D1431704

Decomposition, Combustion, and Detonation Chemistry of Energetic Materials

MATERIALS RESEARCH SOCIETY
SYMPOSIUM PROCEEDINGS VOLUME 418

Decomposition, Combustion, and Detonation Chemistry of Energetic Materials

Symposium held November 27-30, 1995, Boston, Massachusetts, U.S.A.

EDITORS:

Thomas B. Brill
University of Delaware
Newark, Delaware, U.S.A.

Thomas P. Russell
Naval Research Laboratory
Washington, DC, U.S.A.

William C. Tao
Lawrence Livermore National Laboratory
Livermore, California, U.S.A.

Robert B. Wardle
Thiokol Corporation
Brigham City, Utah, U.S.A.

M|R|S

MATERIALS
RESEARCH
SOCIETY

PITTSBURGH, PENNSYLVANIA

This work was supported in part by the Office of Naval Research under Grant Number N00014-96-1-0057. The United States Government has a royalty-free license throughout the world in all copyrightable material contained herein.

Single article reprints from this publication are available through University Microfilms Inc., 300 North Zeeb Road, Ann Arbor, Michigan 48106

CODEN: MRSPDH

This book has been registered with Copyright Clearance Center, Inc. For further information, please contact the Copyright Clearance Center, Salem, Massachusetts.

Published by:

Materials Research Society
9800 McKnight Road
Pittsburgh, Pennsylvania 15237
Telephone (412) 367-3003
Fax (412) 367-4373
Homepage http://www.mrs.org/

Library of Congress Cataloging in Publication Data

Decomposition, combustion, and detonation of energetic materials : symposium
    held November 27-30, 1995, Boston, Massachusetts, U.S.A. / editors Thomas
    B. Brill, Thomas P. Russell, William C. Tao, and Robert B. Wardle.
        p. cm.—(Materials Research Society symposium proceedings ; v. 418)
        Includes bibliographical references and index
        ISBN 1-55899-321-5 (alk. paper)
        1. Combustion—Congresses. 2. Thermodynamics—Congresses. 3.
    Explosives—Congresses. 4. Propellants—Congresses. I. Brill, Thomas B.
    II. Russell, Thomas P. III. Tao, William C. IV. Wardle, Robert B. V. Series:
    Materials Research Society symposium proceedings ; v. 418
QD516.D34 1996
662' .2—dc20

                                                                    96-3930
                                                                    CIP

Manufactured in the United States of America

CONTENTS

*Invited Paper

*Invited Paper

vi

*Invited Paper

*Invited Paper

*Invited Paper

PREFACE

The processes associated with decomposition, combustion, and detonation of energetic materials are immensely complex. Advancements in understanding are hard won because of the dynamic interplay of physical and chemical forces in the heterogeneous state, frequently occurring on short timescales over a short linear dimension. The subject is truly interdisciplinary and broad, as evidenced by the range of topics presented in this symposium. In addition to technical breadth, significant international participation was achieved, especially from Russia.

Progress in the understanding of energetic materials from the initial synthesis, in the nonreacting state, and finally during decomposition on timescales from hours to picoseconds has markedly advanced in recent years. A major factor has been the incorporation of advanced diagnostics, such as AFM, broadband time-resolved spectroscopy, laser methods, high-level theory, and high-speed computing. New and unusual compounds, which have been sought on the basis of theoretically optimized properties, are progressing from imagination to reality. Control over physical properties, such as defect number and crystal growth characteristics, is occurring. Promising new levels of understanding about reaction pathways, kinetics, sensitivity mechanisms, combustion mechanisms, and shock fronts are being gained. The focus on component materials has continued to expand beyond studies of the neat oxidizers and monopropellants to the behavior of binders, metals and metalloids, and whole new classes of compounds. Many novel materials were mentioned throughout this symposium.

This was the second symposium held by the Materials Research Society on the subject of energetic materials. Given the enthusiasm expressed by many of the attendees, we hope that it will be followed by more such symposia.

Thomas B. Brill
Thomas P. Russell
William C. Tao
Robert B. Wardle

January 1996

ACKNOWLEDGMENTS

The organizers acknowledge with gratitude the financial support of the Office of Naval Research (Richard Miller), the Naval Research Laboratory (Thomas Russell), the Army Research Office (Robert Shaw), Lawrence Livermore National Laboratory (Ronald Atkins and William Tao), Los Alamos National Laboratory (Philip Howe), and the Russian Foundation of Basic Research (Vladimir Fortov).

The staff of MRS made their usual efficient and major contribution to the success of this symposium.

The session chairs did an excellent job of keeping the program on schedule and of stimulating discussion. They were Jeff Bottaro, Phil Howe, Bob McKenney, Ross Sausa, Carl Melius, Jim Belak, Carter White, Paul Urtiew, and Jerry Hinshaw.

Finally, we are most appreciative to the community of authors for providing an important set of articles describing current advances in their laboratories.

MATERIALS RESEARCH SOCIETY SYMPOSIUM PROCEEDINGS

MATERIALS RESEARCH SOCIETY SYMPOSIUM PROCEEDINGS

Prior Materials Research Society Symposium Proceedings available by contacting Materials Research Society

Part I

Synthesis and Characterization

Research on New Energetic Materials

R. S. Miller, Office of Naval Research, Arlington VA 22217, Millerr@onrhq.onr.navy.mil

ABSTRACT

The goal in this paper is to review progress in two emerging topics in energetic materials science. These emerging two areas are fluorine and oxygen rich energetic crystals and polymers and environmentally friendly energetic material classes.

INTRODUCTION

Fluorine and oxygen rich energetic crystals and polymers will provide a new approach to increasing composite propellant and explosive energy density and energy release rates. This class of energetic materials will be used to demonstrate that advances in computational chemistry and solid state physics can be used to begin to understand detonation and combustion processes It is anticipated that fluorinated as well as the oxygenated combustion and detonation products will accelerate the rates of metal particle consumption in composite propellants and explosives. Enhanced and tailorable energy release rates and critical diameters of metallized composite explosives will provide new technological opportunities for both military and civilian applications. Environmentally friendly energetic materials are of great current interest to reduce life cycle waste and pollution as well as life cycle cost. Thermoplastic elastomers, which have reversible crosslinking mechanisms, are one of the required keys to the gate and pathway to achieving substantial waste and pollution reduction goals.

EXPERIMENT

The long range goal is to provide polycyclic crystalline explosive and propellant ingredients which will have a maximum density and a tolerable sensitivity to detonation by shock waves and other mechanical and thermal stimuli. Maximizing density of the crystalline oxidizer in a composite propellant or explosive is crucially important because it permits the volume fraction of the particulates within the composition to be minimized. Minimizing the volume fraction of the particulates within the composite energetic solid maximizes the energy absorbing capability or toughness of the explosive or propellant. High energy absorption is important because improvement in the insensitivity of munitions is also required. Polycyclic energetic molecular structures which are inherently dense and contain a variety of energetic functional groups are long term chemical synthesis goals. The increase in density between linear n-hexane (0.66 g/cm3), monocyclic cyclohexane (0.78 g/cm3) and polycyclic adamantane (1.07 g/cm3) vividly illustrates this progression. The carbon to hydrogen ratio also increases in this progression and approaches infinity in the case of diamond. Within a given type of molecular structure the substitution of energetic groups for hydrogen dramatically increases density. Substitution of 3 of the methylene groups with 3 nitramino groups in cyclohexane (0.78 g/cm3) increases density by 1.0 g/cm3 to 1.8 g/cm3 in RDX. Functional groups that contain nitrogen, oxygen and fluorine are inherently dense. These include azido groups ($-N_3$), nitrate ester groups ($-ONO_2$), nitramine groups ($N-NO_2$) and the less common geminal difluoroamino groups ($C-(NF_2)_2$) and geminal dinitro groups ($C-(NO_2)_2$). In energetic molecular families the trend to high density is similar. Within the family of carbon, hydrogen, nitrogen, oxygen explosives and monopropellants the progression to higher density has been maintained. For example, in the nitramine family of explosives, crystalline density increases as the molecular architecture changes from linear nitramines, to monocyclic nitramines (RDX), to the only polycyclic nitramine HexaNitroHexaAzaIsoWurtzitane (HNIW) shown in Figure 1. HNIW has several polymorphs of different density and properties. The density of a polymorph of a crystalline explosive depends upon the conformation of the molecule in the crystal structure. Different polymorphs of the same crystalline material have different densities as well as different sensitivities. Within a single crystal of a given polymorph, the sensitivities to detonation can vary with the direction of the shock waves with respect to the crystallographic axes. Single crystal pentaerythritol tetranitrate (PETN) exhibits these directional properties. The first predictions of the dependence of the detonability of PETN as a function of the crystallographic using solid state qunatum mechanics and band structure theory have neen made.

Mat. Res. Soc. Symp. Proc. Vol. 418 © 1996 Materials Research Society

Figure 1
Molecular Structures of Linear, Monocyclic, and Polycyclic Nitramines

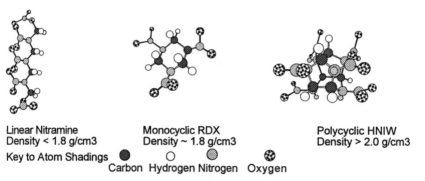

Linear Nitramine
Density < 1.8 g/cm3

Monocyclic RDX
Density ~ 1.8 g/cm3

Polycyclic HNIW
Density > 2.0 g/cm3

Key to Atom Shadings ● ○ ◉ ⊕
Carbon Hydrogen Nitrogen Oxygen

One polymorph of HNIW is the densest and the most energetic explosive known. The high density of HNIW is due to the unusually high number and types of cyclic rings in its molecular structure. HNIW has one six membered ring, two five membered rings and two seven membered rings. The six ring structure of polynitro cubanes is another unique example.

Another approach to increase density and chemical reactivity is to replace fluorine for either hydrogen or nitro groups. In the fluorocarbon family density increases greatly as one replaces fluorine for hydrogen in carbon based polymers. Teflon for example has a density which is much higher than that of polyethylene. Within families of crystalline explosives, density increases as difluoroamino groups replace nitrato or nitramino groups. In a pentaerythritol tetranitrate example, the substitution of two nitrate groups with 8 heavy atoms (6 oxygen and two nitrogen atoms) for two difluoroamino groups with 6 heavy atoms (two nitrogen and four fluorine atoms) results in crystals with essentially equal densities.

Figure 2
Pentaerythritol Tetranitrate and Its Difluoroaminated Analogue

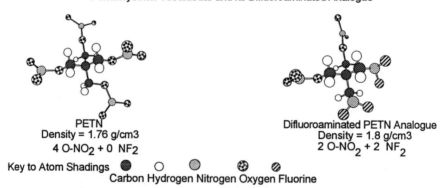

PETN
Density = 1.76 g/cm3
4 O-NO$_2$ + 0 NF$_2$

Difluoroaminated PETN Analogue
Density = 1.8 g/cm3
2 O-NO$_2$ + 2 NF$_2$

Key to Atom Shadings ● ○ ◉ ⊕ ⊘
Carbon Hydrogen Nitrogen Oxygen Fluorine

The difluoroamine group can also be introduced as geminal difluoroamino groups - that is two difluoroamino groups on carbon with SP$_3$ hybridization. In RNFX, the geminal difluoroaminated analogue of RDX, the substitution of two nitrogen, four fluorine and two carbon atoms for two nitrogens and two oxygen atoms results in a substantial increase in the estimated crystal of approximately 0.2 g/cm3 density. The geminal difluoroamino group is not only dense but rich in fluorine. The geminal difluoroamine group is the functional group of choice for substitution into polycyclic analogues of HMX and HNIW.

Figure 3
RDX and its Difluoroaminated Analogue (RNFX)

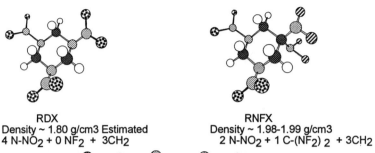

RDX	RNFX
Density ~ 1.80 g/cm3 Estimated	Density ~ 1.98-1.99 g/cm3
4 N-NO$_2$ + 0 NF$_2$ + 3CH$_2$	2 N-NO$_2$ + 1 C-(NF$_2$)$_2$ + 3CH$_2$

Key to Atom Shadings ● ○ ◉ ⊕ ⊘
Carbon Hydrogen Nitrogen Oxygen Fluorine

The next example is in the same monocyclic nitramine family of RDX and HMX explosives. It is the geminal difluoroaminated analogue of HMX (HNFX). HNFX has recently been synthesized. Its crystal structure has not yet been determined by x-ray diffraction but has been estimated to be between 1.980 to 1.999 g/cm3. The estimated densities and crystal structures are a function of the conformation and symmetry of the molecule in the crystal lattice. Solid state quantum mechanics and band structure theory predictions of the electronic structures of the polymorphs indicate that the densest polymorph of HNFX will have a sensitivity to detonation lower than that of RDX as well as lower than that of the other two polymorphs.

Figure 4
HMX and Its Difluoroaminated Analogue (HNFX)

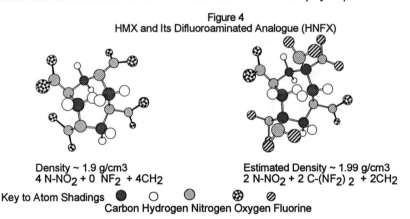

Density ~ 1.9 g/cm3   Estimated Density ~ 1.99 g/cm3
4 N-NO$_2$ + 0 NF$_2$ + 4CH$_2$ 2 N-NO$_2$ + 2 C-(NF$_2$)$_2$ + 2CH$_2$

Key to Atom Shadings ● ○ ◉ ⊕ ⊘
Carbon Hydrogen Nitrogen Oxygen Fluorine

The estimated densities are a function of the conformation and symmetry of the molecule in the crystal lattice. Three isolated molecular conformations for HNFX were predicted and with various symmetries. The estimated densities are a function of the conformation and symmetry of the molecule in the crystal lattice. For HNFX, three polymorphic crystal structures were found with three different types of symmetry.

5

Figure 5
The Three Predicted Conformers of HNFX

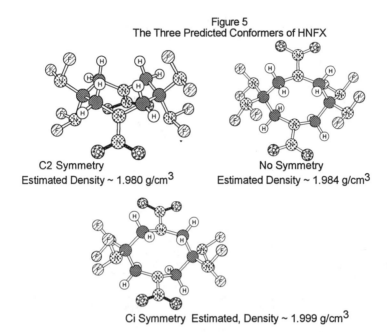

C2 Symmetry
Estimated Density ~ 1.980 g/cm^3

No Symmetry
Estimated Density ~ 1.984 g/cm^3

Ci Symmetry Estimated, Density ~ 1.999 g/cm^3

The fluorine to oxygen balance of the RNFX and HNFX molecules is different. The RDX analogue is richer in oxygen than the HMX analogue. The RDX analogue is balanced to four hydrogen fluoride molecules, one water molecule water, and 3 carbon monoxide molecules. The HMX analogue is a weaker oxidizer. It is balanced to 8 hydrogen fluoride molecules, four carbon monoxide molecules and two atoms of carbon. These approximate product balances assume hydrogen is preferentially oxidized by fluorine in the presence of oxygen. Eventually mixtures of HNFX and RNFX might be desirable to control the overall ratio of fluorine to oxygen.

An imaginary difluoroaminated analogue of HNIW with two geminal difluoroamino groups replacing the nitramine groups on the six membered ring of HNIW is shown in Figure 7. The density of this hypothetical crystalline explosive is predicted to be 2.07 g/cm3; higher than that of HNIW.

Figure 6
Polycyclic Nitramines and Polycyclic Mixed and Difluoroamines/Nitramine

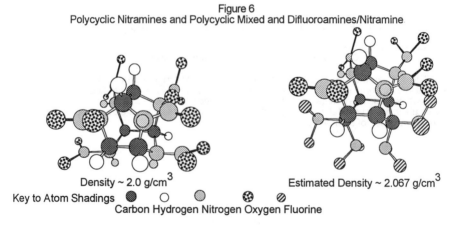

Density ~ 2.0 g/cm^3

Estimated Density ~ 2.067 g/cm^3

Key to Atom Shadings ● ○ ◉ ⊗ ⊘
Carbon Hydrogen Nitrogen Oxygen Fluorine

6

Energetic crosslinkable difluoroaminated hydroxyl terminated polymers have been developed for use in formulations with HNFX and RNFX. These difluoroaminated homopolymers and copolymers are polymerized from functionalized methyl oxetane monomers. The difluoroamino methyl monomers are synthesized by fluorinating one or two methyl carbamate groups situated on the three position of the oxetane ring system. Cationic copolymerization of 3,3-bis(difluoroamino methyl) oxetane with 3-methyl-3-nitrato methyl oxetane is a favored method to control the ratio of nitrato to difluoroamino groups in the polymer. Control of the fluorine to oxygen ratio in the copolymer is controlled by the molar ratio of the 3,3-bis(difluoroamino methyl) oxetane to 3-methyl-3-nitrato methyl oxetane in the mixture of oxetane monomers. Alternatively oxetane monomers with one difluoroamino methyl substituent and one azido methyl or nitrato methyl group can be homo polymerized to form a polymer with a 1:1 ratio of difluoroaminomethyl to nitrato methyl groups.

Figure 7
Chemically Crosslinkable Unsymmetrically Substituted Oxetane Polymers

3-AzidoMethyl, 3-DifluoroaminoMethyl Oxetane
Monomer

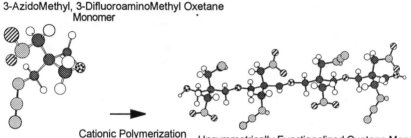

Cationic Polymerization

Unsymmetrically Functionalized Oxetane Monomer
Polymerized to Amorphous Hydroxyl Terminated
Prepolymers Crosslinkable with Isocyanates

Predictions of metal consumption kinetics in gaseous environments composed of prototypical monopropellant combustion products at various fluorine to oxygen ratios show the difluoroaminated analogues to have an enhanced rates of metal consumption. In addition to increasing the rate of metal consumption in composite propellants and explosives, the fluorine in the proper ratio to fluorine to oxygen leads to increases specific impulse. The increase in specific impulse of propellants occurs because of two thermodynamic features. The first factor is a high heat of combustion of the metal and gaseous fluorinated and oxygenated oxidizer and binder decomposition products. The second feature is the nature of the exhaust products. The exhaust products of boronized propellants with both fluorine and oxygen decomposition products contain mostly gaseous OBF species which eliminate two phase flow losses.
Two phase flow losses in aluminized propellants containing only oxygen rich decomposition products produce approximately one percent of flow loss for each 5 weight percent of aluminum in the formulation. Although eliminating oxide particles may be beneficial to performance, it may lead to combustion instability in rocket motor combustion chambers.
Environmentally friendly energetics implies minimizing waste and pollution production and environmental impact through out a munitions life cycle. Increasing emphasis on life cycle cost reduction in defense systems may in the future include all life cycle environmental waste and pollution costs. Environmentally conscientious energetic materials are being preemptively developed and demonstrated to maximize the rate at which environmentally friendly munitions are acquired. The life cycle waste and pollution costs generally start with the production of all the ingredients that will be used in the propellant or explosive that may is utilized in the munition. The life cycle costs may or may not include the life cycle waste and pollution costs of the raw materials, for example nitric acid and sulfuric acids, used in the ingredient manufacturing process. Raw material waste and pollution costs may not be practically important if the volume of use of a particular ingredient is dwarfed by the volume used in civilian application. The life cycle continues through the manufacturing process and utilization of the munitions. The life cycle finishes with demilitirization at the end of the useful life. Environmentally friendly energetic materials and manufacturing processes demand waste minimization throughout the life cycle. Minimizing waste throughout the life cycle implies maximizing the recovery, recycle and reuse of propellant,

7

explosive, and pyrotechnic ingredients of future obsolete weapon systems. Maximum recovery, recycle and reuse of propellant, explosive, and pyrotechnic ingredients will permit (1) avoiding open pit burning or detonation, elimination of the manufacturing waste of replacement ingredients, (2) the recycle of scrap from the manufacturing operations and (3) the elimination of processing solvents and plasticizers from the manufacture of extruded nitrocellulose based gun propellant products. The four key elements of the strategy to recover, recycle, and reuse energetic materials are: (1) to use a reusable binder systems with reversible crosslinking mechanisms; (2) to manufacture recyclable ingredients including binder systems with a minimum of waste; (3) to maximize product quality by using rheological computational tools to design and control mixing, casting and extrusion processes based on recoverable, recyclable, and reusable ingredients; and (4) to demonstrate recovery, recycleability, and reusability of the ingredients used in the munitions before they are manufactured. High energy crystalline melt castable explosives such as trinitroazetidine (TNAZ) may also find use as recoverable, recycleable and reuseable energetic materials.

The irreversible crosslinks in a state-of-the-art conventional chemically crosslinked binder system must be essentially destroyed to even begin to attempt ingredient reclamation. Unfortunately, the strategy to avoid destructive demilitarization had not been developed in the past. Most cast cured chemically crosslinked propellants and explosives in the inventory today will have to be destructively demilitarized. Thermoplastic elastomeric binder systems (TPEs) are more useful because they can avoid this destructive process. TPEs can avoid this destructive reclamation process because they have reversible physical spherulitic crosslinks. The spherulitic physical crosslinks disappear above the melting point of the crystalline "hard" block. The physical crystalline "hard" block crosslinks reform upon cooling.

Figure 8
Molecular Architecture Of A TPE With A Sequential Hard-Soft-Hard- Block Structure

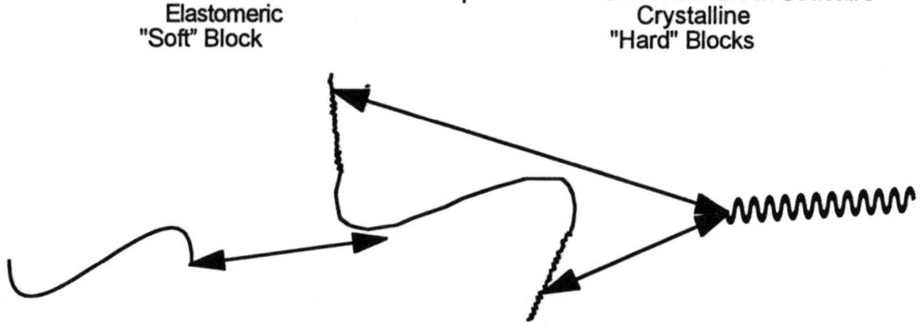

| Elastomeric | Crystalline |
| "Soft" Block | "Hard" Blocks |

The hard blocks act as physical crosslinks because the symmetrical polymer blocks from one polymer molecule cocrystallize with other symmetrical polymer blocks from other polymer molecules. This cocrystallization process forms the necessary three dimensional network of interlinked soft unsymmetrical polymer blocks. The multiblock polymer molecule depicted here is composed of one soft elastomeric block linked at each of its ends with a hard crystalline block. "Hard" blocks are polymerized symmetrical oxetane monomers i.e. those which have two identical functional groups on the "3" position of the oxetane ring system. Azido methyl ($N_3CH_2^-$) groups are the favored functional groups for symmetrical oxetane polymers. The symmetrical nitrato methyl and difluoroamino methyl oxetane monomers do not homopolymerize. Poly(BAMO) or poly(bis(azido methyl) oxetane) is the preferred hard block because of its sharp ~90C° sharp melting point. Crystalline hard blocks are advantageous over glassy hard blocks because the viscosity of the polymer falls rapidly above the crystalline block melt temperature.

8

Figure 9
Symmetrical Oxetane Monomers

Symmetrical BAMO Monomer Is Cationically Polymerized To SemiCrystaline Poly(BAMO)

The "soft" or amorphous block can be any amorphous unsymmetrical difunctional energetic oxetane polymer. Typical "soft" blocks are AMMO and NMMO, i.e. azido methyl methyl oxetane and nitrato methyl methyl oxetane.

Figure 10
Unsymmetrical Oxetane Monomers

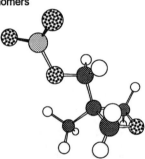

AMMO
3-AzidoMethyl,-3-Methyl Oxetane

NMMO
3-NitratoMethyl,3-Methyl Oxetane

The diversity of unsymmetrical and symmetrical oxetane monomers is high. The newest oxetane monomers are unsymmetrical with two different energetic functional groups, for example nitrato and azido, each bonded to one of two methylene groups which are bonded to the three position on the oxetane ring.

Figure 11
Unsymmetrical Oxetane Monomers with Two Different Energetic Substituents

3-DifluoroaminoMethyl,3-AzidoMethyl
Oxetane

9

The diversity of the oxetane monomer family and control of oxetane polymerization provides the propellant chemist with many possibilities to tailor the energy of the propellant and its combustion at the microstructural level. Additional possibilities are provided by the use of the poly(glycidyl nitate) (PGN) as the soft block of BAMO based thermoplastic elastomers. Control of microstructural combustion may allow manipulation of the pressure exponent and temperature sensitivity of the composite propellant combustion process.

The first successful oxetane TPEs had $(AB)_n$ architectures. "A" represents the crystalline "hard" block and "B" denotes the amorphous "soft" block. The $(AB)_n$ polymer was formed by chain extending isocyanate capped mixtures of BAMO and other soft blocks with butane diol. The important variables are the molecular weights of the isocyanate capped hard and soft blocks and their proportions. Recently, star oxetane TPEs have successfully made which promise a lower melt viscosity for a given polymer molecular weight. Although, ABA TPEs are simpler in concept than either star or $(AB)_n$ architectures, they are more difficult to create because any prematurely terminated "AB" polymer segments have a deleterious effect on elastomers mechanical properties.

Figure 12
Oxetane Based Thermoplastic Elastomeric Binders (TPEs) Architectures
"Hard" Blocks

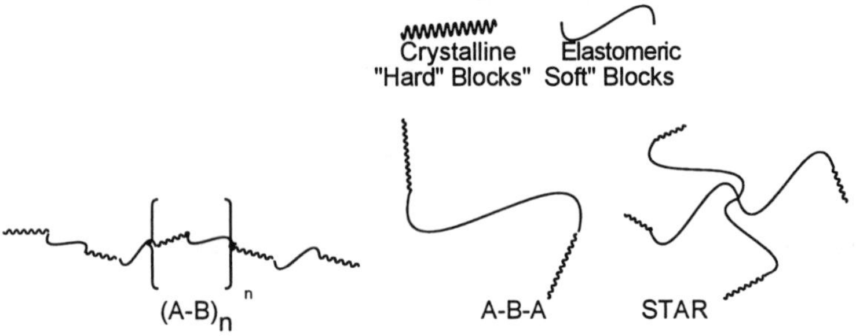

Crystalline   Elastomeric
"Hard" Blocks" Soft" Blocks

$(A-B)_n$ A-B-A   STAR

The mechanical properties degrade because the terminated "AB" segment can not participate in the three dimensional crosslinking network. To participate in the network forming process, each polymer must have at least two hard segments. A minimum of two hard segments from each polymer molecule are needed to cocrystallize with the hard segments from other polymer molecules in order for them to contribute to the elastomers crosslink density. Star and $(AB)_n$ polymer can more effectively form physical crosslinks networks when they are below their melting temperature, as they have more than two hard segments per molecule.

High energy metallized propellants based on oxetane thermoplastic elastomeric binders have been developed. The first rocket motor firing demonstrations are planned. Thermoplastic elastomeric based propellants or explosives may be manufactured by mixing melted TPE with crystalline oxidizer and fuel particles while the polymer is above the melt temperature of the crystalline blocks. Twin screw mixers and extruders specifically designed to accommodate the rheology of the melted particle filled TPE will be utilized to process the higher viscosity TPE based formulations. The molten, explosive particle laden polymer is then cast into a heated rocket motor case, a warhead, or it is extruded into strand or slabs for use in gun propellants. Reversible crosslinking occurs when the temperature composite propellant or explosive is lowered below the melt temperature of the crystalline blocks of the TPE. The TPE based explosive or propellant can be removed from the weapon at the time of demilitirization by heating the unit above the TPE melt temperature. The molten propellant or explosive can then be poured from its containment to await further recovery processing. Oxetane thermoplastic elastomers have the potential to play a major role in the next generation of propellant, explosive and pyrotechnic systems. For this to occur, these materials will have to be manufactured in accordance with future strict environmental regulations and concerns. High on the list of environmental challenges is the elimination of the use of chlorinated solvents in the manufacturing processes of monomer preparation, and monomer polymerization. "Green" chlorinated solvent-free manufacturing technologies are being developed for these materials. The first step to be improved was the synthesis of oxetane monomers. Oxetane monomers are routinely now produced in aqueous solvent systems. Water is the solvent of choice for the preparation of azide based oxetane monomers. Water can also be used to synthesize nitrate ester and difluoroaminated oxetane polymers. The polymerization of the oxetane monomers has normally been carried out in chlorinated organic solvents. However, liquid carbon dioxide has been recently shown useful to replace halogenated hydrocarbon solvents. The hydroxyl oxetane polymers are identical to those produced using normal processing solvents. The final chain linking steps of BAMO and other soft blocks to produce the thermoplastic elastomers are now carried put in liquid CO_2 at a laboratory scale and non-halogenated hydrocarbon solvents at the pilot plant scale. The thermoplastic elastomers produced during these operations are being used in missile and gun propellant development demonstration efforts. Methods to recycle, recover, and reuse the ingredients used in these energetic formulations are being developed. Water soluble oxidizers such as ammonium perchlorate and water insoluble oxidizers such as RDX and HNIW are included in this ongoing effort. Efforts to demonstrate "green" syntheses of oxidizers are also underway.

During the last decade, the science base to understand the rheology of particle laden polymers, to quantify that behavior in constitutive equations, and to efficiently compute particle laden polymer flows in complex geometries have been developed. These experimental and fluid mechanical tools are now being applied to solve energetics processing problems and to help design and control processing operations to insure the uniformity of microstructure. These tools are being first applied to processing design problems at the navy's energetics manufacturing technology center at the Naval Surface Warfare Center, Indian Head Division at Indian Head, Maryland . From a very practical point of view, gradients in explosive crystal, oxidizer crystal or metal particle concentration cause burning rate anomalies, mechanical property variations, and high product rejection rates. The major challenge has been to understand to quantitatively model and control demixing processes when they naturally develops during the manufacturing process. These demixing processes lead to unacceptable changes in the microstructure of the final energetic product.

It was observed that particle migration occurs as particles diffuse from regions of high to low shear rate. To gain a better quantitative understanding of these demixing processes, nuclear magnetic resonance imaging (NMRI) was and continues to be used to investigate these demixing processes. In parallel, Stokesian dynamic simulations of particle motion in newtonian fluids were performed to understand the physics of the demixing process and to develop constitutive equations to describe the complex fluid properties. The NMR images of the sheared suspensions revealed rapid demixing of initially homogeneous suspensions which led to large particulate concentration gradients. Concentration gradients of >40% were established across channels in simple flow fields. NMRI has elucidated the dependency of the migration rate on particle size, concentration, binder viscosity, shear rate, and total shear strain. Particle migration occurs as particles diffuse from regions of high to low shear rate. For example, packed rings of particles migrate to outer stationary cylinder in Couette flow experiments.Gradients in explosive particle concentration cause burning rate and mechanical property variations

A generalized Newtonian fluid theory with migration and a Phillips constitutive model, based on stress induced particle migration, matches the concentration profiles measured by NMRI. It predicts the transient evolution of the particle density profile and it provides a basis to link rheometry with microstructural information. Based on the Stokesian dynamics simulations, alternative constitutive equations are now beginning to be used to improve the description of these microstructural changes. These rheological methods and computational tools can now be used to assist in the design of twin screw extruders. They can also be used to eliminate changes in the microstructure of the propellant or explosive that occur after the mixing process is completed and the particulate composite flows into rocket motor and warhead cases or it is extruded into other shapes.

The thermoplastic elastomers produced during these operations are being used in missile and propellant development demonstration efforts. Methods to recycle, recover, and reuse the ingredients used in these energetic formulations are being developed. Water soluble oxidizers such as ammonium perchlorate and water insoluble oxidizers such as RDX and HNIW are included in this ongoing effort. Efforts to demonstrate "green" syntheses of oxidizers are also underway.

ACKNOWLEDGMENTS

HNIW was first synthesized by Dr. Arnold Nielsen at the Naval Air Warfare Center (NAWC). Characterization of HNIW and improvements in the synthesis approach have been made by many including Dr. Robert Wardle at Thiokol Corporation. The crystal structure and density prediction methods and computations in this were developed and performed respectively by Professor Herman Ammon at the University of Maryland. The experimental studies of the directional sensitivity characteristics of PETN were first made by Dr. Jerry Dick at Los Alamos National Laboratory (LANL). The first predictions of the dependence of the detonability of the polymorphs of HNFX and of PETN as a function of the crystallographic using solid state quantum mechanics and band structure theory have been made by Professor Barry Kunz from Michigan Technological University. The solid state quantum mechanics and band structure theory predictions will be presented at this Symposium. Professor Phillip Eaton at the University of Chicago first synthesized cubane and has discovered along Dr. A. Hashemi at GeoCenters Inc. methods to introduce nitro groups onto six of eight cubyl carbon atoms. Drs. Robert Chapman and Kurt Baum at Fluorochem Inc. synthesized the first HNFX materials but at a low yield. Dr. Chapman and Mr. Mark Welker recently synthesized the first gram scale samples of HNFX at TPL. The modeling of the boron particle consumption has been made by Professor Fred Dryer and Richard Yetter at Princeton University and Dr. Chuck Kolb at Aerodyne Inc.. Drs. Tim and Donna Parr performed the quantitative imaging of aluminum particle combustion.at NAWC. Dr. Richard Gilardi has performed the x-ray diffraction studies on energetic crystals at the Naval Research Laboratory. The difluoroaminated oxetane monomers and polymers were synthesized by Mr. Gerald Manser at GenCorp, Aerojet Division. The first characterized star and A-B-A thermoplastic elastomers as well as the unsymmetrical energetically substituted oxetane monomers and polymers were also synthesized by him. The first successful $(A-B)_n$

thermoplastic elastomers were synthesized by Dr. Robert Wardle at Thiokol Corporation. BAMO monomer was first synthesized by Dr. W. Carpenter at NAWC. The polymerizability of BAMO was first demonstrated by Mr. Gerald Manser at SRI International. Drs. Tom Archibald and Kurt Baum at Fluorochem Inc. conceived the concept of monocyclic RNFX and HNFX molecules and also synthesized TNAZ. The synthesis of PGN was lead by Dr. Rodney Willer at Thiokol Inc.. The experimental and theoretical rheology and flow modeling of highly filled suspensions has been done by Drs. Robert Brown and Robert Armstrong at MIT, Drs. Allan Graham at LANL, Dr. Lisa Mondy at Sandia National Laboratory, Dr. Steven Sinton at Martin-Lockhhed Palo Alto Research Laboratory, Dr. John Brady at California Institute of Technology, Dr. Thomas Stephens at NAWC, and Dr. Dilhan Kalyon at Stevens Institute of Technology.

SELECTED PUBLICATIONS
Oxetane and Poly(Oxetane)s

Manser, G. E.; Newton, L. S.; Cheung, H. W., "Synthesis of Energetic Binders", Aerojet Solid Propulsion Company Final Report, May 1989, Navy Contract No. N00014-86-C-0164.

Manser, G. E., "Cationic Polymerization," U. S. Patent No. 4,393,199; 1984.

Manser, G. E., "Energetic Copolymers and Method of Making Same", U.S. Patent No. 4,483,978; 1984.

Manser, G. E.; Fletcher, R. W., "Nitramine Oxetanes and Polyether formed Therefrom," U. S. Patent No. 4,707,540; 1987.

Manser, G. E.; Shaw, G. C., Internally-Plasticized Polyethers from Substituted Oxetanes," U. S. Patent No. 4,764,586; 1988.

Manser, G. E., Archibald, T. A., "3-Azidomethyl-3-nitratomethyloxetane and Polymers Formed There from, Aerojet Patent Application No. 07-940,269, 1993.

Manser, G. E., Archibald, T. A., "Difluoramino Oxetanes and Polymers Formed There from for Use in Energetic Formulations," U. S. Patent No. 5,272,249; 1993.

Manser, G.E., "Polymers and Copolymers from 3-Azidomethyl-3-Nitratomethyloxetane" US Patent 5,463,019, Oct 31, 1995

Polyoxiranes (PGN)

Willer, R. L.; Day, R. S.; Stern, A. G., "Process for Producing Improved Poly(Glycidyl Nitrate)", U. S. Patent No. 5,017,356; 1992.

Willer, R. L.; Day, R. S.; Stern, A. G., "Isotactic Poly(Glycidyl Nitrate) and Synthesis there of", U. S. Patent No. 5,162,494; 1992.

NF2 Compounds

Manser, G. E., Archibald, T. A., "Difluoramino Oxetanes and Polymers Formed There from for Use in Energetic Formulations," U. S. Patent No. 5,272,249; 1993.

HNIW

Nielson, A. T.; Nissan, R. A.; Vanderah, D. J.; Coon, C. L.; Gilardi, R. D.; George, C. F.; Flippen-Anderson, "Polyazapolycyclics by Condensation of Aldehydes with Amines. 2. Formation of 2,4,6,8,10.12-Hexabenzyl-2,4,6,8,10,12-hexaaza-tetracyclo[5.5.0.05,9.03,11]dodecanes from Glyoxal and Benzylamines," J. Org. Chem., 55, 1459-1466, 1990.

M. Chaykovsky, W. M. Koppes, T. P. Russell, R. Gilardi, C. George, and J. L. Flippen-Anderson, "The Isolation of a Bi(2,4,6,8-tetraazabicyclo[3.3.0]octane) from the Reaction of Glyoxal with Benzylamine", J. Org. Chem., 57, 4295-4297, 1992.

Cubanes

Willer, R. L.; Stern, A.G.,"Diethanolammonium Methyl Cubyl Nitrates - Hydroxylammonium Nitrate Solutions," U. S. Patent No. 5,232,526; 1993.

Hashemi, A. B., Photochemical Carboxylation of Cubanes," Angew. Chem. Int. Ed. Engl., 32 (4), 612-613, 1993,.

Eaton, P. E.; Xiong, Y.; Zhou, J. P., "Systematic Substitution on th Cubane Nucleus: Steric and Electronic Effects," J. Org. Chem.,1992, 57, 4277-4281.

Hashemi, B. A.; Ammon, H. L.; Choi, C. S., "Chemistry and Structure of Phenylcubanes," J. Org. Chem. , 55, 416-420, 1990.

Hashemi, B. A., New Developments in Cubane Chemistry: Phenyl Cubanes," J. Am. Chem. Soc. 1988, 110, 7234-7235.

Eaton, P. E.; Cunckle, G. T.; Marchioro, G.; Martin, R. M., "Reverse Transmetalation: A Strategy for Obtaining Certain Otherwise Difficultly Accessible Organometallics" " J. Am. Chem. Soc., 109 , 948-949, 1987.

Eaton, P. E.; Castaldi, G., "Systematic Substitution on th Cubane Nucleus: Amide Activation for Metalation of "Saturated" Systems," J. Am. Chem. Soc., 107, 724-726, 1985.

Eaton, P. E.; Shankar, B. K. R., Synthesis of 1,4-dinitrocubane," J. Org. Chem. 1984, 49, 185-186.

Advanced Processing & Rheology

Effects of Air Entrainment on the Rheology of Concentrated Suspension during Continuous Processing, "Kalyon, D.M.; Yazici, R.; Jacob, C.; Aral, Bl, Polym. Eng. Sci. , 31, 1386-1392, 1991.

R.J. Phillips, R. C. Armstrong, R.A. Brown, A.L. Graham, and J.R. Abbot "A Constitutive Equation for Concentrated Suspensions that Accounst for Shear-Induced Particle Migration," Physics of Fluids A, 4(1), 30-40 (1992).

J.R. Abbot, L.A. Mondy, A.L. Graham, and H. Brenner, "Techniques for Analyzing the Behavior of Concentrated Suspensions, "Particulate /two Phase Flow (M.C. Roco, Ed.), Chap. 1, Butterworth-Heinamann, Stoneham, MA (1992)

K. E. Newman and T. S. Stephens, "Application of rheology to the processing and reprocessing of plastic-bonded explosives in twin screw extruders," NSWC-IH Technical report 1790, (1995)

N. Phan-Thien, A. L. Graham, S. A. Altobelli, J. R. Abbott, and L. A. Mondy, "Hydrodynamic particle migration in an concentrated suspension undergoing flow between eccentric rotating cylinders", Industrial and Engineering Chemistry, Research, in press, (1995).

D. M. Husband, L. A. Mondy, E. Ganani, and A. L. Graham, "Direct measurements of shear-induced particle migration in suspensions of bimodal spheres." Rheologica Acta, **33**, 185-192, (1994).

T. Stephens and K. Newman , "PEP processing and opportunities for pollution prevention," Proceedings of the Life Cycles of Energetic Materials Conference, Del Mar, CA, Dec. 1994, Los Alamos Report, LA-UR-95-1090, 91-95.

Szady, M.J., Salamon, T. R., Liu, A. W., Bornside, D. E.., Armstrong, R. C.., and Brown, R. A.. (1995) A new mixed finite element method for viscoelastic flows governed by differential constitutive equations., J. Non-Newtonian Fluid Mechanics. 59, 215-243.

P. R. Nott and J. F. Brady, "Pressure driven flow of suspensions: Simulations and Theory," Journal of Fluid Mechanics 275, 157-199, (1994).

J. F. Morris, J. F. Brady, "Self diffusion in sheared suspensions," Journal of Fluid Mechanics, in press (1995).

Explosive Crystal Detonation Theory and Experiment

J. J. Gilman, "Chemical reactions at detonation fronts in solids", Philosophical Magazine B, Vol. 71, No. 6, 1057-1068.

J. J. Dick, Applied Phys. Letters, 44, 859, 1984.

THE DESIGN OF STABLE HIGH NITROGEN SYSTEMS

V.A. TARTAKOVSKY

N. D. Zelinsky Institute of Organic Chemistry, Russian Academy of Sciences
Leninsky prospect 47, Moscow 117913, Russia. Fax: 095 135-5328

ABSTRACT

A general strategy for the design of high nitrogen systems with an adequate degree of stability has been elaborated. The design of nitro compounds in which terminal nitro groups are bonded to the chain of several heteroatoms is a specific case within the strategy. In the process of working out the strategy a number of new high nitrogen systems (dinitrazenic acid or dinitroamide HN_3O_4 and it salts, nitrodiazene oxides RN_3O_3 and tetrazine dioxides) were discovered.
A number of new types of nitro compounds (bicyclo nitro-bis-hydroxylamine, nitrohydrazine, nitrohydroxylamine, sulfo-N-nitroimide and bis-N-nitroimide were synthesized.
This study opens new prospects in the field of the synthesis of high energy materials.

INTRODUCTION

As it is known, energetic properties of a molecule depend, mainly, on density, enthalpy of formation and oxygen balance. In turn, the value of the enthalpy of the formation is considerably affected by the number of N-N bonds in a compound. This idea can be illustrated by four octogen isomers, the real one and three hypothetical, the number of N-N bonds in which increases from 4 to 7. Calculations show that enthalpy of formation of the latter isomer has a twofold increase in comparison with the real octogen isomer. The increase of energetic characteristics proceeds in the same manner.

Table 1. Enthalpies of formation, kcal/mol

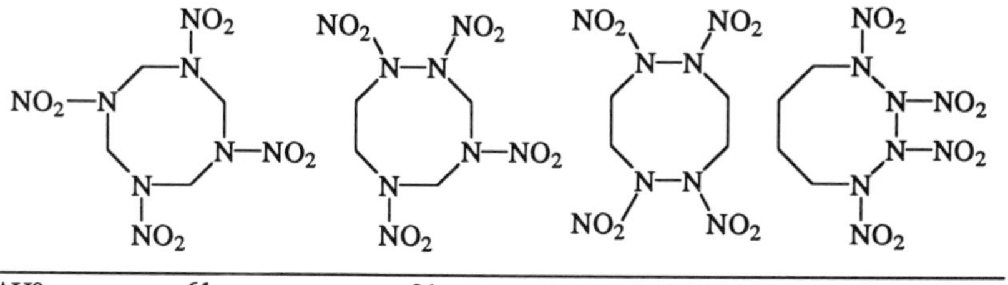

$\Delta H°_f$	61	81	109	133

At the same time it is obvious that to create the latter isomer we should to synthesize a system consisting of 10 heteroatoms directly bonded with each other or (if we divide the system into fragments) to add two nitro groups to terminal atoms of the fragment built of four nitrogen atoms. Is it actually possible to achieve this? What will the stability of such systems depend on?

Which type of nitrogen atoms should be employed for constructing these fragments? These are the questions we would like to find answers to.

So, the first part of our investigation, the general title of which is THE DESIGN OF STABLE HIGH NITROGEN SYSTEMS, was targeted at solving an individual problem: we wanted to elucidate with which fragment, taking into account the number of heteroatoms (mainly of nitrogen), we could bind nitro groups as terminal functions.

EXPERIMENT AND RESULTS

We know a great number of nitro compounds in which the nitro group is bonded with carbon, nitrogen or oxygen atoms. However the consideration of more complicated fragments shows that in more than 99 percent of cases, a carbon atom is located in the β-position relative to the nitro group . We investigated the compounds in which nitrogen, oxygen or sulfur atoms are in a β-position relative to the nitro group. These atoms differ from the carbon ones in having an unshared lone electron pair. Thus this is a study of the influence of an electron pair on β-atoms relative to the nitro group on the nitro compound stability. We shall consider both the basicity of an electron pair and its special position in relation to the nitrogen-α-atom bond.

The term "stability" refers to the possibility of the compound existing under trivial temperatures and its being inert towards common chemical solvents.

Let us consider the nitro group-carbon-nitrogen system [1]. In order to determine how a special position of an electron pair on nitrogen, influences a feasibility of the carbon-nitro group bond breakage, we carried out a comparative study of dioxoazabicyclononane and dioxoazabicyclooctane nitro derivatives. Both bicycles, in the crystalline state, exist in the conformation with a *cis*-position of the nitrogen electron pair and carbon-nitro group bond. In bicyclononane solutions at trivial temperatures a transfer into an antiperiplanar conformation of the electron pair and carbon - nitro group bond takes place. In bicyclooctane solution such a transfer is not observed. It was shown that in the first case an easy substitution of nitro groups for nucleophiles occurred which was preceded by the conformational transfer of the initial nitro compound.

In the case of the bicyclooctane nitroderivative, no nucleophilic substitution takes place. The compound was not changed even when heated with aqueous sodium hydroxide Basicity of an electron pair influences the stability of nitro compounds too. Gem-dinitro compounds also belong to the nitrogen-carbon-nitro group system. However an electron pair of nitrogen is delocalized because of its involvement into the semi-polar bond and a lot of dinitro compounds are known to be sufficiently stable substances. Thus, the removal of the electron pair on nitrogen atom out of the antiperiplanar position, or its delocalization, should increase the stability of nitro compounds of the nitrogen-carbon-nitro group type.

Let us verify these conclusions on some other systems. We were the first to synthesize numerous nitrohydroxylamines and nitrohydrazines, i.e. systems comprising O-N-NO$_2$ and N-N-NO$_2$ fragments. It was established that as the electronegativity of substituents attached to the oxygen atom in nitrohydroxylamines or to β-nitrogen atom of nitrohydrazine increases, stability of nitro derivatives grows [2]. In other words, a decrease of the basicity of an electron pair on β-heteroatoms in such systems increases the compound stability. Although generally both classes of substances possess low stability..

Table 2. Stability of nitrohydrazines

stable at 25°C	stable at 0°C	unstable
$\begin{array}{c}\text{Ac}\diagdown\quad\diagup\text{Ac}\\ \text{N—N}\\ \text{Ac}\diagup\quad\diagdown\text{NO}_2\end{array}$	$\begin{array}{c}\text{Ac}\diagdown\quad\diagup\text{Ac}\\ \text{N—N}\\ \text{Me}\diagup\quad\diagdown\text{NO}_2\end{array}$	$\begin{array}{c}\text{Me}\diagdown\quad\diagup\text{Ac}\\ \text{N—N}\\ \text{Me}\diagup\quad\diagdown\text{NO}_2\end{array}$
$\begin{array}{c}\text{Ac}\diagdown\quad\diagup\text{CO}_2\text{Me}\\ \text{N—N}\\ \text{Ac}\diagup\quad\diagdown\text{NO}_2\end{array}$	$\begin{array}{c}\text{MeO}_2\text{C}\diagdown\quad\diagup\text{CO}_2\text{Me}\\ \text{N—N}\\ \text{Me}\diagup\quad\diagdown\text{NO}_2\end{array}$	$\begin{array}{c}\text{Me}\diagdown\quad\diagup\text{Me}\\ \text{N—N}\\ \text{Me}\diagup\quad\diagdown\text{NO}_2\end{array}$

As oxygen has two mutually perpendicular electron pairs it is apparent that there was no sense in studying the influence of various conformations of nitrohydroxylamines on their stability. In the case of nitrohydrazines it is achievable. We synthesized bicyclic bridging nitrohydrazine, in which an electron pair was in the *cis*-position in relation to the α-nitrogen-NO$_2$ bond and in which the inversion of nitrogen atoms was impossible. This compound proved to be most stable among all nitrohydrazines having been synthesized by us.

decomp. 150°C

Evidently, the most effective way to the stabilize the nitrohydrazines is a complete removal of the electron pair of β-nitrogen. Compounds of this kind are known. They are nitroimides. For example, trimethylnitroimide is a highly stable compound as well as corresponding pyridine nitroimide. An attempt to lower the basicity of the electron pair on β-nitrogen in nitrohydrazines as a result of introducing it into an aromatic system was ineffective.

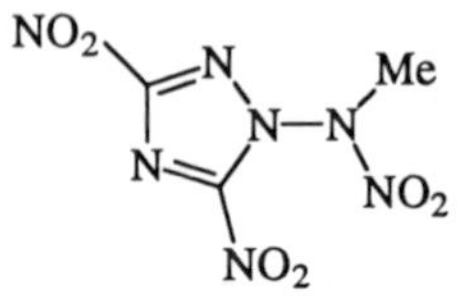

NO₂ N
  \\ // \
   N—N
  /   \ Me
 N   N
  \   NO₂
   NO₂

sta ble at 25°C

Nitrohydrazines themselves, as a rule, display low stability, but their salts are much more stable.

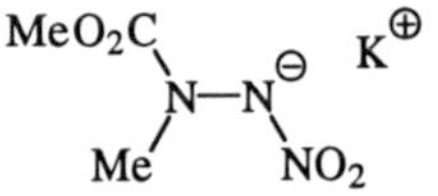

MeO₂C
   \   K⊕
    N—N⊖
   /   \
  Me   NO₂

mp 152°C

Here is another example that shows the efficiency of the liquidation of the electron pair on the β-heteroatom. It is the S-N-NO₂ system. We managed to synthesize N-NO₂ analogs of sulfoxides and sulfones. The experiment demonstrates that they are quite stable compounds. Some of these compounds contain fragments with a relatively high number of heteroatoms directly bonded to each other. Note that these systems can be formally described as systems with the alternating charges [3, 4].

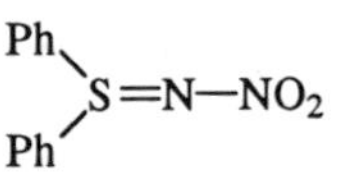

Ph
   \
    S=N—NO₂
   /
Ph

mp 98°C

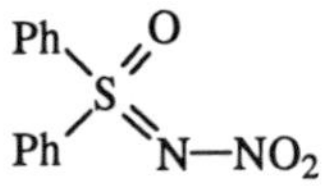

Ph   O
  \ //
   S
  / \
Ph   N—NO₂

mp 96°C

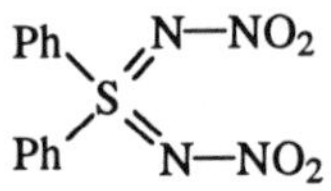

Ph   N—NO₂
  \ //
   S
  / \
Ph   N—NO₂

mp 172°C

O⁻—N⁺—N⁻   N⁻—N⁺—O⁻
      ‖   \   /   ‖
      O   S   O
           /++\
          R   R

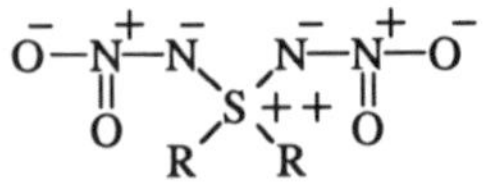

Let us turn back to the most interesting systems consisting of nitrogen chains with terminal nitro groups. We have found out that the most effective way to stabilize such systems is a complete removal of the electron pair from the β-nitrogen atom, i.e. the synthesis of nitroimides. It has also been shown that nitrohydrazine salts are sufficiently stable compounds. Therefore we decided to synthesize molecules that would contain both fragments. The amination of N-aminoimidazole gave bis-aminoimidazole salt and the following nitration leads to bis-(nitroimido)-imidazole K-salt.

mp 230°C   mp 154—161°C   mp 165—167°C

In this compound one nitroamine fragment represents nitroimide, the negative charge of which, is compensated by the positive charge on the nitrogen, being part of the heterocyclic system. The second fragment is nitrohydrazine anion, the negative charge of which, is compensated by outer cation. Actually it is a symmetric system. All these compounds are rather stable ones; their decomposition starts only after melting, that, as a rule, is higher than 150-200°C.

Now, applying this principle we would be able to synthesize N-nitro compounds containing nitro groups bonded to sufficiently long chains of nitrogen atoms.

Thus the application of this method to the 1,2,3-triazole allows to obtain the unique nitrogen-oxygen system consisting of 11 atoms of nitrogen and oxygen directly bonded to each other. Both nitro groups in this compound are bonded with the system containing 5 nitrogen atoms.

mp 226°C

Both K-salts of derivates of unsubstituted 1,2,3-triazole and benzo-1,2,3-triazole obtained are stable in 200°C and start decomposing only after melting [5].

As the result of this part of the work we can formulate two rules. The main reason for the unstability of the Y-X-NO$_2$-type systems is the influence of the electron pair which is especially great when the electron pair is antiperiplanar to the X-NO$_2$ bond. Changes in the molecular geometry or the delocalization of the pair leads to the stabilization of such systems.

It is quite possible to construct nitrogen-oxygen or sufur-nitrogen-oxygen systems with a sufficiently great number of heteroatoms directly bonded with each other. These systems can be formally described as systems in which positively charged heteroatoms are positioned between heteroatoms carrying electron pairs.

Certainly, in the consideration of the stability of each concrete compounds we should take into account the whole complex of factors, that are the mutual influence of various functional groups, conformational peculiarities, etc. However, the rules we could formulate in the result of this study permit us to plan the synthesis of new compounds with a great certainty, rejecting the structures with a low chance of existing beforehand, and at the same time, taking into account, those, which previously seemed unreal. These rules can be also used in the computer search for new structures with the desirable properties.

A few words on the computer approach for the design of polynitrogen-oxygen systems. To solve this problem we suggest a formal method based on the qualitative theory of excitation. The method is based on combining, in a certain way, structural fragments of allyl anion and trimethylene methane dianion through the same atom with a further substitution of all or almost all carbon atoms by nitrogen or oxygen atoms using certain rules.

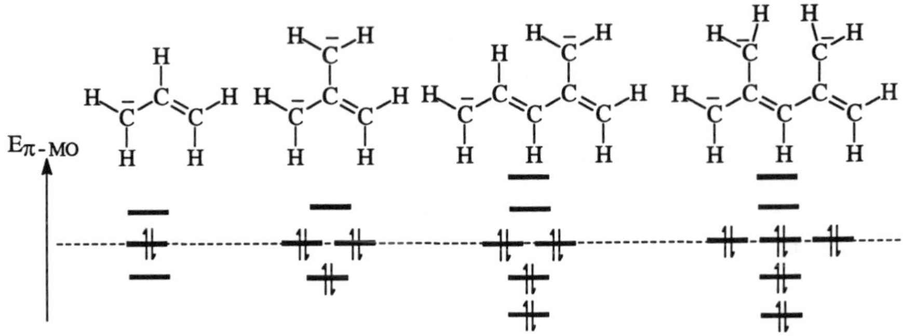

$E\pi\text{-MO}$

$CH_2 =$ fragment was replaced by $O=$, $-CH=$ by $-N=$, $-CH_2^-$ by $-O^-$ and $\diagdown C=$ by $\diagdown N^+=$

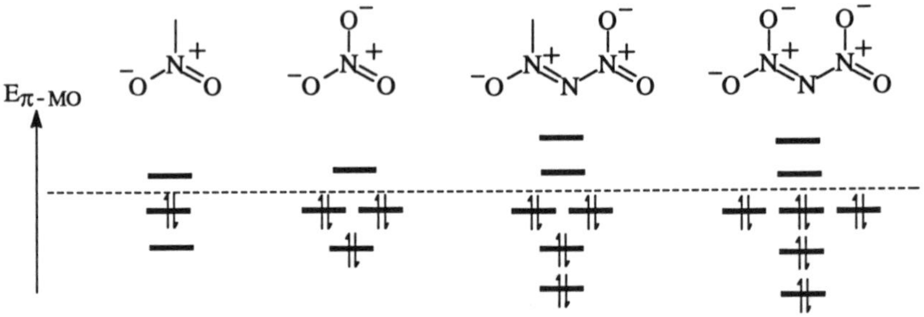

$E\pi\text{-MO}$

In addition to linear combinations, cyclic systems can also be formed. Molecules built according to this principle have a closed binding electronic shell, this being a necessary condition for the existence of the compounds and a main criterion for the selection that lies in the basis of our method. Now we are discussing only an π-electronic shell. However a mighty destabilizing factor in polyheteroatomic systems is the interaction of unshared lone pairs located on neighboring heteroatoms. That is why we introduce a condition stating that the atoms carrying electron pairs should alternate with positively charged atoms. The construction of molecules from elements of the trimethylene methane dianion automatically provides this condition as the central carbon atom of this system can be substituted by a positively charged atom of ammonium-type nitrogen. In the case of molecule building from fragments of allyl anion the middle carbon atom should be substituted for the RN^+ fragment. The compensation of the positive charge can be either external (these will be compounds of the salt-type) or internal, by creating a semipolar bond, as it takes place, for example, in a nitro group. This path is more advantageous.

Let us consider some examples. A combination of the allyl anion with the trimethylene methane dianion leads to a compounds that contain a new functional group, viz, nitrodiazene oxide. It is a very interesting group of the N_3O_3 composition. It contains the same number of nitrogen atoms as an azide group and the same amount of oxygen as a nitrate group. All nitrogen atoms are bonded to each other and all oxygen atoms are active. The addition of such two groups, for example, to furazan would enable one to introduce 6 active oxygen and 6 nitrogen atoms into the molecule, i.e. to obtain a compound of the $C_2N_8O_7$ composition. This new nitrogen -oxygen system undergoes the rule of the charge alternation.

There are two approaches to the synthesis of this type of compounds. Nitroso derivatives are starting compounds in both cases. The first path is a double-step. At the first step the nitroso compound is transformed to that of the ArN(O)=NR type, and at the second step R is substituted by a nitro group. The second variant is one-step, in which a nitroso compound being affected by nitronitrene. We have tried both ways. The first variant was realized via intermediates of the ArN(O)=NCOMe and ArN(O)=NBu^t types by the substitutional nitration with nitronium salts. The second variant was realized by means of electrochemistry. For the one-step synthesis of nitrodiazene oxides a reaction of nitroso compounds with nitramide can also be used [6].

$$\text{Ar—N=O} \xrightarrow{\text{Br}_2\text{NR}} \text{Ar—}\overset{\overset{\text{O}}{\uparrow}}{\text{N}}\text{=N—R} \xrightarrow[\text{CH}_3\text{CN}]{\text{NO}_2\text{X}} \text{Ar—}\overset{\overset{\text{O}}{\uparrow}}{\text{N}}\text{=N—NO}_2$$

$$\overset{\cdot\cdot}{\underset{\cdot\cdot}{\text{N}}}\text{—NO}_2$$

R = Ac, Bu^t,   X = BF$_4$ or SiF$_6$

The replacement of the nitro group by the nitrodiazenoxide is considerably advantageous for molecule energetic characteristics. It is expressed in the increase of the enthalpy of formation (50 kcal/mol with a single replacement), density and detonation rate.

Unfortunately, in spite of the record-breaking energetic characteristics, these compounds are unlikely to be implemented in practice due to their relatively low stability. Melting points of the most nitrodiazene oxides interesting from the practical standpoint are in the 80—120°C range. As a rule, these compounds start todecompose either directly at the moment of melting or 5—10°C after. This proved to be a general feature of this class of compounds and it is connected with the fact that the nitrodiazenoxide group does not belong to a single plane. Due to the repulsion of oxygen atoms the nitro group is almost 90° evolved to the plane in which the diazenoxide fragment atoms lie. This leads to the lengthening of the nitrogen-nitro group bond to 1.47—1.49Å and its weakening.

Now let us discuss a combination of two fragments of the trimethylene methane dianion. The substitution of all carbon atoms by nitrogen and oxygen atoms leads us to the dinitramide anion system. The ammonium salt of this anion, ADN, is widely known at present. This compound is one of the best oxidizers for solid propellants and it considerably exceeds ammonium perchlorate in efficiency. ADN has other positive characteristics: the ecological purity of combustion products and their elevated transparency owing to the lack of chlorine atoms in its composition. Dinatramide salts were for the first time synthesized in my laboratory in May 1971 and later, independently, in the USA in the end of the 80-s. It should be emphasized that the discovery of both dinitramide and its salts, apart from its practical significance, is a breakthrough in chemistry. Dinitramide is one of the strongest inorganic acids and on its basis hundreds of simple and complex salts can be synthesized like on the basis of well-known common inorganic acids.

A few words about the name of the new class of compounds. The name dinitramide is not quite correct. Amide is the name of organic derivites of ammonia which contain only one functional group at the nitrogen atom. Compounds, containing two functional groups at nitrogen atom are called imides. So it will be more correct to call it dinitroimide.

But to choose the name for the first representative of the new class of inorganic compounds we took into account its properties. It is one of the strongest mineral acids ($pK_a = -5.62 \pm 0.04$). That is why we name it "Dinitrazenic acid". But American chemists gave the name DNA and used this name consistanly for long time. I shall use this name.

There are likely to be three different approaches to the synthesis of dinitroamide salts. Two of them are based on the creation of the $N(NO_2)_2$ fragment in the organic molecule with the further elimination of the $N(NO_2)_2$ anion. The principle of the third approach in the direct nitration of ammonia or its simple derivatives. All three approaches to the synthesis of DNA salts were realized in various options both in Russia and the USA [7,8].

$$(NO_2)_2NCH_2CH_2X \xrightarrow{Base} (NO_2)_2N^- + CH_2=CHX$$

$$(NO_2)_2NR \longrightarrow (NO_2)_2N^- + R^+$$

$$NH_3 \text{ or } NH_2X \xrightarrow{NO_2^+} (NO_2)_2NH$$

Dinitroamide in its concentrated state is unstable due to the self-protonation reaction. In dilute water solutions, dinitramide is rather stable. The stability of DNA salts in a solid state varies in very broad ranges and is likely to be connected with the structural symmetry of an anion in the crystalline lattice. The more symmetric the anion and the more equivalent the nitro groups, the more stable are the salts. The anion symmetry is, in its turn, dependent on the cation size. It is probable that the smaller the cation, the higher is its ability to distort the anion symmetry by coordinating mainly with one of oxygens of the anion. Certainly, these speculations are of a qualitative character and the stability of the salts is affected by a whole set of factors, particularly the nature and amount of micro impurities.

DNA salts, specifically Hg salt, can be applied to the organic synthesis and serve as a source of various gem-dinitramines . During alkylation the salts can also form appropriate nitroethers which, as a rule, are rather unstable and decompose in solutions at room temperature.

The last part of my communication will be devoted to the chemistry of a new class of high nitrogen heterocyclic compounds, viz, tetrazine dioxides. From the results discussed above it becomes evident that the stability of compounds is greatly affected by a special position of the heteroatomic fragment and, particularly, by the removal of its individual elements from the plane. It would be logical to suppose that the involvement of such a fragment into an aromatic system should benefit to its stability increase. As mentioned above, our approach to the formal designing of polynitrogen-oxygen systems can be also applied to cyclic structures.We have chosen a more accessible structure, viz, tetrazine derivatives. For the first time tetrazines were synthesized only in the end of the 80-s. They proved to be rather unstable. According to our theory, the introduction of two oxygen atoms into the tetrazine cycle should increase the system stability and tetrazine 1,3-dioxide should be most stable, as this compound completely undergoes the principle of the charge alternation and this isomer is more advantageous by the decomposition energy in comparison with 1,4-isomer.

We proposed a scheme for the synthesis of tetrazine 1,3-dioxides,the key step of which is the reaction of a tert-butyldiazene oxide group with *ortho*-situated N_2O cation. The starting product in this scheme is *ortho*-aminonitrosobenzene. The transformation of a nitroso group to a diazene oxide caused no difficulties but the transformation of an amino group to the $-N=N=O^+$ cation system required the elaboration of new reactions.

We managed to find two reactions of this type and both of them have lead us to the target product. The reaction with the use of N_2O_5, which gave to benzotetrazine 1,3-dioxide, yields a number of its derivatives as a result of by-processes. The route *via* diazo salts leads only to target 1,3-dioxide in good yields. As expected, tetrazine 1,3-dioxides proved to be highly stable compounds, their decomposition starts at above 200°C. Further on a few more tetrazine 1,3-dioxide derivatives were obtained and reactions of electrophilic and nucleophilic substitution on a benzene ring were investigated. All these dioxides are rather stable [9].

CONCLUSION

Apparently, a search of stable high energy compounds should be carried out in various directions. Successes achieved in the chemistry of wuritzitane are a good example to this. Yet, we believe that our approach extends greatly the scope of activities for specialists in the synthesis of high energy compounds and, besides, is of great theoretical interest as the problem of understanding the structure-property link, which is one of the key challenges in the theoretical chemistry.

ACKNOWLEDGMENTS

Dioxoazabicyclononane and dioxoazabicyclooctane nitro derivatives were synthesized and studied by Dr. I. E. Chlenov. He obtained nitrohydroxylamines also. The synthesis of nitrohydrazines was performed by Dr. S. L. Ioffe and Dr. A. V. Kalinin. Sulfo-N-nitroimides were obtained by Dr. O. P. Shitov. He also synthesized bis-(nitroimido)-imidazole K-salt and 1-nitroamino-1,2,3-triazol-3-nitroimide K-salt. The strategy for the design of high nitrogen systems was elaborated by Dr. A. M. Churakov and Dr. I. E. Chlenov. The synthesis of nitrodiazene oxides was lead by Dr. A. M. Churakov and Dr. S. L. Ioffe. Dinitrazenic acid and its salts was first synthesized by Dr. O. A. Luk'yanov. Fused 1,2,3,4-tetrazine 1,3-dioxides were synthesized by Dr. A. M. Churakov and Dr. S. L. Ioffe.

REFERENCES

1. Chlenov, I. E., Morozova, N. S. and Tartakovsky, V. A., Bull. Acad. Sci. USSR, Div. Chem. Sci, 32, 1713, 1983.

2. Kalinin, A. V., Apasov, E. T., Ioffe, S. L. and Tartakovsky, V. A., Bull. Acad. Sci. USSR, Div. Chem. Sci., 40, 988, 1991.

3. Myasnikov, V. A., Vyazkov, V. A., Yudin, I. L., Shitov O. P. and Tartakovsky, V. A., Bull. Acad. Sci. USSR, Div. Chem. Sci., 40, 1116, 1991.

4. Shitov, O. P., Vyazkov V. A. and Tartakovsky V. A., Bull. Acad. Sci. USSR, Div. Chem. Sci., 2440, 38, 1989.

5. Shitov, O. P., Seleznev A. P. and Tartakovsky V. A., Bull. Acad. Sci. USSR, Div. Chem. Sci., 40, 1115, 1991.

6. Rezchikova, K. I., Churakov, A. M., Shlyapochnikov V. A. and Tartakovsky, V. A., Mendeleev Commun, 100, 1994.

7. Luk'yanov, O. A., Gorelik V. P. and Tartakovsky, V. A., Russ. Chem. Bull., 89, 43, 1994.

8. Bottaro, J. C., Schmitt, R. J., Penwell, P. E., Ross D. S., "Dinitramide salts and method of making same", WO 91/19669.

9. Churakov, A. M., Ioffe S. L. and Tartakovsky, V. A., Mendeleev Commun, 101, 1991.

POLYNITROALKANES, NITROAZOLES AND OXIMES IN ELECTROORGANIC SYNTHESIS

Murat NIYAZYMBETOV
Zelinsky Institute of Organic Chemistry, Moscow and University of Delaware, Department of Chemistry and Biochemistry, Newark, DE 19716, murat@brahms.udel.edu

ABSTRACT

Electrochemical oxidation of anions of polynitroalkanes (PNA) proceeds via one-electron transfer and results in the formation of unstable radicals. The radicals undergo fragmentation, abstraction of a hydrogen atom from the solvent, and reaction with aromatic compounds to form aryl-PNA. The yield of aryl-PNA depends on the structure of the radicals and on the electron donating properties of the aromatic substrates. Due to their electrophilic properties, PNA-radicals also react with compounds containing heteroatoms with unshared electron pairs resulting in formation of dinitromethylides. Under anodic electrolysis conditions dinitromethylides which have a carbanionic structure can effectively transfer the dinitromethyl group to other organic substrates. Electrochemical oxidation of anions of nitroazoles also proceeds via one-electron transfer to generate nitroazolyl radicals. Their principal reaction is abstraction of hydrogen from the solvent. However, electrolysis of nitroazoles in the presence of aromatic compounds results in formation of N-aryl derivatives in good yield. It has been shown that electrooxidation of mono oximes leads to formation of unstable nitroso dimers, while electrooxidation of vicinal dioximes results in formation of furoxans in high yield.

INTRODUCTION

Over the past 10-15 years the use of electrochemistry as a synthetic tool in organic chemistry has remarkably increased.[1] An attractive feature of the electrochemical method is that it has the possibility to control the reagent electron over a wide range. For example, electrochemical method enables a wide variety of organic anions to be generated at the cathode under very mild conditions and without using a chemical base. The main advantage of this approach is that it is very easy to control both the solvent and counterion, which can be very useful because the nucleophilic reactivity of anions depends strongly on these factors.[2] The cathodic generation of anions is most useful for strongly nucleophilic anions. By contrast, thermodynamically stable and weakly nucleophilic anions, such as anions of polynitroalkanes or nitroazoles, are useful in anodic (oxidation) processes for generating electrophilic intermediates (radicals and carbenes). Another possibility is to use weak nucleophiles as acceptors of highly-reactive anodically-generated electrophiles (cation-radicals and cations). Traditionally, electrochemistry of polynitro- and nitroheterocyclic compounds was mainly limited to solving analytical and mechanistic problems. Nevertheless the electrochemical method can be applied in the synthesis of a variety of polynitroalkanes and nitroazoles. This article is a review of the use of anions of polynitroalkanes and nitroazoles in electrosynthesis both as precursors for electrophilic radicals and acceptors of electrogenerated cation-radicals and cations. Finally, the anodic behavior of oximes and vicinal dioximes has been studied. A general method for electrosynthesis of furoxans has been developed.

EXPERIMENTAL

The cell for divided electrolyses was a 100-ml glass beaker with a polyethylene cover in which was fixed a platinum gauze anode (35 cm^2) and the cathode compartment which was a polypropylene tube (Nalgene Centrifuge Tube, 28.7 × 103 mm). About 50 6-mm diameter holes were drilled in the side of the tube and it was wrapped with three layers of tracing paper to serve as the separator (diaphragm). The paper was held in place by a few rounds of cotton thread. The cathode was a 10-mm diameter graphite rod.

A 0.1-0.2 N solution of LiClO$_4$ in dry MeCN (60 ml) containing 3-8 mmol of the corresponding salts of polynitroalkanes, nitroazoles or dinitromethylides and 10 vol. % of organic substrate was placed in the anodic compartment of the divided cell. The solution was stirred with a magnetic stirrer. In the cathode compartment was placed 40 ml of supporting electrolyte. The electrolysis was performed at potentials corresponding to oxidation of anions of polynitroalkanes or dinitromethylides. After electrolysis the solution was removed and the solvent was evaporated under vacuum. The products were extracted from the residue by ether (polynitroalkanes) or by ethyl acetate (nitroazoles).

The instrumentation and detailed procedures of the electrolyses, product identification and electroanalytical studies have been described elsewhere.[3-7]

RESULTS

It is known that electrochemical oxidation of anions of mononitro alkanes proceeds via one-electron transfer and formation of radicals. The main reaction of these radicals is a dimerization reaction.[7] Anions of polynitroalkanes (PNA), which are weak nucleophiles, have totally different anodic behavior. Detailed investigation of the electrooxidation of the anions of PNA showed[3] that the reaction proceeds also via one-electron transfer and formation of radicals of PNA. The radicals are unstable and undergo fragmentation and abstract a hydrogen atom from the solvent. However, due to their electrophilic properties, radicals of PNA give no dimeric products[3] but react with aromatic substances[4] on the aromatic ring to form the arylpolynitroalkanes (Scheme 1) in 85-95 yield based on the aromatic substrates.[8]

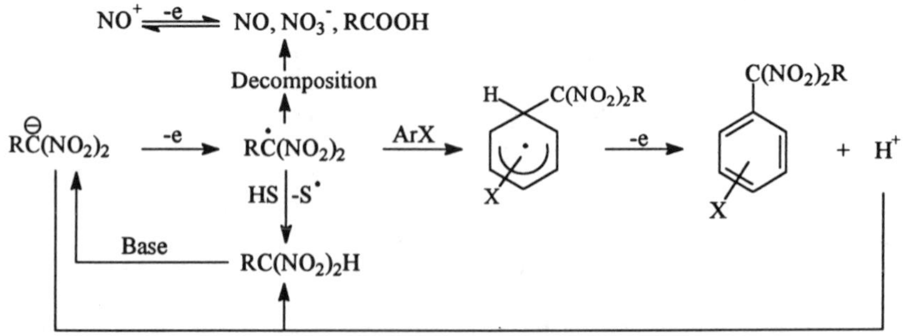

R = H, Ph, Cl, CN, NO$_2$; X = H, Me, OMe, Cl

Scheme 1

The reaction is accompanied by formation of the un-ionized PNA because reaction of the radical with solvent and protonation of starting anion occur. However, un-ionized PNA are relatively strong *C-H* acids (pK_a=0-5), and the addition of the weak base will deprotonate the neutral PNA so formed, allowing the reintroduction of the anion into the reaction.

It is obvious that the electrochemical method gives a short synthetic path to arylpolynitroalkanes. The chemical synthesis of these compounds includes several steps and the overall yields are low.[9]

Due to their strong electrophilic properties, electrogenerated radicals of PNA can react with compounds containing heteroatoms having unshared electron pairs to form the corresponding dinitromethylides.[10] Dinitromethylides have a carbanionic structure and can be oxidized . It was found that oxidation of these compounds proceeds *via* formation of an unstable cation-radical which decomposes with formation of dinitrocarbene.[11] The dinitrocarbene generated thus inserts into a C-H bond of benzene and its derivatives (Scheme 2) to form the corresponding dinitromethylaryl compounds.[12] The carbene mechanism can be considered for products which are formed in low yields during electrolysis of dinitromethylides in the presence of cyclohexane, cyclohexene, pyridine, and fluoride ion.[12,13]

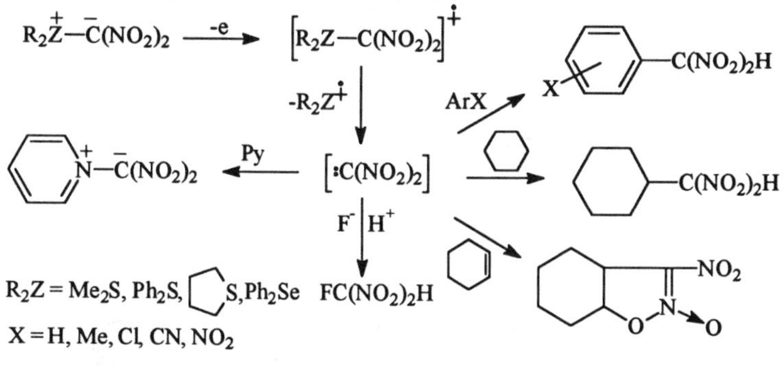

Scheme 2

Anodic oxidation of the anions of nitro derivatives of imidazole, pyrazole, triazole and tetrazole proceeds *via* one-electron transfer and generation of electrophilic nitroazolyl radicals.[5,14] No dimerization products were observed. The principal reaction of these radicals is abstraction of hydrogen from the solvent. But electrolysis in the presence of aromatic compounds results in formation of the corresponding aryl derivatives. So, anodic electrolysis of a mixture of tetraethylammonium salt of 3-nitro-1,2,4-triazole and un-ionized nitrotriazole in the presence of benzene in MeCN in the undivided cell results in formation of the N-phenyl derivatives (Scheme 3) in 67% isolated yield.[5] The anion of nitrotriazole is partially regenerated at the cathode during electrolysis.

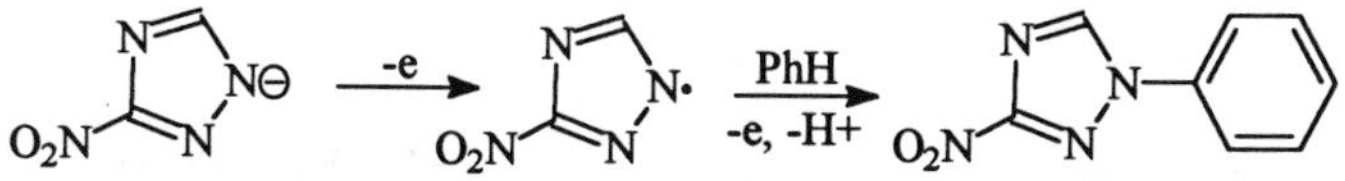

Scheme 3

A wide variety of organic substances such as aromatic compounds, condensed aromatic compounds, aromatic and nonaromatic heterocycles, amines, olefins, thiols, sulfides, alcohol's etc., under anodic electrolysis conditions in nonaqueous media, can be oxidized to cation-radical or cation intermediates. Due to the strong electrophilic character of these intermediates, they are able to couple with nucleophiles, especially anionic nucleophiles, that are present in the solution.

This approach opens excellent possibilities for the synthetic utility of thermodynamically stable anions of PNA and nitroazoles which are usually very weak nucleophiles. Due to the high oxidation potentials of these anions, the electrolysis proceeds by selective oxidation of the organic substrate rather than the anion. However, the cation-radicals or cations so formed can react with unoxidized anions of PNA and nitroazoles at the surface of electrode or in the bulk solution.

Aromatic compounds with electron-donating substitutents or condensed aromatic substances, which are oxidized at relatively low potentials, can serve as a good source for electrogenerated electrophilic intermediates. So, electrooxidation of 1,4-dimethoxybenzene proceeds via one-electron transfer and formation of a relatively stable cation radical. In the presence of the anion of 3-nitro-1,2,4-triazole, which is oxidized at a higher potential ($E_{1/2}$ = 1.77 V vs. Ag/Ag$^+$ in MeCN) than 1,4-dimethoxybenzene ($E_{1/2}$ = 1.0 V vs. Ag/Ag$^+$ in MeCN), complete two-electron oxidation occurs forming the aryl derivative of the nitrotriazole in good yield (75-85%).[15] The cation radical reacts with the anion of nitrotriazole to produce a cyclohexadienyl radical which, in turn, undergoes oxidation and deprotonation (Scheme 4).

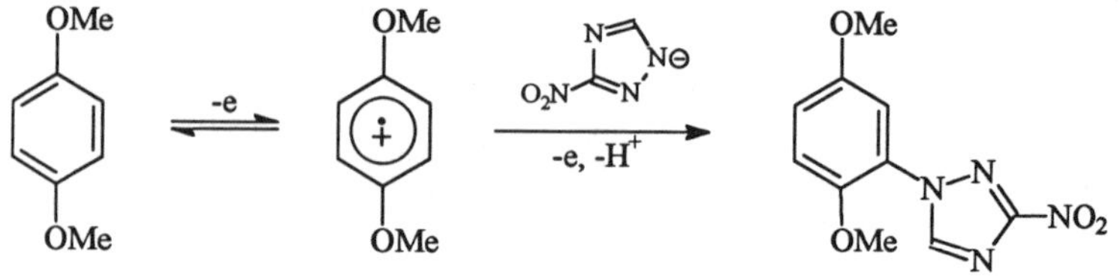

Scheme 4

By analogy, the anion of 3-nitro-1,2,4-triazole reacts with the electrogenerated cation radical of naphthalene at the α-position of the aromatic system.[15b] The formation of α-trinitromethyl- and α-dinitromethylnaphthalenes was observed when naphthalene was oxidized in the presence of the salts of dinitromethane and trinitromethane.[8]

The electrochemical oxidation of sulfides[8,10b] in the presence of the anion of trinitromethane and oxidation of trimethylamine[16] in the presence of the anion of dinitromethane proceed via intermediate formation of cationic intermediates, which subsequently react with the anions to produce the corresponding polynitromethyl derivatives (Scheme 5).

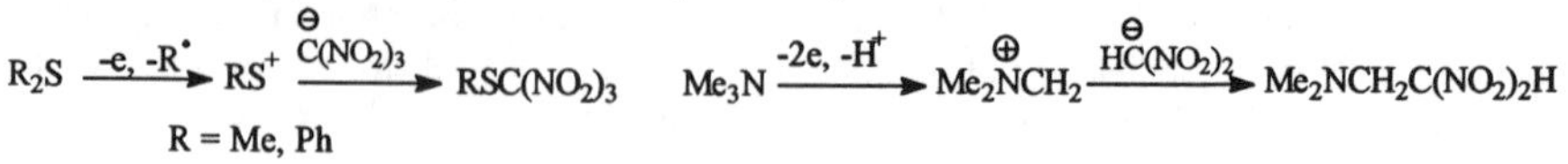

Scheme 5

Anions of PNA are very good acceptors of electrogenerated halogens.[17] It has been found that anions, the oxidation potentials of which are higher than the oxidation potential of chloride anion ($E_{1/2}$ = 0.8 V vs. Ag/Ag$^+$ in MeCN), are chlorinated in quantitative yields in both aqueous and nonaqueous media on Pt, graphite, or ruthenium dioxide-titanium anodes. The formation of chlorinated products of carbanions with oxidation potentials lower than chloride can be achieved in high yield with ruthenium dioxide-titanium anodes, high concentrations of chloride, and high current densities.[17]

It should be noted that electrochemical chlorination allows selective preparation of either monochloro or dichloro derivatives of dinitromethane. Electrolysis of a solution of 0.2 M Et$_4$NCl MeCN in the presence of the anion of dinitromethane with charge consumption of 1 F/mol results in the exclusive formation of monochlorinated product in 99% yield. Under the same conditions when 5 F/mol charge was passed or the electrolysis was carried out in a solution of 3-5 M NaCl in H$_2$O containing 0.5 g/L NaHCO$_3$, the dichlorinated product was formed in 80-93% yield. For comparison it should be noted that the known chemical chlorination of the anion of dinitromethane results in formation of dichlorodinitromethane in 70% yield and chlorodinitromethane can be prepared in 56% yield (total yield 39%) only by dechlorination of dichlorodinitromethane.[18]

Electrosynthesis can be applied in the formation of a variety compounds. In this example, anodic behavior of oximes and vicinal dioximes has been studied.[19] On the basis of voltammetric investigations and potential control electrolyses, it has been found that electrochemical oxidation of aldoximes proceeds via one-electron and proton transfer with formation of iminoxyl radicals, which further dimerize to nitroso dimers. The oxidation of ketoximes, on the other hand, involves further oxidation of intermediate iminoxyl radicals to iminoxonium cations, which react with the starting ketoximes to give also nitroso dimers (Scheme 6). These dimers are not stable and decompose with formation of a number of compounds.

The electrochemical behavior of vicinal dioximes is similar to that of ketoximes. The intermediate iminoxyl radical is oxidized to oxoimmonium cation which reacts with the adjacent oxime group to give a furoxan ring (Scheme 6).[19b]

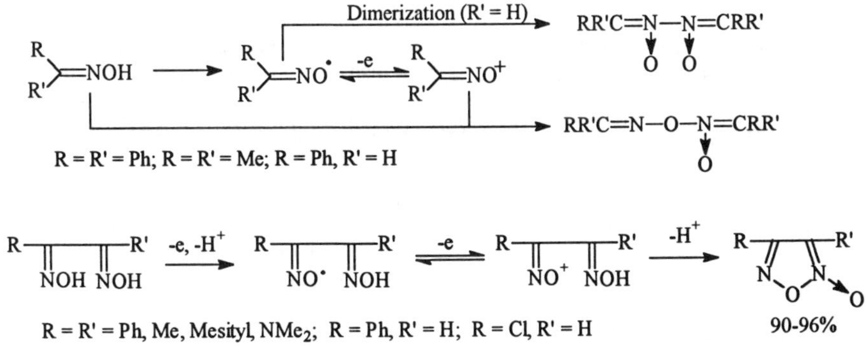

$$R = R' = Ph; \ R = R' = Me; \ R = Ph, R' = H$$

$$R = R' = Ph, Me, Mesityl, NMe_2; \ R = Ph, R' = H; \ R = Cl, R' = H \qquad 90\text{-}96\%$$

Scheme 6

The influence of electrolysis conditions including electrode composition, solvent, electrolysis potential, electrolyte and temperature on the electrosynthesis of furoxans has been studied.[6] It was found that the formation of furoxans with high yields occurs on Pt or graphite anodes at the potentials below oxidation potentials of the furoxans and passing 40-60% of the theoretical amount of electricity at 15-25 °C. In basic media the electrolysis potential should be maintained at the value sufficient to generate oxoimmonium cation. Compare with other chemical methods the electrochemical method allows synthesis of different kinds of mono- and di-substituted furoxans with high purity and yield.[6]

REFERENCES

1. (a) H. Lund, M. M. Baizer, <u>Organic Electrochemistry</u>; 3rd ed. (Marcel Dekker, New York, 1991). (b) T. Shono, <u>Electroorganic Synthesis</u>, (Academic Press, London, 1991). (c) S. Torii (editor), <u>Novel Trends in Electroorganic Synthesis</u>, (Kodansha, Tokyo, 1995).

2. M. E. Niyazymbetov, D. H. Evans, J. Chem. Soc., Perkin Trans. 2, 1333 (1993).

3. V. A. Petrosyan, M. E. Niyazymbetov, Izv. Akad. Nauk SSSR. Ser. Khim. <u>1987</u>, 603.

4. V. A. Petrosyan, M. E. Niyazymbetov, ibid., <u>1988</u>, 368.

5. V. A. Petrosyan, M. E. Niyazymbetov, M. S. Pevzner, B. I. Ugrak, ibid., <u>1988</u>, 1643.

6. M. E. Niyazymbetov, E. V. Ul'yanova, V. A. Petrosyan, Elektrokhimiya 28, 555 (1992).

7. (a) V. A. Petrosyan, M. E. Niyazymbetov (unpublished results, 1991); (b) S. Wawszonek, T.-Y. Su, J. Electrochem. Soc. 120, 745 (1973).

8. M. E. Niyazymbetov, V. A. Petrosyan (unpublished results, 1988).

9. (a) S. S. Novikov, L. I. Khmel'nitskii, O. V. Lebedev, Zh. Obshch. Khim. 28, 2296 (1958); (b) L. I. Khmel'nitskii, S. S. Novikov, O. V. Lebedev, ibid., 28, 2303 (1958); (c) S. S. Novikov, L. I. Khmel'nitskii, O. V. Lebedev, Izv. Akad. Nauk SSSR. Otdel. Khim. Nauk <u>1960</u>, 1783; (d) L. I. Khmel'nitskii, S. S. Novikov, O. V. Lebedev, ibid., <u>1960</u>, 2019.

10. (a) M. E. Nyazymbetov, V. A. Petrosyan, Izv. Akad. Nauk SSSR. Ser. Khim. <u>1985</u>,1213; (b) M. E. Niyazymbetov, V. A. Petrosyan, Elektrokhimiya., 21, 136 (1985).

11. V. A. Petrosyan, M. E. Niyazymbetov, A. A. Fainzilberg, S. A. Shevelev, V. V. Semenov, Izv. Akad. Nauk SSSR. Ser. Khim. <u>1989</u>, 356.

12. M. E. Niyazymbetov, V. A. Petrosyan, , A. A. Fainzilberg, S. A. Shevelev, V. V. Semenov,.ibid., <u>1989</u>, 1094.

13. V. A. Petrosyan, A. A. Fainzilberg, M. E. Niyazymbetov, V. N. Solkan, ibid., <u>1989</u>, 1089.

14. M. E. Niyazymbetov, V. A. Petrosyan, A. A. Gakh, A. A. Fainzilberg, ibid., <u>1987</u>, 2390.

15. (a) M. E. Niyazymbetov, L. V. Mikhal'chenko, V. A. Petrosyan, Abstracts of papers of National Conference on Aromatic Nucleophilic Substitution, USSR, Novosibirsk, <u>1989</u>, 91; (b) M. E. Niyazymbetov, L. V. Mikhal'chenko, V. A. Petrosyan (unpublished results, 1988).

16. M. E. Niyazymbetov, V. A. Petrosyan, Izv. Akad. Nauk SSSR. Ser. Khim. <u>1984</u>, 1676.

17. V. A. Petrosyan, M.E. Niyazymbetov, B. V. Lyalin, ibid., <u>1987</u>, 305.

18. V. I. Erashko, S. A. Shevelev, A. A. Fainzil'berg, ibid., <u>1965</u>, 2060.

19. (a) V. A. Petrosyan, M. E. Niyazymbetov, E.V. Yl'yanova, ibid., <u>1990</u>, 625; (b) M. E. Niyazymbetov, E. V. Yl'yanova, V. A. Petrosyan, ibid., <u>1990</u>, 630.

AMINONITROHETEROCYCLIC *N*-OXIDES --
A NEW CLASS OF INSENSITIVE ENERGETIC MATERIALS

RICHARD A. HOLLINS*, LAWRENCE H. MERWIN*, ROBIN A. NISSAN*, WILLIAM S. WILSON*, RICHARD D. GILARDI**
*Naval Air Warfare Center Weapons Division, 1 Administration Circle, China Lake, CA 93555-6001
**Naval Research Laboratory, 4555 Overlook Ave SW, Washington, D.C. 20375-5000

ABSTRACT

The need continues for new powerful but insensitive explosive ingredients, which match the performance of RDX with the insensitivity of TATB. One approach has been to take the inherent stability of an aromatic heterocycle, combine this with the explosive insensitivity and thermal stability associated with alternating amino and nitro groups, and to supplement the performance of the molecule with an energy contribution from the *N*-oxide functionality. The synthesis, characterization and properties (both sensitivity and performance) of aminonitro-pyridine-1-oxides and aminonitropyrimidine-1,3-dioxides will be described.

INTRODUCTION

The ongoing requirement for ever more powerful energetic materials is well understood. However there is a longstanding parallel requirement for new insensitive energetic materials, and specifically for materials which couple the performance of RDX (VofD 8940 m/s, P_{CJ} 378 kbar, m.p. 204°C, $h_{50\%}$ 22 cm) with the stability and insensitivity of TATB (VofD 7860 m/s, P_{CJ} 277 kbar. m.p. 350°C, 10/10 no fires at 200 cm). Our approach to the synthesis of such materials has been to take the inherent stability of an azaheterocycle, coupled with alternating nitro and amino groups known to confer stability and insensitivity on TATB through intramolecular and inter-molecular hydrogen bonding, and to enhance the oxygen balance by the inclusion of the *N*-oxide functionality, thereby supplementing the explosive energy of the resultant molecule. Specific synthetic targets include 3,5-dinitro-2,4,6-triaminopyridine-1-oxide (1), 3,5-diamino-2,4,6-trinitropyridine-1-oxide (2) and the isomeric 4,6-diamino-2,5-dinitro- and 2,5-diamino-4,6-dinitropyrimidine-1,3-dioxides (3) and (4). (Densities are those calculated by the Holden method [1]; detonation properties are calculated by the method of Rothstein and Petersen [2].)

(1)	(2)	(3)	(4)
Density 1.81 g/cm³	Density 1.90 g/cm³	Density 1.92 g/cm³	
VofD 8010 m/s	VofD 8650 m/s	VofD 8930 m/s	
P_{CJ} 291 kbar	P_{CJ} 351 kbar	P_{CJ} 377 kbar	

EXPERIMENTAL RESULTS AND DISCUSSION

Mixed acid nitration of 2- and 4-aminopyridines occurs smoothly to give the expected aminonitropyridines. In a like manner, treatment of 2,6-diaminopyridine with 90% nitric acid in 96% sulfuric acid at 0-5° and then 60-65°C gave a 60-65% yield of 2,6-diamino-3,5-dinitro-

$$(5) \qquad\qquad (6)$$

pyridine (5), which was then oxidized to 2,6-diamino-3,5-dinitropyridine-1-oxide (6) in 80% yield by heating under reflux with 30% aqueous hydrogen peroxide in glacial acetic acid. (This procedure has also been reported by Ritter and Licht [3].) ^{1}H-n.m.r. spectra of both (5) and (6) in DMSO at ambient temperature showed two two-proton signals for the amine groups, which coalesced to a single four proton signal at elevated temperatures. This behavior was interpreted as encouraging evidence for the desired intramolecular hydrogen bonding with the adjacent nitro groups and N-oxide functionality, even in solution. Single crystal X-ray diffraction studies on (6) confirmed the intramolecular hydrogen bonding, in addition to intermolecular bonding which results in assembly of the molecules in graphitic sheets strongly reminiscent of TATB; the observed crystal density was 1.88 g/cm^3. As expected, the hydrogen bonding also contributes to the stability and insensitivity of (6), which melts with decomposition above 340°C (but without endotherm or exotherm below that temperature, and is insensitive to impact (10/10 no fires at 200 cm), sliding friction(10/10 no fires at 1000 lb) and electrostatic discharge (up to 10/10 no fires at 0.25 J). Finally, preliminary experiments on charges pressed from compositions containing 5% EVA (ethylene vinyl acetate copolymer) indicate that the explosive performance of (6) the predictions for both that material and for TATB. Thus the Rothstein method [2] appears well-suited for prediction of explosive performance of heterocyclic N-oxides.

Two potential routes were considered for the synthesis of (1), the first being oxidation of 3,5-dinitro-2,4,6,triaminopyridine (7). The latter compound has been prepared in a six-step sequence from pyridine-1-oxide [4], but we preferred to use an oxidative amination procedure employing potassium permanganate in liquid ammonia. Treatment of 2-chloro-3,5-dinitro-pyridine under these conditions led initially to a mixture of (5) and (7) [5]. Prolonged reaction gave (7) as the sole product in up to 30% yield, while treatment of (5) in the same manner gave a 61% yield of the desired (7). It was later shown that a 40% conversion of (5) to (7) could also be achieved by a "vicarious nucleophilic amination" procedure employing hydroxylamine in aqueous potassium hydroxide [6]. Oxidation of (7) using 30% hydrogen peroxide in acetic acid under reflux gave the desired 3,5-dinitro-2,4,6-triaminopyridine-1-oxide (1), but only in 9 - 10% yield.

$$(5) \qquad\qquad (7)$$

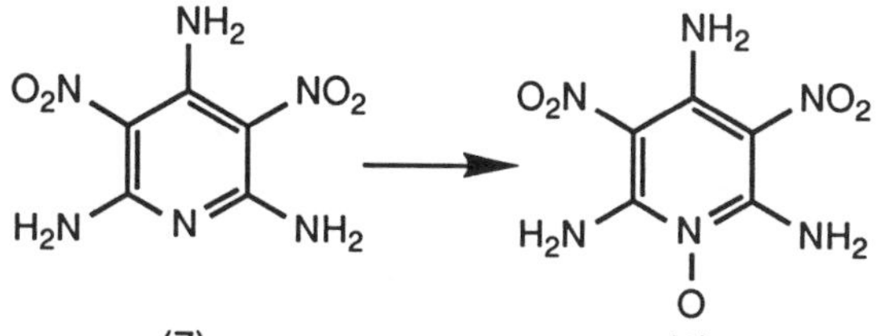

(7) → (1)

The alternative approach to the synthesis of (1) involved amination of (6). Indeed treatment of (6) under the vicarious amination conditions (hydroxylamine in aqueous potassium hydroxide at ambient temperature) gave (1) in 39% yield, while acidification of the mother liquors allowed reclamation of 34% of the starting material. Thus the yield of (1) was 59% based on consumption of (6).

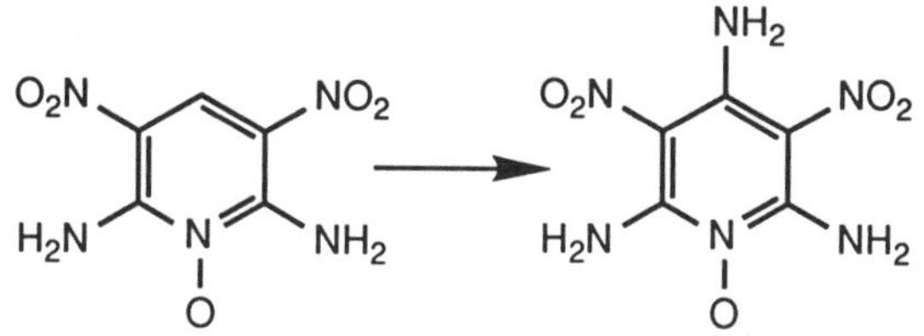

(6) → (1)

^{1}H-n.m.r. spectrum of (7) in DMSO at ambient temperature showed three two-proton signals for the amines, collapsing to two signals (two and four protons respectively) at elevated temperatures, indicative of the desired intramolecular hydrogen bonding. Low solubility prevented observation of the spectrum of (1) at ambient temperature, while at elevated temperatures two signals (two and four protons respectively) were observed. However single crystal X-ray diffraction experiments confirmed the structure of (1), its intramolecular and intermolecular hydrogen bonding and the resultant graphitic crystal structure. The crystal density of 3,5-dinitro-2,4,6-triaminopyridine-1-oxide is 1.88 g/cm^3, and the extensive hydrogen bonding also manifests itself in insensitivity to impact (10/10 no fires at 200 cm) and stability (melts with decomposition above 308°C).

Aminonitropyridines are not accessible by nitration of 3-aminopyridines. However mixed acid nitration of 3,5-dimethoxypyridine at 40°C gave 3,5-dimethoxy-2,6-dinitropyridine (8) in 36% yield, while mixed acid nitration of the N-oxide occurred smoothly at 90°C, giving 3,5-dimethoxy-2,6-dinitropyridine-1-oxide (10) in 42% yield [7]. Heating (8) in a Carius tube at 100°C with ethanolic ammonia gave 3,5-diamino-2,6-dinitropyridine (9) in 75% yield, but ammonolysis of (10) gave 2-amino-3,5-dimethoxy-6-nitropyridine-1-oxide (12) rather than the desired 3,5-diamino-2,6-dinitropyridine-1-oxide (11). Clearly the N-oxide functionality activates the 2-nitro group to nucleophilic displacement.

Crystals of (9) suitable for X-ray diffraction studies have not been obtained. However ^{1}H-n.m.r. spectra of this material in DMSO at ambient temperature reveal only one four-proton signal for the amine protons, suggesting that there is little intramolecular hydrogen bonding, at least in solution. Attempts at further nitration of these compounds in the 4-position have proven unsuccessful, as have attempts to oxidize (9) to (11). Current endeavors are devoted to attempts at "vicarious" amination of 2,4,6-trinitropyridine and its *N*-oxide, which are accessible via an alternative route [8].

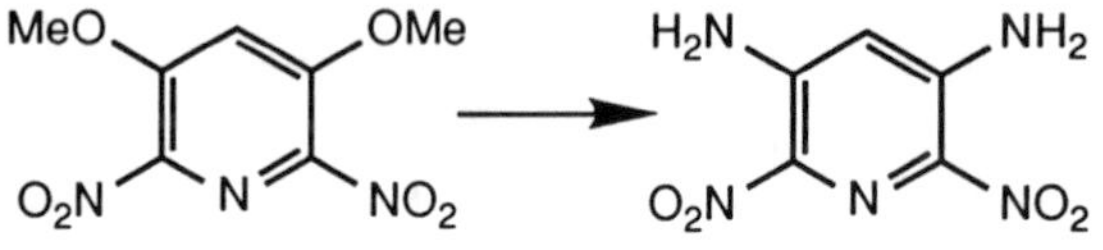

(8)   (9)

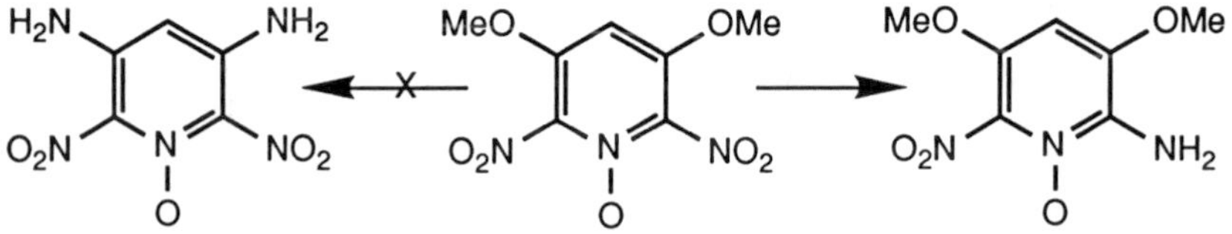

(11)   (10)   (12)

Nitration of pyrimidines occurs readily at the 5-position; 2-nitropyrimidine has been prepared by an indirect route, but 4(6)-nitropyrimidines are unknown. Attention was therefore focussed on 4,6-diamino-2,5-dinitropyrimidine-1,3-dioxide (3) rather than the alternative isomer (4). Mixed acid nitration of 2,4,6-triaminopyrimidine at ambient temperature gives 5-nitro-2,4,6-triaminopyrimidine (13) in 90% yield [9]. Oxidation of (13) with 30% aqueous hydrogen peroxide in trifluoroacetic acid at ambient temperature gives 5-nitro-2,4,6-triaminopyrimidine-1,3-dioxide (14) in 80% yield [10]. Empirical calculations predict that (14) should have a density of 1.84 g/cm^3 [1], velocity of detonation of 8150 m/s and detonation pressure of 304 kbar [2]. It was found to be chemically stable (melts with decomposition above 264°C), and insensitive to impact (10/10 no fires at 200cm), sliding friction (10/10 no fires at 1000 lb) and electrostatic discharge (10/10 no fires at 0.25 J). However although it showed the desired hydrogen bonding, and the now familiar graphitic crystal structure, (14) was found to contain one equivalent of water in the crystal lattice, and had a density of only 1.82 g/cm^3.

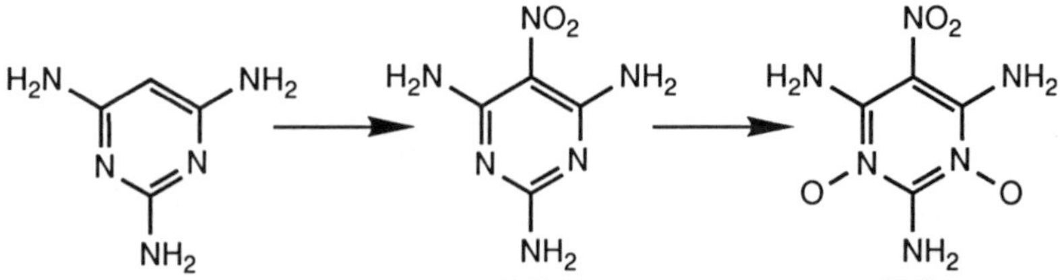

(13)   (14)

Mixed acid nitration of 4,6-diaminopyrimidine at ambient temperature also proceeded smoothly to give 5-nitro-4,6-diaminopyrimidine (15) in 90% yield [11]. Not unexpectedly, prolonged reaction at elevated temperatures failed to afford nitration in the 2-position, effecting instead degradation of the ring system. Nitration of the 1,3-dioxide (16) might be expected to give the desired substitution at the 2-position, but oxidation of (15) with 30% hydrogen peroxide in trifluoroacetic acid gave instead 2-hydroxy-5-nitro-4,6-diaminopyrimidine-1-oxide (17), presumably by rearrangement of (16) under the reaction conditions.

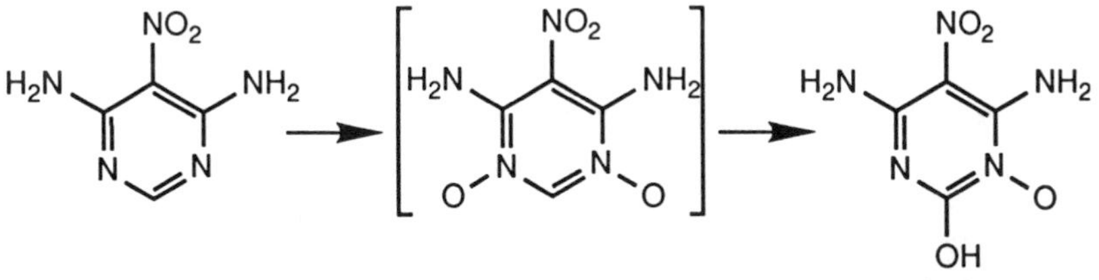

(15)   (16)   (17)

Alternative approaches to the synthesis of (3) are now being sought. 2-Nitropyrimidine was prepared from 2-aminopyrimidine by oxidation of an intermediate sulfilimine [12], and extension of this procedure to 4,6-disubstituted pyrimidines is being examined. Nitration, ammonolysis and oxidation should lead to the desired compound.

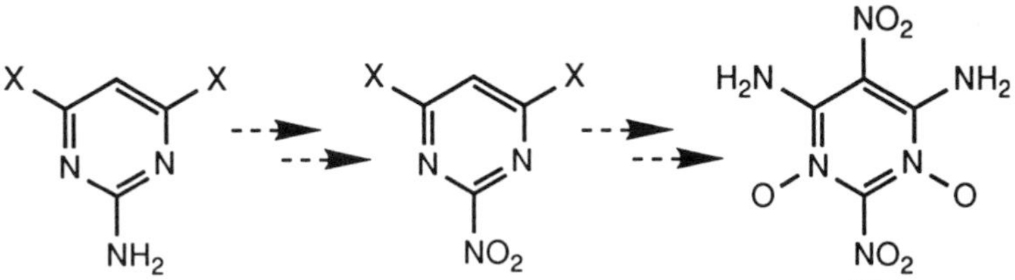

(3)

CONCLUSIONS

Aminonitropyridine-1-oxides and -pyrimidine-1,2-dioxides have been prepared which show stability and insensitivity comparable with those of TATB. Empirical calculations predict incremental improvement over the explosive performance of TATB. Preliminary experiments indicate that the performance of 2,6-diamino-3,5-dinitropyridine-1-oxide matches that predicted, demonstrating the applicability of this empirical method for heterocyclic N-oxides. Thus the feasibility of developing powerful but insensitive explosives from this class of compounds is confirmed. However the specific energetic targets sought have proven elusive, and alternative indirect routes to their synthesis are being explored.

ACKNOWLEDGEMENTS

This research has been supported by the Naval Air Warfare Center Weapons Division Independent Research program, and by various funds administered by the Office of Naval Research.

REFERENCES

1.   Naval Surface Weapons Center. *Estimation of Normal Densities of Explosives from Empirical Atomic Volumes*, by D.A. Cichra, J.R. Holden & C.R. Dickinson. Silver Springs, Md., NSWC, February 1980. 47 pp. (NSWC-TR-79-273, publication UNCLASSIFIED.)

2.   L.R. Rothstein & R. Petersen, *Prop. and Explo.*, **4**, 56 (1979); **6**, 91(1981).

3.   H. Ritter & H.H. Licht, *J. Heterocyclic Chem.*, **32**, 585 (1995).

4.   M.D. Coburn & J.L. Singleton, *J. Heterocyclic Chem.*, **9**, 1039 (1972).

5.   M. Wozniak, A. Baranski & B. Szpakiewicz, *Liebigs Ann.Chem.*, 7 (1993).

6.   J. Meisenheimer & E. Patzig, *Chem.Ber.*, **39**, 2533 (1906); A.R. Katritzky & K.S. Laurenzo, *J.Org.Chem.*, **51**, 5039(1986).

7.   C.D. Johnson, A.R. Katritzky & M. Viney, *J. Chem. Soc. (B)*, 1211 (1967); C.D. Johnson, A.R. Katritzky, N. Shakir & M. Viney, *J. Chem. Soc.*, *(B)* 1213 (1967).

8.   H. H. Ritter & H. Licht, *Prop., Explo. and Pyro.*, **13**, 25 (1988).

9.   J.A. Carbon, *J. Org. Chem.*, **26**, 455 (1961).

10. T.J. Delia, D.E. Portlock & D.L. Venton, *J. Heterocyclic Chem.*, **5**, 449 (1968).

11. D.J. Brown, *J. Soc. Chem. Ind.,* 69, 353 (1950).

12. E.C. Taylor, C.-P. Tseng & J.B. Rampal, *J. Org. Chem.*, **47**, 552 (1982).

SYNTHESIS OF HIGH DENSITY INSENSITIVE ENERGETIC TETRAAZAPENTALENE DERIVATIVES

M.L. TRUDELL *, G. SUBRAMANIAN, G. ECK, J.H. BOYER
*Department of Chemistry, University of New Orleans, New Orleans, LA 70148, mltcm@uno.edu

ABSTRACT

An improved nitration procedure has been developed for the synthesis of y-Tacot (3). The synthesis of y-DBBD (8) has been completed in three steps from 3 in 37% yield. y-DBBD (8) was found to be thermally stable up to 274 °C (decomposed) and insensitive to impact (hammer blow). Nitration of z-BDDB (6) and y-DBBD (8) gave carbonyl derivatives resulting from *in situ* hydroysis/oxidation of the corresponding tetranitro derivatives z-TBBD (7) and y-TBBD (9), respectively.

INTRODUCTION

The compounds triaminotrinitrobenzene (1, TATB)[1], 2,4,8,10-tetranitrobenzotriazolo-[2,1-*a*]benzotriazol-6-ium inner salt (2, z-Tacot)[2,3], 2,4,8,10-tetranitrobenzotriazolo[1,2-*a*]benzotriazol-6-ium inner salt (3, y-Tacot)[4,5], 2,6-dipicrylbenzo[1,2-*d*][4,5-*d'*]bistriazole-4,8-dione (4)[6] and more recently 5-nitro-4,6-bis(5-amino-3-nitro-1*H*-1,2,4-triazol-1-yl)pyrimidine (5, DANTNP)[7] have been developed as insensitive energetic materials for a variety of industrial and military applications (Figure 1). However, despite favorable insensitivity to heat, impact and electric shock, the density and energetic properties (detonation velocity, D; detonation pressure, P_{CJ}) of these compounds are inferior to those observed for more conventional explosives[8,9].

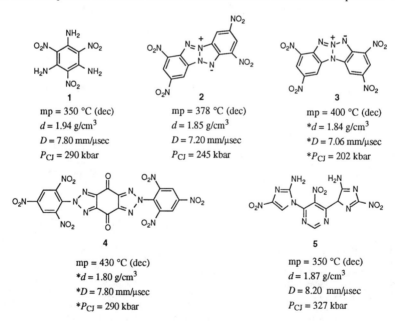

1	2	3
mp = 350 °C (dec)	mp = 378 °C (dec)	mp = 400 °C (dec)
d = 1.94 g/cm³	d = 1.85 g/cm³	*d = 1.84 g/cm³
D = 7.80 mm/μsec	D = 7.20 mm/μsec	*D = 7.06 mm/μsec
P_{CJ} = 290 kbar	P_{CJ} = 245 kbar	*P_{CJ} = 202 kbar

4	5
mp = 430 °C (dec)	mp = 350 °C (dec)
*d = 1.80 g/cm³	d = 1.87 g/cm³
*D = 7.80 mm/μsec	D = 8.20 mm/μsec
*P_{CJ} = 290 kbar	P_{CJ} = 327 kbar

Figure 1. Insensitive Energetic Materials (*computed values[10]).

Because of the inherent thermal stability of the dibenzotetraazapentalene ring system, **2** and **3** were identified as attractive precursors for the development of a new class of high density insensitive energetic materials. Based on computed densities and energetic properties, the nitro and furoxano substituted isomeric derivatives **6 – 9** were envisaged as attractive synthetic targets for development as new high density insensitive energetic compounds (Figure 2)[10]. The synthesis of 4,11-dinitro[1,2,5]oxadiazolo[3,4-*e*][1,2,5]oxadiazolo[3',4':4,5]benzotriazolo[2,1-*a*]benzotriazol-6-ium inner salt 1,8-dioxide (**6**, z-DBBD) from **2** has recently been achieved in three steps in 21% overall yield[11]. z-DBBD (**6**) was found to be thermally stable up to 310 °C and exhibited moderate impact insensitivity (dropweight test (2.5 kg): z-DBBD (**6**) = 19 cm; RDX_{std} = 18 cm)[12].

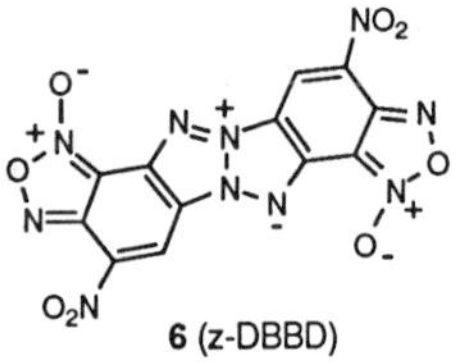

6 (z-DBBD)	7 (z-TBBD)
$(C_{12}H_2N_{10}O_8)$	$(C_{12}N_{12}O_{12})$
*d = 1.96 g/cm^3	*d = 2.06 g/cm^3
*D = 7.52 mm/μsec	*D = 8.03 mm/μsec
*P_{CJ} = 245 kbar	*P_{CJ} = 319 kbar

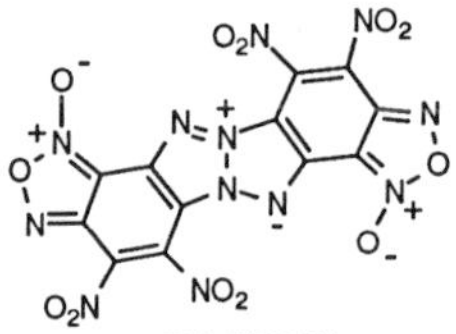

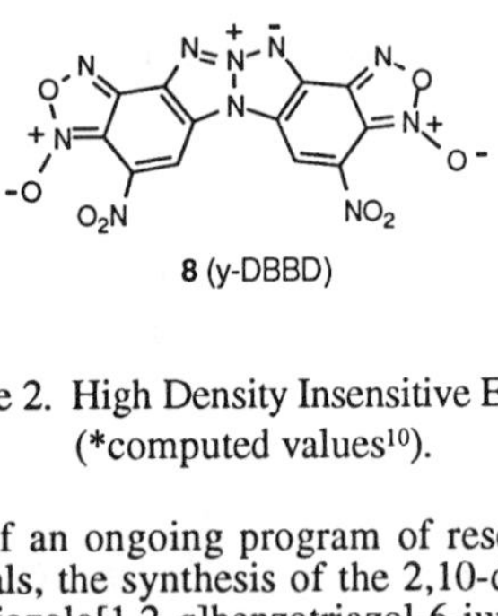

8 (y-DBBD)	9 (y-TBBD)

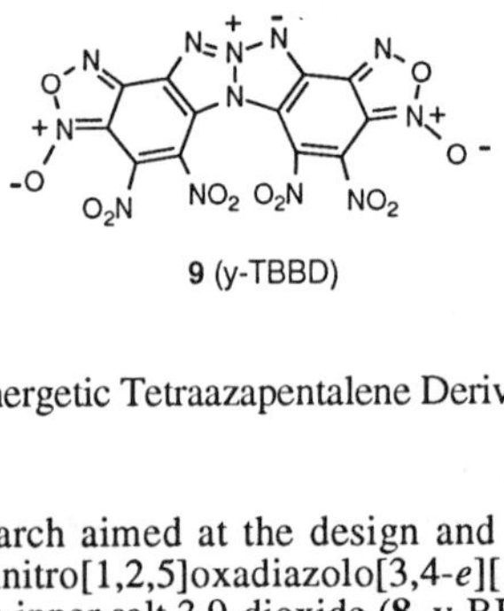

Figure 2. High Density Insensitive Energetic Tetraazapentalene Derivatives (*computed values[10]).

As part of an ongoing program of research aimed at the design and synthesis of new energetic materials, the synthesis of the 2,10-dinitro[1,2,5]oxadiazolo[3,4-*e*][1,2,5]oxadiazolo-[3',4':5,4]benzotriazolo[1,2-*a*]benzotriazol-6-ium inner salt 3,9-dioxide (**8**, y-BBBD) has recently been completed. Herein we wish to describe the synthetic sequence and preliminary insensitivity data for y-BBBD (**8**) as well as report on attempts to prepare z-TBBD (**7**) and y-TBBD (**9**).

EXPERIMENTAL

All chemicals were purchased from Aldrich Chemical Co., Milwaukee, WI. Melting points and decomposition points are uncorrected. NMR spectra were recorded on a Varian Multiprobe 300 MHz spectrometer and IR spectra were recorded on a Perkin-Elmer 1600 FTIR spectrometer. All compounds were homogeneous by TLC and NMR. Elemental analyses were obtained from Galbraith Laboratories, Inc., Knoxville, TN, and Midwest Micro Lab, Indianapolis, IN. All reported compounds gave satisfactory carbon and hydrogen analyses. Due to the high nitrogen content and explosive nature of these compounds, some reported microanalytical data for nitrogen were outside the standard acceptable limit of ±0.4%. However, duplicate and triplicate analyses for nitrogen were usually within ±1% of calculated values and corresponded to the empirical formula of the compound. *Caution!* **Compounds 3, 8, 11, and 12 should be handled as potentially explosive materials!**

y-Tacot (3).

The dibenzotetraazapentalene 10[4] (15.6 g, 0.075 mol) was dissolved in sulfuric acid (195 mL) and the mixture was cooled to 10 °C in an ice-bath. Nitric acid (90%, 300 mL) was then added dropwise, keeping the flask temperature below 25 °C. After the addition was complete, the reaction mixture was stirred for 15 min at room temperature and then heated at 75 °C for 10 min. The mixture was cooled to 20 °C and poured into ice-water (25 L). The yellow precipitate was filtered, washed with water (3 × 100 mL) and dried. The crude compound (25.8 g) was recrystallized from DMF (550 mL) to give 3 (25.1 g, 86%). An analytical sample was prepared by recrystallization from acetone. mp 398 °C (dec) [Lit[4] mp 400 °C (dec)]. IR (KBr) 3097, 1629, 1586, 1536, 1413, 1377, 726 cm^{-1}. ^{1}H NMR (DMSO – d_6) δ 10.6 (d, J = 1.9 Hz, 2H), 9.3 (d, J = 1.8 Hz, 2H). ^{13}C NMR (DMSO – d_6) δ 142.8, 141.2, 135.0, 125.6, 120.7, 116.2. Anal. calcd for $C_{12}H_4N_8O_8$: C, 37.12; H, 1.04; N, 28.86. Found: C, 37.02; H, 1.02; N, 27.82.

2,10-Diazido-4,8-dinitrobenzotriazolo[1,2-a]benzotriazol-6-ium Inner Salt (11).

y-Tacot (3) (17.53 g, 45 mmol) and sodium azide (23.37 g, 360 mmol) in dry DMSO (600 mL) were heated at 70 – 75 °C for 24 hours. The mixture was then cooled at 15 °C for 1.5 h and the yellow-orange solid which separated was collected by filtration and washed with ethyl alcohol (100 mL) and diethyl ether (100 mL) to give 11 (14.3 g, 83%). The crude compound was used directly in the next step without any further purification. A pure sample was prepared for analysis by recrystallization from DMF. IR (KBr) 3072, 2123, 1542, 1522, 1337, 1115, 741 cm^{-1}. ^{1}H NMR (DMSO – d_6) δ 9.8 (d, J = 1.8 Hz, 2H), 8.2 (d, J = 1.7 Hz, 2H). Anal. calcd for $C_{12}H_4N_{12}O_4$•C_3H_7NO: C, 39.76; H, 2.43; N, 40.16. Found: C, 39.71; H, 2.52; N, 39.27.

2,10-Diazido-3,4,8,9-tetranitrobenzotriazolo[1,2-a]benzotriazol-6-ium Inner Salt (12).

Nitric acid (90%, 47.5 mL) was cooled in an ice-bath and 11 (12.6 g, 0.033 mol) was added keeping the temperature below 10 °C. Stirring was continued for 2 h at 0 – 5 °C. The mixture was poured into ice-water (1 L) and the orange-brown precipitate was filtered, washed with water (100 mL) and dried to give 12 (11.9 g, 76%). The material was usually of sufficient purity to go on to the next step. However, an analytically pure sample was obtained by dissolving the crude compound in acetone (12.5 mL) at 40 °C. The insoluble material was removed and the solution was triturated with hexane. The mixture was kept in a freezer overnight and the precipitate was filtered. The material was then recrystallized from acetone to give 12 in pure form. mp 280 °C (dec). IR (KBr) 2144, 1558, 1507, 1339, 1320, 1292, 907, 820 cm^{-1}. ^{1}H NMR (DMSO – d_6) δ 10.1 (s). ^{13}C NMR (DMSO – d_6) δ 140.9, 136.8, 133.2, 124.2, 122.1, 108.7. Anal. calcd for $C_{12}H_2N_{14}O_8$•C_3H_6O: C, 34.10; H, 1.53; N, 37.12. Found: C, 33.78; H, 1.58; N, 34.34.

y-BDDB (8).

The diazidotetranitro derivative 12 (10.0 g, 21 mmol) in 1,2-dichlorobenzene (650 mL) was heated for 1 h at 150 °C. The mixture was cooled in an ice-bath for 2 h and the precipitate was filtered and washed with diethyl ether. The filtrate was triturated with acetonitrile to give 8 (6.0 g, 58%) in pure form. mp 274 °C (dec). IR (KBr) 1654, 1575, 1534, 1414, 1330, 1296, 999, 704 cm^{-1}. ^{1}H NMR (DMSO – d_6) δ 10.4 (s). ^{13}C NMR (DMSO – d_6) δ 146.6, 134.6, 132.5, 123.7, 117.2, 107.5. Anal. calcd for $C_{12}H_2N_{10}O_8$: C, 34.80; H, 0.49; N, 33.82. Found: C, 34.78; H, 0.58; N, 31.23.

z-DTBBD (13).

Under an atmosphere of nitrogen, fluorosulfonic acid (1.35 g, 0.80 mL, 13.5 mmol) was placed in a round bottom flask (25 mL) equipped with magnetic stirrer and nitrogen inlet tube. Nitric acid (100%, 1.19 g, 0.80 mL, 19 mmol) was added in one portion at 25 °C. After 5 min, dry nitromethane (5 mL) was added over a period of 10 min with stirring. The dinitro derivative 6[11] (350 mg, 0.85 mmol) was added in one portion. The temperature was maintained at 80 – 90 °C for two hours, nitromethane was removed under vacuum and the mixture was cooled and added to ice-water (150 mL) to bring about the precipitation of 13. The material was recrystallized from acetonitrile, triturated with boiling ether for 15 min, cooled, filtered and dried to give 13 as a red microcrystalline solid (84 mg, 20%). mp 340 – 341 °C (dec); an irreversible color change from red to yellow was brought about by heating 180 – 185 °C. IR (KBr) 1654 (C = N), 1560 (NO$_2$), 1482 (NO$_2$), 1364 (NO$_2$), 1072, 1031, 985, 899, 774, 504 cm^{-1}. ^{13}C NMR (DMSO–d$_6$) δ 166.65, 163.57, 149.41, 137.51, 113.91, 103.10. Anal. Calcd for C$_{12}$H$_4$N$_8$O$_{10}$: C, 34.30; H, 0.96; N, 26.65; O, 38.09. Found: C, 33.93; H, 0.96; N, 25.27; O, 36.92.

4,5,11,12-Tetraoxo-14H-[1,2,5]oxadiazolo[3,4-e][1,2,5]oxadiazolo-[3',4':4,5]benzotriazolo-[2,1-a]benzotriazol-6-ium inner salt 1,8-dioxide (14).

The compound **13** (420 mg, 1 mmol) was placed in a round bottom flask (10 mL) and heated in an oil-bath up to 140 °C (an irreversible color change from red to yellow was observed). The sample was then kept at this temperature overnight, cooled and recrystallized from dry acetone to give **14** as a yellow solid (130 mg, 34%). mp 330 °C (dec). IR (KBr) 1729 (C = O), 1675 (C = N), 1467, 1383, 1329, 1049, 971, 671 cm^{-1}. Anal. Calcd for C$_{12}$N$_8$O$_8$: C, 37.53; N, 29.16. Found: C, 37.19; N, 28.52.

RESULTS

The benzotriazolo[1,2-a]benzotriazol-6-ium inner salt (**10**) was prepared from o-phenylenediamine and 2-chloronitrobenzene according to the procedure developed by Kauer and Carboni[4]. Nitration of **10** was found to proceed cleanly with 90% HNO$_3$/H$_2$SO$_4$ at 0–5 °C, followed by a brief heating period (10 min) at 75 °C. This gave the 2,4,8,10-tetranitro derivative **3** as the sole product in 86% yield (Figure 3). This modified procedure provided **3** in higher yield and in a higher state of purity than the literature nitration conditions[4]. The orientation of the nitro groups in y-Tacot (**3**) has recently been unequivocally confirmed by X-ray crystallography[13].

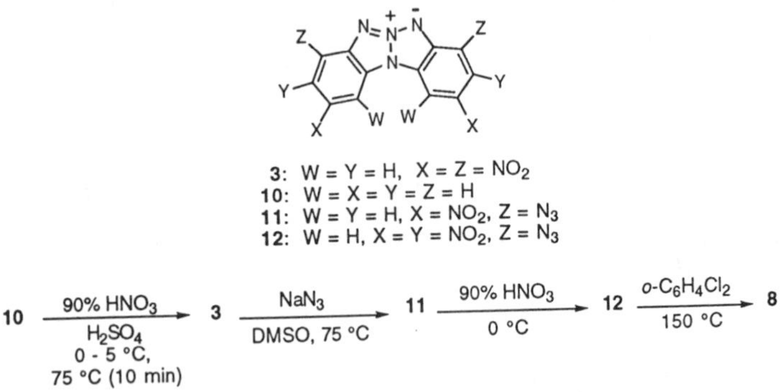

3: W = Y = H, X = Z = NO$_2$
10: W = X = Y = Z = H
11: W = Y = H, X = NO$_2$, Z = N$_3$
12: W = H, X = Y = NO$_2$, Z = N$_3$

Figure 3. Synthesis of y-DBBD (**8**).

With multigram quantities of **3** in hand, attention turned toward further functionalization of the benzo rings of **3** for the construction of the furoxan ring systems of the initial target compound **8**. Treatment of **3** with sodium azide in dimethylsulfoxide at 70–75 °C furnished a single symmetrical diazidodinitro derivative in 83% yield resulting from nucleophilic substitution of an equivalent pair of nitro groups by the azide anion. The structure of the compound, pending confirmation by X-ray crystallographic analysis, is believed to be that of the 4,8-diazido-2,10-dinitro derivative **11** (Figure 3). From previous studies with z-Tacot (**2**), nucleophilic substitution of the 4,10-nitro groups (adjacent to the pyridine-like nitrogen atoms of tetraazapentalene moiety) was found to take place regiospecifically[11]. By analogy, the 4,8-nitro groups are believed to be the site of azide substitution in **3** resulting in the formation of **11**. Unequivocal identification of the structural orientation of the azido and nitro groups of **11** is currently under investigation.

Nitration of the diazidodinitro derivative **11** in 90% HNO$_3$ proceeded easily to give the 4,8-diazido-2,3,9,10 tetranitro derivative **12** in 76% yield (Figure 3). The ease of the nitration of

11 stems from activation of the *C(3)*- and *C(9)*-positions toward electrophilic attack by the *ortho*-directing effect of the azido groups[14]. Despite favorable computed density and improved computed energetic properties for **12** ($d = 1.84$ g/cm^3, $D = 8.12$ mm/µsec, $P_{CJ} = 301$ kbar)[10], the material was considerably more sensitive than y-Tacot (**3**). The diazidotetranitro derivative **12** was found to have good thermal stability (decomposed at 280 °C) but was impact sensitive (violent explosion with flame when struck by a hammer) while **3** was completely stable under these conditions.

Thermolysis (1,2-dichlorobenzene, 150 °C) of **12** furnished the new heterocyclic system 2,11-dinitro[1,2,5]oxadiazolo[3,4-*e*][1,2,5]oxadiazolo[3',4':5,4]benzotriazolo[1,2-*a*]benzotriazol-6-ium inner salt 3,10-dioxide (**8**, y-DBBD) in 58% yield (Figure 3). This served to confirm the presence of two sets of contiguous azido and nitro groups and supported the structural assignment of **11**. The red microcrystalline material **8** was found to be stable up to temperatures of 274 °C at which point the material decomposed. In addition, in these laboratories **8** was found to be insensitive to impact; no detonation was observed when the material was struck by a hammer.

With both z-DBBD (**6**) and y-DBBD (**8**) available, further nitration of these compounds to give the target compounds **7** and **9** has been the subject of recent investigations. Although **7** has not been successfully isolated from the nitration reaction media, an unusual by-product presumably resulting from a hydrolysis-oxidation reaction of **7** was obtained. Treatment of **6** with normal nitration procedures was unsuccessful and gave only recovered starting material[15]. Although there are several reagents available for nitration of deactivated aromatic systems, a number of successful nitrations have been reported in the literature using fluorosulfonic acid in 100% nitric acid[15]. Treatment of **6** with fluorosulfonic acid in 100% nitric acid at 70–80 °C for 2 h and after workup afforded a red crystalline material in 20% yield (Figure 4). The structure of the product of this reaction was unequivocally identified by X-ray crystallography as the hydrated form of the 5,12-dioxo-4,4,11,11-tetrahydroxy-[1,2,5]oxadiazolo[3,4-*e*][1,2,5]oxadiazolo[3',4':4,5]benzotriazolo-[2,1-*a*]benzotriazol-6-ium inner salt 1,8-dioxide (**13**, DTBBD)[16]. The formation of **13** presumably results from the hydrolysis and concomitant oxidation of **7** in the fluorosulfonic acid/100% nitric acid media. Subsequent hydration of one of the carbonyls of each of the 1,2-diketone moieties then results in the observed product **13**. Similar reactivity was observed for y-DBBD (**8**). Treatment of **7** with fluorosulfonic acid/100% nitric acid also afforded carbonyl compounds. However, the structure of these compounds has not been determined at this time.

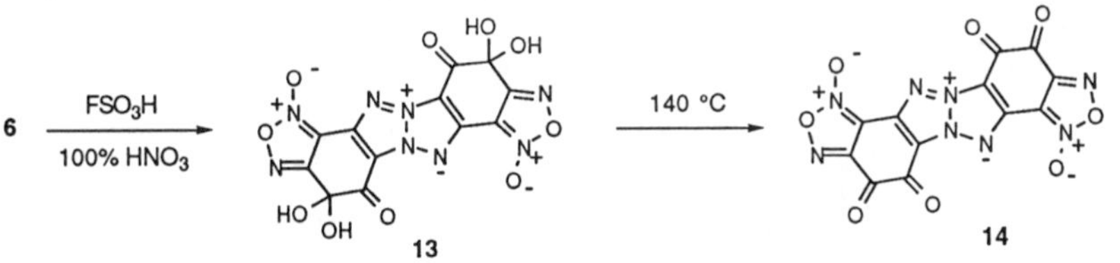

Figure 4. Attempted Nitration of z-BDDB (**6**). A Novel Hydrolysis/Oxidation Reaction.

The hydrate **13** was converted into the tetraoxo derivative **14** by heating at 140 °C overnight (Figure 4). The resultant yellow material **14** was obtained in 34% yield. The zero-hydrogen content tetraoxo derivative **14** was computed to have high density ($d = 2.08$ g/cm^3)[10]. However, **14** was computed to have poor energetic properties ($D = 7.95$ mm/µsec and $P_{CJ} = 286$ kbar) relative to **2**, **6** and **7**. In addition, **14** was found to be thermally stable and insensitive to impact; no detonation was observed when the material was struck with a hammer.

The lability of the *ortho*-dinitro groups of **7** and **9** towards nucleophilic displacement by water is consistent with the chemistry observed for 1,2,3,4-tetranitrobenzene and hexanitrobenzene which has been converted into picric acid and trinitrophloroglucinol respectively upon exposure to moisture[17]. Displacement of the *ortho*-dinitro groups presumably results in the formation of an *o*-dihydroquinone species which then readily undergoes autooxidation to give the observed quinone-like species (**13** and **14**).

CONCLUSIONS

The new tetraazapentalene derivative, y-DBBD (8) was synthesized in three steps from y-Tacot (3) in 37% overall yield. The material was found to have good thermal stability and qualitative measurements have shown 8 to possess good insensitivity toward impact. In addition, we have found that further nitration of 6 and 8 leads to the formation of moisture sensitive nitration products which undergo further oxidation to give o-quinone-like species (13 and 14). The further development of new heterocyclic systems which exploit the insensitivity and thermal stability of the tetraazapentalene ring systems is currently under investigation.

ACKNOWLEDGMENTS

We gratefully acknowledge the financial support of this work by the Office of Naval Research (contract number : N00014-90-J-1661) and Program Officer Dr. Richard S. Miller.

REFERENCES

1. T. Urbanski and S.K. Vasudeva, J. Scient. Ind. Res. **37**, 250 (1978).
2. R.A. Carboni, J.C. Kauer, W.R. Hatchard and R.J. Harder, J. Am. Chem. Soc. **89**, 2626 (1967).
3. R.A. Carboni, U.S. Patent No. 2 904 545 (1959).
4. J.C. Kauer and R.A. Carboni, J. Am. Chem. Soc. **89**, 2633 (1967).
5. J.C. Kauer, U.S. Patent No. 3 262 943 (26 July 1966).
6. J.K. Berlin and M.D. Coburn, J. Heterocycl. Chem. **12**, 235 (1975).
7. C. Wartenberg, P. Charrue and F. Lavel, Prop. Expl. Pyro. **20**, 23 (1995).
8. A.T. Nielsen, in *Chemistry of Energetic Materials*, edited by G. Olah and D.R. Squire (Academic Press Inc., New York, 1991), pp. 95-124.
9. R. Meyer, *Explosives*, 3rd ed. (VCH, Weinheim, 1987), p. 150 and 202.
    (RDX; mp 204 °C; d = 1.81 g/cm^3; D = 8.85 mm/sec)
    (HMX; mp 282 °C; d = 1.9 g/cm^3; D = 9.1 mm/sec)
10. The density d (g/cm^3), detonation velocity D (mm/sec) and detonation pressure P_{CJ} (kbar) were computed with a program obtained from the Naval Weapons Center, China Lake, CA.
11. G. Subramanian, J.H. Boyer, D. Buzatu, E.D. Stevens, and M.L. Trudell, J. Org. Chem. **60**, 6110 (1995).
12. W. Koppes, Naval Surface Warfare Center, Indian Head, MD 20640–5035 (private communication).
13. E.D. Stevens, University of New Orleans (private communication).
14. M.E. Biffin, J. Miller, and D.B. Paul, in *The Chemistry of the Azido Group*, edited by S. Patai (Wiley and Sons, New York, 1971), pp. 209-212.
15. G.A. Olah, R. Malhorta and S.C. Narang, *Nitration*, (VHC Publishers Inc., New York, 1976), pp. 9-83.
16. G. Subramanian, J.H. Boyer, W. Koppes, R. Gilardi, and M.L. Trudell, J. Org. Chem. (submitted).
17. T. Urbanski, *Chemistry and Technology of Explosives*, (Pergamon Press, Oxford, 1964), pp 258–259.

ANTA AND ITS OXIDATION PRODUCTS

Kien-Yin Lee*, Richard Gilardi**, Michael A. Hiskey* and James R. Stine*
* MS C920, Los Alamos National Laboratory, PO Box 1663, Los Alamos, NM 87545
** Laboratory for the Structure of Matter, Naval Research Laboratory, Washington D.C. 20375

ABSTRACT

5-Amino-3-nitro-1H-1,2,4-triazole (ANTA) is a molecule with high stability. Aside from being an insensitive high explosive (IHE), it is also used as a synthon for other potential new IHEs. The crystal structure of ANTA was resolved by X-ray crystallography. However, when ANTA was recrystallized from 2-butanone, crystals with molecular packing characterized by extended planar sheets were obtained (ß-ANTA). The crystal density of ß-ANTA is 1.73 g/cm^3, which is less dense than α-ANTA (ρ = 1.82g/cm^3).

The high-nitrogen molecule, 5,5'-dinitro-3,3'-azo-1,2,4-triazole (DNAT) was calculated to have a high density and a positive heat of formation (ΔH_f). In an attempt to prepare DNAT, we have studied the oxidation of ANTA with different oxidizers. It was found that DNAT is the reaction product when the potassium salt of ANTA was oxidized with potassium permanganate. However, when ANTA was oxidized with ammonium persulfate in aqueous medium, one of the reaction products obtained was the azoxy moleclue of DNAT (DNAzT). We are unable to determine the crystal densities of either DNAT and DNAzT because the crystals obtained were solvated with the crystallization solvents.

BACKGROUND

ANTA molecule was first prepared by Pevzner et al, as an end product from acid hydrolysis of 5-acetamido-3-nitro-1,2,4-triazole, but no properties were reported[1]. In 1982, we proposed to evaluate ANTA as a potential insensitive high explosive (IHE) for nuclear weapons application. Because of the extreme low overall yield of preparing ANTA by existing process, the study was discontinued. In 1989, we improved the synthesis of ANTA to give more than 95% yield[2], and consequently several ANTA-based IHEs were prepared[3,4]. In 1992, ANTA was produced in large quantity by Lawrence Livermore National Laboratory for a high explosive performance study[5].

The structure of ANTA, grown from ethanol and chloroform was determined by X-ray crystallography, and the molecules of ANTA were found to form ribbons which are twisted in orientation[6]. If ANTA was recrystallized from 2-butanone, crystals with molecular packing characterized by extended planar sheets were obtained (ß-ANTA).

High-nitrogen molecules offer unique features which are suitable for both explosives and propellants application. Data from calculated burn performance of high-nitrogen content nitroheterocyclic energetic materials indicate that the flame temperatures of those molecules are low, and they produce low-molecular-weight gases as the burn products[7]. In 1983 we proposed to prepare 5,5'-dintro-3,3'-azo-1,2,4-triazole (DNAT) because it was predicted to have both a high density and a positive heat of formation. We attempted to prepare DNAT by nitration of the parent compound, 3,3'-azo-1,2,4-triazole with acetyl nitrate. However, the nitration product isolated was found to be the nitramine isomer of DNAT[8]. We then studied the oxidation of ANTA with different oxidizers, and evaluated the properties of DNAT.

In this paper, the crystal structure of ß-ANTA will be reported. The syntheses of both DNAT and 5,5'-dinitro-3,3'-azoxy-1,2,4-triazole (DNAzT) by oxidation of ANTA will also be described.

CRYSTAL STRUCTURE OF ß-ANTA

<u>Crystallization of ß-ANTA</u>   Samples of ANTA was dissolved in 2-butanone at temperature 45°C. The solution was then cooled to 25°C at a rate of 1°C per hour. Clear yellow irregular crystals of ß-ANTA was then obtained after the solution was evaporated to dryness at 25°C.

<u>Summary</u>   ANTA, $C_2H_3N_5O_2$, M_r = 129.1, monoclinic, $P2_1/n$, a = 6.517 (1), b = 9.691 (1), c = 8.410 (1) Å, β = 111.10 (1)°, V = 495.5 (1) Å^3, Z = 4, Dx = 1.730 g/cm^3, λ(Mo Kα) = 0.71073 Å, μ = 0.142 mm^{-1}, F (000) = 264, T = 294 K, final R = 0.048, wR = 0.043 for 750 independent observed reflections.

The polymorph is planar to within ± 0.006 Å (±0.05 Å including hydrogen atoms) and forms hydrogen-bonded sheets. In these almost-planar sheets, the non-hydrogen atoms lie less than ±0.17 Å from the x = 0.25 and 0.75 planes. The interplanar spacing, [1/(2aSinθ)], is 3.040 Å.

<u>Experimental</u>   A clear yellow irregular 0.12 x 0.20 x 0.25 mm crystal was used for data collection on an automated Siemens *R3m*/V diffractometer with incident beam monochromator. 25 centered reflections within 25.4 ≤ 2θ ≤ 41.1° used for determining lattice parameters. $(\sin(\theta)/\lambda)_{max}$ = 0.65 Å^{-1}, range of *hkl* : 0 ≤ h ≤ 8, -12 ≤ k ≤ 0, -10 ≤ l ≤ 10.   Standards 4,0,0, 0,-2,0, 0,-1,-3, monitored every 97 reflections with random variation of 2.0 % over data collecion, θ/2θ scan mode, scan width [2θ($K_{\alpha1}$) - 1.0] to [2θ($K_{\alpha2}$) + 1.0]°, a constant ω scan rate of 6.0° min^{-1}, 1321 reflections measured, 1146 unique, R_{int} = 0.017, 750 observed with $F_0>3\sigma(F_0)$. Data corrected for Lorentz and polarization effects. The structure solution, by direct methods, and the full-matrix least-squares refinement used programs in SHELXTL (Sheldrick 1980). $\Sigma w(|F_0| - |F_c|)^2$ minimized where $w = 1/[\sigma^2(|F_0|) + g(F_0)^2]$, g = 0.000225. Secondary extinction parameter p = 0.0027(5) in $F_c = F_c/[1.0 + 0.002(p)F_0^2/\sin(2\theta)]^{0.25}$. 95 parameters were refined: atomic coordinates and anisoptropic thermal parameters for all non-H atoms, H atoms included using riding model [coordinate shifts of C applied to attached H atoms, C-H distance set to 0.96 Å, H angles idealized, U_{iso}(H) set to 1.1 U_{eq}(C) or, if methyl, 1.2 U_{eq}(C)]. $(\Delta/\sigma)_{max}$ = 0.001, ratio of observations to parameters 7.9:1, R = 0.048, wR = 0.043, S = 1.27. R = 0.080 for all data. Final difference map excursions 0.24 and -0.24 eÅ^{-3}. Atomic scattering factors from *International Tables for X-ray Crystallography* (1974).

   A thermal ellipsoid plot of ß -ANTA showing the refined molecule imbedded in one of the hydrogen-bonded layers that occur in the crystal is shown in Figure 1. Figure 2 is the thermal ellipsoid drawing of two α-ANTA ribbons showing some of the secondary NH...N links and the pronounced twist.

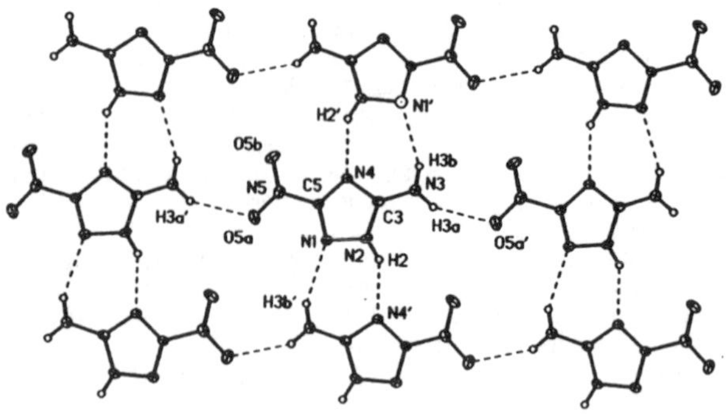

Figure 1. A thermal ellipsoid plot of ß -ANTA showing the refined molecule imbedded in one of the hydrogen-bonded layers that occur in the crystal with ellipsoids drawn at the 20% probability level

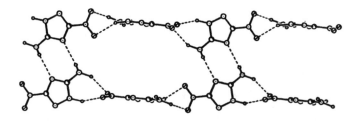

Fig. 2. A thermal ellipsoid drawing of two α-ANTA ribbons showing some of the secondary
NH....N links and the pronounced twist

The bond angles and bond lengths of ß -ANTA are listed in Table I.

Table I.   Bond Angles (°) and Lengths (Å) of ß -ANTA

N(1) - N(2)	1.367	(3)		N(1) - C(5)	1.301	(3)	
N(2) - C(3)	1.351	(3)		C(3) - N(3)	1.333	(4)	
C(3) - N(4)	1.341	(3)		N(4) - C(5)	1.341	(3)	
C(5) - N(5)	1.445	(4)		N(5) - O(5a)	1.223	(3)	
N(5) - O(5b)	1.225	(3)					

N(2) - N(1) - C(5)	100.3	(2)		N(1) - N(2) - C(3)	110.7	(2)	
N(2) - C(3) - N(3)	124.8	(3)		N(2) - C(3) - N(4)	109.0	(2)	
N(3) - C(3) - N(4)	126.2	(2)		C(3) - N(4) - C (5)	101.6	(2)	
N(1) - C(5) - N(4)	118.4	(2)		N(1) - C(5) - N(5)	120.2	(2)	
N(4) - C(5) - N(5)	121.4	(2)		C(5) - N(5) - O(5a)	117.5	(2)	
C(5) - N(5) - O(5b)	117.6	(2)		O(5a) - N(5) - O(5b)	124.9	(2)	

OXIDATION OF ANTA WITH AMMONIUM PERSULFATE

Oxidation Procedure

    The oxidation reaction was carried out by a modification procedure by Solodyuk et al.[9] A
solution of ammonium persulfate [$(NH_4)_2S_2O_8$] and ANTA was stirred at 70°C for 1.5 hours.
After cooling to room temperature, the aqueous mixture was extracted with ethyl acetate, followed
by drying with magnesium sulfate. The reaction products obtained after evaporating the ethyl
acetate on a rotary evaporator were collected and analyzed by ^{13}C NMR spectroscopy. It was
found that the oxidation product contains more than one compound. Attempts to separate the
oxidation products by recrystallization of the final mixture were unsuccessful. Hence, a different
approach for the crystallization of the reaction products was carried out.

Crystallization of Oxidation Products

The crystals used for structure determination was obtained by solvents interface technique. Thus, onto the surface of the oxidation products solution in ethyl acetate in a test tube was carefully pipetted a small amount of dimethylformamide (DMF). The test tube was then covered with a rubber stopper and left in the refrigerator. After a period of about a week, yellow crystals were observed.

Structure of DMF-solvated 5,5'-Dinitro-3,3'-Azoxy-1,2,4-Triazole (DNAzT)

The crystals obtained by this appproach were found to be DMF-solvated DNAzT. There are two molecules of DMF for each DNAzT molecule. The yellow parallelepiped crystals, $C_4H_2N_{10}O_5$ * $2[C_3H_7NO]$, $M_r = 416.3$, are monoclinic, space group Pn, with a = 11.963(2), b = 6.057(2), c = 13.616(2) Å, $\beta = 105.16(2)°$, V = 952.3(2)Å^3, Z = 2, Dx = 1.452g/cm^3, λ(Cu Kα) = 1.54178 Å, μ = 1.074mm^{-1}, F(000) = 432 and T = 294 K.

The thermal ellipsoid drawing of DMF-solvated DNAzT showing the atom numbering is shown in Figure 3. The dotted [O] atom in the center of the molecule represents the alternative location for oxygen atom of the -N=N- bridge, i.e.the oxygen atom of the bridge sometimes stacks in one orientation, sometimes in the other.

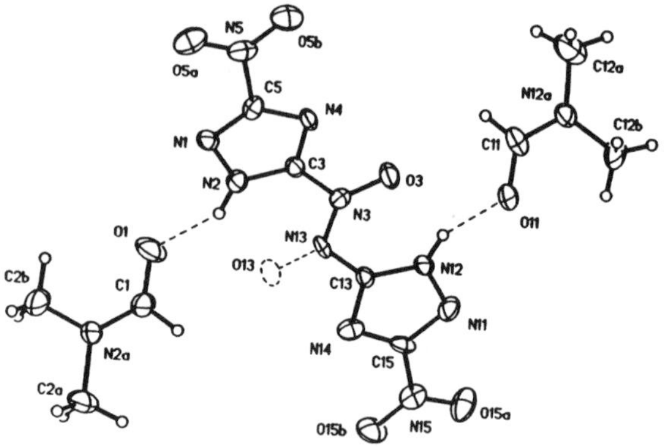

Fig. 3. A thermal ellipsoid drawing of DMF-solvated DNAzT showing the atom numbering

Calculated Physical properties and Specific Impulse (Isp) of DNAzT

Because of the low yield of DNAzT obtained by oxidation of ANTA with ammonium persulfate, we have not yet determined the explosive and sensitivity properties of DNAzT. However, data from calculation indicate that DNAzT is a dense molecule. Based on calculated heat of formation of 100 kcal/mol, the specific impulse (Isp) of DNAzT was calculated to be 244.8 sec, Table II.

Table II.   Calculated Properties of DNAzT

Molecular Formula	$C_4H_2N_{10}O_5$
Wt. %N	52
Density (g/cm^3)	1.885
Heat of Formation (kcal/mol)	100
Isp, sec	244.8

OXIDATION OF ANTA WITH POTASSIUM PERMANGANATE

Oxidation Procedure

A solution of potassium salt of ANTA (KANTA) was mixed with potassium permanganate (KMnO$_4$) at O °C. After stirring at this temperature for about an hour, the compound obtained was the potassium salt of DNAT, which was then acidified to yield the free acid form, Reaction 1. The oxidation product was confirmed by ^{13}C NMR spectroscopy to be the 5,5-dinitro-3,3'-azo-1,2,4-triazole (DNAT).

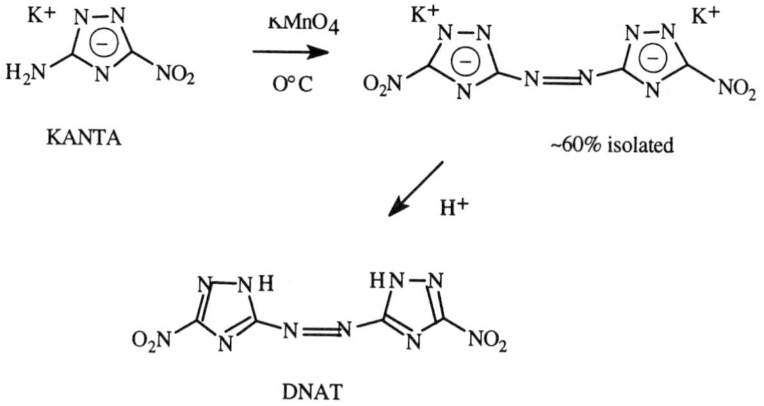

KANTA

~60% isolated

H$^+$

DNAT

Reaction 1. Preparation of 5,5'-Dinitro-3,3'-Azo-1,2,4-Triazole

Properties of DNAT

DNAT is lemon in color, and is soluble in most polar solvents. When recrystallized from water, hydrated DNAT is obtained, as confirmed by ^{1}H NMR spectroscopy. The water molecule can be removed by drying the molecule in the oven at 110°C for overnight or longer. The molecule burns without smoke and solid residues. It is less sensitive to impact than HMX, and has a positive heat of formation (78 kcal/mol), Table III. However, because of its acidic nature, DNAT becomes less attractive in some applications.

Table III.   Physical and Explosive Properties of DNAT

Molecular Formula	C$_4$H$_2$N$_{10}$O$_4$
Wt %N	55
Density (g/cm^3)	1.88 (calc.)
DTA Exotherm (°C)	175
Impact Sensitivity (Type 12)(cm)	69
Heat of Formation (kcal/mol)	78

DNAT crystals of suitable size were obtained by recrystallization from N-methyl pyrrolidone (NMP) or dimethyl sulfoxide (DMSO). However, we are unable to determine the density of DNAT because the crystals were analyzed to be solvated with the crystallization solvents.

RESULTS AND DISCUSSIONS

A different polymorph of ANTA was observed by X-ray crystallography when ANTA was recrystallized from 2-butanone, instead of from a mixture of ethanol and chloroform. The β-ANTA is less dense ($\rho = 1.730$ g/cm^3) than α-ANTA, and the molecular packing is characterized by extended planar sheets.

In an attempt to prepare the high-nitrogen molecule, 5,5'-dinitro-3,3'-azo-1,2,4-triazole, we have studied the oxidation of ANTA with different oxidizers. The high-nitrogen molecule, 5,5'-dinitro-3,3'-azoxy-1,2,4-triazole (DNAzT) was obtained when ANTA was oxidized with ammonium persulfate. No explosives properties of DNAzT were determined due to the low yield of DNAzT by the present oxidation process. However, data from calculated properties indicate that DNAzT is a dense molecule, and is a molecule for potential propellants applicaiton. We are in the process of looking into other oxidation process to prepare DNAzT.

Since the oxidation product of ANTA with ammonium persulfate is the azoxy compound of DNAT, we then studied the oxidation of ANTA with potassium permanganate. It was found that a better yield of DNAT was obtained if the oxidation reaction was carried out with the potassium salt of ANTA, instead of ANTA. Although the yield for the preparation of the potassium salt of DNAT is about 60%, the conversion of the salts to the free DNAT form is essentially 100%. DNAT burns without smoke and solid residues. It appears that DNAT may have application as a low signature burning rate modifier. However, the two protons on the DNAT molecule are acidic, and hence it becomes less attractive in some applications.

We were unable to determine the crystal densities of both DNAT and DNAzT because the crystals obtained from recrystallization are solvated.

ACKNOWLEDGEMENT

The authors are grateful to Dr. May Chan, NAWC, China Lake for her specific impulse (Isp) calculations. This work was performed under the auspices of the US Department of Energy, and was supported, in part, by the Office of Naval Research, Mechanics Division.

REFERENCES

1. M. S. Pevzner, T. N. Kulibabina, N. A. Povarova and L. V. Kilina, Khim. Geterotsikl. Soedin. 8, 1132 (1979).

2. K.-Y. Lee and C. B. Storm, US patent 5,110,380, May (1992).

3. K.-Y. Lee, E. Garcia and D. Barnhart, LA-12248-MS, Los Alamos National Laboratory Report, March (1992).

4. K.-Y. Lee, M. D. Coburn and M. A. Hiskey, LA-12582-MS, Los Alamos National Laboratory Report, June, (1993)

5. R. L.Simpson, P. F. Pagoria, A. R. Mitchell and C. L. Coon, Propellants, Explos.,and Pyrotech., 19(4), 174 (1994).

6. E. Garcia and K.-Y. Lee, Acta Cryst. C **48**, 1682 (1992)

7. M.M. Stinecipher, K.-Y. Lee and M. A. Hiskey, Proceedings ot 31st AIAA/ASME/SAE/ASEE Joint Propulsion Conference, July 10-12 (1995).

8. K.-Y. Lee, U.S. patent 4,623,409, Nov. (1986).

9. G. D. Solodyuk,M. D. Boldyrev, B. V. Gidaspov and V. D. Nikolaev, Zhur. Organ. Khimi, 17, No 4, 861, April (1981).

CHARGE DENSITIES AND ELECTROSTATIC POTENTIALS FOR ENERGETIC MATERIALS

A.A. PINKERTON, A. MARTIN
Department of Chemistry, University of Toledo, Toledo, OH 43606

ABSTRACT

High resolution ($\sin\theta /\lambda < 1.34$ Å^{-1}), low temperature (85 K) X-ray diffraction data has been used to map the deformation density and the derived electrostatic potential for three dinitramide salts. The traditional presentation of contour maps has been replaced with 3D views of the molecule. A comparison of the dinitramide ions from each salt is presented.

INTRODUCTION

Charge density studies of energetic materials are not new. However, no intensive study on the effects of the geometry on the charge density and the derived electrostatic potential for any such material has been carried out. A recently reported class [1] of energetic materials is composed of the salts of the dinitramide anion, $N(NO_2)_2^-$, DN. A number of papers have described the characterization and thermal decomposition of the ammonium salt [2-5] and there have been reports of theoretical studies of the anion and the parent acid to gain insight into the possible mechanism of decomposition [6-8]. It is known from room temperature X-ray diffraction studies [9] on a variety of salts that the anion has much structural flexibility, in particular with respect to the amount of molecular twist (from 0 to 45°). At the same time, there are distinct differences in chemically equivalent bond lengths (up to 0.045 Å for bonds to the bridging nitrogen).

In order to obtain further insight into the bonding and perhaps reactivity of these compounds, we have carried out more detailed X-ray diffraction studies on three such derivatives - the mono- and di-protonated biguanidinium (BIGH$^+$ and BIGH$_2^{2+}$) and ammonium salts which span the complete range of torsion angles (5.1, 28.9 and 42.7° respectively). We have mapped the electron density distribution (reported as the deformation density, i.e. the difference between the total observed electron density and that obtained by the overlap of neutral spherical atoms) for all three compounds. We have also derived the electrostatic potentials for these systems to give insight into the solid state interactions and perhaps a starting point for molecular dynamics calculations

EXPERIMENT

X-ray diffraction data were obtained at 85 K for crystals of (BIGH)(DN), (BIGH₂)(DN)₂ and ADN using an Enraf-Nonius CAD4 automatic diffractometer and an Oxford Cryostream nitrogen gas cooling device. Data were obtained to the limits of the diffractometer ($\sin\theta/\lambda < 1.34$ Å^{-1}). The electron density was modeled using atom centered multipoles [10] and refined to conventional R values of 0.028, 0.026 and 0.022 respectively. The contribution from neutral spherical atoms was subtracted, resultant structure factors calculated and Fourier transformed to obtain deformation density maps. The results of the multipole refinements were also used to derive electrostatic potential maps. These numerical maps were used as input to obtain various 3D color coded maps, ray-traced maps and virtual reality representations of the results.

RESULTS

We now have the data to compare and contrast the deformation densities of the same ion in three different geometries. In order to compare and contrast these large amounts of data we use a combination of color coded 3D cutaway isosurfaces and virtual reality techniques. As this is only meaningful on a computer screen, for the purposes of this paper, we have prepared gray scale ray-traced images to prepare the reader for comparing the data across the experiments. Figures 1, 3 and 5 show the deformation density in the N-N-N plane of the dinitramide anion as a semi-transparent map, + signs indicating a build up of electron density and - signs indicating depletion. The white contour represents a deformation density of 0.0 e / Å³. The atomic positions and bonds have been marked using 0.2 Å spheres and 0.02 Å cylinders respectively. Each image is centered in the same place with the longer N-N bond at the top.

In the three figures we see the expected build up of electron density in the bonding and lone pair regions, coupled with depleted regions necessary to obtain charge balance. Strain in the anion is indicated by the fact that the maximum bond density lies off the bond vector of the N-N-N bonds (bent bonds). The deformation density distribution between the 'equivalent' N-N bonds in dinitramide becomes more unsymmetrical as the torsion angle increases. The effect is small in the almost flat anion in (BIGH)(DN) but becomes more pronounced for the severely twisted anion in ADN. There is, however, no correlation with the difference in the two bond lengths (Δ N-N 0.023, 0.038 and 0.016 Å respectively). We also note that the central lone pair in the N-N-N plane becomes more diffuse as the torsion angle increases.

Because of the different torsion angles, no meaningful comparison of the N-O bonding regions nor of the oxygen lone pair regions can be obtained from this representation, however, from the complete 3D maps we observe little difference across the series in these regions.

In figures 2, 4 and 6, we show gray scale cutaway isosurfaces representing the electrostatic potentials for (BIGH)(DN), (BIGH₂)(DN)₂ and ADN. In these figures the longer N-N bond is at the bottom. Although the isolated atom model that we have used should produce an electrostatic potential isolated from the crystal lattice, the influence of the lattice still resides in the electron density and is thus not completely removed from the

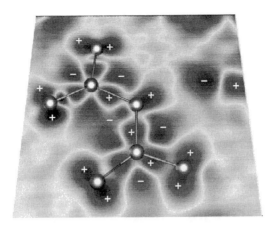

Figure 1: Deformation Density of Dinitramide in (BIGH)(DN)

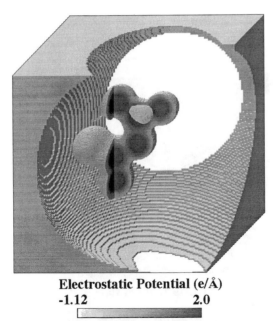

Electrostatic Potential (e/Å)
-1.12   2.0

Transparent from -1.02 to -0.48

Figure 2: Electrostatic Potential of Dinitramide from (BIGH)(DN)

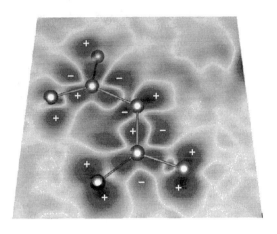

Figure 3: Deformation Density of Dinitramide in $(BIGH_2)(DN)_2$

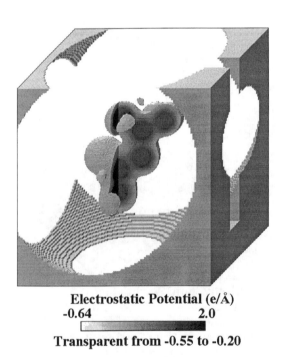

Electrostatic Potential (e/Å)
-0.64   2.0

Transparent from -0.55 to -0.20

Figure 4: Electrostatic Potential of Dinitramide from $(BIGH_2)(DN)_2$

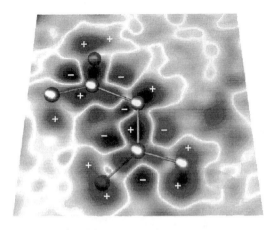

Figure 5: Deformation Density of Dinitramide in ADN

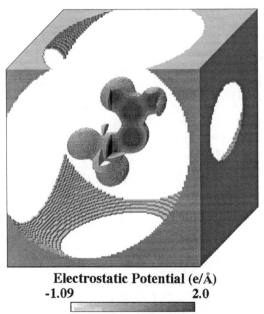

Electrostatic Potential (e/Å)
-1.09   2.0

Transparent from -1.02 to -0.34

Figure 6: Electrostatic Potential of Dinitramide from ADN

computed potential. Again, we observe a trend with respect to the torsion angle of the anion. In (BIGH)(DN) which has the smallest torsion angle, the most negative region (-1.12 e/Å) lies above the N-N-N plane over a terminal nitrogen atom with another smaller negative region over the other terminal nitrogen atom. A similar situation is observed for $(BIGH_2)(DN)_2$ with a minimum value of -0.636 e/Å, however, the relationship between the magnitudes of the two minima is reversed with respect to the N-N bond asymmetry. Finally for ADN which has the largest torsion angle, we observe the same asymmetry in the potential with respect to the bond asymmetry as for $(BIGH_2)(DN)_2$ but now with a minimum value of -1.09 e/Å.

Interestingly, both *ab-initio* and semi-empirical calculations [9] produce electrostatic potentials for the gas phase ions that have the most negative region in the N-N-N plane near to the bridging nitrogen. Thus, we suggest that the use of experimental electrostatic potentials obtained from X-ray crystallography should give a more reliable starting point for understanding solid state reactivities, e.g. by the use of molecular dynamics calculations.

ACKNOWLEDGMENTS

The authors would like to thank the Office of Naval Research for financial support (Contract Nos. N00014-93-1-0597 and N00014-95-1-0013).

REFERENCES

1.   J.C. Bottaro, R.J. Schmitt, P.E. Penwell and D.S. Ross, U.S. Pat. 5,198,204 and 5,254,324 (1991).
2.   T.B. Brill, P.J. Brush and D.G. Patil, Combust. Flame, 92, 7788 (1991).
3.   R.J. Doyle, Org. Mass Spectrom. 28, 83 (1993).
4.   M.J. Rossi, J.C. Bottaro and D.F. McMillen, Int. J. Chem. Kinet. 25, 549 (1993).
5.   R.J. Schmitt, M. Krempp and V.M. Bierbaum, Int. J. Mass Spectrom. Ion Processes 117, 612 (1992).
6.   H.H. Michels and J.A. Montgomery Jr., J. Phys. Chem., 97, 6602 (1993).
7.   P. Politzer and J.M. Seminario, Chem. Phys. Lett., 216, 348 (1993).
8.   P. Politzer, J.M. Seminario, M.C. Concha and P.C. Redfern, J. Mol. Struct. THEOCHEM, 287, 235 (1993).
9.   A. Martin, A.A. Pinkerton, R.D. Gilardi and J.C. Bottaro, to be published.
10.   N.K. Hansen and P. Coppens, Acta Cryst., A34, 909 (1978).

COMPUTATIONAL DETERMINATION OF HEATS OF FORMATION
OF ENERGETIC COMPOUNDS

PETER POLITZER, JANE S. MURRAY AND M. EDWARD GRICE
University of New Orleans, Department of Chemistry, New Orleans, Louisiana
70148 USA

ABSTRACT

A recently-developed density functional procedure for computing gas phase heats of formation is briefly described and results for several categories of energetic compounds are summarized and discussed. Liquid and solid phase values can be obtained by combining the gas phase data with heats of vaporization and sublimation estimated by means of other relationships. Some observed functional group effects upon heats of formation are noted.

INTRODUCTION

The heat of formation is frequently viewed as a measure of the energy content of a compound. As such, it is an important factor to consider in designing new energetic materials or evaluating existing ones. For example, the heat of formation enters into the calculation of such key explosive and propellant properties as detonation velocity [1,2], detonation pressure [1,2] and specific impulse [3].

We have recently developed a nonlocal density functional procedure for computing gas phase heats of formation [4]. The advantages of a computational approach are of course that it can be applied to suggested target molecules prior to undertaking syntheses, and to newly-prepared compounds even when the available amount is insufficient for laboratory characterization. We have now applied our technique to a variety of molecules that are of potential interest as energetic materials [5-8]. These as well as more recent results will be summarized and discussed.

THEORY

Our procedure [4] involves calculating ΔE for the formation of the molecule from its elements at 0 K followed by conversion to ΔH at 298 K by assuming ideal behavior and adding the translational, rotational and vibrational energies. Empirical correction terms corresponding to the various coordination states of the carbons, nitrogens and oxygens are also added.

ΔE is computed by a nonlocal density functional procedure (Gaussian 94 [9], Becke exchange and Perdew correlation functionals [10,11], 6-31G** basis set), using optimized geometries. The vibrational energy was originally determined from the normal mode frequencies, the calculation of which can be quite demanding in terms of computer resources. We have found, however, that this energy can be estimated accurately from stoichiometry-based relationships [12]; accordingly we now compute the frequencies only when we wish to confirm that the geometry corresponds to a local minimum in the energy (indicated by the absence of imaginary frequencies [13]).

In Table I, our calculated heats of formation are compared to the experimental values for some compounds relevant to the area of energetic materials. Additional comparisons have been presented elsewhere [4]. The average absolute error in the test cases is 3.4 kcal/mole, indicating very satisfactory agreement.

The procedure that has been described yields gas phase heats of formation, and these will be the primary subject of this paper. For practical purposes, however, liquid and solid phase values are often of greater interest; these can be obtained if the heats of vaporization and sublimation are known, using eqs. (1) and (2):

Table I. Comparison of Some Calculated and Experimental
Gas Phase Heats of Formation

	Molecule	Heat of formation, kcal/mole	
		Calculated	Experimental
1	$(H_3C)_2N-NO_2$	-2.7[a]	-1.1[b]
2	pyridine	31.4[c]	33.56[b]
3	1,3,5-trinitrobenzene (NO_2, O_2N, NO_2)	18.2[c]	14.9[b]
4		124.3[d]	130[e]
5		125.1[d]	124[g]

[a]Reference 6.   [b]Reference 45.
[c]Reference 4.   [d]Reference 8.
[e]Extrapolated from the experimental values for the dibenzo- and monobenzo-derivatives, which are 142.8 and 136.4 kcal/mole, respectively.[f]
[f]Y. T. Chia and H. E. Simmons, J. Am. Chem. Soc. 89, 2638 (1967).
[g]Extrapolated from the experimental values for the dibenzo- and monobenzo-derivatives, which are 132.1 and 128.2 kcal/mole, respectively.[f]

56

$$\Delta H_f^{298}(\text{liquid}) = \Delta H_f^{298}(\text{gaseous}) - \Delta H_{vap}^{298} \qquad (1)$$

$$\Delta H_f^{298}(\text{solid}) = \Delta H_f^{298}(\text{gaseous}) - \Delta H_{sub}^{298} \qquad (2)$$

We have recently developed quantitative representations of both ΔH_{vap}^{298} and ΔH_{sub}^{298} in terms of computed quantities related to electrostatic potentials on molecular surfaces [14,15]; the approach is one that we have used successfully to relate a variety of liquid, solid and solution properties to quantities calculated for individual molecules [14,16,17]. These expressions for ΔH_{vap}^{298} and ΔH_{sub}^{298} can accordingly be used in conjunction with $\Delta H_f^{298}(\text{gaseous})$ to obtain liquid and solid phase heats of formation, via eqs. (1) and (2).

RESULTS AND DISCUSSION

We will present and discuss our computed gas phase heats of formation for several classes of compounds that are currently of interest as potential energetic materials. Some of these data have been published elsewhere, as will be indicated. We will also mention two solid phase heats of formation that we have recently obtained, using eq. (2) and our expression for ΔH_{sub}^{298}.

1. Tetraazapentalenes

The presence of several linked nitrogens in a molecule (nitrogen catenation) is frequently associated with instability [18]. It is therefore notable that certain derivatives of the tetraazapentalene 4 show unexpected stability [19-21]; of particular interest in the present context is TACOT, 6, which has a melting/decomposition temperature of 378° C [22]. We have recently speculated that the surprising stability that has been observed experimentally for the molecular framework 4 may be due to the relatively positive character that we have found for the two triply-coordinated nitrogens [8].

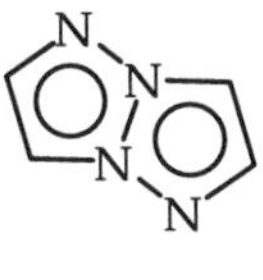

4

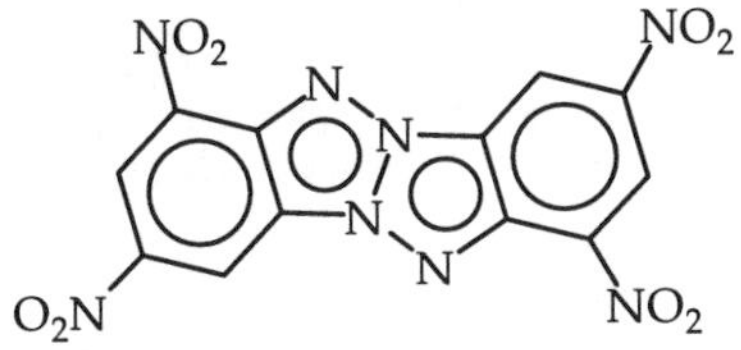

6

In Table II are listed our computed gas phase heats of formation for the two isomeric tetraazapentalenes 4 and 5 and for three of their derivatives. None of these compounds are known at present, although various other derivatives have been prepared [19-21,23,24], including 6 [22]. 7 [25] and 8 and 9 [26] have been proposed as target energetic materials, and synthesis efforts have been undertaken.

Table II. Calculated Gas Phase Heats of Formation of Some Tetraazapentalenes[a]

	Molecule	ΔH_f^{298} kcal/mole	ΔH_f^{298} cal/g
4		124	1151
5		125	1158
7		253	1318
8		426	1439
9		442	1493
10		247	630
11 RDX		45.7 (experimental value[b])	206

[a]The values for all of the tetraazapentalenes except **10** are from reference 8.
[b]References 27 and 28.

The data in Table II are given in both kilocalories/mole and calories/gram; it is the latter that is relevant for predicting detonation performance [1-3]. In order to provide a basis for comparison, the experimentally-determined gas phase heat of formation of RDX, **11**, a prominent military explosive, is also included [27,28].

Most of the tetraazapentalenes in Table II have quite large positive heats of formation, greater by at least a factor of five, in calories/gram, than that of RDX; the high values obtained for the unsubstituted molecules **4** and **5** are particularly striking. The relatively low ΔH_f^{298} found for **10** may indicate that

Table III. Calculated Gas Phase Heats of Formation of Some Furazans and Triazoles[a]

	Molecule	ΔH_f^{298} kcal/mole	ΔH_f^{298} cal/g
12		43	330
13 NOTO		202	903
14 DNAF		169	621
15 NTO		−5.7	−44
16 ADNT		87	502
11	RDX	45.7	206
		(experimental value[b])	

[a]The value for NOTO, **13**, is from reference 8.
[b]References 27 and 28.

framework nitrogens are more effective than nitro groups in producing a large heat of formation. The diazide 8 and the ditetrazole 9 are tautomers, which are expected to be related through a tautomeric equilibrium, 8 ⇌ 9. Our results show the diazide to be the more stable, by about 16 kcal/mole. This could be a matter for concern, from the standpoint of an energetic material, because of the tendency of many azides toward facile decomposition [29].

2. Furazans and Triazoles

Like the tetraazapentalenes, the furazans and triazoles in Table III fit into the category of high-nitrogen compounds. However these have actually been synthesized: 12 [30,31], NOTO, 13 [32], DNAF, 14 [33], NTO, 15 [34,35], ADNT, 16 [36]. NOTO has the interesting feature of being a liquid at room temperature [32]. Table III shows that all of these compounds except NTO have positive heats of formation which are considerably larger, on a calories/gram basis, than that of RDX but much smaller than those of most of the tetraazapentalenes in Table II.

3. Nitramines and Difluoramines

The difluoramino group, $-NF_2$, continues to be of interest as an ingredient of energetic materials, particularly propellants [2,3,37,38]. A judicious combination of nitro and difluoramino substituents offers the possibility of enhanced propellant performance due to an increased number of moles of gaseous combustion products per gram of material (provided that some hydrogens are present) [3,6,39].

In conjunction with recent computational studies of various existing or proposed nitramine and difluoramine systems [6,40], we have calculated the gas phase heats of formation of the compounds in Table IV. While most of them are positive, only 18 has a gas phase value (in calories/gram) greater than that of RDX.

A consistent pattern in Table IV is that ΔH_f^{298} is lower (less positive) for the difluoramine than for its nitramine analogue. (Leroy et al have observed the same to be true when the $-NF_2$ and $-NO_2$ are attached to carbons [41], as we do also for the pair 20, 22.) However the difluoramine may still have a higher specific impulse (a measure of propellant thrust [3,22,39]) due to producing more moles of gaseous products per gram upon combustion. Thus the specific impulse of 23 is predicted to be higher than that of RDX, 11, despite having a lower estimated heat of formation [3].

23
$$NF_2$$

11, RDX
$$NO_2$$

Table IV. Calculated Gas Phase Heats of Formation of Some
Nitramines and Difluoramines[a]

	Molecule	ΔH_f^{298} kcal/mole	ΔH_f^{298} cal/g
1	$(H_3C)_2N-NO_2$	−2.7	−30
17	$(H_3C)_2N-NF_2$	−20	−206
18	$O_2N-N\diamond N-NO_2$	44	297
19	$O_2N-N\diamond N-NF_2$	29	186
20 TNAZ	$\begin{matrix} O_2N \\ O_2N \end{matrix}\diamond N-NO_2$	31	160
21	$\begin{matrix} O_2N \\ O_2N \end{matrix}\diamond N-NF_2$	16	79
22	$\begin{matrix} O_2N \\ F_2N \end{matrix}\diamond N-NO_2$	7.4	38
11	RDX	45.7 (experimental value[b])	206

[a]All of these calculated results except that for **22** are from reference 6.
[b]References 27 and 28.

4. Some C, N, O, F Compounds

We have investigated computationally a group of 19 unsaturated molecules that contain only the elements C, N, O and F (Table V) [7]. Their high N/C and O/C ratios suggest that these are potential energetic molecules or precursors. To the best of our knowledge, most of these molecules are presently unknown, an exception being **26** [42,43].

The calculated gas phase heats of formation are given in Table V. These were obtained by a slightly modified version of the method described earlier, in that *ab initio* HF/6-31G* optimized geometries and vibrational frequencies were used and ΔE was computed with the density functional program DeMon [44]. Structures **26A**, dinitrosoacetylene, and **26B**, which is more accurately designated as dinitrile-di-N-oxide, were both found to correspond to energy minima at the *ab initio* HF/6-31G*

Table V. Calculated Gas Phase Heats of Formation of Some C, N, O, F Compounds[a]

	Molecule	ΔH_f^{298} kcal/mole	ΔH_f^{298} cal/g
24	$O_2N-C\equiv C-NO_2$	89	763
25	$O_2N-C\equiv C-NO$	100	1004
26A	$ON-C\equiv C-NO$	130	1550
26B	$ON\equiv C-C\equiv NO$	73	868
27	$O_2N-C\equiv C-NF_2$	74	606
28	$ON-C\equiv C-NF_2$	95	901
29	$N_3-C\equiv C-NO_2$	144	1287
30	$OCN-C\equiv C-NO_2$	54	486
31	$(O_2N)_2C=C\begin{smallmatrix}NO_2\\CN\end{smallmatrix}$	80	426
32	$\begin{smallmatrix}O_2N\\CN\end{smallmatrix}C=C(NO_2)_2$	108	572
33	$(O_2N)_2C=C=O$	3	22
34	$O_2N-N=C=N-NO_2$	87	659
35	$(O_2N)_2C=C=C=O$	39	274
36	$(O_2N)_2C=C=C=N-NO_2$	115	613
37	$O_2N-N=C=C=N-NO_2$	128	888
38	$O_2N-N=C=C=C=N-NO_2$	136	870
39		102	465
40		135	716
41		130	559

(continued)

Table V. Calculated Gas Phase Heats of Formation of Some C, N, O, F Compounds[a]
(continued)

	Molecule	ΔH_f^{298} kcal/mole	ΔH_f^{298} cal/g
42	$(O_2N)_2C$⬡$C(NO_2)_2$	9	35
11	RDX	45.7	206
		(experimental value[b])	

[a]The results for **24 - 30, 33-35, 37** and **38** are in reference 7.
[b]References 27 and 28.

level [7]. Table V shows that **26B** is the more stable, which is consistent with experimental evidence [42,43].

Most of these compounds are predicted to have rather large positive heats of formation (calories/gram). One of the highest, not surprisingly, is the azide **29**. The pairs **24, 27** and **25, 28** show again that replacing –NO$_2$ by –NF$_2$ lowers ΔH_f^{298}, as observed above when these groups were on nitrogens.

Particularly striking, especially on a calories/gram basis, is the *increase* in ΔH_f^{298} that occurs when the nitro group is replaced by the nitroso. This can be seen in **24, 25** and **26A** and again in **27** and **28**. (**26B** is more nitrile-N-oxide than nitroso in character, as mentioned above.) This effect has also been observed experimentally; the heat of formation of **43** is 46.44 kcal/mole (322.5 cal/g) vs. 13.9 kcal/mole (79.0 cal/g) for **44** [45]. In view of these findings, it is interesting to note evidence

| **43** | **44** | **45** | **11, RDX** |

suggesting the possibility of another desirable consequence of substituting –NO for –NO$_2$; an experimental study showed that the impact sensitivity of **45** is significantly less than that of RDX, **11**, while the explosive power (measured by dent depth) is essentially the same [46].

A further comparison of interest is between =C(NO$_2$)$_2$ compounds and their =N(NO$_2$) analogues. The latter have the greater heats of formation, as can be seen from the pairs **36, 37** and **40, 41**.

5. Solid Phase Heats of Formation

As was pointed out earlier in this paper, we now have the means for reliably estimating heats of vaporization and sublimation [14,15]. These permit us to convert our computed gas phase heats of formation to liquid and solid phase values, using eqs. (1) and (2), respectively. We have thus far applied this procedure to two compounds, 1,3,3-trinitroazetidine (**20**, TNAZ) and DNAF (**14**).

For TNAZ, we find $\Delta H_{sub}^{298} = 22$ kcal/mole; when combined with ΔH_f^{298}(gaseous) = 31 kcal/mole (Table IV) via eq. (2), the result is ΔH_f^{298}(solid) = 9 kcal/mole. This is in excellent agreement with the experimental value, 8.7 kcal/mole [38]. For DNAF, we obtain $\Delta H_{sub}^{298} = 32$ kcal/mole. Using ΔH_f^{298}(gaseous) = 169 kcal/mole (Table III), eq. (2) gives ΔH_f^{298}(solid) = 137 kcal/mole.

CONCLUSIONS

We have presented and discussed the results of density functional calculations of gas phase heats of formation for several categories of energetic compounds. We have also demonstrated the feasibility of converting these to solid phase values.

For the systems investigated, some key observations are:

(1) The tetraazapentalenes have particularly large positive heats of formation (calories/gram).
(2) Replacement of a nitro by a difluoramino group, whether on a carbon or on a nitrogen, lowers the heat of formation.
(3) Replacement of a nitro by a nitroso group increases the heat of formation.
(4) Replacement of $=C(NO_2)_2$ by $=N-NO_2$ increases the heat of formation.

ACKNOWLEDGMENT

We greatly appreciate the financial support of the Office of Naval Research, through contract No. N00014-95-1-0028 and Program Officer Dr. Richard S. Miller.

REFERENCES

[1] M. J. Kamlet and S. J. Jacobs, J. Chem. Phys. **48**, 23 (1968).

[2] T. Urbanski, *Chemistry and Technology of Explosives* (Pergamon Press, New York, 1984).

[3] P. Politzer, J. S. Murray, M. E. Grice and P. Sjoberg, in *Chemistry of Energetic Materials*, edited by G. A. Olah and D. R. Squire (Academic Press, New York, 1991), ch. 4.

[4] D. Habibollahzadeh, M. E. Grice, M. C. Concha, J. S. Murray and P. Politzer, J. Comp. Chem. **16**, 654 (1995).

[5] P. Politzer, P. Lane, P. Sjoberg and M. E. Grice, Technical Report No. 72, Contract No. N00014-91-J-4057, Office of Naval Research, Arlington, VA, October 20, 1994.

[6]P. Politzer, P. Lane, M. E. Grice, M. C. Concha and P. C. Redfern, J. Mol. Struct. (Theochem) **338**, 249 (1995).

[7]P. Politzer, P. Lane, P. Sjoberg, M. E. Grice and H. Shechter, Struct. Chem. **6**, 217 (1995).

[8]M. E. Grice and P. Politzer, J. Mol. Struct. , in press.

[9]M. J. Frisch, G. W. Trucks, H. B. Schlegel, P. M. W. Gill, B. G. Johnson, M. A. Robb, J. R. Cheeseman *et al.*, Gaussian 94 (Revision B.3) (Gaussian, Inc., Pittsburgh, PA, 1995).

[10]A. D. Becke, Phys. Rev. A **38**, 3098 (1988).

[11]J. P. Perdew, Phys. Rev. B **33**, 8822 (1986).

[12]M. E. Grice and P. Politzer, Chem. Phys. Lett. **244**, 295 (1995).

[13]W. J. Hehre, L. Radom, P. v. R. Schleyer and J. A. Pople, *Ab Initio Molecular Orbital Theory* (Wiley-Interscience, New York, 1986).

[14]J. S. Murray and P. Politzer, in *Quantitative Treatments of Solute/Solvent Interactions*, edited by J. S. Murray and P. Politzer (Elsevier, Amsterdam, 1994), ch. 8.

[15]M. DeSalvo, E. Miller, J. S. Murray and P. Politzer, unpublished work.

[16]J. S. Murray, T. Brinck, P. Lane, K. Paulsen and P. Politzer, J. Mol. Struct. (Theochem) **307**, 55 (1994).

[17]P. Politzer, J. S. Murray, T. Brinck and P. Lane, in *Immunoanalysis of Agrochemicals; Emerging Technologies*, edited by J. O. Nelson, A. E. Karu, and R. B. Wong (ACS, Washington, 1995), ch. 8.

[18]F. R. Benson, *The High Nitrogen Compounds* (Wiley-Interscience, New York, 1984).

[19]R. A. Carboni and J. E. Castle, J. Am. Chem. Soc. **84**, 2453 (1962).

[20]R. Pfleger, E. Garthe and K. Raner, Chem. Ber. **96**, 1827 (1963).

[21]R. A. Carboni, J. C. Kauer, J. E. Castle and H. E. Simmons, J. Am. Chem. Soc. **89**, 2618 (1967).

[22]J. Köhler and R. Meyer, *Explosives*, 4th ed. (VCH Publishers, New York, 1993).

[23]R. A. Carboni, J. C. Kauer, W. R. Hatchard and R. J. Harder, J. Am. Chem. Soc. **89**, 2626 (1967).

[24]J. C. Kauer and R. A. Carboni, J. Am. Chem. Soc. **89**, 2633 (1967).

[25]J. H. Boyer and M. L. Trudell, private communication.

[26]M. L. Trudell, private communication.

[27]D. R. Stull, E. F. Westrum and G. C. Sinke, *The Chemical Thermodynamics of Organic Compounds* (Wiley, New York, 1969).

[28]J. M. Rosen and C. Dickinson, J. Chem. Eng. Data **14**, 120 (1969).

[29]J. H. Boyer and F. C. Canter, Chem. Rev. **54**, 1 (1954).

[30]G. D. Solodynk, M. D. Boldyrev, B. V. Gidaspov and V. D. Nikolaev, Zh. Org. Khim. **17**, 861 (1981).

[31]A. Gunasekaran, private communication.

[32]A. Gunasekaran and J. H. Boyer, Heteroatom Chem. **4**, 521 (1993).

[33]A. Gunasekaran, M. L. Trudell and J. H. Boyer, Heteroatom Chem. **5**, 441 (1994).

[34]K.-Y. Lee and M. D. Coburn, U. S. Patent , 4,733,610 (1988).

[35]K.-Y. Lee and R. Gilardi, in *Structure and Properties of Energetic Materials*, edited by D. H. Liebenberg, R. W. Armstrong and J. J. Gilman (Materials Research Society, Pittsburgh, 1993).

[36]R. Schmitt and J. Bottaro, SRI International, private communication.

[37]R. F. Gould, in *Advances in Chemistry Series, No. 54* (American Chemical Society, Washington, 1966).

[38]T. G. Archibald, L. C. Garver, A. A. Malik, F. O. Bonsu, D. D. Tzeng, S. B. Preston and K. Baum, Report No. ONR-2-10, Office of Naval Research, Arlington, VA, Contract No. N00014-78-C-0147, February, 1988.

[39]R. T. Holzmann, in *Advanced Propellant Chemistry*, edited by R. F. Gould (American Chemical Society, Washington, 1966), Ch. 1.

[40]P. Politzer and M. E. Grice, J. Chem. Res. (S) , 296 (1995).

[41]G. Leroy, M. Sana, C. Wilante, D. Peeters and S. Bourasseau, J. Mol. Struct. (Theochem) **187**, 251 (1989).

[42]C. Grundmann, Angew. Chem. **75**, 450 (1963).

[43]C. Grundmann, V. Mini, J. M. Dean and H.-D. Frommeld, Liebigs. Ann. Chem. **687**, 191 (1965).

[44]D. R. Salahub, R. Fournier, P. Mlynarski, I. Papai, A. St. Amant and J. Ushio, in *Density Functional Methods in Chemistry*, edited by J. K. Labanowski and J. W. Andzelm (Springer-Verlag, New York, 1991), ch. 6.

[45]J. B. Pedley, R. D. Naylor and S. P. Kirby, *Thermochemical Data of Organic Compounds*, 2nd ed. (Chapman and Hall, London, 1986).

[46]S. Iyer, Propell. Expl. Pyrotech. **7**, 37 (1982).

MONTE CARLO SIMULATIONS OF CRYSTALLINE TATB

THOMAS D. SEWELL
Theoretical Division, Los Alamos National Laboratory, Los Alamos, New Mexico 87545

ABSTRACT

We are performing constant-*NPT* Monte Carlo calculations of the physical properties of crystalline TATB. Our approach is to employ an atomistic model in which the individual molecules are treated as semi-rigid entities. Each molecule is allowed to undergo rigid translations and rotations, and in some cases limited intramolecular flexibility is conferred on the molecules *via* exocyclic torsions. Additionally, the size and shape of the simulation box is allowed to vary. Our immediate interest is in computing the density, lattice energy, lattice constants, and other structural parameters as a function of temperature. Preliminary results indicate that simulations involving only two molecules suffice for calculations of the energy and density, but that more molecules are required to compute the lattice constants. Intramolecular flexibility is important, particularly at higher temperatures.

INTRODUCTION

1,3,5-triamino-2,4,6-trinitrobenzene (TATB) is the primary energetic material in the plastic-bonded high explosive known as PBX-9502 (see Fig. 1). TATB is considered an insensitive high explosive, meaning that initiation culminating in full detonation is relatively difficult to achieve *via* common mechanisms such as mechanical shock, spark initiation, cook-off, or deflagration. As such, TATB is a relatively safe material and is usually used as a secondary explosive in a detonation chain. However, recent studies have revealed that samples of TATB become more sensitive at high temperatures. This is a potential cause for concern since explosives are sometimes stored in relatively harsh thermal environments and any degradation in the safety characteristics of the material needs to be at the very least examined, and hopefully understood from a scientific perspective.

Figure 1. Structure of TATB. The chemical formula is $C_6N_6O_6H_6$. Black atoms: C; large white atoms: N; shaded atoms: O; small white atoms: H. The structure is planar with D_{3h} symmetry.

We are developing codes for Monte Carlo modeling of the pressure and temperature dependence of the physical properties of organic molecular crystals. As we are employing atomistic treatments of the materials, our calculations provide information about the microscopic details of how the crystals change with temperature and/or pressure. The goal of the work is to develop a predictive capability that will enable us to make purely theoretical statements concerning the nonreactive behavior of a wide range of materials under conditions not readily amenable to experimental measurement. Of particular interest are the density, energy, elastic constants, and crystal packing at elevated temperatures and pressures. However, at the present time we are still working to build confidence in our methods and models at the lower ends of the *P-T* domain.

Mat. Res. Soc. Symp. Proc. Vol. 418 ® 1996 Materials Research Society

In this report we present preliminary results of Monte Carlo calculations of the physical properties of TATB as a function of temperature. All of the results were computed in the *NPT* ensemble at a pressure of 1.0 bar. It is not our aim at this point to reproduce the physical properties to high accuracy; rather, our present goal is to assess the effects of simulation size, treatment of the electrostatic potential, and intramolecular flexibility on our results. To facilitate this goal, we have employed simple, "generic" forms and parameterizations for all of the non-electrostatic interactions and have used an idealized model of the TATB molecule. More realistic treatments will be used in future work.

THEORY

Monte Carlo Method

The Monte Carlo approach is useful for studying the physical properties of molecular crystals [1,2]. A useful introduction to the method has been given by Kalos and Whitlock [3]. Monte Carlo moves in the *NPT* ensemble were performed in four distinct stages: independent linear translations of the molecular centers of mass, three dimensional rotations about the molecular centers of mass, changes in the size and shape of the simulation box, and variations of the dihedral angles associated with exocyclic groups. A Monte Carlo step was defined to be the result of $N_{com}=10$ attempted translations of each molecule, $N_{rot}=10$ rotations of each molecule, $N_{tor}=5$ torsional displacements of each dihedral angle, and $N_{vol}=10$ attempted shape/size moves. For clarity, the translation operation was performed once for each molecule in turn, with the cycle being repeated N_{com} times; and similarly for the other operations. The step sizes for each kind of move were adjusted so as to achieve an acceptance/rejection ratio near unity.

A warm-up walk comprised of at least 100 Monte Carlo steps was performed to relax the geometry away from the initial structure prior to the accumulation of Monte Carlo observations for subsequent analysis. Longer warm-up segments were used at higher temperatures. The random walk was continued subsequent to the warm up until 2000 observations were obtained. The individual observations were coarse grained to obtain statistically independent observations of the system based on the battery of tests described by Hald [4]. Further details concerning the Monte Carlo algorithm used in this work are being presented elsewhere [2].

Potential-Energy Surface

We write the potential energy as a sum of repulsion, dispersion (van der Waals), and electrostatic intermolecular terms plus exocyclic torsional intramolecular terms. The intermolecular repulsions were represented by simple exponential functions,

$$V_{rep} = A_{ij} \exp[-b_{ij}R_{ij}], \tag{1}$$

and dispersion (van der Waals) interactions were taken to be

$$V_{vdW} = -C_{ij} R_{ij}^{-6}. \tag{2}$$

Here, R_{ij} is the internuclear separation of atoms i and j on different molecules. Parameters for these interactions were taken from the work of Williams and coworkers [5].

The electrostatic contributions to the energy were expressed in terms of atom-centered multipole expansions [6],

$$V_{ele} = E_{00} + E_{01} + E_{11} + E_{02} + E_{12} + E_{22} + E_{03} + E_{13} + E_{23} + E_{33}, \tag{3}$$

where we identify the individual terms in the electrostatic potential as E_{kl}; k and l refer to particular atom-centered multipole moments: 0, 1, 2, and 3 denote charges, dipoles, quadrupoles, and octupoles, respectively. The first term, $E_{00} = q_i q_j R_{ij},^{-1}$ is just the familiar charge-charge interaction used in most simulations. The use of higher-order multipoles is likely to be important for molecules having lone pairs of electrons, particularly in systems where intermolecular hydrogen bonding occurs, as they confer an explicit directionality on the potential interactions. Expressions for the E_{kl} are provided in Table 1 of Ref. [6]. Descriptions of how the atom-centered multipoles are generated may be found in [6] and references therein.

Rigid exocyclic torsions of the nitro groups were included in some of the calculations. (The effects of including amino rotations and possibly out-of-plane wags for both kinds of groups will be considered in future work.) We used a simple, zeroth-order description of the torsions in which each exocyclic group was taken to be independent of the others. Specifically, we employed the following function,

$$V_{tor} = V_0[1-\cos^2\phi], \tag{4}$$

where ϕ is the NO_2 dihedral angle and V_0 is the barrier determined from the quantum chemical calculations described in the next section.

Molecular Geometry and Barrier to Nitro Rotation

The equilibrium geometry of TATB is predicted to be planar and to possess D_3 symmetry at the HF/3-21g level of *ab initio* quantum theory. The calculated structure is quite close to D_{3h} symmetry, however; a systematic, few-degree distortion of the C-N bonds away from the bisectors of the adjacent C-C-C angles in the benzene ring results in the lower symmetry point group. Likewise, the transition state for NO_2 rotation is close to C_{2v} symmetry. The energetic difference between these two stationary points is 71.2 kJ/mol; inclusion of the MP2 correction for electron correlation lowers the barrier to 68.7 kJ/mol. This relatively high barrier is due to significant intramolecular hydrogen bonding in the equilibrium structure.

We have ignored the slight distortions of TATB away from D_{3h} and C_{2v} symmetry at the equilibrium and transition-state geometries, respectively. Thus, at this early stage of our work, TATB is taken to be a planar molecule having D_{3h} symmetry at the equilibrium structure, with a barrier of 68.7 kJ/mol to rotation of a single nitro group about its torsional axis. The structure shown earlier (Fig. 1) corresponds to the equilibrium geometry used.

RESULTS

Fits to the Electrostatic Potential

A summary of the fits to the electrostatic potential surrounding an isolated TATB molecule is

TABLE I. Fit of the electrostatic potential.

	Range (Hartrees)					
	-0.0600 <E< -0.00159		-0.00159 <E< 0.00159		0.00159<E <0.08	
Model	RMS	RRMS	RMS	RRMS	RMS	RRMS
PDQ[a]	0.7085E-03	0.9196E-01	0.1559E-03	0.1117E+02	0.1472E-02	0.8588E-01
ME 0[b]	0.4305E-02	0.4288E+00	0.5678E-03	0.3113E+02	0.6006E-02	0.5633E+00
ME 1	0.1528E-02	0.1933E+00	0.4313E-03	0.3295E+02	0.2275E-02	0.2746E+00
ME 2	0.3059E-03	0.4042E-01	0.7766E-04	0.1049E+01	0.1177E-02	0.4807E-01
ME 3	0.1157E-03	0.1185E-01	0.2951E-04	0.4032E+00	0.7754E-03	0.2967E-01

a. CHELPG potential-derived charges obtained using Gaussian 92 [7].
b. Atom-centered multipole expansions carried out to order "*l*".

provided in Table I. The data correspond to the RMS and relative RMS errors in the potential energy obtained from the atom-centered multipole expansion (carried out to order l) compared to that computed rigorously using Gaussian 92 [7]. The results for potential-derived "CHELPG" charges [7] are also included. The data are sorted into regions of negative, near zero, and positive potential. The order at which the multipole expansion yields a better fit is indicated in the table. From the information in the table it is obvious that the atom-centered multipole expansion carried out through quadrupoles gives a significantly better overall fit to the electrostatic potential surrounding the molecule than does a simple potential-derived charge treatment.

Density, Energy, and Lattice Constants

Plots depicting the temperature dependence of the density and energy as functions of simulation size, electrostatic model, and molecular flexibility are presented in Fig. 2. Error bars represent one standard deviation. Table II contains numerical data for the density, energy, and lattice constants. The experimental results of Cady and Larson [8] are included for reference. It is seen in panel a of Fig. 2 that the CHELPG charges (upper trace) give the best agreement to the experimental result (about 4.6% error). Atom-centered multipoles yield a somewhat larger error (6.25%). It is important to recall, however, that the repulsion and dispersion parameters have not been optimized for any of the electrostatic models used here; thus, the results could probably be adjusted to yield better agreement with experiment. (The fact that the computed densities are too low is thought to result from the use of repulsion and dispersion parameters which were optimized using energy minimized structures, thereby overestimating the "size" of the atoms.)

Significantly, the size effects seem to be fairly small (open symbols at T=300K) and, based on the limited results shown here, there does not seem to be a strong effect in passing from quadrupoles to octupoles in the electrostatic treatment. We seem to be able to compute reasonable values for the density and expansion coefficients using as few as two molecules. As would be expected, there is a direct correlation between the energy and density; the higher densities associated with use of potential-derived CHELPG charges in panel a of Fig. 2 are consistent with the lower energies seen in panel b.

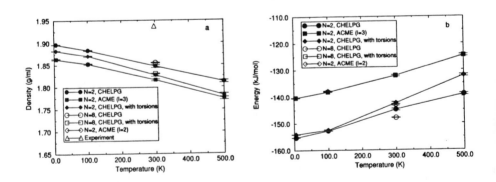

Figure 2. Temperature dependence of the density (panel a) and energy (panel b) of crystalline TATB as a function of electrostatic model, simulation size, and presence or absence of NO_2 torsions. Descriptions of the data points are given in the figure legends.

The role of the torsional degrees of freedom is clear in Fig. 2. Note the increase in (negative) curvature in the density (panel a) as the temperature is increased. (Compare the trace denoted by filled circles -- no torsion -- to the one denoted by filled diamonds -- same electrostatic model, but with torsion included.) This is accompanied by a similar increase in the (positive) curvature for the corresponding energy trace in panel b. Although it is possible to account for intramolecular modes in the heat capacity using statistical mechanical arguments, it is not obvious how to include the effects of such modes on the density without explicit incorporation into the simulations.

The lattice constants are not as yet under good statistical control in our calculations (Table II). It appears that simulations involving moderate numbers of molecules (N=8 may suffice) will be necessary in order to obtain reliable results at noncryogenic temperatures. In high-temperature simulations using N=2, the molecules are able to "slide" in the a-b crystallographic plane; larger simulations prevent this to some extent. It does appear from our limited data, however, that the main mode of expansion in the crystal is in the direction parallel to the c axis (i.e., perpendicular to the molecular planes), consistent with experiment.

TABLE II. Density, energy and lattice constants as a function of temperature, simulation size, electrostatic model, and presence or absence of flexible nitro groups. Densities in g/ml, energies in kJ/mol, distances in Å, and angles in degrees.

T (K)	N[a]	l[b]	Tors[c]	Density	Energy	a	b	c	α	β	γ
5	2	0	no	1.8965(2)[d]	-155.295(2)	9.17	9.18	6.21	90.0	89.9	120.1
100	2	0	no	1.8831(8)	-152.76(6)	9.18	9.19	6.24	89.4	90.5	120.1
300	2	0	no	1.847(3)	-144.8(4)	e					
300	8	0	no	1.8546(8)	-147.88(7)	9.20	9.21	6.30	89.6	90.1	120.0
500	2	0	no	1.812(3)	-139.1(4)						
5	2	3	no	1.863(1)	-140.39(6)	9.24	9.18	6.27	96.7	85.6	119.3
100	2	3	no	1.852(1)	-138.19(7)	9.25	9.19	6.33	97.5	81.2	119.3
300	2	3	no	1.816(2)	-132.1(3)	9.28	9.22	6.34	91.3	86.5	119.2
500	2	3	no	1.774(3)	-124.5(4)						
100	2	2	no	1.852(1)	-137.89(9)	9.25	9.20	6.27	93.6	84.8	119.4
5	2	0	yes	1.8823(2)	-154.11(6)	9.17	9.17	6.85	110.9	89.4	120.0
100	2	0	yes	1.870(1)	-152.5(2)						
300	2	0	yes	1.827(3)	-142.4(4)						
300	8	0	yes	1.8302(8)	-143.0(3)	9.20	9.20	7.21	114.0	89.3	120.0
500	2	0	yes	1.781(4)	-132.1(4)						
300	Expt [8]			1.937	-----.----	9.01	9.03	6.81	108.6	91.8	120.0

a. Number of molecules in the simulation.
b. Order of the atom-centered multipole expansion; $l=0$ corresponds to CHELPG charges.
c. Indicates whether torsional motion of nitro groups was included in the calculations.
d. Numbers in parentheses represent the uncertainty in the last digit shown.
e. Gaps in the table indicate that the numbers were not sufficiently well determined.

CONCLUSIONS

Monte Carlo methods can be used to predict the temperature dependence of the physical properties of polyatomic molecular crystals. Although the computational cost rises quickly with the number of molecules considered in such simulations, reasonably good results for properties such as density and energy (and the associated derivatives) can be obtained from simulations involving only a few independent molecules (approximately the number of molecules in the crystallographic unit cell). Based on our limited results, it appears that bigger calculations are required in order to predict reliable values for lattice constants.

Atom-centered multipole treatments of the electrostatic interaction potential are a viable means for properly taking into account the nonspherical nature of the electrostatic potential surrounding atoms which contain lone pairs of electrons. Given that most high-explosive molecules contain such atoms, the use of atom-centered multipole expansions needs to be considered when performing simulations of crystal structures if high accuracy is desired. However, in order for this approach to be used to best advantage, it will be necessary to reparameterize the potential functions used to describe the repulsion and dispersion interactions.

An atom-centered multipole expansion carried out through quadrupoles yields a better fit to the electrostatic potential than is obtained using CHELPG potential-derived charges. However, for the TATB simulations performed to date, there do not seem to be strong effects on either the energy or density associated with the use of potential-derived charges versus atom-centered multipole treatments of the electrostatics. (Although the magnitudes differ somewhat, the slopes of the curves in Fig. 2 are similar.) Investigations are underway to determine whether differences exist for other, more incisive, measures of the crystal structure.

Inclusion of limited torsional motion leads to an increase in the heat capacity of TATB. It also leads to an increase in the coefficient of volumetric expansion (particularly parallel to the c axis). These effects become more important at higher temperatures. Thus, it is necessary to include such motions when performing calculations at elevated temperatures. The degree to which additional "floppy" modes must be included is not clear at this point and remains a topic for future research.

Much work remains to be done. Our immediate goals are to perform simulations using optimized potential parameters, larger numbers of molecules, increased intramolecular flexibility, and extensions to points along the pressure axis; and to consider other molecules such as TNT and nitroguanidine to obtain a better sense of how generally reliable these kinds of calculations can be.

ACKNOWLEDGMENT

This work was supported by the U. S. Department of Energy.

REFERENCES

1. S. Yashonath, S. L. Price, and I. R. McDonald, Mol. Phys. **64**, 361 (1988).

2. T. D. Sewell, *Monte Carlo Simulations of Crystalline Benzene* (manuscript in preparation).

3. M. H. Kalos and P. A. Whitlock, *Monte Carlo Methods. Volume I: Basics* (John Wiley & Sons, New York, 1986).

4. A. Hald, *Statistical Theory with Engineering Applications* (John Wiley & Sons, New York, 1952), Ch. 13.

5. D. E. Williams and S. R. Cox, Acta Cryst. **B40**, 404 (1984).

6. J. P. Ritchie and A. S. Copenhaver, J. Comp. Chem. **16**, 777 (1995).

7. Gaussian 92/DFT, Revision G.1, M. J. Frisch, G. W. Trucks, H. B. Schlegel, P. M. W. Gill, B. G. Johnson, M. W. Wong, J. B. Foresman, M. A. Robb, M. Head-Gordon, E. S. Replogle, R. Gomperts, J. L. Andres, K. Raghavachari, J. S. Binkley, C. Gonzalez, R. L. Martin, D. J. Fox, D. J. Defrees, J. Baker, J. J. P. Stewart, and J. A. Pople, Gaussian, Inc., Pittsburgh PA, 1993.

8. H. H. Cady and A. C. Larson, Acta Cryst. **18**, 485 (1965).

CRYSTAL GROWTH OF ENERGETIC MATERIALS DURING HIGH ACCELERATION

M. Y. D. LANZEROTTI*, J. AUTERA*, L. BORNE**, J. SHARMA***
*U. S. ARMY ARDEC, Picatinny Arsenal, NJ 07806 5000
**French-German Research Institute of Saint-Louis (ISL), France
***Naval Surface Warfare Center, Silver Spring, MD 20903

ABSTRACT

Studies of the growth of crystals of energetic materials under conditions of high acceleration in an ultracentrifuge are reported. When a saturated solution is accelerated in an ultracentrifuge, the solute molecules move individually through the solvent molecules to form a crystal at the outer edge of the tube if the solute is more dense than the solvent. Since there is no evaporation or temperature variation, convection currents caused by simultaneous movement of solvent and solute are minimized and crystal defects are potentially minimized. Crystal growth is controlled by the g-level of the acceleration. In addition, solution inclusions and bubbles migrate out of the saturated solution as a result of the pressure gradient induced by the g-force. We present results of TNT, RDX, and TNAZ grown at high g from various solutions.

INTRODUCTION

Crystal growth from a solution can be considered a heterogeneous chemical reaction of the type where a portion of the liquid goes into crystal form[1,2]. In the laboratory at 1 g, crystal growth methods include solvent evaporation at constant temperature and slow cooling. Crystal growth occurs when the solution becomes supersaturated. The crystal growth is controlled by simultaneous movement of solute and solvent in convection currents.

Supersaturation can also occur in an initially saturated solution during high acceleration[3-6]. If the solution is initially saturated, then under acceleration the solution at the outer edge of the accelerating tube becomes supersaturated. A density gradient is established. Thus at high g (above 1000 g), the crystal growth mechanism is different. The solute molecules individually move through the solvent molecules to form a crystal if the density of the solute is greater than the density of the solvent. Crystal growth is controlled by the g-force. In this new method crystal defects caused by temperature variation or evaporation are potentially minimized. Two international conferences on crystal growth at high g levels addressed these issues[7,8], although only a few of the reports addressed g-levels above 1000 g[4,9].

In addition, solution inclusions and voids in the accelerated saturated solution migrate out of this saturated solution as a result of the pressure gradient induced by the g-force. Thus, voids and solution inclusions are less likely to form in a crystal grown under high acceleration. Since the mechanical sensitivity of an explosive is significantly influenced by defects in the crystal[10-14] this feature of the crystal is important for numerous applications utilizing energetic materials[15-20]. The long term objective of this program [6,9,21] is to understand the fundamentals of the crystal growth process and thereby to reduce the formation of defects in crystals of energetic materials so that they will be less sensitive to mechanical shock.

TECHNIQUE

The experiments on crystal growth are performed using saturated solutions. The samples of saturated solutions can be filtered prior to insertion into the centrifuge tube in order to remove seed crystals. A Beckman preparative ultracentrifuge model L8-80 with a swinging bucket rotor model SW 60-Ti is used to accelerate the saturated solution up to 500,000 g. Polyallomer centrifuge tubes with hemispherical ends are used in the experiments. After an experimental run the centrifuge tube with saturated solution sample is removed from the bucket and the saturated solution is poured off if a crystal has formed. If necessary, the polyallomer tube is cut lengthwise

73

with a razor blade to remove the crystals and to study the physical features and habit of the crystal and polycrystal materials without damaging the crystals.

RESULTS

Polycrystalline materials are found on the curved interior surface of the polyallomer centrifuge tube. In the experiments performed to date, the curved surface appears to inhibit single crystal formation. A hemispherical Teflon insert with a flat surface interfacing with the saturated solution is inserted into the tube to provide a flat surface that yields single crystal growth.

A number of experimental runs have been made for TNT (trinitrotoluene), RDX (cyclotrimethylene-trinitramine), and TNAZ (1,3,3-trinitroazetidine). These runs have been made for various values of temperature, time and acceleration. The results are shown in Table I and Table II for TNT and RDX, respectively. The results are shown in Tables III through V for TNAZ. Pressure in units of Pascals divided by 6.894757 x 10^3 equals pressure in units of psi.

Table I. TNT Crystal Growth Experiments at High g at 25°C in Ethyl Acetate Solution

Acceleration (x 10^3 g)	Pressure (x 10^6 Pa)	Growth Surface	Time (hr)	Filter	Results
13	6	Curved	16	No	No crystals
29	13	Flat	64	Yes	No crystals
50	24	Curved	17	No	Polycrystalline
50	24	Flat	15	Yes	Individual crystals aligned parallel to acceleration, 2-mm length, habit is coffin-like[22]
50	24	Flat	92	Yes	Polycrystalline
50	24	Flat	16	Yes	Individual crystal aligned perpendicular to acceleration, 5-mm length, habit is coffin-like[22]
50	24	Flat	21	No	Many individual crystals, 2-mm length, habit is coffin-like[22]

The results of Table I show that 2-5-mm size TNT crystals have been grown from TNT saturated ethyl acetate solution at 50,000 g at approximately 24 x 10^6 Pa and 25°C for 16 hours. The pressure at the growth surface depends on the density and the height of the saturated solution and the acceleration. The density of the TNT saturated ethyl acetate solution (≈ 1.44 g/cc) is estimated from the solubility of TNT in ethyl acetate (59.8 g/100 g ethyl acetate at 21°C[23], the density of TNT (1.65 g/cc)[24], and the density of ethyl acetate (0.9 g/cc)[25]. Acceleration at 50,000 g at 25°C for 92 hours results in polycrystalline TNT. The crystal structure of the 5-mm size TNT crystal has been determined to be monoclinic by x-ray analysis[23,26].

The results of Table II show that 2-mm size RDX crystals have been grown from RDX saturated solution at 200,000 g at approximately 56 x 20^6 Pa and 25°C for 17 hours. The density of the RDX saturated acetone solution (0.85 g/cc) is estimated from the solubility of RDX in acetone (7.3 g RDX/100 g acetone at 20°C)[23], the density of RDX (1.81 g/cc)[24], and the density of acetone (0.79 g/cc)[25].

The RDX crystals grown at 200,000 g and at 1 g have been studied in fluids of matching refractive index to reveal defects and solution inclusions in the crystals. Optical micrographs of RDX crystals grown at 200,000 g and at 1 g are shown in liquids of matching refractive index (1-bromonaphthalene, R. I. = 1.600) in Figure 1. The growth surface of the RDX crystal grown at 200,000 g has been identified by the dark spots across the lower edge of the crystal in Figure 1. Insoluble impurities present in the saturated solution have reached the high g growth surface before solute molecules have reached the growth surface to begin crystal growth. A comparison of the two crystals shows clearly that individual RDX crystals grown at 200,000 g have far fewer voids than the RDX crystals grown at 1 g.

Table II. RDX Crystal Growth Experiments at High g at 25°C in Acetone Solution

Acceleration (x 10^3 g)	Pressure (x 10^6 Pa)	Growth Surface	Time (hr)	Filter	Results
50	14	Curved	17	No	No crystals
200	56	Flat	17	No	Individual crystals, 2-mm length, orthorhombic[27]
200	56	Flat	23	No	Individual crystals, 3-mm length, orthorhombic[27]
200	56	Flat	118	No	Polycrystalline

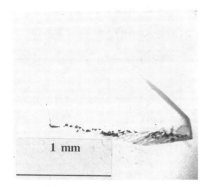

Figure 1. Optical micrograph of RDX crystal grown at 200,000 g (left) and at 1 g (right)

The results of Table III show that 2-mm x 3-mm TNAZ[28] crystals have been grown from TNAZ saturated acetone solution at 50,000 g at approximately 20 x 10^6 Pa during a 19 hour run. The density of the TNAZ saturated acetone solution (1.23 g/cc) is estimated from the solubility of TNAZ in acetone (0.44 g/cc)[29], the density of TNAZ (1.84 g/cc)[30], and the density of acetone[25].

Table III. TNAZ Crystal Growth Experiments at high g on a Flat Growth Surface and 25°C in Acetone Solution

Acceleration (x 10^3 g)	Pressure (x 10^6 Pa)	Time (hr)	Filter	Results
13	5	15	Yes	No crystals
50	20	19	Yes	Crystals, 2 mm x 3 mm
200	82	17	No	Polycrystalline

The results of Table IV show that TNAZ crystals up to 2 mm x 5 mm in size have been obtained under a 50,000 g acceleration at approximately 15 x 10^6 Pa during a 64 hour run. The results of Table IV also show that 2-mm x 6-mm TNAZ crystals have been grown at 200,000 g at approximately 59 x 10^6 Pa for a 19 hour run. The density of TNAZ saturated methyl alcohol solution (0.88 g/cc) is estimated from the solubility of TNAZ in methyl alcohol (0.09 g/cc), the density of TNAZ, and the density of methyl alcohol (0.79 g/cc)[26].

Table IV. TNAZ Crystal Growth Experiments at High g on a Flat Growth Surface at 25°C in Methyl Alcohol Solution

Acceleration (x 10^3 g)	Pressure (x 10^6 Pa)	Time (hr)	Filter	Results
50	15	64	No	Crystals, 2 mm x 5 mm
200	59	93	No	Polycrystalline
200	59	19	No	Crystals, 2 mm x 6 mm

The results of Table V show that 0.5-mm x 3-mm TNAZ needles have been grown under a 200,000 g acceleration at approximately 58 x 10^6 Pa for a 23 hour run. Further Table V also shows that 2-mm x 6-mm and 1-mm x 4-mm TNAZ crystals have been grown under 50,000 g at approximately 14 x 10^6 Pa for runs of 96 hours and 64 hours, respectively. The density of the TNAZ saturated ethyl alcohol solution (0.85 g/cc) is estimated from the solubility of TNAZ in ethyl alcohol (0.06 g/cc)[29], the density of TNAZ, and the density of ethyl alcohol (0.79 g/cc)[26].

Table V. TNAZ Crystal Growth Experiments at High g on a Flat Growth Surface at 25°C in Ethyl Alcohol Solution

Acceleration (x 10^3 g)	Pressure (x 10^6 Pa)	Time (hr)	Filter	Results
200	58	23	No	Needles, 0.5 mm x 3 mm
50	14	96	No	Crystal, 2 mm x 6 mm
50	14	64	No	Crystal, 1 mm x 4 mm

DISCUSSION

The objectives of this investigation are to understand the fundamental chemistry and physics of crystal growth during high acceleration and to make explosives more insensitive to mechanical shock by reducing the formation of defects in the crystals[16-21]. For the first time TNT, RDX, and TNAZ crystals have been grown by this new method to control crystal growth by the g-force in an ultracentrifuge. TNT crystals as large as 5-mm in size have been grown at 50,000 g in saturated ethyl acetate solution. RDX crystals as large as 3-mm in size have been grown at 200,000 g from saturated acetone solution. TNAZ crystals as large as 3 mm in size have been grown at 50,000 g in saturated acetone solution. TNAZ crystals as large as 6 mm in size have been grown at 200,000 g in saturated methyl alcohol solution. TNAZ crystals as large as 6 mm in size have been grown at 50,000 g in saturated ethyl alcohol solution.

The crystals grown at high g have been compared with crystals grown at 1 g in the laboratory by slow evaporation. Under optical microscopy the crystals grown at high g appear to be free of the voids and of the solution inclusions that are found in crystals grown at 1 g. The RDX crystals grown at high g have been studied in a fluid of matching refractive index to reveal defects and solution inclusions in the crystals. Figure 1 shows that individual RDX crystals grown at 200,000 g have far fewer voids than RDX crystals grown at 1 g.

Since this is the first investigation to explore the possibility of producing improved crystals of energetic materials, a wide range of parameter space has been explored in these initial experiments. Therefore a wide range of acceleration has been used. For example, Table I shows that no crystals have been obtained in TNT saturated ethyl acetate solution accelerated at 13,000 g for 16 hours or 29,000 g for 64 hours. When the acceleration is increased to 50,000 g individual crystals are obtained. These results show that the acceleration must be sufficient that a sufficient number of molecules reach the end of the tube during the run to form a crystal. Unfiltered solutions at approximately the same conditions result in many small individual crystals. Suspended particles may have provided nuclei for the growth of many small individual crystals.

When the duration of a 50,000 g run is increased to 92 hours so many molecules reach the end of the tube that polycrystalline material is formed.

Table II shows that no crystals have been obtained in RDX saturated acetone solution at 50,000 g for 17 hours. When the acceleration is increased approximately four times to 200,000 g individual crystals are obtained. When the duration of a 200,000 g run is increased to 118 hours so many molecules reach the end of the tube that polycrystalline material is formed.

Table III shows that no crystals have been obtained in TNAZ saturated acetone solution accelerated at 13,000 g. When the acceleration is increased approximately four times to 50,000 g, crystals are obtained. When the acceleration is increased an additional four times to 200,000 g, polycrystalline material is obtained. The run times are approximately constant for the three experiments. When the acceleration is too high so many molecules reach the end of the tube that polycrystalline material is formed.

Solubility and solvent molecular structure can also influence crystal growth. TNAZ is approximately equally soluble in methyl alcohol and ethyl alcohol. Table IV shows that TNAZ crystals have been grown at 200,000 g from methyl alcohol. Table V shows that only very small TNAZ needles have been grown at 200,000 g from ethyl alcohol. The run times are approximately constant for the two experiments. The difference in the molecular structure of methyl alcohol and ethyl alcohol may be the reason. It may be easier for the TNAZ molecules to move individually through the methyl alcohol molecules than the ethyl alcohol molecules.

This initial investigation has only begun to elucidate the crystal physics and chemistry of TNT, RDX, and TNAZ. Further work in these directions will provide more understanding and will likely produce important results that can be used in applications of energetic materials.

ACKNOWLEDGMENTS

We thank Drs. J. Lannon, R. Surapaneni, C. Choi, S. Iyer, Messrs. B. Travers, and M Joyce, all at U. S. Army ARDEC, and Mr. W. Lukasavage, GEO-CENTERS, Inc. for helpful comments.

REFERENCES

1. R. A. Laudise, The Growth of Single Crystals, Prentice-Hall, Inc., New Jersey, p. 39, 1970.

2. A. Holden and P. Singer, Crystals and Crystal Growing, Doubleday & Company, Inc., New York, 1960.

3. P. J. Shlichta and R. E. Knox, J. Crystal Growth 3, pp. 808-813 (1968).

4. P. J. Shlichta, J. Crystal Growth 119, pp. 1-7 (1992).

5. I. Amato, Science 253, pp. 30-32 (1991).

6. M. Y. D. Lanzerotti, J. Autera, J. Pinto and J. Sharma in High Pressure Science and Technology - 1993, edited by S. C. Schmidt, J. W. Shaner, G. A. Samara, and M. Ross (AIP Conference Proceedings 309, Part 2, American Institute of Physics, New York, NY, 1994), pp. 489-491.

7. L. L. Regel, M. Rodot and W. R. Wilcox, eds. J. Crystal Growth 119, pp. 1-175 (1992).

8. L. L. Regel and W. R. Wilcox, eds., Materials Processing In High Gravity, Plenum Press, New York, 1994.

9. M. Y. D. Lanzerotti, J. Autera, J. Pinto and J. Sharma, ibid, pp. 181-184.

10. F. Baillou, J. M. Dartyge, C. Spyckerelle, and J. Mala in Proc. Tenth Symposium (International) on Detonation, 1993, pp. 816-823..

11. L. Borne, ibid, pp. 286-293.

12. A. Van Der Steen, H. J. Verbeek, and J. J. Meulenbrugge in Proc. Ninth Symposium (International) on Detonation, 1989, pp. 83-88.

13. I. B. Mishra and L. J. Van de Kieft, Proc. 19th International Annual Conf. of ICT, Karlsruhe, 1988, pp. 25-1 to 25-21.

14. L. Borne, Proc. Europyro 95, 6eme Cong. Intl. de Pyro, Tours-France, pp. 125-131, 1995.

15. J. J. Dick, J. App. Phys. **53**, pp. 6161-6167 (1982).

16. J. J. Dick, J. App. Phys. Lett. **44**, pp. 859-861 (1984).

17. J. J. Dick in Shock Waves In Condensed Matter, edited by Y. M. Gupta (Plenum Press, New York, NY, 1986), pp. 903-907.

18. J. J. Dick, R. N. Mulford, W. J. Spencer, D. R. Petit, E. Garcia, and D. C. Shaw, J. App. Phys. **70**, pp. 3572-3587 (1991).

19. J. J. Dick, E. Garcia, and D. C. Shaw in Shock Compression of Condensed Matter-1991, edited by S. C. Schmidt, R. D. Dick, J. W. Forbes, D. G. Tasker (Elsevier Science Publishers, The Netherlands, 1992) pp. 349-352.

20. J. J. Dick, E. Garcia, and D. C. Shaw in Shock Compression Of Condensed Matter - 1993 edited by S. C. Schmidt, J. W. Shaner, G. A. Samara, and M. Ross (AIP Conference Proceedings 309, Part 2, American Institute of Physics, New York, NY, 1994), pp. 1373-1376.

21. M. Y. D. Lanzerotti, J. Autera, J. Pinto, and J. Sharma in Army Science Conference Proceedings, 1994, Vol. I, pp. 69-75.

22. H. G. Gallagher and J. N. Sherwood in Structure and Properties of Energetic Materials, edited by D. H. Liebenberg, R. W. Armstrong, and J. J. Gilman (Mater. Res. Soc. Proc. 296, Pittsburgh, PA 1993), pp. 215-219.

23. S. Morrow, U. S. Army ARDEC, 1989.

24. B. M. Dobratz and P. C. Crawford, Properties of Chemical Explosives and Explosive Simulants, UCRL-52997, Lawrence Livermore National Laboratory, University of California, Livermore, CA, 1985.

25. R. C. Weast, Handbook of Chemistry and Physics, CRC Press, Cleveland, OH, 1975-1976.

26. S. M. Kaye, Encyclopedia of Explosives and Related Items, Picatinny Arsenal Technical Report 2700, Vol. 9, p. T263 (1980).

27. J. T. Rogers, Physical and Chemical Properties of TDX and HMX, Control No. 20-P-26, 1962.

28. T. G. Archibald, R. Gilardi, K. Baum, and C. George, J. Org. Chem. **55**, p. 29920 (1990).

29. D. Stec, GEO-CENTERS, INC., 1992.

30. S. Iyer, S. Eng, M. Joyce, R. Perez, J. Alster, D. Stec, Proc. of the Joint International Symposium on Compatibility of Plastics and Other Materials with Explosives, Propellants, Pyrotechnics and Processing of Explosives, Propellants, and Ingredients, pp. 80-84, 1991.

PHASE RELATIONSHIPS INVOLVING RDX AND COMMON SOLID PROPELLANT BINDERS

E. Boyer*, P.W. Brown**, and K.K. Kuo*
*Department of Mechanical Engineering, The Pennsylvania State University, University Park, PA 16802
**Department of Materials Science and Engineering, The Pennsylvania State University, University Park, PA 16802

ABSTRACT

The solubilities of the common propellant ingredients acetyl triethyl citrate (ATEC) and cellulose acetate butyrate (CAB) and their effects on RDX (cyclotrimethylenetrinitramine) unit cell dimensions were investigated. If there is appreciable solid solubility, solutions will form and will have enthalpies of fusion, melting temperatures, and other characteristics different from those of pure RDX. It is desirable to establish the properties of such mixtures when designing new propellant formulations. Samples were aged at an elevated temperature to speed the formation of solid solutions. A least-squares analysis of X-ray diffraction data was used to obtain the lattice parameters from which unit cell volume and solubility was deduced. Both ATEC and CAB caused an expansion of the unit cell, indicating the formation of a solid solution. The limit of solubility in the ATEC/RDX mixture appeared to be approximately 13 wt% ATEC, while the CAB/RDX limit is above 16 wt% CAB. In both mixes, the cell volume expanded linearly with increasing proportion of binder.

INTRODUCTION

Because of many favorable characteristics, RDX (cyclotrimethylenetrinitramine) has been extensively utilized as an ingredient in both gun and solid rocket propellants. RDX is used in Low Vulnerability Ammunition (LOVA) propellant formulations due to its relative insensitivity to accidental energy stimuli. The large amount of gas generated and high energy released during combustion make it very attractive for both rocket and gun propulsion applications. The absence of HCl in the combustion products makes RDX desirable on an environmental basis as well. It is useful to characterize the condensed-phase pyrolysis behavior and achieve better understanding of the combustion processes of RDX and RDX-based propellants so that detailed simulation of the burning phenomena of RDX-based propellants can be achieved.

Due to the fact that the regression rate of solid propellants can be greatly influenced by surface and subsurface reactions, better knowledge of condensed phase phenomena is vital to gain a more in-depth understanding of the combustion of nitramine solid propellants. Much work has been done on the decomposition and subsequent reactions involving RDX and other propellant ingredients [1-7], but very little attention has been given to the effects of solid-state interactions in the condensed phase [8,9]. Because propellants consist of a combination of oxidizer, binders, plasticizers, and other ingredients it may not be adequate to model a composite propellant by treating it merely as a physical mixture. Interactions between the components must be understood in order to create an accurate model. Changing intermolecular forces due to ingredient interactions can affect density, enthalpy of fusion, melting temperature, stability, aging characteristics, and general combustion behavior. Effects of changing intermolecular forces due to transitions of pure phases of HMX (cyclotetramethylenetetranitramine), a high-energy material very similar to RDX, have been investigated previously and shown to be important [10], but little work has been done on the effect of solid solutions between crystalline nitramines and binder materials. Formulations of two common RDX-based propellants, XM39 and M43, are shown in Table 1.

X-Ray Diffraction

A method useful in quantifying the effect of ingredient interactions is X-ray diffraction (XRD) analysis. In crystalline materials, unit cells combine into a lattice forming planes of molecules that

Table 1. Ingredients in XM39 and M43 propellants.

Propellant	Material	%
XM39	RDX Cellulose Acetate Butyrate (CAB) Acetyl Triethyl Citrate (ATEC) Nitrocellulose (NC) (12.6% N) Ethyl Centralite (EC)	76.0 12.0 7.6 4.0 0.4
M43	RDX Cellulose Acetate Butyrate (CAB) Energetic Plasticizer Nitrocellulose (NC) (12.6% N) Ethyl Centralite (EC)	76.0 12.0 7.6 4.0 0.4

scatter incident X-rays (see Fig. 1). A strong beam is diffracted when Bragg's Law satisfied:

$$n\lambda = 2d \sin \theta \tag{1}$$

where n represents any positive integer, λ the wavelength, d the interplanar spacing, and θ the angle of incidence. X-rays diffracted from different planes in the material are in phase, constructively reinforcing one another and leading to a strong beam being returned [11,12]. A diffractometer used to measure this effect is generally constructed so it scans through a range of 2θ angles in order to measure peaks formed by the diffraction of X-rays off differently-spaced planes present in the crystalline lattice of the material under study. Through careful analysis of the position and intensity of these peaks, one can reconstruct the form and shape of the unit cell of the sample. When samples are in powdered form, such as in this study, reflections from all planar spacings present in the material will be detected during a 2θ scan.

Solid solutions are formed by the introduction of one type of molecule into a lattice made up of another. There are two possible forms; interstitial, in which case molecules are inserted into existing spaces between molecules in the lattice, or substitutional, where the molecules replace original molecules in the lattice. The presence of substituents in the lattice affects intermolecular forces, leading to a change in dimensions of the unit cell of the lattice that can be quantitatively measured using XRD techniques.

EXPERIMENTAL

Motivation

In a previous study [13,14], XRD analysis of quenched samples of XM39 and M43 propellants showed evidence of the formation of RDX/binder solid solutions in both virgin and extinguished samples. Therefore, it is important to gain a better understanding of the effects of solid solution formation on the combustion and other characteristics of propellant formulations.

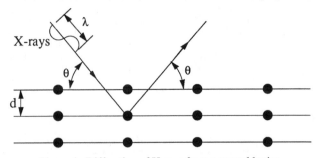

Figure 1. Diffraction of X-rays from a crystal lattice

<u>Sample Preparation</u>

To isolate the effects of individual ingredients, binary mixtures of RDX and common binder materials were made. Samples formulated to contain up to 16 wt% of binder allowed the variation in unit cell parameters with percentage substituent and the limit of solid solubility to be determined. Because military grade RDX may contain a significant portion of HMX, the RDX used for this study was purified using an acetone recrystallization process developed by Schroeder [15]. After purification, measured proportions of the RDX crystals and binder were co-dissolved in sufficient quantities of acetone to form a clear liquid. The resulting solutions were then shock-precipitated into large volumes of water forming fine particles of the mixtures. Particle sizes formed an approximately normal distribution around 10 μm mean diameter. The samples of varying percentages of propellant ingredients and RDX were aged at 80° C for 1 week to allow solid solutions to form.

<u>Data Acquisition and Analysis</u>

A Scintag/USA PAD-V automated diffractometer interfaced to a VAX 3100 was used for data acquisition and analysis. Before scanning, 5 wt% of a silicon standard [16] was added to each sample and mixed thoroughly to enable calibration of peak positions so instrument systematic error could be eliminated. The powdered samples were packed into glass slides and scanned at the following conditions: 1.54059 cm wavelength Cu X-rays, 10-48 degrees 2θ scanned at a scan rate of 0.5 degrees 2θ per minute with a 0.02 degree chopper increment. Lattice parameters were determined using a least-squares fit computer program.

RESULTS

Diffraction patterns for the RDX/ATEC mixtures are shown in Fig. 2 [17]. Examination of the peaks showed a systematic shift with increasing concentration of ATEC. This is clearly seen with the peaks near 13 and 18 degrees 2θ, indicating a change in lattice spacing. Using peaks determined from these patterns, the Argonne program [18] was employed to generate unit cell parameters. Figures 3a and b show the cell parameters and resulting cell volumes, respectively. From these figures, it can be seen that there is very little change with small percentages of ATEC added, but the unit cell begins to expand significantly at ATEC concentrations greater than 4 wt%. The expansion is due to primarily to the elongation of the a lattice parameter, although the c lattice parameter expands as well, while the b lattice parameter actually contracts slightly. With these effects combined, the unit cell volume has expanded by about 5 % at 14 wt% ATEC. The lattice parameter and cell volume growth appears to be linear between about 4 and 13 wt% ATEC, then ceases to increase. However, the data are not precise enough to draw a firm conclusion on the extent of solubility. More experiments at higher percentages of ATEC need to be performed before it can be definitively determined that 13 wt% is the limit of ATEC solubility in RDX.

Much less change in the RDX unit cell resulted from the addition of CAB. The use of a Si standard eliminated most systematic errors, so the simpler Appleman program [19] was used to fit lattice parameters to the CAB data. All three lattice parameters (shown in Fig. 4a) increased continuously with increasing concentration of CAB. This yielded a definite linear, but less than 1 % increase in unit cell volume (Fig. 4b) between pure RDX and 16 wt% CAB. No plateau in cell volume was seen, indicating that extent of solid solubility of CAB in RDX appears to be greater than 16 wt%.

CONCLUSIONS

From this work, it can be seen that there is a significant solid-solution effect between RDX and common solid propellant binders as shown by the expansion of the RDX unit cell. The extent of solid solubility in the ATEC/RDX system appears to be about 13 wt% ATEC, while for CAB/RDX it may be greater than 16 wt% CAB. Due to the relative size of the molecules involved, the solid solutions of both ATEC and CAB in RDX are believed to be substitutional. Although the size of the ATEC molecule is much smaller than CAB, it has a stronger effect on the expansion of the unit cell of RDX. These changes in material and combustion properties due to

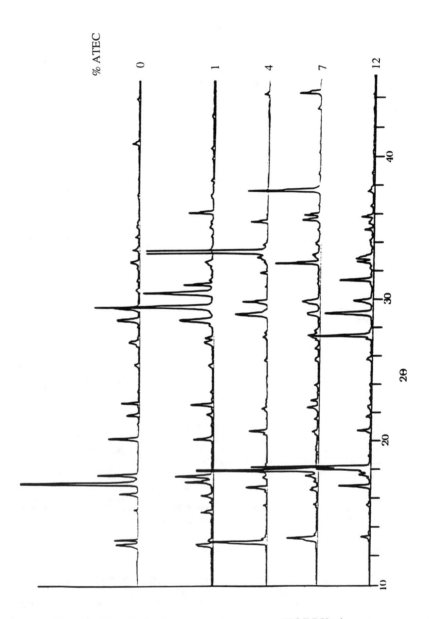

Figure 2. X-ray diffraction patterns for several ATEC/RDX mixtures.

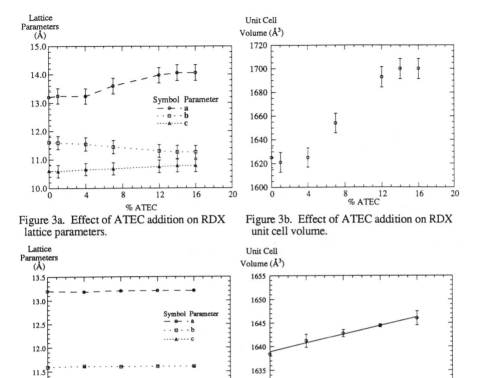

Figure 3a. Effect of ATEC addition on RDX lattice parameters.

Figure 3b. Effect of ATEC addition on RDX unit cell volume.

Figure 4a. Effect of CAB addition on RDX lattice parameters.

Figure 4b. Effect of CAB addition on RDX unit cell volume.

the solid solution formation should be closely examined in the modeling of combustion processes of RDX-based composite propellants.

Future Work

Work is currently underway to extend the present solid solution characterization study to systems of RDX and other propellant ingredients. In addition, it will be useful to examine HMX mixtures. Studying the HMX/RDX system can assist in better understanding the difference between military grade RDX, used in actual propellants, and pure RDX as commonly used in combustion models. Determining the effect of temperature on solid solubility and constructing phase diagrams will greatly aid in integrating these results into combustion models.

ACKNOWLEDGEMENTS

This work has been sponsored by the Army Research Office under Contract Nos. DAAH04-93-G-0364 and DAAL03-92-G-0118. The program manager is Dr. Robert W. Shaw. His interest and support are gratefully acknowledged. The assistance of Mr. Sean J. Wilson of the Pennsylvania State University in the early start of this work is greatly appreciated.

REFERENCES

1. T. L. Boggs, in Fundamentals of Solid-Propellant Combustion, edited by Kenneth K. Kuo and Martin Summerfield (Progress in Astronautics and Aeronautics 90, American Institute of Aeronautics and Astronautics, New York, 1984) pp. 121-175.

2. R. A. Fifer, in Fundamentals of Solid-Propellant Combustion, edited by Kenneth K. Kuo and Martin Summerfield (Progress in Astronautics and Aeronautics 90, American Institute of Aeronautics and Astronautics, New York, 1984) pp. 177-237.

3. R. Behrens, Jr. and S. Bulusu, J. Phys. Chem. **96**, pp. 8877-8891 (1992).

4. Michael A. Schroeder, Robert A. Fifer, Martin S. Miller, Rose A. Pesce-Rodriguez, and Gurbax Singh, Technical Report BRL-TR-3337, Ballistic Research Laboratory, Aberdeen Proving Ground, Maryland, May 1992.

5. T. B. Brill, P. J. Brush, D. G. Patil, and J.K Chen, Twenty-Fourth Symposium (International) on Combustion, The Combustion Institute, Pittsburgh, PA, 1992, pp. 1907-1914.

6. T. B. Brill and P. J. Brush, Phil. Trans. R. Soc. London **339**, pp. 377-385 (1992).

7. C. F. Melius, in Chemistry of Physics and Energetic Materials, edited by S. N. Bulusu (Kluwer Academic Publishers, The Netherlands, 1990) pp. 51-78.

8. Gasper J. Piermarini, Stanley Block, and Philip J. Miller, in Chemistry of Physics and Energetic Materials, edited by S. N. Bulusu (Kluwer Academic Publishers, The Netherlands, 1990) pp. 51-78.

9. R. J. Karpowicz and T. B. Brill, J. Phys. Chem. **88**, pp. 348-352 (1984).

10. T. B. Brill and R. J. Karpowicz, J. Phys. Chem. **86**, pp. 4260-4265 (1982).

11. Raymond P. Goehner and Monte C. Nichols, in ASM Handbook Volume 10: Materials Characterization (American Society for Metals, 1986) pp. 332-343.

12. William D. Callister, Jr., Materials Science and Engineering: An Introduction, 2nd ed. (John Wiley & Sons, Inc., New York, 1991) pp. 54-59.

13. Sean J. Wilson, Master's Thesis, The Pennsylvania State University, 1995.

14. S. J. Wilson, P. W. Brown, and K. K. Kuo (in preparation).

15. Michael A. Schroeder (private communication).

16. National Bureau of Standards Standard Reference Material 640b Silicon Powder.

17. S. J. Wilson, P. W. Brown, E. Boyer, and K. K. Kuo (in preparation).

18. M. H. Mueller and L. Horton, Argonne National Laboratory Report ANL6176, January 1969.

19. Daniel E. Appleman and Howard T. Evans, Jr., USGS Report GD-73-003, United States Geological Survey, Washington, D. C., 1973.

OPTICAL PROPERTIES OF RDX AND HMX

R. A. ISBELL, M. Q. BREWSTER
Department of Mechanical and Industrial Engineering
University of Illinois
Urbana, Illinois 61801

ABSTRACT

Optical properties of RDX (cyclotrimethylene-trinitramine) and HMX (cyclotetramethylene-tetranitramine) were obtained from 2.5 to 18 μm using scattering-corrected KBr pellet-FTIR transmission spectrometry. Absorption index (k) was measured directly and refractive index (n) was deduced using dispersion theory. At 10.6 μm the absorption coefficients were RDX, 2800 cm^{-1} and HMX, 5670 cm^{-1}.

INTRODUCTION

Optical properties of energetic materials are of interest for a variety of reasons. Infrared transmission spectra are commonly used to identify chemical structure by correlating absorption peaks with certain bonds and lattice vibrations. For these purposes qualitative (or relative) information about absorption strength as obtained from conventional FTIR spectroscopy is sufficient. Other studies require quantitative (or absolute) optical property information, *i.e.*, the optical constants, n (refractive index) and k (absorption index). Investigative methods that are benefitted by knowledge of the material optical constants include laser pyrolysis, for determining energetic material decomposition chemistry, and laser-augmented combustion and ignition. Knowledge of the optical constants allows estimation of surface (Fresnel) reflection losses and spatial distribution of radiation absorption. While reports containing relative absorbance are plentiful, optical constant data for energetic materials are scarce.

This paper reports the results of a technique developed to determine optical constants for energetic materials. Data are reported for RDX and HMX. The method uses KBr pellet-FTIR transmission spectrometry in conjunction with dispersion theory and accounts for radiation scattered by the KBr pellets. Advantages of the FTIR/KBr pellet method are that optical constants can be obtained even in strongly absorbing regions with sufficient dilution and fine enough particles and they can be obtained over a broad spectral range.

PROCEDURE AND ANALYSIS

RDX and HMX samples were prepared by dispersing their respective powders in KBr powder and pressing the mixture into pellets. Before mixing, both energetic material and KBr powders were ground until most of the particles were submicron in size (see next section for particle size requirement). Transmission spectra were measured for samples with various optical thickness using an FTIR spectrometer over the range of 2.5 to 18 μm. The RDX and HMX contained only trace amounts of impurities which did not affect the optical properties. Optical properties were determined by measuring the normal, spectral transmissivity of the samples. KBr was taken to be non-absorbing in the spectral range of interest, although scattering by KBr was accounted for as described below.

Theoretical Considerations

The transmissivity of a homogeneous (non-scattering) plane layer in air subject to normal incident monochromatic radiation is given by

85

$$T = \frac{I_t}{I_o} = \frac{(1-R_n)^2 e^{-\tau}}{1 - R_n^2 e^{-2\tau}} \tag{1}$$

where K_a is the spectral absorption coefficient, L is the thickness of the sample, $\tau = K_a L$ is the optical thickness, $I_{t,o}$ are the transmitted and incident intensities, respectively, and R_n is the normal Fresnel interface reflectance given by

$$R_n = \frac{(n-1)^2 + k^2}{(n+1)^2 + k^2}. \tag{2}$$

The absorption coefficient and absorption index (k) of a homogeneous material are related by

$$K_a = \frac{4\pi k}{\lambda_o} \tag{3}$$

where λ_o is the vacuum (~air) wavelength of the incident radiation. If the layer consists of a non-absorbing, non-scattering matrix (KBr) and an absorbing sample material of the same refractive index, extinction of incident radiation by scattering will be negligible. Furthermore, if the particle size, d, of the absorbing component is small such that $K_a d \ll 1$, then the effective absorption coefficient of the layer is the intrinsic absorption coefficient of the absorbing constituent weighted by its volume fraction $K_a f_v$. (In particle light scattering theory this approximation is known as Rayleigh-Gans scattering [1], and carries with it the formal restrictions $2x|\tilde{n}-1| \ll 1$ and $|\tilde{n}-1| \ll 1$, where $x = \pi d/\lambda$ and $\tilde{n} = n - ik$.) If the condition $K_a d \ll 1$ is not satisfied, more complicated particle light scattering theory would have to be used to relate the pellet volumetric optical properties to those of the absorbing constituent particles and the technique would lose its advantage of accuracy. If the condition $K_a d \ll 1$ is not satisfied but the data are analyzed assuming that the pellet absorption coefficient is the volume fraction weighted absorbing constituent value $K_a f_v$ anyway, the material absorption coefficient K_a will be underpredicted. (Ostmark, et al., [2] reported an absorption coefficient of 875 cm^{-1} at 10.6 μm for RDX but did not report particle size whereas our measurements yielded 2800 cm^{-1}.)

In KBr pressed pellets there are discontinuities in refractive index at the boundaries between KBr particles and between KBr and sample particles[1] which cause scattering of radiation. Scattering can be accounted for by including a scattering coefficient K_s which is additive with the absorption coefficient. Thus the optical thickness of the layer is based on the extinction coefficient which is the sum of the scattering and absorption coefficients.

$$\tau = K_e L; \quad K_e = K_a f_v + K_s \tag{4}$$

Substituting Eq. (4) into (1) and rearranging gives

$$A = \log\left(\frac{1}{T}\right) = \log\left[(1-R_n)^{-2}\right] + 0.4343L[K_a f_v + K_s] + \log\left(1 - R_n^2 e^{-2\tau}\right) \tag{5}$$

where A is the absorbance. Equation (5) is the appropriate relation to apply to uncorrected KBr pellet transmission data if quantitative absorption information is desired. One difficulty in doing so, particularly in the near infrared and shorter wavelength regions, is the fact that the scattering

---

[1] The latter contribution, scattering by sample particles, is usually negligible because their concentration is so low, $f_v \ll 1$.

coefficient can be of non-negligible magnitude compared with the volume fraction weighted absorption coefficient, making an independent measurement of the scattering coefficient or an equivalent scattering correction necessary. In the FTIR/KBr pellet technique the scattering contribution was removed from the transmission measurement data by baseline shifting. This eliminates K_s in Eq. (5). To account for the (typically very small) volume fraction in the pellets occupied by voids due to imperfect packing the actual sample thickness L was reduced slightly, $L_{eff}=L(\rho_{act}/\rho_{theo})$ where ρ_{act} is actual pellet density, and ρ_{theo} is theoretical pellet density (no voids). The sample particle volume fraction f_v now becomes the (more easily measured) value based on the assumption of no voids (*i.e.*, volume fraction of sample particles relative to KBr and oxidizer only). Except for moderate to high values of R_n and very small L, the last term in Eq. (5) is negligible.

$$A = \log\left(\frac{1}{T}\right) = \log\left[(1 - R_n)^{-2}\right] + 0.4343 K_a f_v L_{eff} \tag{6}$$

This is the form of Beer's Law used in the analysis. It indicates a linear relationship between absorbance and the product $f_v L_{eff}$ with the slope of A versus $f_v L_{eff}$ being the absorption coefficient. Both f_v and L_{eff} were varied independently. Linear regression was applied to absorbance data on a spectral basis and absorption coefficients were extracted from the slopes.

The measured spectra for absorption index $k(\lambda)$, as determined by the FTIR-spectrometry technique, were curve fit using an optimization routine and dispersion theory relations [1].

$$n^2 - k^2 = n_e^2 + \sum_j \frac{\omega_{p,j}^2 \left(\omega_{o,j}^2 - \omega^2\right)}{\left(\omega_{o,j}^2 - \omega^2\right)^2 + \gamma_j^2 \omega^2} \tag{7}$$

$$2nk = \sum_j \frac{\omega_{p,j}^2 \gamma_j \omega}{\left(\omega_{o,j}^2 - \omega^2\right)^2 + \gamma_j^2 \omega^2} \tag{8}$$

Here ω_p, ω_o, and γ represent the oscillator strength (plasma frequency), oscillator frequency, and line width of the "jth" oscillator, respectively. The residual value of refractive index n_e occurs at the short wavelength end of the spectral interval of interest, which is the visible/near infrared region in this case. Its value results from electronic transitions in the ultraviolet region. This procedure resulted in values for the oscillator parameters (ω_p, ω_o, and γ) and the refractive index spectrum $n(\lambda)$.

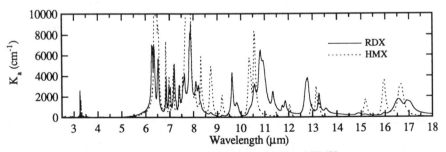

Fig. 1 Absorption coefficient for RDX and HMX

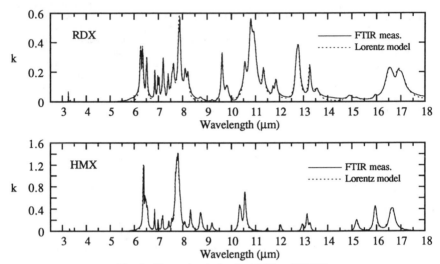

Fig. 2 Absorption index for RDX and HMX

RESULTS

Figure 1 shows the absorption coefficient spectra. The spectra are active over various regions corresponding to the molecular or lattice structures of the individual materials. Some of the peaks are truncated due to insufficient transmission at these wavelengths. At 10.6 μm (CO_2 laser wavelength) the absorption coefficients are $K_{a,RDX} = 2803$ cm^{-1} and $K_{a,HMX} = 5672$ cm^{-1}. The absorption index (k) is calculated from absorption coefficient using Eq. (3) and the results are shown in Figure 2. RDX has a complicated molecular structure as indicated by numerous, closely spaced bands ($k \sim O(10^{-1})$). These bands exhibit some overlap and are moderately strong. k_{RDX} approaches 0.6 near 7.8 μm and 10.8 μm. The complex spectrum arises from a complex molecular structure consisting of potentially 21 bond stretches, 36 bond-angle bends, 3 wag-angle bends, and 9 Leonard-Jones non-bonded interactions. CH participates over 2700 to 3100 and 0 to 300 cm^{-1}. No other bonds contribute appreciably to these regions. The CN bond has many bands over 400 to 1600 cm^{-1}. NO participates strongly over 1500 to 1600 cm^{-1} with weaker peaks over 500 to 1300 cm^{-1}. N-C-N bending contributes weakly over 350 to 843 cm^{-1}. H-C-H bending has a strong peak at 1234 cm^{-1} with weaker bands over 1039 to 1311 cm^{-1} [3]. HMX has a structure similar to that of RDX with numerous, closely-spaced, moderately strong bands in k_{HMX} ($k \sim O(10^{-1})$). k_{HMX} has values up to 1.4 at 7.88 μm. HMX bonds participate in similar spectral regions as in RDX but the spectra show shifted peak positions and strengths. Detailed discussions of molecular structure and bond-peak assignments are available for RDX [3-5] and HMX [6]. At 10.6 μm the absorption index values are $k_{RDX} = 0.236$ and $k_{HMX} = 0.478$.

The effect of spectral resolution in the FTIR measurements was also considered. Most measurements were conducted with a resolution of 2 cm^{-1}. To check this effect, relative absorbance spectra for each material were also measured at 0.125 cm^{-1} (0.0028 μm at 10.6 μm) resolution over the spectral range 1000 - 900 cm^{-1} (10.0 - 11.1 μm). No noticeable differences were observed.

To determine the refractive index (n), dispersion theory was applied to the measured absorption index spectra. Refractive index values in the visible region were assumed as follows: $n_{e,RDX} = 1.597$ and $n_{e,HMX} = 1.594$ [5-6]. k was modeled by varying the oscillator parameters of

Eqs. (7) and (8). RDX was modeled with 40 oscillators and HMX with 39. The Lorentzian modeled k spectra are shown in Fig. 2 with the experimentally determined values. Excellent agreement with the experimental spectra demonstrates the appropriateness of the Lorentzian model. The refractive index and normal reflectance, n and R_n, were calculated from the oscillator parameters and Eqs. (2, 7-8). Figures 3 and 4 show the Lorentzian modeled refractive index and normal reflectivity spectra. $R_{n,RDX}$ fluctuates rapidly throughout the spectrum with values ranging from 2% to 11%. For HMX, strong bands near 6.4 and 7.75 μm result in reflectances varying from near zero to 23% and 28%, respectively. At 10.6 μm, $R_{n,RDX} = 4.70\%$, and $R_{n,HMX} = 15.0\%$.

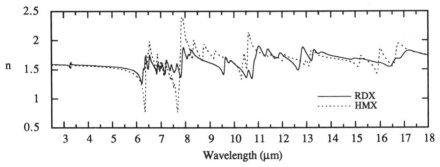

Fig. 3 Refractive index for RDX and HMX from Lorentz model

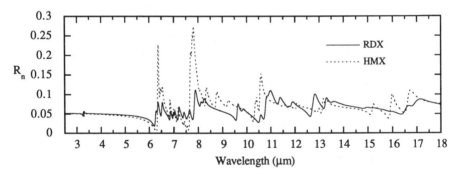

Fig. 4 Normal reflectivity for RDX and HMX from Lorentz model

DISCUSSION

The KBr/FTIR technique used in this study is dependent on the assumption of Rayleigh-Gans scattering to obtain accurate absorption coefficient data. The key condition that must be satisfied is that the particles be optically thin $K_a d << 1$. In this study particles were ground to the sub-micron size range, as verified by optical microscopy, but no attempt at a more precise size determination was made. Assuming an effective average absorbing particle size of 0.5 μm and using 0.3 as the maximum limit of $K_a d$ gives a maximum absorption coefficient of 6000 cm^{-1}. Thus there is high confidence in the accuracy of the indicated absorption coefficients below this value and an increasing (although as yet unquantified) tendency for underestimation in the values reported above 6000 cm^{-1}. Except for the highest absorption peaks, most of the values were under this value, including those at 10.6 μm. Previously reported values for RDX absorption coefficient [2] are lower than those reported here. Failure to satisfy the Rayleigh-Gans scattering limit would result in underprediction of the absorption coefficient.

SUMMARY

The KBr-pellet/FTIR transmission spectroscopy technique has been developed to obtain the optical constants of relatively strongly absorbing energetic materials, such as HMX and RDX, over the spectral range 2.5 to 18 μm. Absorption coefficient and absorption index were measured directly and refractive index was determined from dispersion theory. Efforts to satisfy rigorously the necessary Rayleigh-Gans scattering limit resulted in significantly higher absorption coefficients than reported previously for RDX.

ACKNOWLEDGMENTS

Support for this work from the Office of Naval Research (N00014-91-J-1977), Richard S. Miller scientific officer, is gratefully acknowledged. The support of the Naval Air Warfare Center in providing RDX and HMX samples is also greatly appreciated.

REFERENCES

1. Brewster, M. Q. (1992), Thermal Radiative Transfer and Properties, John Wiley & Sons, New York.

2. Ostmark, H., M. Carlson, and K. Ekvall (1994), "Laser Ignition of Explosives: Effects of Laser Wavelength on the Threshold Ignition Energy," *J. Energetic Materials,* Vol. 12, pp. 63-83.

3. Sewell, T.D., Chambers, C.C., and Thompson, D.L. (1993), "Power spectral study of the classical vibrational dynamics of RDX," *Chemical Physics Letters*, Vol. 208, No. 1,2, pp. 125-134.

4. Alix, J. and Collins, S. (1991), "The photochemistry of RDX in solid argon at 10K," *Canadian J. Chemistry*, Vol. 69, pp. 1535-1538.

5. Military Specifications (Sept. 1991), *RDX Properties*, Unit 16, Unclassified CPIA/M3, pp. 1-21.

6. Military Specifications (Sept. 1991), *HMX Properties*, Unit 15, Unclassified CPIA/M3, pp. 1-22.

OVERVIEW ON ENERGETIC POLYMERS

JACQUES BOILEAU
Scientific Expert - Direction de la Recherche et de la Technologie (DRET)
Délégation Générale pour l'Armement - Paris - France

ABSTRACT

Energetic materials for missiles, gun munitions or pyrotechnic devices often are mixtures in a biphasic form, with a filler and a binder. To satisfy the user needs, an analysis of functional requirements together with constraints (safety, vulnerability, ageing, environment, disposal, price...) is useful to choose a convenient binder. From this point of view numerous synthetic energetic polymers proposed or developed as binders are reviewed with regard to their syntheses, processing, properties and possible uses. These polymers contain explosophore groups : $C-NO_2$ aliphatic or aromatic, ONO_2, NNO_2, NF_2 and N_3.

Some research projects are suggested. Among them in the list of published polymers, following a NIMIC (NATO) suggestion, note the reason of a development interruption. Some dinitropolystyrene - polyvinyl nitrate mixtures or copolymers could exhibit interesting properties. For unknown reasons, some mixtures of crystalline filler · with polymeric binder, generally in a biphasic form, may also be monophasic for a same composition. What properties are modified between the both forms (e.g. combustion mechanisms, erosion, ideal character of the detonation) ? It is also interesting to pursue into a newly open route to thermo-plastic elastomers.

INTRODUCTION

This paper concerns a short overview on energetic polymers used in the binders of energetic material formulation, excluding nitrocellulose, nitrostarch and also the plasticizers. It is divided into three parts : (1) a short presentation of a methodology by a systemic approach from the customers needs to the characteristics of the materials. (2) a listing of the requirements, coming from the uses (propellants for rockets and missiles, powders for guns and small arms, high explosives for ammunitions) and from the processes. (3) a review on synthetic energetic polymers containing explosophore groups NO_2, NF_2, N_3. In the conclusion, some recommendations are given.

1. A methodology by a systemic approach

It may be interesting to start from the needs, the goals expressed by the users. From this expression, it is possible to imagine solutions, namely systems, which have to satisfy these needs; that is to fulfil the wanted functions in the constraints of safety, stability, storage, environmental problems, disposal ... and cost. It shall be expressed as requirements, for which it is necessary to find representative measurement methods and specifications, to give a guarantee to the customer that the system shall satisfy the required need.

The development of nitrocellulose for gunpowders last century and of the castable explosive TNT, was followed by introduction of ingredients to improve the performances. These include nitroglycerine for gunpowders and nitramines for high explosives. But there were limitations due to erosion of tubes, safety (burning and explosions) and vulnerability. Problems became acute for propellants, where mechanical properties are very important. The progress in polymers satisfied the required needs with biphasic mixtures, i.e. fillers (nitramines, ammonium perchlorate ...) and binders; and the differences between propellants, gunpowders and high explosive disappeared.

But the request for better performances, higher safety, lower vulnerability, lower aggressivity towards environment, and, important, lower prices, pushed the study of new binders particularly the energetic ones [1].

Binders are composed of polymers, often plasticizers, and other additives. They are studied as component of propellants in rocket motors and missiles, in guns and small arms, and as high explosives in warheads (artillery, mines, bombs, nuclear weapons ...), in pyrotechnical devices, and recently as active armor. Each use generates its own specifications, the economic criteria introduce other imperative factors. Another consequence of specific and more numerous requirements is that the possible solutions are more specific for a need.The universal energetic polymer for all needs is not in sight.

2. Requirements

A tentative listing of requirements for propellants, powders and high explosives is given here [2].

a) Functional requirements :

For propellants, the requirements involve specific impulse Isp, volumic specific impulse Isp x d (d is density), burning rate, pressure exponent, sensitivity of burning rate to the initial temperature, ease of initiation, flame temperature, burning stability and mechanical properties at various temperatures (e.g. between -40 and +70°C and during the missile launch). For some special uses, possibility to extinguish and to reinitiate during the launch.

For powders for guns and small arms the requirements are burning rate under high pressures, burning rate versus pressure, impetus, ballistic properties to obtain a given projectile velocity under a maximum pressure, maximum temperature of burning gases, temperature coefficient and brittleness.

For high explosives the requirements are heat of formation, energy release, density and detonation velocity D (the value of dD^2 is often used but sometimes misused [3]), Gurney coefficient, critical diameter, initiation and pop-plot. The mechanical properties are less important than for propellants, not functional but only as constraints, e.g. for safety.

b) Requirements due to the constraints :

Generally the safety during fabrication, loading, transportation, storage, the ageing, the toxicity of exhaust gases, the environmental problems, the demilitarization and disposal are the main concerns..

More specifically discretion of missiles during launch and flight, muzzle flash of guns, erosion of tubes in guns and small arms, thermostability for high explosives in air-carried ammunitions or in nuclear weapons are of concerns.

All these requirements are important in the choice of processing. For energetic mixtures, these processings may be classed in three groups.

P1 - Dissolution of the polymeric binder in a solvent to coat hard particles followed by compression of the coated particles, or, sometimes by impregnation with a plasticizer is used. This kind of processes is convenient for small or medium size pieces, and when a high volume proportion of filler is required.

P2 - The melt /cast process : mixing of the filler with an unreactive thermoplastic binder, and heating until a pourable mixture is obtained, which is then poured into a mould. A high temperature must be maintained during the process, but recycling, repair and dismantling are possible [4]. Cracks by shrinkage must be avoided.

P3 - The cast-cure processes [5]. In a first step, fillers and two liquid components are mixed thoroughly. Some additives and catalysts are added to initiate a relatively slow reaction between the liquid components. The mixture is quickly poured into a mould. Functionality of components and mixture viscosity are essential factors. This process is used for large missiles, and also with a mixer-screw extruder for small devices. The most classical reaction occurs between diols and diisocyanates to form the urethane linkage. Trifunctional additives allow the binder to crosslink giving the final mechanical properties. But in the P3 processes, recycling of the binder is impossible.

The energetic polymers often fit into one of these processes P1, P2, P3 (see Table 1). Each process introduces its own requirements. It is necessary to translate all these above requirements to the material components : filler, polymeric binder, plasticizer, surface agents as additives, with the mutual interactions between these components, as parts of a system. For the polymers, viscosity, compatibilities, and interactions with the filler (surface tension) and with the plasticizers (miscibility, exudation, Tg ...) are imperative requirements. A knowledge of these interactions is an important target for predictive reasons. See for instance a report about the prediction by modelization of the surface tension of thermoplastic polymers (e.g. PVN) in contact with HMX [9], or of mechanical properties [10,11].

Table 1

Raw Material	First reaction	Energetic reaction product	Processing
Non energetic polymer	Nitration	Energetic polymer *PVN DNPS*	P1
	Exchange	Precursor or prepolymer (A) *GAP*	Adding (B) P3
Energetic polymer PVN	Exchange	Other energ. polymer *PV Azide*	P1 (?)
Energetic monomer *(GLYN, NIMMO BAMO, AMMO...)*	Polymerization or copolymerization	Energ. Prepolymer (A) *Poly BAMO... Some acrylates*	Adding (B) P3
		Energetic polymers *(acrylates)*	P1
	Block copolymerization	Thermoplastic elastomer (TPE) *BAMO/AMMO...*	P2
Energetic molecule	Polycondensation	Energ. polymer *PNP*	P1
	Reaction with HCHO	Energ. polyformals prepolymers	Adding (B) P3
Molecule with $C = C$ or $C = O$	Reaction with HNF_2 or $N_2 F_4$	Fluoraminated binders	P1 or (adding B) P3

A = diol (or diisocyanate)
B = diisocyanate (or diol)
P1, P2, P3 : processings (see text)

3. Synthetic energetic polymers

Many papers have been published recently [1, 4, 6, 7, 8]. With a few exceptions, these polymers contain "explosophore" chemical groups such as NO_2, linked to a C, O or N atom, NF_2 and N_3. A more exotic polymer with a tetrazene group was published [12]. Because it is difficult to have a description according to these chemical groups, we have tried to put together some polymers used by a same process P1, P2 or P3 (Table 1), knowing that some of them can be processed by two different ways.

(a) Dinitropolystyrene (DNPS)

The 4-nitro, 2,4-dinitro (MP 56°C) and 2,4,6-trinitrostyrene monomers are known and studied essentially in Russia. Polymerization and copolymerization are possible, between di-and trinitrostyrene in cyclohexanone, giving low molecular weight products [13].

But the most interesting method of preparation is direct nitration by mixed acids [14]. Molecular weights may be higher than 100 000. DPNS is a yellow powder with low sensitivity, very high thermal stability (over 280° C), and chemical stability of years. It is soluble in cyclohexanone, but very brittle without additives. It is used as a binder in gun propellants in mixture with polyvinylnitrate (PVN), by coating gunpowder by a solvent process P1 with cyclohexanone, and subsequently pressing it between 100 and 140° C [15]. Its cost is fairly low. It is interesting to observe that the mechanical properties and the thermostability of mixtures DNPS-PVN are not linear versus mixture proportions.

(b) Polynitropolyphenylene (PNP)

Developed by a German Company (Dynamit Nobel) [16], PNP is a very thermostable amorphous (> 95%) polymer (MW ~ 2000). Obtained by Ullmann's reaction of 1,3-dichloro-2,4,6-trinitrobenzene with copper powder in nitrobenzene at about 150°C . It has no melting point and is soluble in some solvents. Coating of fillers is obtained with an acetone-ethanol solution. It can be used as binder in thermostable pyrotechnic compositions [17] by process P1.

In the nitroaromatic series are also described polynitropoly-(p.xylene) and polynitroindene (stability > 300° C) obtained by nitration of the polymers, and poly (2,4,6-trinitro-phenylglycidylether), by polymerization of the monomer [18], with a softening range 64-70° C.

(c) Polyvinylnitrate (PVN)

The monomer $CH_2=CHONO_2$ is practically unknown, but perhaps fugitively isolated once [19] as a liquid. The polymer is known since 1929 [20, 21]. Synthesis by nitration of polyvinyl-alcohol with mixed acids (ICT process [22]), $HNO_3 + Ac_2O$ [23] or nitric acid in presence of a chlorinated solvent (the process used in France). Molecular weight exceeds 100 000 and density is about 1.5g/cm3. Softening temperature is very low, about 40° C but higher with isotactic PVN [23]. It is stabilized by 2-NDPA. Pure PVN (15.7% N) is more stable than NC by vacuum stability test, and the gas evolution vs time is linear. It is compatible with DNPS which improves its mechanical properties and its thermostability (see above). It is soluble in many solvents.

It can be processed by coating nitramines (process P1) and then pressing the granulates. PVN was patented in cast-cured double-base composite propellants as an additive. In many propellant or gunpowder compositions, some part of NC may be replaced by PVN. It is also possible to obtain monophase mixtures PVN-RDX, PVN-HMX, [24] and PVN-PETN.

Another energetic nitrate ester monomer was isolated in 1950 [25,8], e.g. the acrylate nitrate of 1,2-ethanediol which gives a soft polymer. Perhaps by copolymerization, it is possible to obtain TPE (thermoplastic elastomer). The starting glycol mononitrate can be easily obtained by reaction of N_2O_4 with ethylene oxide [19].

(d) Polyvinylazide

The monomer is known as a very sensitive liquid (BP 30° C). The polymer was recently prepared by reaction of a metallic azide with polyvinylnitrate in an inert organic solvent at 20 - 80° C. The substitution may be partial [26].

(e) C - NO2 aliphatic and N - NO2 derivatives

Among the mononitroaliphatic derivatives C - NO2, the nitroethylenic monomers (nitroethylene, 1-nitropropene) may polymerize, but thermostability is poor.

For gem-dinitro derivatives (not fluorinated) the most used starting molecules are 2,2-dinitropropanol (DNP), 2,2-dinitro-1,3propanediol (DNPD) and 3,3-dinitro-1,5pentanediol. With the exception of DNP acrylate (DNPA) [27, 28] and AFNOL [28], these dinitrodiol derivatives were transformed into prepolymeric diols or diisocyanates (process P3).

The plastic bonded LX-09 using DNPA seems to be better than PBX 9404, with a better stability, but some poorer safety properties, perhaps due to the hardness of this binder. It seems to be now abandoned.

For P3 processing, nitrated diisocyanates are obtained by following sequence :

$$CO_2H \longrightarrow COCl \longrightarrow CON_3 \longrightarrow NCO$$

But the most used is a polymeric diol. To obtain it with a precise functionality, some different ways are used : for instance, reaction of formaldehyde with an excess of the starting diol or a mixture of diols. Many monomers and prepolymers were synthesized up to 1982 [27] but it seems that none of them was found convenient because poor physical properties were obtained. The same situation exists with some N-NO$_2$ or mixed N-NO$_2$ and C(NO$_2$)$_2$ molecules.

Another route was recently described (1991) involving acylation of DNPD with adipic acid chloride in excess, followed by a reaction with a short diol [29]. In propellant formulation, Isp x d is equivalent to products with GAP (see later) and the vulnerability is low.

Other nitramines include a polymer -[-N(NO$_2$)-CH$_2$-]$_n$ which was described in the 1950's and obtained during hexamine nitrolysis, but seems to be forgotten. The trimer and the tetramer (BSX and AcAn, with acetyl groups at the ends) have been used as starting materials to obtain the diol, which is linked by reaction with HCHO and a fluorinated diol, giving a mixed formal terminated by two hydroxyl groups. This last diol is a prepolymer for a polyurethane (process P3). This cured polymer is able to retain a large quantity of FEFO [bis-(2-fluoro-2,2-dinitroethyl) formal] as a plasticizer [30]. Another starting molecule is EDNA (ethylene-dinitramine).

The fluorodinitrocarbon group is fairly stable, and the 2-fluoro-2,2-dinitroethyl group was introduced into many molecules, giving plasticizers and sometimes monomers or polymers. The patented binders have the property of absorbing the fluorodinitroplasticizers, giving to the propellant more energy and density. For instance, the fluorodinitrocarbon group is linked laterally to a carbon of a diol, which reacts after with a diisocyanate during the processing (P3) [31]. Some of these polymers have poor mechanical properties e.g. the poly(2-fluoro-2,2-dinitro-ethylglycidyl)ether.

(f) Fluoroaminated polymers - NF$_2$

NF$_2$ is an energetic functional group and some polymers were synthesized containing adjacent NF$_2$ groups by reaction of N$_2$F$_4$ with C=C [32] or gemdifluoroamino groups by reaction of HNF$_2$ with C=O [33]. These polymers are able to absorb an important quantity of difluoroamino energetic plasticizers.

Mixed polymers containing both C(NO$_2$)$_2$ and -C(NF$_2$)-C(NF$_2$)- groups were also recently patented [34]. This mixture with acrylates and HMX is melt/cast. Price is probably a limiting factor.

(g) Oxiranes and oxetanes as raw materials

The advantages of process P3 (diol + diisocyanate) pushed the research in the direction of energetic diols synthesis via oxiranes and oxetanes. These compounds can be cationically polymerized (e.g. with BF$_3$/Et$_2$O as catalyst) in presence of a diol, in order to obtain a OH functionality near 2 and a molecular weight between about 1000 and 5000. Polymerization conditions are adjusted to obtain a liquid with low Tg. The diol reacts after mixing with the filler with added isocyanates following conventional methods. An appropriate quantity of triol or triisocyanate and a catalyst to control the pot life are added to the mixture.

Two major energetic groups are introduced (ONO_2 or N_3) into an oxirane or oxetane molecule.

In the first family, an important breakthrough was made in UK with the use of N_2O_5 as nitrating agent, dissolved in a chlorinated solvent [8, 35]. It is possible to obtain glycidyl nitrate (GLYN) and nitratomethyl-methyloxetane (NIMMO)

$$H_2C\text{-}CH\text{-}CH_2ONO_2$$
$$\diagdown O \diagup$$

GLYN

NIMMO

Polymerization of NIMMO is carried out using BF_3/Et_2O and 1,4 butanediol. The product has an onset of decomposition at 170°C by DSC. It is usable as binder after curing in propellants and explosives giving reduced vulnerability. To overcome the relatively high Tg (-25°C) and to use its good mechanical properties, copolymers were prepared with GLYN or with BAMO (see later) - NIMMO is now in development. Polymerization of GLYN seems to be more difficult, but density of poly GLYN (1.43) is higher than for poly NIMMO (1.26).

The most important compound of the N_3 oxirane and oxetane family is the GAP glycidyl azide polymers of $H_2C\text{-}CH\text{-}CH_2N_3$
$$\diagdown O \diagup$$

It was found some years ago that, contrary to many metallic azides, some organic azides are shock-insensitive. They are obtained by an exchange reaction between NaN_3 with an halogenated compound.

Polymerization of the GAP monomer seems to be too slow. It is necessary to start from poly-(epichlorhydrin) (PECH), reacting with NaN_3, either in a solvent or in water with a phase transfer catalyst. PECH can be prepared as a low molecular weight polymer with an adequate structure chosen to optimize binder mechanical properties (Rockwell process). Another way is to start from commercial high molecular molecular weight product, giving a branched GAP [36]. An important review on GAP is given [37][38]. Japanese teams, under the direction of N.Kubota, have made and published numerous studies on GAP and its properties and combustion mechanisms [39, 40].

GAP seems to have no special advantage for gunpowders [41] or for high explosives, but it has a large development in propellant formulations. They provide a reduced vulnerability, and a higher combustion rate. The relatively low price is also an advantage. A drawback to solve is the temperature sensitivity of the GAP combustion rate.

Some other azides are obtained starting from bis-2,2-(bromomethyl)-1,3-propanediol giving esters by reaction with a dicarboxylic acid, followed by reaction with NaN_3 [42]. But the two most important azides after the GAP are two oxetane derivatives, the bis(azidomethyl)(BAMO) and the methylazidomethyl (AMMO) oxetane. Poly BAMO is a relatively stiff solid. In order to have a liquid, BAMO is copolymerized with tetrahydrofuran (THF). AMMO gives a soft and elastic polymer with density 1.06. These azides thermally decompose over 200° C in two steps. The first occurs with N_2 evolving. For an overview on these azides, see [43, 44, 45, 46].

New polymers are recently synthesized by the team of G. Manser, starting from oxetanes substituted in the position 3 by CH_2NF_2 or on the same molecule, by CH_2N_3 and CH_2ONO_2 [47].

To obtain better mechanical properties at low temperatures, two kinds of copolymers were studied : BAMO-AMMO and BAMO-NIMMO, the first ones are more thermostable, but give less energy release.

But a very interesting discovery was made in these copolymers by Wardle (Thiokol)[48, 49] concerning the association BAMO-AMMO. This team solved the problem of preparation of a TPE polymer by block-copolymerization which often gave a non-reproducible product. They studied the chemical mechanism and perform the reaction in two steps, with a monomer activation approach ("pseudo-living" polymerization).

CONCLUSIONS

(1) Following NIMIC suggestions [2], it is interesting to record, through the literature, energetic polymers or theoretically polymerizable monomers, the development of which was stopped or not undertaken, giving if it is possible reasons for this stop. For example the reason might by unsatisfactory properties, difficulties in the synthesis, poor stability, expense, or fair properties but no need at that time, no more money, research team disappeared or moved and so forth.

(2) For the polymers an effort needs to be sustained to understand and to predict surface properties, compatibility, interaction with plasticizers,mechanical and thermal properties.

(3) More knowledge is needed on the combined effect between filler particles size and kind of the binder (inert or energetic) on the energetic properties such as ballistics for propellants, erosivity for gunpowders, detonics (e.g. ideality) for high explosives. In some cases, the biphasic composition becomes monophasic when the filler has nanosize particles; the crystalline phase disappears in mixtures of RDX-PVN, PETN-PVN, NC-RDX. What are the structural conditions for the filler and the binder to obtain this monophasic product ? Is it a stable or metastable situation?

(4) More understanding is needed on the properties, the polymerization and copolymerization of ethyleneglycol nitrate-acrylate and vinyl nitrate.

(5) The relationship of structure to properties of nitrated products of polystyrenes or their copolymers needs to be known. There seems to be a way to obtain relatively cheap binders for

pressable material or for TPE (melt cast process P2).

(6) To fulfill simultaneously the mechanical properties requirements and reprocessing, the effort to manage the polymeric binders structure which was begun by the Wardle team to obtain TPE by combining stiff and flexible segments, is very interesting. It is also interesting to explore the field of interpenetrated polymers [50] with polymers containing energetic moieties.

ACKNOWLEDGMENTS

The assistance of the French DRET is acknowledged.

REFERENCES

1 -   J. M. Decore. Internat - Défense et Technologie (Sept. 1994) (Connection International Publishers - Brussels) p. 16-20; R. S. Miller, ibid pp. 21-27.

2 -   A. Sanderson, P. Kernen, B. Stokes, J. Fitzgerald-Smith. Proceedings of the 1994 NIMIC Workshop (NATO) (22 - 24 June) Brussels - p. 15-44.

3 -   M. Défourneaux - 25 th Intern. Ann. Conf. ICT 1994 p. 2/1-14.

4 -   D. M. Hoffman, T. W. Hawkins, G. A. Lindsay, R. B. Wardle, G. E. Manser - Life Cycle of Energetic Materials - Conf. Dec. 1994 - Proceedings p. 75-90; L. F. Cannizzo, R. B. Wardle, R. S. Hamilton, W. W. Edwards - Ibid pp. 221-225.

5 -   A. Davenas (Ed.), Solid Rocket Propulsion Technology - Pergamon Press - Oxford, New-York (1992).

6 -   K. Klager - 13 th Int. Ann. Conf. ICT 1982, p. 11-30, L. R. Rothstein, ibid pp. 245-256.

7 -   M. B. Frankel, L. R. Grant, J. E. Flanagan - J. Prop. Power 8 (3) 560-3 (1992).

8 -   M. E. Colclough, H. Desai, R. W. Millar, N. C. Paul, M. J. Stewart, P. Golding, Polym. Advanced Technology 5 554- 560 (1994).

9 -   R. Mevrel, S. Poulard, J. Dumiel, M. Vignollet - Rheol. Acta 13 (2) 318-22 (1974).

10 -   R. Reed, V. Brady - ADPA - Proceedings of Joint Int. Symp. 22-24 April 1991 San Diego p. 492 - 500

11 -   R. G. Stacer, D. M. Husband, Prop. Expl. Pyro. 16 167-176 (1991).

12 -   I. Mishra, L. J. van de Kieft - 19 th Int. Ann. Conf. ICT 1988, 25/2-9

13 -   K. Kalnins, A. D. Kutsenko, Yu E. Kirsh, N. A. Barba, I. D. Korzh - Vysokomol. Soedin. Ser. A 32 (6) 1268-75 (1990) - CA 113 132878 f.

14 - A. M. Pujo, J. Boileau, F. M. Lang - Mém Poudres 35 41-50 (1953)

15 - J. Boileau, L. Leneveu (SNPE), French Patent N° 2658505 (21.02.1990).

16 - K. H. Redecker, R. Hagel - 18 th Int. Ann. Conf. ICT 1987 p. 26/1-10 ; Prop. Expl. Pyro. 12 196-201 (1987).

17 - B. Berger, B. Haas, G. Reinhard - 26 th Int. Ann. Conf. ICT 1995 p. 2/1-14.

18 - S. Oinuma - Tokyo Kogyo Shik. Hokoku 65 (2) 52-6 (1970) - C. A. 74 141409e

19 - A. M. Pujo, J. Boileau,Mem Poudres 37, 35-48 (1955).

20 - W. Diepold,Explosivst. 18 2 (1970).

21 - T. Urbanski - Chemistry and Technology of Explosives - Pergamon Press New York. II, (1965) p. 173 -4; IV (1984) p. 413 - 21.

22 - E. Backof - 12 th Int. Ann. Conf. ICT 1981 p. 67-84.

23 - R. A. Strecker, F. D. Vanderame - US Pat 3965081 (22/06/1976).

24 - N. A. Messina, L. S. Ingram, G. F. Konig, M. Summerfield, W. Crawford - Report 1987 PCRL - FR - 87004 - AD - 184178 - CA 108 223920 z.

25 - N. S. Marans, R. P. Zelinski - JACS 72 5330 (1950).

26 - E. E. Gilbert US Pat. 4839420 (13/06/89) - CA 111 157031u.

27 - E. E. Hamel - 13 th Int. Ann. Conf. ICT 1982 p. 69 - 84.

28 - B. Dobratz,LLNL Explosives Handbook UCRL 52997 - 16/03/1981.

29 - M. Piteau, A. Becuwe, B. Finck - ADPA Int. Meeting San Diego (22 - 24/04/1991) Proceedings p. 69-79.

30 - H. G. Adolph, D. M. Cason-Smith,US Pat. 5266675 (30/11/1993).

31 - M. B. Frankel, E. F. Witucki (Rockwell),US Pat 3832390 (27/08/1974).

32 - R.M. Price - US Pat - 3781252 (25/12/1973); US Pat 3829336 (27/08/1974).

33 - Jiang Zhirong, Xu Guilin, Chen Yuanfa - 21 st Int. Ann. Conf. ICT 1990 p. 52/1-6.

34 - B. A. Scott, L. E. Koch,US Pat 5092944 (03/03/1992).

35 - R. W. Millar, M. E. Colclough, P. Golding, P. J. Honey, N. C. Paul, A. J. Sanderson, M. J. Stewart, Phil. Tran. R. Soc. London A. 339 305-19 (1992) (Edited by J. E. Field and P. Gray).

36 - E. Ahad - 21 st Int. Ann. Conf. ICT 1990, p. 5/1-13; E. Ahad and 15 coll. - 24 th Int. Ann. Conf. ICT 1993 p. 75/1-12.

37 - M. B. Frankel, L. R. Grant, J. E. Flanagan, J. Prop. Flame $\underline{8}$ (3) 560-3 (1992).

38 - AA. N. Nazare, S. N. Asthana, H. Singh, J. Energ. Mat. $\underline{10}$ 43-63 (1992).

39 - N. Kubota, T. Sonobe - 19 th. Int. Ann. Conf. ICT 1988 p. 2/1-12; G. Nakashita, N. Kubota, Prop. Exp. Pyro. $\underline{16}$ 177-81 (1991).

40 - N. Kubota - 6e Congrès Internat. de Pyrotechnie - 5-9 June 1995 Tours (France). Proceedings p. 13-17.

41 - F. Schedlbauer, Prop. Expl. Pyro $\underline{17}$ 164-71 (1992).

42 - T. Keicher, F. W. Wasmann - Prop. Expl. Pyro. $\underline{17}$ 182-4 (1992).

43 - G. E. Manser (Aerojet), 21 st Int. Ann. Conf. ICT (1990) p. 50/1-13.

44 - Young-Gu Cheun, Jin-Seuh Kim, Byung-Wook Jo (South Korea), 25 th Int. Ann. Conf. ICT (1994) p. 71/1-14.

45 - Ying-Ling Liu, Ging-Ho Hsine, Yie-Shun Chiu (Taiwan), J. Appl. Polym. Sci $\underline{58}$ 579-86 (1995).

46 - J. K. Chen, T. B. Brill, Comb. Flame $\underline{87}$ 157-168 (1991); T. B. Brill, Chemistry in Britain Jan. 1993 p. 34-36.

47 - G. Manser, A. Malik, T. Archibald - Patent PCT. 17/03/1994. C. A. $\underline{122}$, P 82365x; T. Archilbald, R. Carlson, A. Malik, G. Manser, US Pat. 5362848 (08/11/1994).

48 - R. B. Wardle, J. H. Hinshaw, W. W. Edwards, Joint Int. Symp. ADPA 22-24 April 1991. San Diego. Proceedings p. 89-96.

49 - L. F. Cannizzo, R. B. Wardle, R. S. Hamilton, W. W. Edwards, Life Cycles of Energ. Mat. 1994 Conf. Del Mar. Proceedings p. 221-225.

50 - S. Parthiban, B. N. Raghunandan, S. R. Jain, Def. Sci. J. $\underline{42}$ (3) 147-156 (1992).

Part II

Thermal Decomposition

KINETIC ANALYSIS OF SOLID-STATE REACTIONS:
AN IMPROVED ANALYSIS METHOD.

D.P. SMITH, M.M. CHAUDHRI
Materials Group, PCS, Cavendish Laboratory, University of Cambridge, Madingley Road, Cambridge, CB3 0HE, UK. Correspondence to dps13@cus.cam.ac.uk

ABSTRACT

Many reactions of interest occur in the solid state, or with low mobility of the reactants or products, with the result that the physical mechanisms operative during the reaction determine the shape of the reaction rate-time profile; i.e. *topochemistry* is important. However, often improper account is taken of the physical mechanisms, resulting in erroneous values for the activation energy and kinetic constants.

An improved method of analysis is suggested, allowing the mechanisms of reaction to be determined explicitly. This thus allows the activation energy and kinetic constants to be determined with good theoretical justification. The method has been verified by studying the thermal decomposition of barium azide, where the activation energy was determined to better than 1%. The operative mechanisms so determined are in agreement with visual observations and the literature.

INTRODUCTION

Many reactions of interest occur in the solid state, such as decomposition, dehydration, dehydroxylation, calcination, reduction, oxidation, ageing, tarnishing, polymeric inversion, crystalline transitions, or any activated process. Even reactions occurring in the liquid or vapour phase may be thought to follow solid state-type mechanisms, with reactions occurring on vessel walls or at foreign particles. These reactions may be monitored by a number of methods [1], such as differential scanning calorimetry (DSC), thermogravometry (TG), thermomanometry, quantitative IR, quantitative X-ray analysis, dilatometry, conductivity measurements (electrical or thermal), optical properties, etc.. The base criterion is the ability to monitor the temporal variation of the global amount of material decomposed.

For reactions occurring in solution or inviscid liquids, high mobility allows the reactants and products to be well mixed, and the concentration of any species is uniform and a state variable. However, in the solid state, the mobility of reactants is absent (or very much reduced), and the concentration of reactants and products can vary locally through the sample. Alternatively, reaction is limited to an interface, and the concentration is again no longer a state variable. Thus we have *topochemistry*, where shape, phase of product and reactant, the action of products (such as the formation of eutectics and catalysis), porosity, and factors such as strain introduced by the formation of the product must be taken into account. In summary, due to the lack of (or reduction in) mobility of the reactants compared to the inviscid case, the physical mechanisms occurring during a reaction *must* be taken into account in any attempt to characterise the reaction.

During a reaction there exists a rate-limiting step which determines the kinetics. The form of the rate equation used to model this rate-limiting step, and hence the progress of the reaction, is almost invariably given by equation (1):

$$\underbrace{\frac{d\alpha}{dt}}_{\substack{rate\ of \\ conversion}} = \underbrace{K(T)}_{\substack{function\ of \\ T\ only}} \cdot \underbrace{f(1-\alpha)}_{\substack{function\ of \\ \alpha\ only}} \quad \text{where} \quad K(T) = K_\infty e^{\frac{-E_{act}}{kT}} \tag{1}$$

where α is the global degree of conversion of reactant, t is time, T temperature, k the Boltzmann constant, E_{act} the activation energy, and K_∞ the pre-exponential factor. The global rate of reaction, $\frac{d\alpha}{dt}$, is inferred from macroscopic measurements, such as heat flow $\frac{dQ}{dt}$ in DSC experiments, for instance. The above relationship is, however, empirical, and has no theoretical derivation. The use of the Arrhenius equation, seen in equation (1) as $K(T)$, to solid state reactions has recently been justified theoretically [2]. The applicability of equation (1) has been shown in the case of isothermal experiments by the fact that (α,t) graphs are isomorphic on scaling the time axis [3], al least within a temperature range. This suggests that the rate equation is separable into a temperature-dependent part and an α-dependant part, which determine the shape of the (α,t) curve, expressed in equation (1) as $K(T)$ and $f(1-\alpha)$ respectively.

Dynamic experiments appear attractive, as in a single experiment an entire temperature range of interest can be covered. However, it has been doubted [4] whether equation (1) is applicable in a dynamic run. While alternative forms of equation (1) have been proposed for a dynamic experiment, it has been argued [5] that equation (1) is still valid. However, there are many problems inherent with the analysis of dynamic data. It can be shown [6] that the effect of the operative physical mechanism, $f(1-\alpha)$, on the dynamic results is inherently masked by the temperature being a function of time, with the result that from a single dynamic experiment it is not possible to determine uniquely the operative mechanism, the activation energy or the pre-exponential factor. Performing multiple dynamic runs may allow these to be determined through some non-trivial analysis, though most of the benefits of performing dynamic experiments over isothermal experiments are lost. Hence the analysis method proposed in this paper is based on isothermal experiments.

The functional form, $f(1-\alpha)$, determines the shape of the normalised (α,t) curve, and is derived theoretically for the appropriate operative mechanisms (see table I). Reviews discussing the theory and derivation of the different mechanisms are given in references [3, 7, 8]. In the case of well-stirred, inviscid liquids and gases, the rate of reaction is proportional to the concentration of each species involved in the rate determining step raised to some power, n. This is often written in the form of the functional form given in equation (2):

$$\frac{d\alpha}{dt} = K(T)(1-\alpha)^n \qquad (2)$$

where n should be an integer. This rate equation forms the basis of many popular analysis methods and software supplied by many thermal analysis equipment manufacturers. However, as explained above, equation (2) is *not* applicable to reactions in the solid state.

The functional forms listed in table I include reaction mechanisms based on nucleation, phase boundary movement, unimolecular decay and diffusion. The list is in not exhaustive, and one may derive a new theoretical functional form which describes the specific physical processes occurring in an experiment. The list merely presents 14 functional forms commonly met in the literature, and which are most likely to be met in an experiment.

The interpretation of $K(T)$ in equation (1) is dependent on the operative mechanism. The activation energy E_{act} represents the energy barrier to the rate determining step, and has units of energy. The pre-exponential factor K_∞ must have units of s^{-1}, and may include diffusion coefficients, nucleation rates, probability of branching or termination of a process, initial interfacial area, size, geometric factors, transport coefficients etc. from the theoretical derivation of the functional forms. It also includes any normalisation factors and numerical constants. Hence even for a given mechanism, K_∞ may vary by orders of magnitude from material to material, even from sample to sample. For this reason, sample consistency is important, e.g. in particle size distribution, else scatter will be seen in the results.

EXPERIMENTAL

To illustrate the method, experiments were performed on anhydrous barium azide in a Mettler DSC30. This material was chosen because its energetics and mechanisms of decomposition are well documented in the literature, and under an optical microscope it can be *visually* observed to form compact nuclei during thermal decomposition.

Barium azide was synthesised in the laboratory by the double decomposition of hot aqueous sodium azide and barium iodide. Unsaturated solutions of the reactants at ~90°C were mixed, resulting in the formation of a thick precipitate of *anhydrous* barium azide, which has a much lower solubility than the reactants. The solid was then filtered and washed with copious amounts of 96% ethanol, in which the reactants are again much more soluble than the product (hence the choice of the iodide). The solid was then dried under vacuum in a desiccator in the presence of phosphorus pentoxide. This method of preparation is different from methods proposed in the literature; these methods involve hydrazoic acid or were deemed dangerous. Standard qualitative chemical tests and X-ray energy dispersive analysis using a scanning electron microscope showed there to be no iodide or sodium present in the sample, and X-ray

Table I - 14 commonly met physical reaction mechanisms and their functional forms.

$\Phi_i(1-\alpha)$ = integrated, normalised form of $f_i(1-\alpha)$. m = slope of log(-ln(1-α)) vs. log(t) graph.

i	$f_i(1-\alpha)$	$\Phi_r(\alpha)$	m	Mechanism
1	$\alpha^{1-\frac{1}{n}}$	$(2\alpha)^{\frac{1}{n}}$	1.08n	Power law nucleation.
2	α	$\dfrac{\ln(\alpha)+C_0}{\ln(0.5)+C_0}$	—	Linear branching chains of nuclei, no overlap during growth.
3	$\alpha(1-\alpha)$	$1-C_0\ln\left(\dfrac{1-\alpha}{\alpha}\right)$	—	Branching chain nucleation, interference during growth.
4	$(1-\alpha)\alpha^2$	$\dfrac{1}{C_0-2}\left[C_0-\ln\left(\dfrac{1-\alpha}{\alpha}\right)-\dfrac{1}{\alpha}\right]$	—	Cubic auto-catalysis.
5	$[-\ln(1-\alpha)]^{1-\frac{1}{n}}(1-\alpha)$	$\left[\dfrac{\ln(1-\alpha)}{\ln(0.5)}\right]^{\frac{1}{n}}$	n	Random nucleation, growth accompanied by ingestion.
6	$\alpha^{\frac{1}{2}}$	$\dfrac{\alpha_0^{-\frac{1}{2}}-\alpha^{-\frac{1}{2}}}{\alpha_0^{-\frac{1}{2}}-(0.5)^{-\frac{1}{2}}}$	—	Instantaneous nucleation, time dependent growth.
7	1	2α	1.12	1-D phase boundary growth.
8	$(1-\alpha)^{\frac{1}{2}}$	$\dfrac{1-(1-\alpha)^{\frac{1}{2}}}{1-(0.5)^{\frac{1}{2}}}$	1.07	2-D phase boundary growth.
9	$(1-\alpha)^{\frac{2}{3}}$	$\dfrac{1-(1-\alpha)^{\frac{1}{3}}}{1-(0.5)^{\frac{1}{3}}}$	1.04	3-D phase boundary growth.
10	$(1-\alpha)$	$\dfrac{\ln(1-\alpha)}{\ln(0.5)}$	1	Unimolecular decay.
11	$\dfrac{1}{\alpha}$	$4\alpha^2$	0.56	1-D diffusion controlled growth.
12	$[-\ln(1-\alpha)]^{-1}$	$\dfrac{\alpha+(1-\alpha)\ln(1-\alpha)}{(0.5)+0.5\ln(0.5)}$	0.54	2-D diffusion controlled growth.
13	$\left[(1-\alpha)^{-\frac{2}{3}}-(1-\alpha)^{-\frac{1}{3}}\right]^{-1}$	$\dfrac{\left[1-(1-\alpha)^{\frac{1}{3}}\right]^2}{\left[1-(0.5)^{\frac{1}{3}}\right]^2}$	0.54	3-D diffusion controlled growth.
14	$\left[(1-\alpha)^{-\frac{1}{3}}-1\right]^{-1}$	$\dfrac{\frac{3}{2}\left[1-(1-\alpha)^{\frac{2}{3}}\right]-\alpha}{\frac{3}{2}\left[1-(0.5)^{\frac{2}{3}}\right]-0.5}$	~0.5	3-D diffusion controlled growth (alternative derivation).

diffraction studies confirmed that the solid was anhydrous form of barium azide. The crystals had the appearance under the microscope of platelets of typical dimension ~0.1mm.

Mettler gold plated steel high pressure pans (part no. ME-26732) were used for the study, which can support an internal pressure of 15MPa. A typical sample size of ~5mg was sealed in the pan in an argon atmosphere, as oxygen was found to affect the sample. ($\frac{dQ}{dt}$,t) (α,t) and data were displayed graphically on a printer, then electronically scanned and digitised. Isothermal scans were performed in the temperature range 162.5-182.5°C, a range which produced traces of good quality and high resolution. The experiment used to illustrate the analysis was run BAN973, an isothermal at 170°C.

ANALYSIS

Since the experiments in this paper were performed on a DSC, analysis is focused around DSC results, though data collected by other methods only need to be processed to produce (α,t) and ($\frac{d\alpha}{dt}$,t) data. In the case of raw DSC data, it must first be processed to remove the shift in the baseline, which necessarily occurs due to the difference in heat capacity between reactant and products. It is better to work from raw DSC data, since the DSC signal only needs normalising to give ($\frac{d\alpha}{dt}$,t) data, which may be integrated to give (α,t), also needed in the analysis. Obtaining ($\frac{d\alpha}{dt}$,t) data by numerically differentiating (α,t) data often leads to banding due to

Figure 1: The log graph indicates by its linearity that a single mechanism is dominant up to $\alpha\approx0.75$; the slope m of the linear section is 2.7, which suggests a nucleation mechanism. Note banding at low α, due to digitisation of the signal.

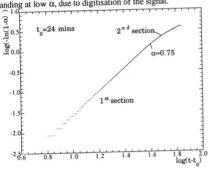

Figure 2: Fit of Avrami-Erofeev nucleation, Φ_5, up to $\alpha=0.75$, as dictated by figure1, providing such a good fit that the data can hardly be distinguished from the line.

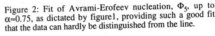
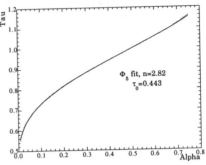

digitisation, thus degrading the quality of the data. A final preparation is to remove data outside the range $0.005<\alpha<0.995$, since data at $\alpha=0$ and 1 are not true dynamic data.

An indication of the operative mechanism is given by the graph of $\log[-\ln(1-\alpha)]$ versus $\log(t-t_0)$, where t_0 is the time origin of the operative mechanism (see figure 1). Such plots were first employed by Avrami [9], as they yielded directly the numerical index in her nucleation mechanism. Hancock and Sharp [10] observed that such plots of many of the forms in table I were linear in the range $0.15<\alpha<0.5$; however, plotting computer generated data for all the forms in table I over the whole range of α showed that all except $\Phi_{2,3,4}$ and Φ_6 had long linear sections, some curving up gently only at high α. Forms $\Phi_{2,3,4}$ and Φ_6 (see table I) continuously curve up with increasing α over the whole range of α, though all the functional forms are smooth and continuous. The slope 'm' of the linear section is listed in table I, and falls into three groups: $m>1$ is indicative of a nucleation mechanism, $m\approx1$ of unimolecular decay or a phase boundary controlled mechanism, and $m\approx\frac{1}{2}$ of a diffusion controlled mechanism.

It must be stressed that such plots are insensitive, and are only used to give an indication of which mechanism may be operative, and also the location of any changes in the mechanism, indicated by a change in slope. Such a change is usually seen as a bend down, in the opposite direction to the deviation seen in computer-generated data of a single mechanism, making changes to the mechanism indisputable. The shift in the time origin, t_0, may be interpreted as a combination of the uncertainty in time origin always present in isothermal experiments, and any induction period which may exist. The value of t_0 is that which makes the $\log[-\ln(1-\alpha)]$ versus $\log(t-t_0)$ graph linear, as in Figure 1, and is found by trial and error.

As can be seen from figure 1, the $\log[-\ln(1-\alpha)]$ versus $\log(t-t_0)$ is linear up to $\alpha\approx0.75$, with a slope of $m\approx2.7$, suggesting that a single mechanism is dominant up until $\alpha\approx0.75$, and that the mechanism must be a nucleation mechanism. The linearity of the graph also means that the only possible alternatives are Φ_1 or Φ_5 (table I). Above $\alpha\approx0.75$, a second mechanism is dominant, which can be analysed in a similar manner, as described later.

To determine precisely which mechanism is operative, the possible functional forms as dictated by the $\log[-\ln(1-\alpha)]$ versus $\log(t-t_0)$ must be fitted to (α,t) data, up to the value of α dictated by the log plot, by non-linear regression, which is routinely possible on desk-top computers. To do this, (α,t) data must be normalised to produce (α,τ) data. From equation (1):

$$\dot{\alpha}=K(T)f(1-\alpha) \quad \text{then} \quad \int\frac{d\alpha}{f(1-\alpha)}=K(T)t, \quad \text{or} \quad F(1-\alpha)=K(T)t$$

Now, at $\alpha=\frac{1}{2}$, $F(1-\alpha)=F_{\frac{1}{2}}=K(T)t_{\frac{1}{2}}$. Then $\Phi(1-\alpha)=\dfrac{F(1-\alpha)}{F_{\frac{1}{2}}}=\dfrac{t}{t_{\frac{1}{2}}}=\tau$,

defining Φ (see table I) However, we also have to allow for the shift in the time origin, t_0:

108

$$\Phi(1-\alpha) = \frac{F(1-\alpha)}{F_{\frac{1}{2}}} = \frac{t-t_0}{t_{\frac{1}{2}}-t_0}, \quad \text{so that if } t_0 = 0, \quad \Phi = \tau \text{ as before. With } \tau = \frac{t}{t_{\frac{1}{2}}} \text{ and}$$

$\tau_0 = \dfrac{t_0}{t_{\frac{1}{2}}}$, then
$$\Phi(1-\alpha) = \frac{\tau - \tau_0}{1 - \tau_0} \qquad (3)$$

The integrated functional forms listed in table I may now be fitted with τ_0 as a variable. The goodness of fit may then be judged by eye. In the case of BAN973, the fit of Φ_1 is relatively poor, especially at high α. The fit may be improved by fitting only up to a lower value of α, though we know from figure 1 that we must fit all the data up to $\alpha \approx 0.75$ as a single section. Thus the fit is rejected. Figure 2 shows the fit of Φ_5, which can be seen to be so good that the data can hardly be distinguished from the curve. Note the slight deviations beginning to occur at very high α due to competition from a second mechanism, which dominates for $\alpha > 0.75$. The interpretation of the non-integral numerical constant n is complex, and will not be discussed here. The value for τ_0 corresponds to a value of $t_0 = 24$ min, which is in excellent agreement with figure 1 and the 'foot' of the DSC exotherm, giving confidence in the fit.

Having identified the operative mechanism and functional form, the temperature-dependent part of equation (1), $K(T)$, may be determined by plotting the rate of reaction versus the functional form $f(1-\alpha)$ over the same range of α, yielding $K(T)$ as the slope (figure 3). The goodness of the fit again gives confidence in the analysis method. The slope from such a plot forms a single point at a discrete temperature on the Arrhenius plot.

In the Arrhenius plot of *all* the barium azide experiments, at a given isothermal temperature scatter occurs, which is probably due to a variation in particle size, as discussed above. A variation in particle size will present a different surface area, hence may affect the reaction rate profile, and hence $K(T)$. Shaking the sample container will cause separation of particle sizes, hence samples should be of the same particle size to avoid scatter. Figure 4 shows an Arrhenius plot for selected data, from experiments which were performed consecutively with very similar particle size distributions. By using the suggested analysis method, and given a uniform sample set, the activation energy (given by the slope in figure 4) may be determined to a precision as high as 0.6%, as obtained in figure 4, compared to typical values of 5-15% in the literature. The value found for the activation energy for the thermal decomposition of barium azide was $E_{act} = 37.3 \pm 0.3$ kcal/mol, and it was found that $\ln(K_\infty/s^{-1}) = 40.3 \pm 0.3$. Units of kcal are used due to their almost universal use in the literature.

To analyse the second section of figure 1, the process is very similar, with α and t reset thus:

$$\alpha' = \frac{\alpha - \alpha_1}{1 - \alpha_1} \qquad \text{and} \qquad t' = t - t_1 \qquad (4)$$

where the subscript '1' refers to the start point of the new mechanism; in the case of BAN973 this is $\alpha_1 \approx 0.78$. The analysis procedure described above is then repeated with the new data set.

Figure 3: A plot of rate of reaction versus $f(1-\alpha)$ yields the Arrhenius term $K(T)$ as the slope, which forms one point in the Arrhenius plot. The linearity of the graph gives confidence in the method.

Figure 4: Arrhenius plot for selected data, which come from consecutive experiments with similar sample characteristics. Note that the slope gives an error of only 0.6% for the activation energy.

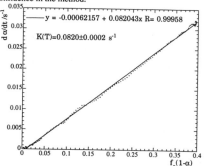

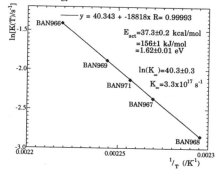

The decision as to *whether* to reset the data or leave it unreset depends on the nature of the new mechanism, and whether the new mechanism has been operative since the start of the reaction, merely dominated previously by another rate limiting process, or whether it only began part-way through the reaction. This decision must be made by the user, based on the results, and values such as the new t_0 for the time origin of the new mechanism in real time must be self-consistent.

In the case of BAN973, the data needed resetting, and to make the $\log[-\ln(1-\alpha)]$ versus $\log(t-t_0)$ plot linear, a value of $t_0=65.5$ min was obtained, which corresponds to $\alpha=0.78$ in the original DSC data, which is in excellent agreement. The slope of the new $\log[-\ln(1-\alpha)]$ versus $\log(t-t_0)$ plot was found to be $m'\approx0.95$, suggesting that the new dominant mechanism is probably unimolecular decay. This is confirmed by non-linear regression, with no other mechanisms providing a better fit. The rate of reaction versus $f(1-\alpha)$ plot (c.f. figure 3), as expected, provides a different slope, though when the Arrhenius plot is made for all the barium azide experiments, the same values for activation energy and pre-exponential factor are obtained.

DISCUSSION AND CONCLUSIONS

In the literature there are many reported values for the 'activation energy' of various processes in the decomposition of barium azide, ranging from 5-166 kcal/mol. Even within a single paper, authors report several activation energies based on different measurements. Discussion and interpretation of the activation energy in terms of physical processes in the solid is beyond the scope of this paper, for which the reader is referred to [3, 7, 11]. Whilst there are many values given for the 'activation energy' of the decomposition of barium azide, many are for observable events which are *not* the rate determining step. One energy which is relatively consistent and agreed upon is the excitation of the azide ion at ~37 kcal/mol. Hence it is suggested that this energy relates to the activation energy found in this study, which relates to the rate determining step. The mechanisms found to be operative during the decomposition were nucleation followed by unimolecular decay, in excellent agreement with the literature and visual observations [3, 7, 11], in that nuclei could be *seen* to form and grow.

The analysis method used in this paper has provided a sound theoretical basis for thermal analysis, and the results quoted have good theoretical justification. The effect of the physical mechanism operative during a reaction or transition may be isolated from the effect of the term K(T), which contains temperature, activation energy and pre-exponential factor, and is determined explicitly. This then allows the activation energy and pre-exponential factor to be determined without the interference from the functional form, with a precision much higher than generally achieved in the literature. Whilst the process is more laborious than previous methods, this inconvenience is outweighed by the improved results, and the sound theoretical justification and understanding of the results.

REFERENCES

1. W.W. Wendlandt, Thermal Analysis, 3rd ed., (Chemical Analysis 19), Wiley, 1985.

2. A.K. Galwey and M.E. Brown, Proc. Roy. Soc. London A450 p.501-512 (1995).

3. D.A. Young, Decomposition of Solids, Pergamon, Oxford, 1966.

4. T.B. Tang and M.M. Chaudhri, J. Thermal Anal. 18 p.247-261 (1980)

5. J. Šesták and J. Kratochvil, J. Thermal Anal. 5 p.193-201 (1973)

6. T.B. Tang, Thermochim. Acta 58 p.373-377 (1982);
   D.P. Smith, "Characterisation of Peracids", PhD Thesis, University of Cambridge (1995)

7. W.E. Garner, Chemistry of the Solid State, Butterworths, London, 1955.

8. T.B. Tang and M.M. Chaudhri, J. Thermal Anal. 17 p.359-370 (1979)

9. M.J. Avrami, J. Chem. Phys. 8 p212-224 (1940)

10. J.D. Hancock and J.H. Sharp, J. Amer. Ceram. Soc. 55 p.74-77 (1972)

11. H.D. Fair and R.F. Walker, Energetic Materials, Vols I & II, Plenum, New York, 1977

SOLID-PHASE THERMAL DECOMPOSITION OF 2,4-DINITROIMIDAZOLE (2,4-DNI)

LEANNA MINIER,* RICHARD BEHRENS, JR.* AND SURYANARAYANA BULUSU**
*Sandia National Laboratories, Combustion Research Facility, Livermore, CA 94551
** Energetic Materials Division, U.S. Army, ARDEC, Dover, NJ 07801-5001

ABSTRACT

The solid-phase thermal decomposition of the insensitive energetic aromatic heterocycle 2,4-dinitroimidazole (2,4-DNI: mp 265-274°C) is studied utilizing simultaneous thermogravimetric modulated beam mass spectrometry (STMBMS) between 200° and 247°C. The pyrolysis products have been identified using perdeuterated and ^{15}N-labeled isotopomers. The products consist of low molecular-weight gases and a thermally stable solid residue. The major gaseous products are NO, CO_2, CO, N_2, HNCO and H_2O. Minor gaseous products are HCN, C_2N_2, NO_2, $C_3H_4N_2$, $C_3H_3N_3O$ and NH_3. The elemental formula of the residue is C_2HN_2O and FTIR analysis suggests that it is polyurea- and polycarbamate-like in nature. The rates of formation of the gaseous products and their respective quantities have been determined for a typical isothermal decomposition experiment at 235°C. The temporal behaviors of the gas formation rates indicate that the overall decomposition is characterized by a sequence of four events; 1) an early decomposition period induced by impurities and H_2O, 2) an induction period where CO_2 and NO are the primary products formed at relatively constant rates, 3) an autoacceleratory period that peaks when the sample is depleted and 4) a final period in which the residue decomposes. Arrhenius parameters for the induction period are $E_a = 46.9 \pm 0.7$ kcal/mol and $Log(A) = 16.3 \pm 0.3$. Decomposition pathways that are consistent with the data are presented.

INTRODUCTION

The relatively new energetic aromatic heterocycle, 2,4-dinitroimidazole (2,4-DNI), has performance characteristics between that of HMX and TNT and has better thermal stability and shock and impact insensitivity than these two commonly used materials [1]. The crystal structure of 2,4-DNI (Figure 1) is thought to contribute to the enhanced stability observed for 2,4-DNI [2]. Since there is potential use for 2,4-DNI as an insensitive high explosive, characterizing the solid-phase thermal reaction chemistry is important to gain insight into the reactions controlling the release of energy from this material. Such insight can provide information required to probe correlations between molecular structure and performance, to evaluate the reactions that may affect the long-term aging of systems using 2,4-DNI, and to build predictive models for assessing the long-term stability of 2,4-DNI. To our knowledge there is little information pertaining to 2,4-DNI thermal decomposition chemistry. We have utilized simultaneous thermogravimetric modulated beam mass spectrometry (STMBMS) to characterize the mass spectrum of 2,4-DNI [3] and are now applying this method to characterize the reaction mechanisms and chemical kinetics associated with the thermal decomposition of 2,4-DNI in the solid phase. The identities and rates of formation of the gaseous reaction products are reported as well as characteristics of the solid residue formed during pyrolysis.

EXPERIMENTAL

Instrument description and data analysis.

The STMBMS apparatus used to conduct the thermal decomposition experiments and the basic data analysis procedures have been described previously [5-7]. The sample sizes used for this study are approximately 10 mg. The alumina reaction cell used to contain the samples during the experiments is typically fitted with a 25μ-diameter orifice. A nominal electron energy of 20 eV is used to minimize fragmentation of the reaction products in the mass spectrometer.

Unlabeled 2,4-DNI samples were obtained from P. Pagoria at LLNL (~ 98% pure) and used as received. 2,4-DNI isotopomers were synthesized as published.[8] Pyrolytic residues are analyzed using FTIR spectroscopy and elemental analysis. The FTIR spectra are recorded from pressed KBr pellets with 8 cm⁻¹ resolution using a Bio-Rad FT-40 FTIR spectrometer. Elemental analyses are performed by Huffman Laboratories, Inc., Boulder, CO.

RESULTS AND DISCUSSION

Solid-Phase Thermal Decomposition Characteristics

The identities of the reaction products and the temporal behaviors of their rate of formation from the isothermal decomposition of 2,4-DNI in the solid phase between 200° and 247°C can be characterized by four

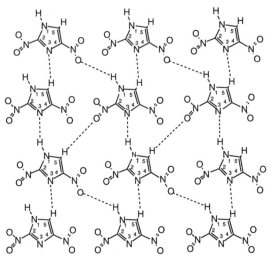

Figure 1. The hydrogen bonding in a ribbon layer of 2,4-DNI molecules. The crystal structure consists of stacked layers that are not linked by hydrogen bonds.[4]

different periods as illustrated in Figure 2. The major gaseous products formed are NO, CO_2, H_2O, HNCO, N_2, CO and HCN (Table I). Minor amounts of other gaseous products with similar temporal behaviors include C_2N_2, NO_2, NH_3, imidazole and $C_3H_3N_3O_3$. The four characteristic events of the thermal decomposition of 2,4-DNI at 235°C, illustrated in Figure 2, are: 1) an early decomposition [I]; 2) an induction period [II]; 3) an autoacceleratory period that peaks when the unreacted 2,4-DNI [III] is depleted; and 4) a slow decomposition of the polymeric residue that continues until the temperature is decreased [IV]. Absorption bands at 3296, 3156, 1748, 1628, 1572 and 1304 cm⁻¹ in the infrared spectra of the nonvolatile polymeric residue suggests that it contains urea and carbamate functionalities.

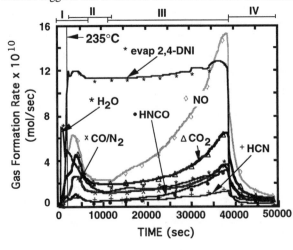

Figure 2. Gas formation rates of products from the isothermal decomposition of 10.3 mg of 2,4-DNI at 235°C. Four periods characterize the decomposition: I) early thermal decomposition, II) an induction period, III) an autoacceleration period and IV) decomposition of the residue.

Table I. Solid-phase 2,4-DNI decomposition products and product distribution at 235°C.

mass (m/z)	Product Formula (and Source)	Product (mg)	Molar Ratio[a]
17	NH_3	< 0.01	< 0.01
18	H_2O	0.17	0.45
27	HCN (N from ring)	0.06	0.10
28	CO	0.06	0.11
	N_2 (N from ring)	0.13	0.22
30	NO (N from C-2 and C-4 nitro group in ~50:50 ratio)	0.74	1.18
43	HNCO (N from ring)	0.18	0.21
44	CO_2	0.55	0.60
46	NO_2 (NO_2 mostly from C-2)	0.01	0.01
52	C_2N_2 (N from ring)	0.02	0.02
68	imidazole	< 0.01	< 0.01
129	$C_3H_3N_3O_3$ (N from ring)	< 0.01	< 0.01

2,4-DNI evaporated (mg)   7.04
2,4-DNI decomposed (mg)   3.28
residue (mg)   1.33
total mol gas/mol 2,4-DNI decomp   2.9

[a]Ratio of moles of product formed to moles of decomposed 2,4-DNI for an unlabeled sample at 235°C. Orifice diameter is 25μ.

The early decomposition period (I) involves the effect of impurities on the decomposition of 2,4-DNI whereas the induction and autoacceleratory periods (II and III) result from only the decomposition of 2,4-DNI. These three periods are illustrated in the data shown in Figure 3 for decomposition at 235°C. The temporal behaviors of the ion signals representing the gaseous decomposition products HNCO (m/z=43), H_2O (m/z=18), CO_2 (m/z=44), and a residual impurity from synthesis, 4-nitroimidazole (4-NI; m/z=113), are typical of all the solid-phase decomposition experiments. The decomposition is characterized by the five following events: 1) An immediate evolution of contaminants and impurities occurs as the sample approaches the isothermal temperature. In this case, 235°C was obtained at approximately 2200 seconds. 2) Simultaneous with 4-NI evolution is the desorption of H_2O from the 2,4-DNI. 3) An early decomposition occurs from near the onset of the isothermal temperature to approximately 4000 seconds. 4) An induction period ensues that is determined from the onset of isothermal temperature to the time that an increase in the HNCO is observed after the early decomposition is complete (~9000 seconds). 5) Following the induction period, the HNCO signal shows an accelerating rate of evolution. The initial portion of the accelerating rate can be approximated by a linear rate of increase up to approximately 15000 seconds. Then other product ion signals, such as H_2O and CO_2, start to show autoaccelerating rate as well.

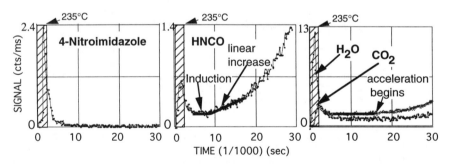

Figure 3. Ion signals representing the early evolution of the products HNCO, H_2O, CO_2 and the impurity 4-NI from the thermal decomposition of 2,4-DNI at 235°C.

Early decomposition period. The early decomposition period (I) observed near the onset of the isothermal temperature is due to many factors such as the presence of impurities and the condition of the crystal lattice. Imperfections within the lattice caused by the presence of contaminants, including adsorbed and occluded H_2O molecules, can initiate the decomposition of 2,4-DNI molecules not stabilized by the lattice. Separate studies of 4-NI thermal decomposition show that under similar experimental conditions, 4-NI undergoes substantial evaporation with little decomposition. However, the dinitroimidazole isomers of 1,4-DNI and 4,5-DNI decompose and melt at temperatures below 200°C, and the temporal behaviors of ion signals associated with the decomposition products from these dinitroimidazole isomers indicate their presence during the early decomposition period. However, these contaminants are typically less than ~1% (w/w) of the 2,4-DNI sample and cannot account for the entire early decomposition. The major contributor to the early decomposition is H_2O. The end of the early decomposition stage, shown in Figure 3, coincides with the decrease in the H_2O signal at about 4000 seconds. The H_2O signal observed during the initial stages of 2,4-DNI degradation is suspected to originate from adsorbed and occluded H_2O, not from H_2O as a thermal reaction product.

Effect of H_2O on 2,4-DNI thermal decomposition. The effect of H_2O on 2,4-DNI decomposition is important to understand since it has a measurable affect on the early decomposition process and may affect the thermal stability and long-term aging of a system that contains 2,4-DNI. To determine if H_2O is a factor in the early decomposition, the 235°C isothermal experiment was repeated with a 1.5 hour hold at 120°C to reduce the amount of any adsorbed H_2O. The resulting strong H_2O signal and absence of signals from other products during the 120°C hold, shown in Figure 4, supports the origin of water from adsorbed H_2O and not from the decomposition of 2,4-DNI at this temperature. As 235°C is reached (~6000 seconds), the H_2O signal again increases as remaining H_2O desorbs from the sample, indicating that water is still present with the material. The peak widths of the decomposition products HNCO and CO_2 are narrow, relative to the corresponding peak widths in the experiment without the 120°C hold (Figure 3), suggesting a lesser amount of these products has evolved. Similar to the evolution of products shown in Figure 3, the early HNCO and CO_2 signals decrease with the decrease in the H_2O signal. This again suggests that the early decomposition is correlated to the presence of H_2O.

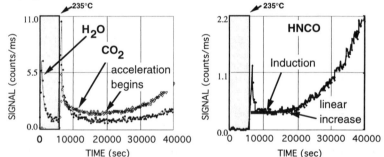

Figure 4. Ion signals representing the early evolution of H_2O, CO_2 and HNCO during the thermal decomposition of 2,4-DNI with an isothermal hold at 120°C until ~5000 seconds and a ramp to a second isothermal hold at 235°C (reached at about 6000 seconds).

We have deduced a general thermal decomposition mechanism for 2,4-DNI (Figure 5) that involves the formation of an isocyanate intermediate. The development of this mechanism is based on our decomposition studies conducted using 2,4-DNI isotopomers. Initially a nitro-nitrite rearrangement occurs yielding NO and an oxy radical. This mechanism is first order with respect to 2,4-DNI and probably occurs with the nitro group in the 2,4-DNI molecule that is not hydrogen bonded. As can be seen from Figure 1, half of the nitro groups from the C-2 position and half of the nitro groups from the C-4 position do not participate in hydrogen bonding. The equal contribution of the NO product from both positions is observed experimentally from thermal decomposition of 2,4-DNI-(2-$^{15}NO_2$) where the product ion signals for NO (m/z=30) and ^{15}NO (m/z=31) are observed to have similar temporal behaviors and intensities.

114

Furthermore, quantification of the early decomposition period at 235°C results in a molar ratio for NO to 2,4-DNI consumed of 1.05, indicating that one mole of NO is formed (two are available) per one mole of 2,4-DNI consumed. The oxy radical remaining after the formation of NO can then rearrange to form an intermediate with an isocyanate functionality (R-N=C=O). The isocyanate functionality is stable, but reactive, and readily reacts with H_2O to yield CO_2. As the integrity of the crystal structure is disturbed from the early events, the H_2O trapped in the crystal lattice has an increased opportunity to interact with available isocyanates. After the impurities are depleted, including the desorbed and occluded H_2O, the early decomposition event ceases and decomposition of pure 2,4-DNI in the induction period continues. Based on initial evaluations, the duration of the induction period does not appear to be changed significantly by the amount of the early decomposition that occurs, being approximately 7500 seconds for the 235°C isothermal experiments represented in Figures 3 and 4. Further experiments to evaluate the effect of H_2O on the early decomposition of 2,4-DNI are presently being conducted.

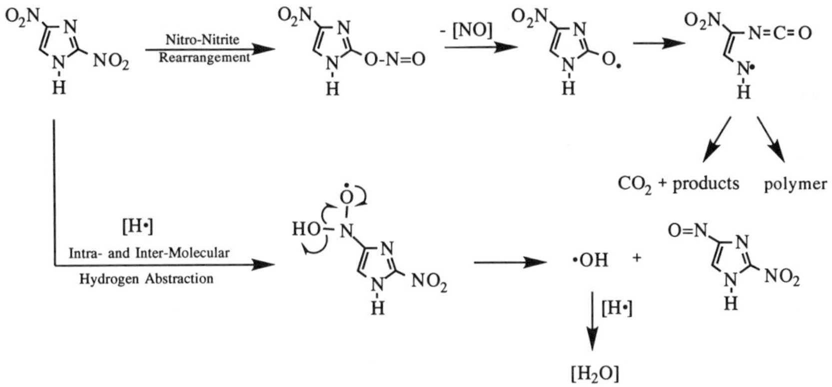

Figure 5. Initial mechanisms for 2,4-DNI condensed-phase thermal decomposition.

Induction and autoacceleration period characteristics. Decomposition intermediates accumulate during the induction period. These intermediates appear to be prepolymeric "subunits" that readily combine with one another at an accelerating rate as the crystal lattice weakens. The low-molecular weight gaseous products are produced as the prepolymeric subunits form and combine with each other. A number of different reactions are likely to occur during this period and two examples are presented in Figure 6. Formation of the isocyanate intermediate (Figure 5) is a likely principal intermediate. However, other intermediates that have amino and hydroxy functionalities, as well as radical intermediates, are likely to form simultaneously and consecutively with the isocyanate intermediate. The intermediates either decompose further or combine with another intermediate. The isocyanate functionality readily reacts with hydroxy and amine groups to form predominately carbamates (R-NH-C(O)-OR') and ureas (R-NH-C(O)-NH-R'), respectively. During the early stages of the decomposition the intermediates can not readily interact with each other due to constraints of the lattice. As dissolution of the lattice increases, more nitro groups are free from hydrogen bonding and can undergo nitro-nitrite rearrangement and form the intermediates that become prepolymer subunits. Consequently, higher concentrations of intermediates lead to increased rates of interaction, which result in the observed autoacceleratory behavior and the formation of the polymeric residue. From the FTIR analysis, the polymeric residue is found to contain functional groups that show absorption bands at frequencies that are characteristic of compounds containing urea and carbamate functionalities. The autoacceleration period continues until one of the primary building blocks, 2,4-DNI, is depleted. As the polymerization occurs, gaseous products with low molecular weights (see Table I) are released. If the experiment is allowed to continue after the 2,4-DNI is depleted, the polymeric residue continues to rearrange in some fashion while continuously releasing low-molecular weight products.

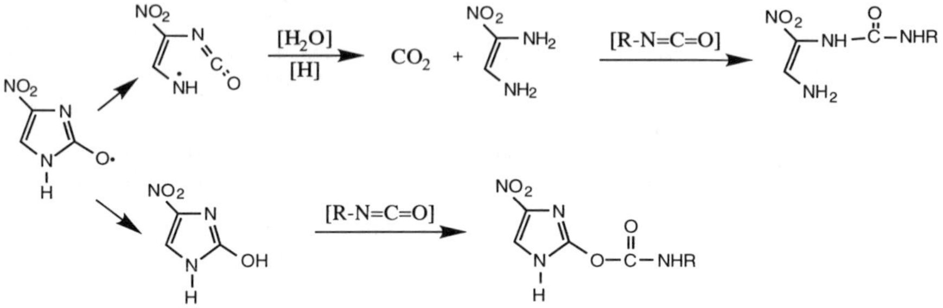

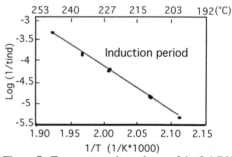

Figure 6. Example of likely reactions occurring during the induction and autoacceleratory periods of 2,4-DNI solid-phase decomposition.

Induction Period Kinetics

Assuming that the early decomposition does not affect the normal thermal degradation of pure 2,4-DNI, the induction period, obtained from the temporal behavior of HNCO as a function of temperature (Table II), may be used to obtain kinetic parameters. The temperature dependence is shown in Figure 7. The Arrhenius parameters are Ea = 46.9 ± 0.7 kcal/mole and Log(A) = 16.3 ± 0.3. These parameters can be used to evaluate the duration of the induction period at specific temperatures. For example, at the upper limit of the storage temperature of the stockpiled munitions, 70°C, the duration of the induction period for 2,4-DNI is 1×10^6 years, as extrapolated from the data. This value reflects the high thermal stability of 2,4-DNI. However, the duration of the induction period may be influenced by other factors, such as material compatibility or containment issues, not considered in this study.

Table II. 2,4-DNI decomposition data.

Number	Temp (°C)	Orifice Diameter (μm)	Induction time (sec)
1	200.4	10	212000
2	209.8	25	70000
3	210.5	25	68000
4	225.4	25	17100
5	225.0	25	16000
6	235.2	25	6850
7	235.2	25	6600
8	246.9	25	2250

Figure 7. Temperature dependence of the 2,4-DNI induction period.

CONCLUSIONS

The thermal decomposition of 2,4-DNI in the solid-phase between 200° and 247°C is characterized by the following four events. 1) An early decomposition period occurs near the onset of the isothermal temperature that is attributed to the presence of impurities remaining from the synthesis and from adsorbed and occluded H_2O in the 2,4-DNI crystal lattice. 2) When the impurities are depleted, an apparent induction period is observed that is defined to be from the onset of the isothermal temperature to time of the increase of the HNCO signal. 3) The induction period is followed by an autoaccelerating rate period. The initial accelerating rate of this period can be approximated with a linear rate of increase. The maximum autoaccelerating sample. 4) The solid residue that forms during the decomposition continues to release low-molecular weight gases until the experiment is stopped.

The presence of adsorbed and occluded H_2O in the crystal structure of 2,4-DNI increases the rate of thermal decomposition of 2,4-DNI. The increase of the ion signals from the early decomposition products , CO_2 and HNCO, occurs shortly after the increase of the ion signal from adsorbed H_2O . These ion signals then decrease with the decreasing H_2O ion signal. Partial dehydration of the 2,4-DNI crystal before pyrolysis results in reduction in the amount of the early decomposition products. However, the presence of H_2O impurity does not significantly affect the induction time at 235°C. Further studies are being conducted to evaluate the effect of H_2O in detail.

A set of reaction pathways have been developed that are consistent with the data. Initially a nitro-nitrite rearrangement leads to the formation of NO and an oxy radical intermediate. This reaction is first order relative to 2,4-DNI and the rearrangement occurs at the nitro group that is not hydrogen bonded in the 2,4-DNI lattice. The oxy radical intermediate rearranges to form an isocyanate functionality. The isocyanate functionality is stable but reactive and, in the presence of H_2O, readily reacts to produce CO_2. This pathway dominates the early decomposition observed for 2,4-DNI and ceases when adsorbed and occluded H_2O is depleted. During the induction period the isocyanate intermediates and other intermediates with amine and hydroxy functionalities accumulate, producing low-molecular weight gases as they form and react. The intermediates readily combine with isocyanate functionalities and result in the formation of a polymeric residue that is polyurea- and polycarbamate-like in character. The buildup of the polymeric residue occurs at an increasing rate due to dissolution of the crystal lattice, thus increasing the interaction frequency between the intermediates, and accounting for the observed increase in reaction rate.

The Arrhenius parameters for the induction period are Ea = 46.9 ± 0.7 kcal/mole and Log(A) = 16.3 ± 0.3. These parameters are based on the assumption that the early decomposition does not affect the kinetics of pure 2,4-DNI thermal decomposition. Using the parameters to extrapolate to 70°C, an induction time of 1×10^6 years is obtained. The long induction time is a reflection of the high thermal stability of 2,4-DNI. However, the induction time may be affected by other factors relating to compatibility and containment issues and by factors influencing the early decomposition period.

Further studies are being conducted to evaluate the early decomposition in detail. A more detailed analysis of 2,4-DNI solid-phase decomposition will be presented in a future paper.

ACKNOWLEDGMENTS

The authors thank D.M. Puckett for experimental assistance and K. Erickson for private discussions. This work is supported by a Memorandum of Understanding between the U.S. Department of Energy and the Office of Munitions, by the U.S. Department of Energy under contract DE-AC04-94AL85000 and by the U. S. Army, ARDEC.

REFERENCES

1. Jayasuriya, K.; Damavarapu, R.; Simpson, R. L.; Coon, C. L.; Coburn, M. "2,4-Dinitroimidazole: A Practical Insensitive High Explosive?," Lawrence Livermore National Laboratory, UCRL-ID-1133G4, March 12 (1993).
2. R. Simpson, private communication.
3. Minier, L. M..; Behrens, R., Jr.; Bulusu, S. *J. Mass Spec.*, in print.
4. Gilardi, R. and Foltz, F., private communication on the crystal structure of 2,4-DNI.
5. Behrens, R., Jr. *Rev. Sci. Instrum.*, *58*, 451-461 (1987).
6. Behrens, R., Jr. *Inter. J. Chem. Kinetics*, *22*, 135-157 (1990).
7. Behrens, R., Jr. *Inter. J. Chem. Kinetics*, *22*, 159-173 (1990).
8. Bulusu, S.; Damavarapu, R.; Autera, J. R.; Behrens, R. Jr.; Minier, L. M.; Villaueva, J.; Jayasuriya, K.; Axenrod, T. *J. Phys. Chem.* , *99*, 5009-5015 (1995).

THERMAL DECOMPOSITION OF HMX: LOW TEMPERATURE REACTION KINETICS AND THEIR USE FOR ASSESSING RESPONSE IN ABNORMAL THERMAL ENVIRONMENTS AND IMPLICATIONS FOR LONG-TERM AGING

RICHARD BEHRENS* AND SURYANARAYANA BULUSU**
*Sandia National Laboratories, Combustion Research Facility, Livermore, CA 94551
** Energetic Materials Division, U.S. Army, ARDEC, Dover, NJ 07801-5001

ABSTRACT

The thermal decomposition of HMX between 175°C and 200°C has been studied using the simultaneous thermogravimetric modulated beam mass spectrometer (STMBMS) apparatus with a focus on the initial stages of the decomposition. The identity of thermal decomposition products is the same as that measured in previous higher temperature experiments. The initial stages of the decomposition are characterized by an induction period followed by two acceleratory periods. The Arrhenius parameters for the induction and two acceleratory periods are (Log(A)= 18.2 ± 0.8, Ea = 48.2 ± 1.8 kcal/mole), (Log (A) = 17.15 ± 1.5 and Ea = 48.9 ± 3.2 kcal/mole), (Log (A) = 19.1 ± 3.0 and Ea = 52.1 ± 6.3 kcal/mole), respectively. The data can be used to calculate the time and temperature required to decompose a desired fraction of a test sample that is being prepared to test the effect of thermal degradation on its sensitivity or burn rates. It can also be used to estimate the extent of decomposition that may be expected under normal storage conditions for munitions containing HMX. The data, along with previous mechanistic studies conducted at higher temperatures, suggest that the process that controls the early stages of decomposition of HMX in the solid phase is scission of the $N-NO_2$ bond, reaction of the NO_2 within a "lattice cage" to form the mononitroso analogue of HMX and decomposition of the mononitroso HMX within the HMX lattice to form gaseous products that are retained in bubbles or diffuse into the surrounding lattice.

INTRODUCTION

Understanding the thermal decomposition of HMX (I) is important for developing models to predict the response of munitions in abnormal thermal environments and to obtain an understanding of the chemical processes that control its long-term degradation.

The objective of this work is twofold. First, we seek to develop a model that can be used to estimate the extent of thermal decomposition of HMX as a function of time and temperature. This model can be used to guide experiments that are aimed at determining the response of degraded explosives or propellants that contain HMX. For example, to predict the violence of reaction that occurs after ignition in a 'slow cookoff' event, it is important to know the high pressure burn rate characteristics of the degraded materials. These results will provide information on the state of the thermally degraded material just prior to the ignition event. Second, we wish to ascertain the extent and characteristics of decomposition that may occur at lower temperatures which could affect the long-term aging of components containing HMX. In this work the temperature dependence of the global rates of decomposition of δ-HMX will be determined and the results will be used to determine the lower limit on the rate of decomposition at storage conditions assuming that the reaction kinetics of δ-HMX are similar to that of β-HMX.

INSTRUMENTATION AND METHODS

A description of the STMBMS apparatus and the experimental procedures that are used in this study have been described previously.[1,2] For this study, HMX was decomposed in reaction cells using orifice diameters ranging from 10 to 100 μ and 5 to 10 mg samples of HMX that have a purity of greater than 99%. The global rate of decomposition was extracted from the weight loss data measured with a microbalance after correction for evaporation of HMX.

HMX THERMAL DECOMPOSITION PROCESS

Using the described method, we have previously conducted extensive studies on the thermal decomposition of HMX[3,4] and RDX[5,6] in an effort to determine the reaction mechanisms that control the decomposition of these two cyclic nitramines.

PREVIOUS RESULTS

For RDX in the liquid phase (m.p. 200°C), the gaseous products have been identified and their rates of formation have been measured and from these results a reaction mechanism consisting of four parallel reaction pathways were found to control the decomposition of RDX in the liquid phase. Details of the RDX decomposition mechanism have been published.[5,6] In the solid phase, RDX was found to decompose through an intermediate that is the mononitroso analogue of RDX, hexahydro-1-nitroso-3,5-dinitro-s-triazine (ONDNTA).

In contrast to RDX, HMX liquefies between 270° and 280°C and undergoes substantial decomposition in the solid phase. The identity and rates of formation of the products formed in the thermal decomposition of HMX have been determined and a qualitative model describing the process has been developed.[3] The thermal decomposition process of HMX is illustrated with the data shown in Figure 1. The main decomposition products are H_2O, N_2O, CH_2O, NO and CO. The temporal behavior of the gas formation rates of each of the products is characterized by the following sequence: 1) an induction period, 2) an increasing rate of formation between 0 and 10% decomposition, 3) a faster increasing rate of decomposition between 10% and 30% decomposition and a decreasing rate of gas formation after 40% decomposition. Products formed at smaller rates of formation but with similar temporal behaviors include: $(CH_3)_2NNO$, the mononitroso analogue of HMX (1-nitroso-3,5,7-trinitro-1,3,5,7-tetrazocine) and several formamides. The temporal behavior of the decomposition process during the induction period is illustrated with data from the decomposition of a deuterium labeled analogue of HMX, as shown in Figure 2. From this figure one observes a period of time in which the rates of formation of N_2O and CD_2O are low and constant with time. The time at which the gas formation rates first reach a constant value coincides with attaining an isothermal temperature of 235°C in the sample. The time period between the time that the isothermal temperature is reached and when the N_2O signal first starts to increase is referred to as the <u>induction</u> period. The gas formation rates of N_2O and CH_2O during the induction period arise from the decomposition of HMX in either the gas phase or on the surface of the particles.

During the induction period of HMX the material is somehow transformed to a new state that undergoes more rapid decomposition. This transformation could be a nucleation process as observed in the decomposition of many inorganic salts in which nuclei are formed and destroyed

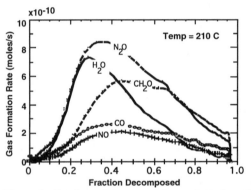

Figure 1. Gas formation rates of the products from the thermal decomposition of HMX at 210°C. The duration of the decomposition process is about 37000 seconds.

while those that reach a certain critical diameter form the basis for the accelerating rate of decomposition as the nuclei grow. The transformation may also be due to the slow decomposition of the HMX within the lattice and the permeation of its gaseous products through the crystal structure, weakening the crystal structure, and leading to the release of gaseous products and the more rapid decomposition of noncrystalline material. This latter mechanism is consistent with the observation of the mononitroso analogue of RDX in the decomposition of RDX in the solid phase and the delay time observed for the onset of rapid decomposition during its decomposition as discussed in a previous paper.[5] It is important to note that the HMX must be undergoing some significant change during the induction period. The acceleratory phase of the decomposition process marks a transition to a more rapid decomposition process. During this phase the HMX undergoes a more rapid rate of decomposition.

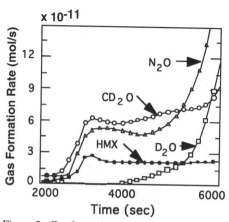

Figure 2. Gas formation rates of thermal decomposition products from HMX-d8 during the induction period and first stage of the acceleratory period.

Our previous work on the decomposition of HMX in the solid phase was conducted over a temperature range from 210° to 235°C. This work was extended to higher temperatures to investigate the qualitative behavior of HMX decomposition in the liquid phase.[7] The work covered in this paper extends the thermal decomposition measurements to lower temperatures. Since the objectives of the study are to obtain data to characterize the state of HMX under abnormal thermal conditions and to investigate the behavior of HMX during long-term aging, it is necessary to measure only the induction period and its early stages in particular, in these studies.

NEW HMX LOWER TEMPERATURE DATA

Experiments on the decomposition of HMX between 175° and 200°C have been conducted to add to the database on its decomposition behavior. The general behavior is quite similar to that observed in the previous work as can be seen from Figure 3 for the decomposition of HMX at 182°C. The ion signals representing the evolution of N_2O and CH_2O are very similar to the data collected at higher temperatures. In this case the induction period, as determined by the time at which the N_2O signal first increases, is approximately 60,000 seconds. As the N_2O signal increases, it is characterized by two different sets of increasing rates, with the first set being less than the second. This is typical of all of the data collected in the different experiments at lower temperatures.

Comparing the induction time of N_2O to CH_2O (Figure 3) shows that the increase in the CH_2O rate of evolution lags that of N_2O by about 40,000 seconds. This behavior is similar to that observed at higher temperatures. Several other products also observed in the higher temperature experiments are dimethylnitrosamine ($m/z=74$), formamide ($m/z=45$), H_2O and CO. The temporal behaviors of these products are similar to those observed at higher temperatures.

Since there is a strong correlation between the identities and temporal behaviors of the products in the new lower temperature experiments and those of the previous higher temperature experiments, it is possible to determine the temperature dependence of the first stages of the thermal decomposition of HMX. The experimental parameters and data collected from experiments with HMX are listed in Table 1. The experiments range in temperature from 175° to 200°C (data from reference 4 is also included in Table 1). HMX samples with mean particle diameters of 150 and 600 μm were used in the experiments. The diameter of the reaction cell orifice was also varied from 10 μm to 100 μm in different experiments. Although there may be an effect due to the extent of gas product containment, which is controlled by the orifice diameter, and particle size, these effects are small compared to the variation in the induction times and rates of reaction with temperature.

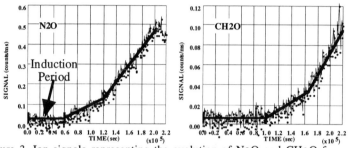

Figure 3. Ion signals representing the evolution of N_2O and CH_2O from the thermal decomposition of HMX at 182°C.

Table I. HMX THERMAL DECOMPOSITION DATA

	Temp (°C)	Particle diameter (μm)	Orifice Diameter (μm)	Induction time (sec)	Decomposition rate constant (sec^{-1}) First	Second
1/2*	175	600	25	270000	2.85E-07	5.09E-07
3	182	600	10	64000	3.97E-07	2.30E-06
4	182	150	10	88000	4.94E-07	9.29E-07
5	182	150	100	62000		
6	200	150	100	9000	3.83E-06	8.77E-06
7	210	150	100	2600	1.30E-05	4.50E-05
8	226	150	100	800		
9	235	150	100	370		

* Experiment carried out in two parts. Sample was heated and held at 175°C for 220,000 seconds, cooled and removed from the apparatus for 5 days, and then again heated and held at 175°C for 220,000 seconds.

INDUCTION PERIOD. The temperature dependence of the induction period is shown in Figure 4. The Arrhenius parameters calculated from the data are Log(A)= 18.2 ± 0.8 and Ea = 48.2 ± 1.8 kcal/mole. The variation in the results from the three experiments conducted at 182°C using different particle sizes and different orifice diameters is shown by the data at 1/T = 0.0022 in Figure 4. The two points shown at 1/T=0.00223 sec^{-1} are two different estimates for the induction time from experiment 1/2 in which the sample was heated and held at 175°C for 220000 seconds, cooled and removed from the apparatus for five days, and then reheated and held at 175°C for another 220000 seconds. The shorter induction time was taken from the data collected in the first heating cycle and the longer induction time was taken from the second heating cycle. The acceleratory phase of the decomposition was not readily apparent in the first heating cycle and the data point was based on the first apparent deviation of the N_2O signal from a constant value. The data from the second heating cycle showed a clear acceleratory period and the induction time is calculated as the sum of the time heated in the first cycle and the normal induction time measured in the second cycle.

The data from the experiment with the two heating cycles clearly show that the HMX sample retains the degradation effects of prior heating. Thus, from this initial data, it appears that the effects of the degradation process that occur during the induction period are irreversible. However, further experiments need to be performed to test this more thoroughly.

ACCELERATORY PERIOD. The acceleratory period that commences at the end of the induction period is characterized by two different acceleration rates as illustrated by the N_2O data shown in Figure 3. The rate constants, derived from averaging the weight loss data measured with the microbalance of the STMBMS over the duration of each acceleratory period, are listed in Table I. There is some variation in the rate constant calculated for the 2nd acceleratory phase due to the measurement being made on different segments of the acceleratory curve. The weight loss data were corrected for evaporation of HMX in the experiments using the reaction cell with the 100 μm diameter orifice. This correction was not necessary in the experiments using the smaller diameter orifices due to the small contribution from evaporation.

The temperature dependence of the 1st and 2nd acceleratory periods is shown in Figure 5. The Arrhenius parameters for the 1st acceleratory period are Log (A) = 17.15 ± 1.5 and Ea = 48.9 ± 3.2 kcal/mole. The Arrhenius parameters for the 2nd acceleratory period are Log (A) = 19.1 ± 3.0 and Ea = 52.1 ± 6.3 kcal/mole.

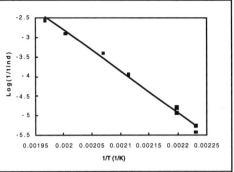

Figure 4. Temperature dependence of the induction period for the thermal decomposition of HMX.

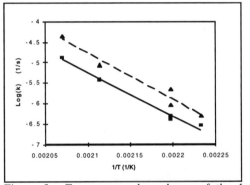

Figure 5. Temperature dependence of the 1st (squares, solid line) and 2nd (triangles, dashed line) acceleratory periods of the HMX decomposition.

MODEL TO PREDICT LOWER TEMPERATURE BEHAVIOR FOR HMX

The data from the induction and the two acceleratory periods can be used to estimate the extent of decomposition as a function of sample temperature and time. It can also be used to estimate the extent of degradation at lower temperatures over long periods of time.

The fraction, Fr, of HMX that has decomposed up to a given point in time, t, when held at a temperature, T, is given by

$$Fr = \left[1 - Exp(-k_1 t_{d1})\right] + \left[1 - Exp(-k_2 t_{d2})\right]$$

where the rate constants for the 1st (k_1) and 2nd (k_2) acceleratory periods are given by

$$Log(k_1) = 17.14 - \frac{10664}{T}$$ and $$Log(k_2) = 19.12 - \frac{11374}{T},$$

the time in each acceleratory period is given by

$$t_{d1} = Min(t - t_{ind}) t_f$$ and $$t_{d2} = Min(t - t_{ind} - t_f, t_{fm}),$$

the induction time is given by

$$Log \, (1/t_{ind}) = 18.23 - \frac{10522}{T},$$

and t_f is the time of the transition from the first to the second acceleratory region and is given by

$$t_f = -Ln \, (1 - f_t) / k_1$$

where f_t is the fraction of decomposition at which there is a transition from the 1st to the 2nd acceleratory phase, and t_{fm} is the time associated with the maximum fraction of decomposition, f_m, that this model is developed to predict and is given by

$$t_{fm} = -Ln \, (1 - f_m + f_t) / k_2$$

The transition fractions used in the model are $f_t = 0.05$ and $f_m = 0.2$. The value for f_t is determined from the break points between the 1st and 2nd acceleratory periods measured in the experiments. The value of 0.2 for f_m is selected because this approximates the amount of decomposition that occurs through the acceleratory phase.

The fraction of decomposition as a function of time at several different temperatures is shown in Figure 6. For example, to achieve 5% decomposition requires the sample to be maintained at 190°C for approximately 19 hours.

HMX LONG-TERM DECOMPOSITION

The length of time predicted for the duration of the induction period and the point at which 0.01% of the sample has decomposed based on the rate constants for the first and second acceleratory periods is shown in Figure 7 for a set of temperatures varying between 20° and 130°C. The data in the table show that if the sample is maintained at 120°C the induction period will last for 10 years before gas starts to be released in the first acceleratory period. Whereas, if the reaction kinetics that control the decomposition were based on

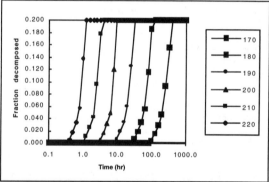

Figure 6. Fraction of HMX decomposed as a function of time and temperature (°C).

the reaction rates that characterize the two acceleratory periods, then the sample only needs to be maintained between 80° and 90°C to undergo 0.01% decomposition in ten years. This illustrates the importance of the process that occurs in the induction period in contributing to the long-term stability of HMX. For example, at 70°C the induction period would last 85,000 years, whereas, decomposition controlled by solely the acceleratory period reactions would occur over 200 to 300 years.

Clearly, the data suggest that under normal storage conditions (< 70°C) HMX should be stable over the lifetime of any component in which it is used, based on these decomposition measurements of pure HMX. However, there are several other points that should be considered regarding the decomposition of HMX. First, it is possible that reactions may occur between HMX and other materials that may come in contact with the particle surface. This may occur at lower temperatures than the decomposition of HMX by itself.

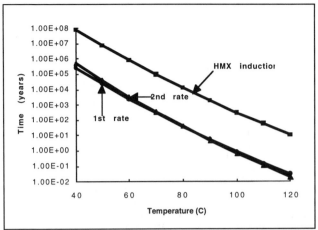

Figure 7. Modeled thermal stability curves of the induction period and constant rate periods for HMX.

These interactions may present a long-term aging issue. Second, during the induction period HMX must undergo some significant changes. These changes may be chemical, physical or both. It is unlikely that they would affect the combustion or detonation behavior of these materials, since no significant change in the energy content of the material is observed. However, these changes may affect the thermal or impact sensitivity if they cause significant changes in the way "hot spots" are formed in the material. Third, the decomposition processes that we have studied above 210°C also occur at lower temperature. Thus, it is reasonable to assume that the details of the reaction processes that we have observed at higher temperatures also occur at lower temperatures (reaction in the lower temperature ß-HMX may differ from the higher temperature ∂-HMX measured in these studies). This process which was observed in the solid phase decomposition of RDX and HMX is the formation of the mononitroso analogue of the parent compound. From our previous data we believe that the mononitroso analogue is formed within the solid lattice. The conversion of HMX to its mononitroso analogue occurs through scission of the $N-NO_2$ bond and reaction of the NO within the surrounding "lattice cage". This is consistent with the results of our isotopic crossover experiments using ^{15}N labeled HMX, which showed that 75% of the mononitroso analogue of HMX formed in the experiment did not participate in isotopic scrambling. It is also consistent with the activation energy of 48.2 ± 1.8 kcal/mole for the induction period measured in this study which is very close to the $N-NO_2$ bond energy of approximately 45 kcal/mole. From our thermal decomposition studies of the mononitroso analogue of RDX we have found that it is significantly less stable than RDX and its main decomposition products are N_2O and CH_2O. This is quite consistent with the products observed in the decomposition of HMX in this study. Thus, from these results, it appears that the process that occurs in the solid phase at low temperature may also be scission of the $N-NO_2$ bond, reaction of the NO_2 within a "lattice cage" to form the mononitroso analogue of HMX, decomposition of the mononitroso HMX within the HMX lattice to form gaseous products that are retained in bubbles or diffuse into the surrounding lattice. These products may affect the sensitivity of HMX significantly.

CONCLUSIONS

The decomposition of HMX between 175° and 210°C exhibits the same behavior as HMX above 210°C. The identity and temporal behavior of the products formed in the lower and higher temperature ranges are similar. Thus, the results of mechanistic studies conducted at the higher temperature range are applicable to the reactions at lower temperatures.

The early stages of the decomposition of HMX can be divided into three segments, an induction period followed by two acceleratory periods. The end of the induction period is characterized by the increase in the rate of N_2O evolving from the sample. The increase in the rate of evolution of CH_2O always lags that of N_2O, suggesting that CH_2O is initially retained in the HMX lattice. Other products such as formamide and dimethylnitrosamine are also observed.

The A-factors and activation energies for the induction period and two acceleratory periods are (Log(A)= 18.2 ± 0.8, Ea = 48.2 ± 1.8 kcal/mole), (Log (A) = 17.15 ± 1.5 and Ea = 48.9 ± 3.2 kcal/mole), (Log (A) = 19.1 ± 3.0 and Ea = 52.1 ± 6.3 kcal/mole), respectively. These activation energies are similar to $N-NO_2$ bond energy in HMX.

Using the Arrhenius parameters determined from the measurements, estimates of the length of time that an HMX containing sample must be held at a given temperature to achieve a desired fraction of decomposition can be determined. This data can be used for preparing samples for testing the effects of thermal degradation on sensitivity and burn rates.

Extrapolating the HMX thermal decomposition results to lower temperatures shows that the degradation is not extensive at the upper temperature limit of the normal stockpile to target sequence, 70°C. However, raising the temperature to 120°C would have a significant effect on the material if it behaves as measured in the present experiments. If for some reason the stability attributed to the induction period were lost and the rate of reaction were controlled by the reactions occurring in the acceleratory phase, then raising the temperature to 85°C would cause significant changes in the material.

The similarity of the results of the lower temperature experiments carried out in this study to our previous mechanistic studies of HMX and RDX at higher temperature allows us to suggest that the following process probably occurs in the early stages of the solid phase decomposition of HMX: scission of the $N-NO_2$ bond, reaction of the NO_2 within a "lattice cage" to form the mononitroso analogue of HMX and decomposition of the mononitroso HMX within the HMX lattice to form gaseous products that are retained in bubbles or diffuse into the surrounding lattice.

Further studies are needed to accurately characterize the effects of particle size, gas containment, and cycling the sample temperature on the thermal decomposition process. However, several experiments conducted in the present study, in which these parameters were varied, indicate that these effects are secondary compared to the variation due to temperature.

ACKNOWLEDGMENTS

The authors wish to thank D.M. Puckett for assistance in collecting the mass spectrometry data. This work was supported by the Memorandum of Understanding between the Office of Munitions and the U.S. DOE, the U.S. Department of Energy under Contract DE-AC04-94AL85000, and the U.S. Army, ARDEC.

1 R. Behrens, Jr.; Rev. Sci. Instrum. **58**, 451-461, 1987.
2 a) R. Behrens, Jr.; Int. J. Chem. Kin. **22**, 135-157, 1990. b) R. Behrens, Jr.; Int. J. Chem. Kin. **22**, 159-173, 1990.
3 R. Behrens, Jr.; J. Phys. Chem. **94**, 6706-6718, 1990.
4 R. Behrens, Jr. and S. Bulusu; J. Phys. Chem. **95**, 5838-5845, 1991.
5 R. Behrens, Jr. and S. Bulusu; J. Phys. Chem. **96**, 8877-8891, 1992.
6 R. Behrens, Jr. and S. Bulusu; J. Phys. Chem. **96**, 8891-8897, 1992.
7 R. Behrens, Jr. and S. Bulusu; Mat. Res. Soc. Symp. Proc., **296**, p. 18-24, 1993.

FREE RADICALS FROM PHOTOLYSIS OF (NTO) 5-NITRO-2,4-DIHYDRO-3H-1,2,4-TRIAZOL-3-ONE STUDIED BY EPR SPIN TRAPPING

M. D. Pace[*], L. Fan, and T. J. Burkey
[*]Code 6120, Naval Research Laboratory, Washington, DC 20375
Department of Chemistry, The University Of Memphis, Memphis, TN 38152

ABSTRACT

NTO (5-Nitro-2,4-dihydro-3H-1,2,4-triazol-3-one) is shown to form a photochemical free radical in alkaline aqueous solution containing nitromethane (CH_3NO_2). The NTO free radical can also be produced by chemical reduction of NTO with hydroquinone in alkaline solution. Details of radical formation and assignment of the NTO radical to a nitro anion radical, having the NTO ring intact, are presented.

INTRODUCTION

NTO (5-Nitro-2,4-dihydro-3H-1,2,4-triazol-3-one) is a new energetic material which is of interest as an insensitive munition, burn-rate additive, and as an explosive mixture with TNT (trinitrotoluene).[1,2,3] Free radicals produced by photochemical and thermal NTO decomposition have been studied by Menapace and coworkers.[4] Their findings show that NTO abstracts hydrogen from solvent molecules or other NTO molecules during ultraviolet photolysis to produce hydroxy-4-oxo-2,3,4-triazolyl nitroxide radicals. NTO synthesis has been carried out by Burkey.[5] This study investigates the photochemistry of NTO in strong alkaline solution. Strongly basic solution has been shown to decompose nitramine explosives such as cyclotrimethylenetrinitramine. For example, the nitromethane *aci*-anion EPR (electron paramagnetic resonance) spin trapping technique has proven effective in detecting nitrogen dioxide ($NO_2^\bullet$) radicals derived from photodecomposition of nitramines in aqueous solutions at high pH.[6] From the spin concentrations of radical adducts, the percent decomposition of nitramines by strong base can be estimated. The nitromethane *aci*-anion (*aci*-nitromethane) spin trapping reaction is as follows[7]:

$$CH_3\text{-}NO_2 \quad \overset{OH^-}{\rightleftarrows} \quad CH_2=NO_2^- \ + \ H_2O \tag{1}$$
$$\text{nitromethane} \qquad \textit{aci}\text{-nitromethane}$$

$$CH_2=NO_2^- \ + \ R^\bullet \ \rightarrow \quad R\text{-}HC^\bullet\text{-}NO_2^- \tag{2}$$
$$\text{adduct radical}$$

This reaction has been applied to trap many radical transients ($R^\bullet$) other than $NO_2^\bullet$.[8] In this study, we have applied reactions (1) and (2) to study NTO. The results are surprising because an NTO ring-intact anion radical, rather than a simple radical adduct, is formed by reaction of NTO with *aci*-nitromethane. The free radical product is shown to be an NTO reduction product from electron transfer between *aci*-nitromethane and NTO.

EXPERIMENTAL

NTO was received from the Naval Ordnance Center, Indian Head, Maryland; ^{15}N-labeled NTO and TO (2,4-dihydro-3H-1,2,4-triazol-3-one) were synthesized by T. Burkey and coworkers. The ^{15}N-labeled compound is labeled on the two adjacent ring nitrogens as shown in Fig. 3. Nitromethane (CH_3NO_2), deuteronitromethane-d$_3$ (CD_3NO_2), sodium hydroxide (NaOH), dimethylsulfoxide ($(CH_3)_2SO$), and hydroquinone ($C_8H_6O_2$) were purchased (Aldrich Chemicals and Norell Chem. Co.). A detailed explanation of the spin trapping reactions has been given.[6] The procedures are as follows. A solution of NaOH (pH > 12) is prepared and the NTO is dissolved into strong alkaline aqueous solvent with the NTO concentration $ca.$ 10^{-4} mol L^{-1}. To 1 mL of this solution is added $ca.$ 1.6×10^{-4} mol of nitromethane which forms the aci-nitromethane as in equation 1. The solution is pipetted into a 250 μL flat cell and photolyzed with ultraviolet (UV) light in $situ$ in the EPR cavity. Unfiltered UV light from a continuous wave lamp source (Hanovia) was used to initiate free radical reactions. All experiments were carried out at room temperature. EPR spectra were recorded using a Bruker ER 200 spectrometer. Signal averaged scans, of typically 20 s per scan, were accumulated to improve spectral signal-to-noise. The EPR signals were observed within minutes of the start of photolysis. The EPR signal of the free radical photochemical reaction product was detected only in solutions containing NTO and nitromethane. Uv/vis spectra were recorded using a Spectronic 2000 (Bausch and Lomb) spectrometer.

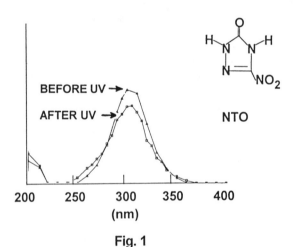

Fig. 1

UV/Vis Spectra Of NTO/Nitromethane

RESULTS

NTO/Aci-nitromethane

Evidence of photochemical NTO reaction is given in Fig. 1 which shows UV/vis spectra recorded before and after photolysis of a 1 mL alkaline solution containing NTO and nitromethane. After 10 minutes of UV photolysis, the solution shows a proportionate decrease of the NTO peak (305 nm) and the nitromethane peak (near 200 nm). This is consistent with reaction of NTO to form a new product. EPR shows that a *free radical* product is photochemically generated.

In Fig. 2a the EPR spectrum from the photolysis of NTO in aqueous NaOH with *aci*-nitromethane is shown. This spectrum has a width of 4.1 mT (1mT = 10 gauss) and is composed of a triplet of quartets. The triplet splitting is due to a nitrogen hyperfine coupling of $a(^{14}N) = 1.495$ mT. The quartet spectral components are not resolved in Fig. 2a due to the broad sweep width necessary to record the entire spectrum. The low field quartet is shown on an expanded scale in Fig. 2b. A modulation amplitude of 0.012 mT was required to resolve the spectral lines. This complicated spectrum is due to three inequivalent nitrogen couplings. The assignments of the three couplings are $a(^{14}N) = 0.31$ mT, $a(^{14}N) = 0.13$ mT, and $a(^{14}N) = 0.05$ mT. The simulated fit is demonstrated using these hyperfine couplings as shown in Fig. 2c. The presence of *four* nitrogen couplings and no proton hyperfine couplings was surprising because *aci*-nitromethane usually forms a radical adduct of type R-HC$^{\bullet}$-NO$_2^-$ as given in equation 2. ^{15}N-labeled NTO was used to investigate this and to determine two of the coupling assignments.

The ^{15}N-labeled NTO compound was labeled on the two adjacent nitrogens of the NTO ring. The EPR spectrum from UV photolysis of this sample with *aci*-nitromethane in alkaline solution is shown in Fig. 3a. This spectrum has a width of *ca.* 3.62 mT. With a 10 mT sweep the entire spectrum appears as a triplet of quartets as in Fig. 2a. However, the pattern differs from that of Fig. 2. The low field set of lines of Fig. 3a is shown on the expanded scale of Fig. 3b. This spectrum identifies the hyperfine couplings which are assigned to the ^{15}N-labeled positions. The pattern of Fig. 3b is assigned to three inequivalent nitrogen couplings, but with two of the couplings due to the ^{15}N nitrogens and the third coupling due to ^{14}N. For ^{15}N the nuclear spin is I = 1/2 instead of I = 1 as with ^{14}N. Each ^{15}N hyperfine coupling therefore contributes a doublet splitting to the spectral pattern, rather than a triplet splitting as with ^{14}N. The two ^{15}N hfcs have different magnitudes. The pattern in Fig. 3a is assigned as $a(^{15}N) = 0.416$ mT, $a(^{15}N) = 0.179$ mT, and $a(^{14}N) = 0.0456$ mT. The accuracy of these assignments is verified by the simulated spectrum shown in Fig. 3c.

The EPR signals as in Figs. 2 and 3 *were not detected* using only NTO in alkaline solution without adding *aci*-nitromethane. The same experiment was conducted with *aci*-nitromethane and alkaline solution, but using TO instead of NTO. In this case, no EPR signals were observed. Experiments with hydroquinone show that the observed radical is a photochemical reduction product of NTO.

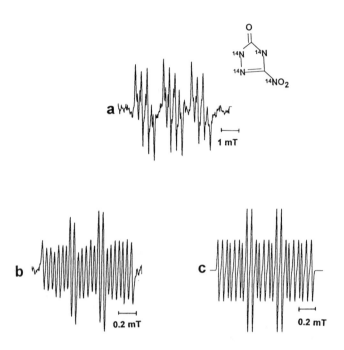

Fig. 2 (a) EPR of ¹⁴N-NTO radical; (b) Low field lines on expanded scale; (c) EPR simulated spectrum of low field lines

NTO/Hydroquinone

Two experiments were conducted using hydroquinone (HQ) which is an electron donor compound. In the first experiment, hydroquinone was added to an alkaline solution (pH 13) of ¹⁵N-NTO. No *aci*-nitromethane was added. EPR spectra were recorded *without* UV photolysis. The resulting spectra are shown in Fig. 4. There is an intense peak observed near the g = 2.002 region of the spectrum. This peak is assigned to an anion radical of hydroquinone as determined by control experiments without ¹⁵N-NTO. A much weaker spectrum is also observed. The weaker spectrum is of an NTO radical product. For example, the low field lines are recorded on expanded scale using high gain as shown in Fig. 4b. This spectrum is the same as that observed in photolysis experiments (compare Fig. 4b to Fig. 3b), confirming that the observed free radical spectra in Figs. 3 and 4 are due to a reduction of NTO.

In a second experiment, a solution of NTO and hydroquinone was prepared in DMSO solution. Before adding NTO the EPR spectrum shows an intense doublet peak near the center of the spectrum which is attributed to a hydroquinone radical. After NTO was added no

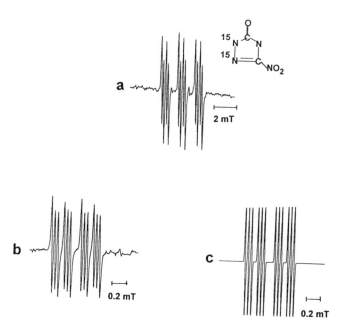

Fig. 3 (a) EPR of ^{15}N-NTO radical; (b) Low field lines on expanded scale;
(c) EPR simulated spectrum of low field lines

new peaks were observed. This indicates that alkaline solution is required to form the NTO radical. The conclusion from these experiments is that the structure of the new radical results from reduction of NTO by electron transfer from *aci*-nitromethane during photolysis or reduction of NTO by hydroquinone without photolysis. The NTO ring remains intact and NO_2 is not lost during radical formation.

DISCUSSION

Photochemical decomposition of energetic materials in alkaline solution has been considered for environmentally safe disposal of these materials. This method is already utilized for nitramines.[9] If NTO were sufficiently decomposed by reaction with strong base, the disposal of NTO by alkaline solution might be considered. Since NTO has a single NO_2 group, the dissociation of the $>C-NO_2$ bond to give NO_2^- (nitrite ion) or $NO_2^\bullet$ (nitrogen dioxide radical) is of particular interest, because $NO_2^\bullet$ is toxic. The radical NO_2-HC$^\bullet$-NO_2^- which would indicate formation of $NO_2^\bullet$ was not observed. The results of this study show that NTO is *not* decomposed significantly by alkaline solution or by the combined effects of short exposure to

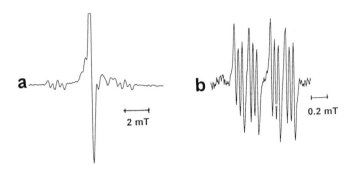

Fig. 4
NTO/Hydroquinone

UV light and alkaline solution. The TO and NTO studies strongly suggest that the $>C-NO_2$ bond remains intact and that the NTO molecule produces a ring-intact free radical intermediate.

Based on the results, a probable assignment of the NTO radical structure can be made. The formation of the radical requires NTO reduction as determined by hydroquinone, and requires basic solution, as determined by the negative result of the NTO/DMSO/hydroquinone experiment. Therefore, the hydroxyl ion is directly involved in the formation of the radical. A possible reaction scheme would lead to an NTO nitro anion radical as shown in Fig. 5.

The resulting anion radical would be formed due to the strong alkalinity of the solution. In previous photolysis and thermal studies of NTO in acetone, Menapace et. al. have detected an NTO hydroxy nitroxide radical formed by hydrogen abstraction from solvent.[4] The hyperfine couplings of the NTO hydroxy radical are typically $a(^{14}N) = 0.8467$ mT (nitroxide nitrogen), $a(^{14}N) = 0.4728$ mT (ring position 2), and $a(^{14}N) = 0.1356$ mT (ring position 4). A hydroxy proton (attached to the nitro group) also gave a hyperfine coupling for this radical with $a(^1H) = 0.7021$ mT. These hyperfine couplings are much different from the NTO radical observed by this study; however, there are some similarities between these radicals. The NTO ring remains intact for both radicals, and there are no resolved hyperfine couplings from the ring protons. The assignment of the NTO reduction product to a nitro anion radical is based on the magnitude of the largest coupling (1.495 mT) which is consistent with many other cyclic nitroxide radicals.

CONCLUSIONS

A reduction product from UV photolysis of NTO in alkaline aqueous solution containing nitromethane is assigned as an NTO nitro anion free radical. This NTO radical was confirmed to be a reduced species of NTO by producing the radical by chemical reduction of NTO with hydroquinone. Further study to confirm the assigned structure will include electron nuclear double resonance experiments to attempt to detect the ring protons, variations of experimental conditions, such as pH, in order to gain further insight into the nature of the reduction reactions of the NTO molecule, and studies of compounds with structures similar to NTO.

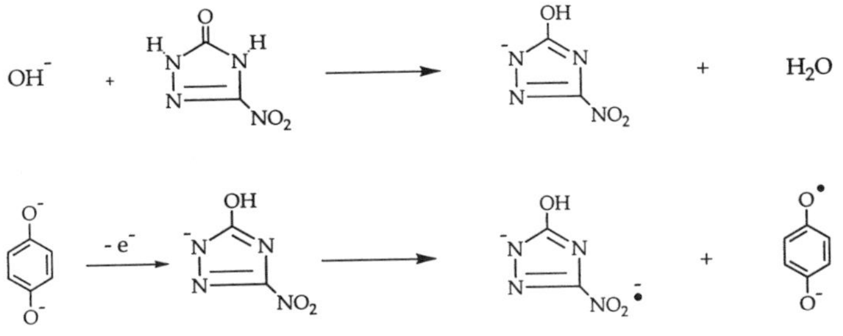

Fig. 5
Possible Scheme of NTO Radical Formation

ACKNOWLEDGEMENTS

This research was sponsored by the Office of Naval Research. Dr. A. W. Snow of the Naval Research Laboratory (NRL), Chemistry Division, is acknowledged for helpful discussion. Dr. T. Russell (NRL) is acknowledged for providing some of the NTO used in this study.

REFERENCES

1. J. C. Oxley, J. L. Smith, Z. Zhou, and R. L McKenney, J. Phys. Chem. **99**, 10383 (1995).
2. G. K. Williams and T. B. Brill, J. Phys. Chem. **99**, 12536 (1995).
3. J. C. Oxley, J. L. Smith, H. Ye, R. L. McKenney, and P. R. Bolduc, J. Phys. Chem. **99**, 9593 (1995).
4. J. A. Menapace, J. E. Marlin, D. R. Bruss, and R. V. Dascher, J. Phys. Chem. **95**, 5509 (1991).
5. L. Fan, C. Dass, and T. J. Burkey, J. Labelled Compounds and Radiopharm. (1995), in press.
6. M. D. Pace, J. Phys. Chem. **98**, 6251 (1994).
7. P. Bilski, C. F. Chignell, J. Szychlinski, A. Borkowski, E. Olesky, and K. Reszka, J. Am. Chem. Soc. **114**, 549 (1992).
8. D. Behar and R. W. Fessenden, J. Phys. Chem. **76**, 1710 (1972).
9. J. D. Corley, Treatment of Wastewaters Containing Energetic Materials, Proceedings of Life Cycles Of Energetic Materials, Del Mar, California, December 11-16, 1994, p. 96.

NTO DECOMPOSITION STUDIES

J.C. OXLEY; J.L. SMITH; K.E.YEAGER; E. ROGERS; X.X. DONG
University of Rhode Island; GRC, P.O. Box 984, West Kingston, RI 02892

ABSTRACT

To examine the thermal decomposition of 5-nitro-2,4-dihydro-3H-1,2,4-triazol-3-one (NTO) in detail, isotopic labeling studies were undertaken. NTO samples labeled with ^{15}N in three different locations [N(1) and N(2), N(4), and N (6)] were prepared.[1] Upon thermolysis, the majority of the NTO condensed-phase product was a brown, insoluble residue, but small quantities of 2,4,-dihydro-3H-1,2,4-triazol-3-one (TO) and triazole were detected.[2] Gases comprised the remainder of the NTO decomposition products. The analysis of these gases is reported herein along with mechanistic implications of these observations.

INTRODUCTION

NTO is a candidate for a effective insensitive high explosive. For this reason we have examined its thermal stability, studying the condensed-phase products and kinetics of NTO decomposition.[2] Over the temperature range 220° to 280 °C, neat NTO exhibited a global activation energy of 78.6 kcal/mol with frequency factor of 2.53 x10^{29} sec^{-1}; NTO in water or methanol (4%) showed lower activation energies 48.8 kcal/mol (6.52x10^{17} sec^{-1}) and 38.4 kcal/mol (1.05 x10^{13} sec^{-1}), respectively. At the low end of the temperature range, the kinetics of NTO in solution and neat were first-order, but above about 255°C decomposition of neat NTO was autocatalytic. Decomposition of NTO in solution or in open thermolysis tubes and decomposition of TO were strictly first order. Build-up of NO$_2$ was speculated to cause autocatalytic decomposition of NTO by interaction with the NTO carbonyl; a survey of ten triazole-like rings, including NTO, showed that those with a carbonyl or a NH$_2$ substituent experienced enhanced decomposition under exogenous$_2$ NO . Decomposition of deuterated NTO, neat and in solution, was noticeably slower than proteo-NTO, indicative of an intramolecular deuterium kinetic isotope effect (DKIE). An intermolecular DKIE was also observed when NTO was decomposed in deuterated solvent. Examination of ten triazole-like rings showed only rings substituted with either NO$_2$ or NH$_2$ exhibited an intermolecular DKIE. From this observation it was concluded that hydrogen and NO$_2$ must be involved in the rate-determining decomposition step.

Since NTO does not melt, 2,4,6-trinitrotoluene (TNT) is contemplated as a molten matrix for it. In comparing the thermal stability of NTO to that of TNT, it was found that at low temperatures (below 230°C) NTO decomposed more slowly than TNT, but that at higher temperatures, it decomposed faster. This was attributed to the autocatalytic nature of the NTO decomposition, which became marked above 255°C. In turn, the autocatalytic nature of the NTO at high temperature was attributed to a change in the decomposition mechanism to one which generated NO$_2$. It was speculated that the rate-determining step in at least one NTO thermolysis pathway involved hydrogen transfer to the nitro group followed by subsequent loss of HONO. This mechanism would be

135

dominant at low temperatures, while at high temperatures C-NO$_2$ homolysis was speculated to be a competitive decomposition pathway.

EXPERIMENTAL SECTION

A series of NTO compounds specifically labeled with ^{15}N at the N(1)-N(2) positions, the N(4) position and the N(6) position were synthesized and authenticated.[1] Melting point glass capillary tubes were used to contain the NTO--unlabeled, labeled, and mixtures consisting of equal masses of N(1)-N(2)/unlabeled, N(1)-N(2)/N(4), N(1)-N(2)/N(6) and N(4)/N(6). Since gas analysis was the goal, care was taken to exclude all air.[3] Samples were heated at 270°C for 20 minutes to ensure greater than 99% decomposition of NTO. Total composition of the decomposition gas was determined using unlabeled NTO. Samples were thermolyzed, opened, and injected into a Varian 3600 gas chromatograph (GC) equipped with Supleco Porapak Q and Molecular Sieve 5A columns connected in series across a thermal conductivity detector (TCD). When tracking the ^{15}N label, it was necessary to avoid contamination by air when sample tubes were opened; thus, each capillary tube was placed in an apparatus consisting of a two way valve connected to a septum adapter. The system was evacuated and then flushed with argon gas three times. When nitrogen was analyzed, it was necessary to quantify the amount of air typically remaining after this procedure. A 7-8 μL aliquot of the argon gas was withdrawn from the apparatus and analyzed to check for residual air contamination in the system. Since O$_2$ was not a NTO decomposition product, the mass intensity of its molecular ion was used to determine ^{28}N$_2$ background. This was accomplished by using the ratio of ^{28}N$_2$/^{32}O$_2$ mass intensities found in the chamber background before the capillary tube was broken; this ratio ranged from about 2.3 to 2.5. After breaking the sample tube and analyzing the sample gas, the intensity of the ^{28}N$_2$ peak was corrected by multiplying the ^{32}O$_2$ peak intensity in the sample gas by the measured background ratio of ^{28}N$_2$/^{32}O$_2$ and subtracting this resultant value from the ^{28}N$_2$ peak intensity. The procedure corrected for any contamination from residual air in the system. A similar correction for background ^{29}N$_2$ from air contamination was obtained using the ratio of ^{29}N$_2$/^{28}N$_2$ in air and multiplying this value by the previously determined ^{28}N$_2$ contamination. For N$_2$ analysis, a Varian 3400 gas chromatograph equipped with a Chrompack PLOT fused silica column with Molecular Sieve 5A stationary phase connected to a Finnigan TSQ 700 tandem mass spectrometer (GC/MS/MS) was used. For N$_2$O analysis, 100 μL of gas was injected into a Hewlett Packard 5890 Series II gas chromatograph equipped with a J&W GS-Q column and interfaced to a Hewlett Packard Series 5971 Mass Selective Detector (MSD). For both N$_2$ and N$_2$O analysis only selected ions were monitored.

RESULTS

Manometric analysis for the total gas produced by NTO decomposition yielded 2.24 moles gas per mole NTO as an average value for 11 analyses (standard deviation SD=0.32). The average value as percent, by weight, of the insoluble solid residue remaining after thermal decomposition of NTO was 30.9% (SD= 1.9% of 14 replicates). The analysis of NTO decomposition gases at equilibrium required use of three different gas chromatographic systems. Gas chromatography (GC) with thermal conductivity detection was used to determine the total composition of the decomposition gas to be 43% N$_2$, 6% N$_2$O, 8% NO, 37% CO$_2$, and 6% CO. To track the fate of the ^{15}N-labeled samples, three to ten samples were run per compound, and the gaseous products of each sample were analyzed twice--once to determine the isotopic content of the nitrogen gas, using GC/MS/MS, and once to determine the isotopic content of the nitrous oxide, using GC/MSD.

Table I accounts for the ^{15}N associated with the generation of N_2 and N_2O from the thermal decomposition of three pure ^{15}N labeled NTO compounds. The mass spectral data is expressed as the ^{15}N isotopic distribution for 100 molecules of N_2 gas. A perfect mass balance would necessitate that the sum of these three percentages equal 100%. However, the total was slightly greater than 100% (105.3 for N_2 and 101.5 for N_2O), and accordingly, these values were used as normalization factors. More complete details can be found in reference 3. We attribute the slight discrepancy to uncertainties associated with obtaining independent mass spectral measurements. The double labeled Sample I had ^{15}N atoms at both the 1 and 2 position. For every 100 molecules of N_2, 52.8 molecules were double labeled, 16.3 had only one ^{15}N and 30.9 molecules contained no nitrogen atoms from the 1 or 2 position of NTO. This interpretation means that 61% (52.5 + 16.3/2) of the nitrogen atoms generated as N_2 gas by NTO decomposition came from the nitrogen atoms at the 1 or 2 position. A comparable interpretation is used for Samples II and III--when the 4 position was labeled 14.7% of the nitrogen atoms from the generated N_2 gas were from the labeled position; when N(6) was labeled, 24.4% of nitrogen atoms came from the labeled position. A similar analysis is given for the isotopic distribution of nitrogen atoms in N_2O. The results indicate that 39.4% of the nitrogen atoms in the N_2O gas from NTO decomposition were from the 1 and 2 positions, 5.6% was from the 4 position and 55.2% was from the 6 position. Table I also shows the computed percentage of each type of N/N reaction necessary to form the observed labeling patterns in N_2 and N_2O.

Tables II and III use the same interpretation as Table I except that Samples IV-VI represent roughly equal molar mixtures of labeled N(1)-N(2) NTO with unlabeled NTO and NTO labeled at positions 6 or 4. For Sample VII equal amounts of NTO labeled at the 4 and at the 6 positions were used. Comparison of Sample I with the labeled sample diluted in half with unlabeled NTO (Sample IV) shows the numbers of double labeled and single labeled molecules of N_2 were reduced by about half while the number of unlabeled N_2 molecules was doubled. This indicates that the double labeled N_2 molecules form intramolecularly from nitrogen atoms N(1) and N(2) since an intermolecular reaction would require the amount of doubly labeled N_2 to be reduced by one fourth. It should be noted that while the gases are products of equilibration, once the permanent gases are formed isotopic scrambling is unlikely.

The complete analysis of the fate of the ^{15}N label during the thermal decomposition of NTO is shown in Table IV. When the label is on N(1) or N(2), 6.5 out of a possible 10 ^{15}N labels are accounted for. However, when the label is on N(6) or N(4), only 2.8 out of 10 or 1.5 out of 10, respectively, are found in the decomposition gases N_2 or N_2O. Some of the missing label may be found in NO. Though small amounts of NO were observed with the GC/TCD system, we were unable to detect low NO levels with either of the GC/ mass spectrometer systems. However, most of the missing ^{15}N must be in the polymeric residue. This means nitrogen atoms from all four positions in the NTO molecule sometimes are found in the polymeric residue.

DISCUSSION

We have previously reported the condensed-phase products of NTO decomposition.[2] Although TO and 1,2,4,-triazole were observed, neither were major products. The major condensed-phase product was an insoluble residue about 31% in mass of the original NTO. Complete elemental analysis of the insoluble residue indicated an elemental ratio of roughly $C_2H_2N_3O$, a formula consistent with "polymerized TO." However, it could be consistent with a number of species, including a one to one mixture of cyclic azines such as melamine [$C_3N_6H_6$] and cyanuric acid [$C_3N_3O_3H_3$]. The analysis of the NTO decomposition gases in this study, as well as the residue, allows us to speculate as to the stoichiometric equation for NTO decomposition at 270°C. Most simply, the reaction can be approximated by production of polymer precursor CN_2 (40g/mol vs. 130 g/mol) and water, along

TABLE I. NORMALIZED N_2 AND N_2O MASS SPECTRAL DATA. ASSIGNMENT AND EVALUATION OF INTERACTION PARAMETERS.

SAMPLES	LABEL	NORMALIZED N_2	NORMALIZED N_2O	ASSIGNMENTS	EVALUATIONS N_2	EVALUATIONS N_2O
I	•—•	52.8	2.5	A	A=53	A=2.5
	•—○	16.3	73.8	D+E	B=0.2	B=1.8
	○—○	30.9	23.7	B+C+F	C=4.5	C=21
II	•—•	0.2	1.8	B	D=14	D=67
	•—○	28.9	7.5	E+F	E=2.7	E=6.5
	○—○	70.9	90.7	A+C+D	F=26	F=1.0
III	•—•	4.5	21.0	C		
	•—○	39.7	68.3	D+F		
	○—○	55.8	10.7	A+B+E		

TABLE IV. Summary of Dinitrogen & Nitrous Oxide Results

Type of N/N Interaction	A N12	B N44	C N66	D N126	E N124	F N46	Sum
Percentage of Each Interaction Represented in Gas Formation							
N2	53%	0.2%	4.5%	14%	2.7%	26%	100%
N2O	2.5%	1.8%	21%	67%	6.5%	1.0%	100%
Atoms N out of 10 NTO from Each Type of Interaction Forming 10 N2 + 1 N2O							
N2	5.30	0.02	0.45	1.40	0.27	2.60	10
N2O	0.03	0.02	0.21	0.67	0.07	0.01	.1

Fate of each type of nitrogen		Atoms of N per 10 NTO N2	N2O	Sum
N12	A+D/2+E/2	6.14	0.39	6.52
N4	B+E/2+F/2	1.46	0.06	1.51
N6	C+D/2+F/2	2.45	0.36	2.81

A= [N(1,2)]=INTERACTIONS THAT SPECIFICALLY INVOLVE NITROGENS AT POSITIONS 1 & 2

B= [N(4,4)]=INTERACTIONS THAT SPECIFICALLY INVOLVE NITROGENS AT POSITION 4

C= [N(6,6)]=INTERACTIONS THAT SPECIFICALLY INVOLVE NITROGENS AT POSITION 6

D= [N(1,2,6)]=INTERACTIONS THAT SPECIFICALLY INVOLVE NITROGENS AT POSITIONS 1,2 & 6

E= [N(1,2,4)]=INTERACTIONS THAT SPECIFICALLY INVOLVE NITROGENS AT POSITIONS 1,2 & 4

F= [N(4,6)]=INTERACTIONS THAT SPECIFICALLY INVOLVE NITROGENS AT POSITIONS 4 & 6

TABLE II. MASS SPECTRAL DATA FOR THE ISOTOPIC DISTRIBUTION OF ^{15}N EXPRESSED AS 100 MOLECULES OF N$_2$ (50:50 MIXTURES)

ISOTOPE	●—●	●—○	○—○	TOTAL N ATOMS	TOTAL ^{15}N	% ^{15}N ATOMS
SAMPLE IV	29.9	7.9	62.2	200	67.7	33.9
SAMPLE V	30.4	22.4	47.2	200	83.2	41.6
SAMPLE VI	36.8	22.5	40.6	199.8	96.1	48.1
SAMPLE VII	7.3	24.2	68.5	200	38.8	19.4

TABLE III. MASS SPECTRAL DATA FOR THE ISOTOPIC DISTRIBUTION OF ^{15}N EXPRESSED AS 100 MOLECULES OF N$_2$O (50:50 MIXTURES)

ISOTOPE	●—●—○	●—○—○	○—○—○	TOTAL N ATOMS	TOTAL ^{15}N	% ^{15}N ATOMS
SAMPLE IV	0.7	43.9	55.5	200.2	45.3	22.6
SAMPLE V	0.5	44.6	54.9	200	45.6	22.8
SAMPLE VI	24.2	49.7	26.1	200	98.1	49.1
SAMPLE VII	5.4	49.8	44.8	200	60.6	30.3

with appropriate amounts of N_2 and CO_2 and N_2O and CO. To generate the observed ratio of gases would require 90% of the decomposition proceed by route (1) and 10% by route (2).

$$NTO \rightarrow CN_2 \quad + \quad H_2O \quad + \quad N_2 \quad + \quad CO_2 \qquad (1)$$
$$NTO \rightarrow CN_2 \quad + \quad H_2O \quad + \quad N_2O \quad + \quad CO \qquad (2)$$

A more detailed stoichiometric reaction for NTO is shown below:

$$10\ NTO \rightarrow C_{11}H_8N_{16}O_4 + 6\ H_2O + \quad 10\ N_2 + 1\ N_2O + 2\ NO + 8\ CO_2 \ + 1\ CO \qquad mol\ gas/\ NTO$$

	residue						
calculated	33wt%	45%	5%	9%	36%	5%	2.20
observed	31wt%	43%	6%	8%	37%	6%	2.24

As a result of our previous study, examining only the decomposition kinetics and condensed-phase products of NTO, we proposed a mechanism (Scheme I) which is initiated by HONO or NO_2 loss.[2] This mechanism is postulated for a number of reasons. (1) NTO decomposed with autocatalytic kinetics attributed to evolved NO_2. (2) Both intra- and intermolecular deuterium kinetic isotope effects were observed for NTO, and other triazole-like rings containing NH_2 or NO_2 groups also exhibited an intermolecular DKIE. (3) TO (effectively NTO minus NO_2) was a minor, but observed, NTO decomposition product. (4) Homolysis of $X-NO_2$ is a decomposition mode common to many classes of explosives-- nitroarenes, nitramines, and nitrate esters. A number of other researchers have cited evidence for loss of HONO or NO_2 from NTO.[4-6] Rothgery et al. observed m/z of 46 among NTO decomposition products but did not detect NO_2 by infrared (IR) until after an equilibration period. They concluded that NO_2 was formed by oxidation of NO during the equilibration of the NTO decomposition products.[7] Recently, Wight et al.[8] have examined the decomposition of NTO in thin films and reported that if N_2O_4 were formed under their experimental conditions (heating rates 10^7 to 10^9 K/s and laser fluences ranging from 0.41 to 3.2 J/cm²), it must have remained in the film in concentrations less than 5% that of CO_2. Using exclusively IR detection, they observed CO_2 (formed by intramolecular oxidation of the carbonyl by the nitro group) as the first decomposition product. The mechanism they proposed (Scheme II) results in the formation of one mole each of N_2 and CO_2 per mole NTO and produces a nitrosoisocyanide radical which would serve as the precursor to the polymeric residue. Direct observation of N_2, however, could not be accomplished by their experimental method; therefore, it is uncertain whether CO_2 or N_2 is actually the first decomposition product. Wight's mechanism (Scheme II)[8] would require that all N_2 gas would be formed from N(1) and N(2). This is not the case. Furthermore, it would not account for the observed intramolecular DKIE,[2,4] unless the scheme were modified, for example, to include an initial $C-NO_2$ to $N-NO_2$ migration. One might speculate that the input of thermal energy results in the formation, however briefly, of the less thermodynamically stable $N-NO_2$ from $C-NO_2$. Nitroarenes,[9] pyroazoles, nitroindazoles, imidazoles, and trizoles[10] all exhibit nitro migration on the ring. Such an initial step would result in an observed DKIE (Scheme III).

Analysis of the data gathered from our labeling studies allows us to determine which nitrogen atoms form which nitrogen-containing gas. At most, 60% of the N_2 gas is from N(1) and N(2); the number may be as low as 50% (Table IV). Interaction of N(6) with N(1), N(2), or N(4) accounts for most of the other 40% N_2, with N(6)/N(6) reaction producing a small amount (about 5%). This means that the mechanism of Wight cannot be the only mode of NTO decomposition. Furthermore, experiments represented in Table II by Samples VI [N(1) or N(2) with N(6)] and VII [N(4) with N(6)] indicate intermolecular reactions do occur between N(6), the nitro nitrogen, and the other ring

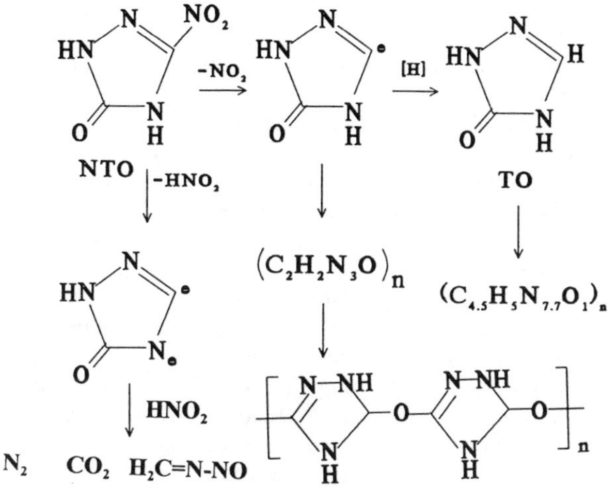

Scheme I (from reference 2)

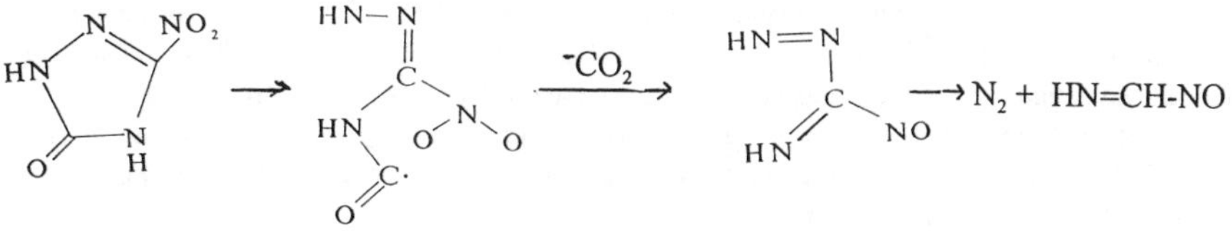

Scheme II (from reference 8)

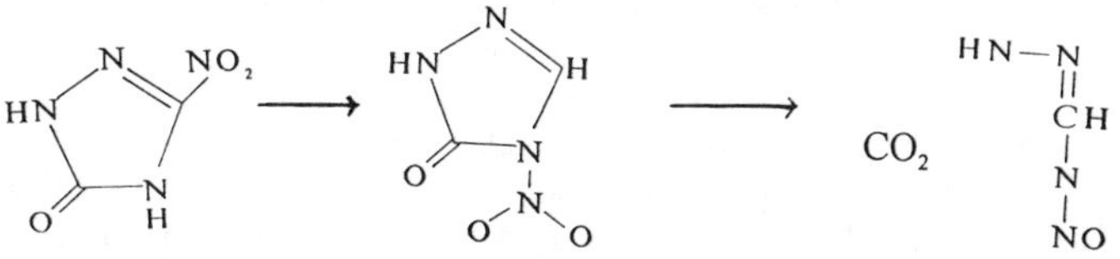

Scheme III

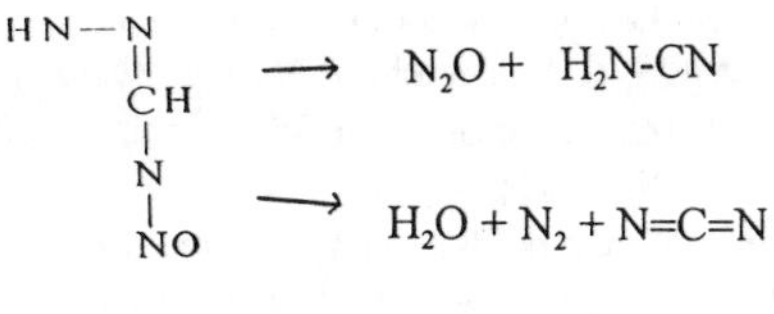

Scheme IV

nitrogen atoms. Considering Sample VI, if the doubly labeled nitrogen were strictly a result of the N(1)/N(2) intramolecular interaction, then the percentage could not be more than 29.9% (Sample IV); instead, it is 36.8%. Moreover, if the interaction of N(4) with N(6) were strictly intramolecular, then there would be no possibility of doubly labeled nitrogen forming in sample VII. These reactions are not minor; 40% of the nitrogen gas is formed by this route. Accounting for labeled nitrogen atoms, as shown in Table IV, indicates that out of 10 NTO molecules, 3.5 atoms of N(1) and of N(2) form neither N_2 nor N_2O gas.

Scheme I is a decomposition route which accounts for the acceleratory effect of NO_2, the DKIE, the 31wt% residue, and most of the decomposition gases. Unlike Scheme II, Scheme I would not require that nitrogen gas come exclusively from N(1) and N(2). Our results show that at least 40% of the observed N_2 is generated by attack of NO_2 on N(1), N(2), or N(4). In the case of N(1) or N(2), attack by N(6) also generates N_2O (Table IV). It is difficult to reconcile these observations with those of Wight other than to claim that experimental conditions were dramatically different, and this may be the case. Otherwise, one can speculate that the amount of free NO_2 is indeed quite low or that NO performs the intermolecular attack on the ring nitrogen atoms. The latter postulate would have NO_2 perform an intramolecular oxidation of the carbonyl to form CO_2; then, before the residue could decompose to form N_2 from N(1) and N(2), N(6) would be lost as NO, to randomly attack N(1), N(2), or N(4). However, NO should have been easily observed by IR. In any case, formation of the unsaturated fragments shown in Schemes I, II, or III can easily be envisioned to result in the observed cyclic azines. Scheme IV illustrates a route to cyanamide, a species known to cyclize to melamine, a NTO decomposition product observed by Brill.[11]

SUMMARY

A series of NTO compounds specifically labeled with ^{15}N at the N(1)-N(2) positions, the N(4) position and the N(6) position were synthesized and thermolyzed at 270°C. Decomposition gases were quantified and the ^{15}N label was traced. It was determined that about half of the reaction proceeds via evolution of CO_2 and N_2 --the former from C(3), the latter from an intramolecular evolution of N(1) and N(2). The remainder of the decomposition results from loss of N(6) in the form of HNO_2, NO_2 or NO and subsequent intermolecular attack on N(1), N(2), or N(4). This latter reaction could account for the DKIE.

ACKNOWLEDGMENTS

The authors thank the Armament Division of Eglin Air Force Base for funding through the Office of Naval Research and Dr. Carlyle Storm for helpful discussions.

REFERENCES

1. Oxley,J.; Smith,J.; Yeager,K.; Coburn,M.; Ott,D. *J. Energetic Materials* **1995**, *13(1&2)*, 93-105.
2. Oxley, J.C.; Smith, J.L.; Zhou, Z.; McKenney, R.L. *J.Phys.Chem*, **1995**, *99(25)*, 10383-10391.
3. Oxley, J.C.; Smith, J.; Yeager, K; Roger, E.; Dong, X.X. submitted *J.Phys.Chem.*
4. Menapace, J.A.; Marlin, J.; Bruss D.; Dascher, R. *J. Phys. Chem.*, **1991**, *95*, 5509-5517.
5. Beard B.C.; Sharma, J. *J. Energetic Materials*, **1989**, *7(3)*,181.
6. Ostmark, H., "Thermal Decomposition of NTO" FOA Rep. D-201782.3, Nov. 1991.
7. Rothgery,E.F.; Audette,D.E.; Wedlich,R.; Csejka,D. *Thermochim Acta*, **1991**, *185*, 235.
8. Botcher, T.R.; Beardall, D.J.; Wight, C.A. submitted *J.Phys.Chem.*
9. "Supplement F: The Chemistry of Amino, Nitroso, and Nitro Compounds and Their Derivatives" Part 1; Patai, S. Ed. Wiley, NY, 1982 p146.
10. Janssen, J.W.A.M.; Habraken, C.L.; Louw, R. *J.Org.Chem.*, **1976**, *41(10)*, 1758-1762.
11. Williams, G.K.; Palopoli, S.F.; Brill, T.B. *Combust. Flame,* **1994**, *98*, 197.

STUDIES OF PYROLYSIS PRODUCTS OF EXPLOSIVES IN SOILS USING INFRARED TUNABLE DIODE LASER DETECTION

J. WORMHOUDT, J. H. SHORTER AND C. E. KOLB
Aerodyne Research, Inc., 45 Manning Road, Billerica, MA 01821

ABSTRACT

We report on laboratory studies of the thermal decomposition of TNT under a variety of conditions. Our goal is a better understanding of the soil contamination detection technique used in site characterization using cone penetrometers, in which hot filament pyrolysis results in small gaseous molecules, such as NO, which can be detected by a variety of methods to indicate the presence of explosives in the soil. Our laboratory studies use a long-path sampling cell and tunable infrared diode laser absorption, which provide sensitive, specific, time-resolved detection of several of the key decomposition products. In this paper, we present both the initial results of laboratory experiments on mixtures of soil and TNT, designed to simulate the operation of a cone penetrometer probe, and also results of a series of heated tube experiments which serve to relate the first type of experiment to previously reported TNT decomposition studies.

INTRODUCTION

This paper differs from the typical contributions to this symposium, which might report fundamental parameters of energetic materials such as Arrhenius expressions for decomposition rates or fractions of gaseous decomposition products, or might apply these parameters in a computer model of a decomposition process. Instead, it describes an area in which those basic parameters and some level of predictive modeling could lead to significant improvements in an important process. The area is the detection of energetic materials such as TNT and RDX in soils, and the process of interest is the thermal decomposition of those materials to yield characteristic and easily detectable small gaseous molecules. Our research program involves the application of infrared tunable laser/multipass absorption cell technology, in two ways: as a sensitive, selective detection method in laboratory experiments designed to clarify the mechanisms involved in pyrolysis techniques for soil contamination detection, and as an *in situ* technique for use in advanced site characterization techniques.

Such techniques are needed because of the magnitude of the soil contamination problem faced by the Department of Defense, the Department of Energy and the Environmental Protection Agency. In particular, the Army must characterize and map soil and groundwater contamination by a range of toxic materials at a large number of current and former bases, munitions plants, storage depots and other facilities. The Army has taken a leading role in the development, testing, and fielding of the Site Characterization and Analysis Penetrometer System (SCAPS), consisting of a hydraulically operated cone penetrometer test unit mounted in a custom-engineered 20-ton truck. The penetrometer unit is equipped with an instrumented probe that collects geotechnical information and provides continuous soil classification and stratigraphy during the penetrometer push.

In addition, the Army, Navy and Air Force have cooperated in the development of various sensors for specific classes of contaminants, including laser-induced fluorescence probes for hydrocarbon contaminants[1] and a probe using an electrochemical sensor to detect the products of pyrolysis of explosives.[2] This latter probe includes an electrochemical cell which can detect NO and some related species which are thermal decomposition products of energetic materials such as TNT, as well as a pyrolyzer unit and provision for gas flows to carry pyrolysis products to the sensor. Our initial goal has been to gain an understanding of the processes underlying the operation of this device. We have also begun laboratory simulations of the operation of a pyrolysis probe, and intend to correlate our observations using a quantitative computational model, finally applying this understanding to the development of improved detection techniques.

To provide an idea of the system whose operation we are trying to simulate and predict, we conclude this introduction with a brief description of the current penetrometer probe for explosives

Mat. Res. Soc. Symp. Proc. Vol. 418 ©1996 Materials Research Society contamination[2]. A schematic diagram is shown in Fig. 1 which focuses on a few of characteristics of the probe that figure importantly in the investigations reported here. The present pyrolyzer unit consists of a 20 cm length of 0.025 cm platinum wire wound in two loops around a ceramic insulator sleeve, typically operated near 900 °C. A sacrificial sleeve protects the gas flow ports and the pyrolyzer unit as the probe is advanced into the soil, then is left behind as the probe is withdrawn. The distance between the pyrolyzer filament and the soil surface is about 0.25 cm. The withdrawal of the probe is halted at desired depths and the pyrolyzer is activated for periods of order 1 minute. Air is continuously pumped from the surface through the air supply ports of the probe, collected through the air return or vapor sampling ports, and directed over the electrochemical sensor located in the probe. The analog signal from the electrochemical sensor, which is proportional to NO concentration, is transmitted to the data acquisition system located in the cone penetrometer truck.

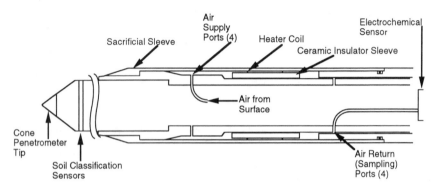

Fig. 1. Schematic drawing of pyrolyzer region of the current cone penetrometer explosives probe. Air pumped from the surface passes out through four air supply ports and between the pyrolyzer and heated soil before being drawn back into probe and sent to electrochemical sensor.

EXPERIMENT

The main section of Fig. 2 shows the components of an initial laboratory experiment which simulates pyrolysis of TNT in soil. It involves heating of a contaminated soil sample and sampling of the resulting gaseous products into a multipass cell where their laser absorption is measured as a function of time. The optical layout for the tunable infrared diode laser system used in the present work, and general features of this type of instrument, have been presented elsewhere.[3,4] The particular heater shown schematically in the figure is planar, with the heater wire wrapped in a spiral below a disk of machinable ceramic. The ring shaped heating surface has a 2 cm o.d. and 0.6 cm i.d. and is formed from nickel-chrome thermal coaxial cable inside a 0.2 cm o.d. inconel sheath. Other heaters have been used or are planned; these differ from the one shown by being closer in design to the penetrometer device, but less well described by one-dimensional heat transfer models.

Decomposition gases are sampled through a 0.25 cm i.d. ceramic tube located in the center of the ring heater at the same distance from the soil surface. After passing through some 0.3 cm o.d. Teflon tubing (to allow motion of the heating element over the surface of the soil), the flow of pyrolysis products passes through a filter constructed from an 50 cm section of 1.9 cm o.d. stainless steel tubing packed with fiberglass filter material, then through a valve and into the multipass cell.

The inset of Fig. 2 shows an alternative source of TNT pyrolysis gases, used in a second series of experiments with the goal of connecting the time and temperature regime of the simulation experiment with the regimes of literature studies of TNT decomposition. This source is a simple heated quartz tube in which a small amount (0.004-0.006 g) of pure TNT is placed after a steady temperature and air flow rate into the multipass cell have been established.

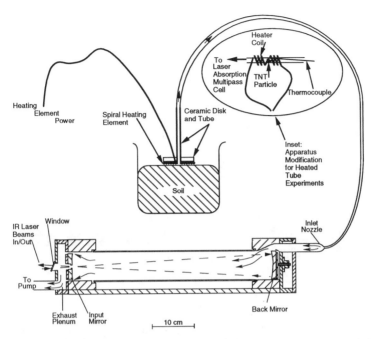

Fig. 2. Schematic drawing of key elements of laboratory apparatus for heating of explosives-contaminated soil, including soil sample, planar spiral heating element with central sampling tube, and laser absorption multipass cell. Inset shows schematic of heated tube experiments.

RESULTS

Fig. 3 shows an example observation from the planar spiral heater configuration in Fig. 2. The two traces are cell concentrations of NO and NO_2, derived from specific absorption lines in the spectral scans of two laser diodes. In this example, the ring heater was about 0.25 cm from the surface of the soil, and was heated to about 900 °C for about 3.5 minutes. The soil sample in this case was sand contaminated with about 100 ppm of TNT by weight. The sample was made up by grinding a weighed amount of TNT using a mortar and pestle, then mixing it with some sand in the mortar, and finally adding that mixture to the remaining sand. It can be seen that the NO and NO_2 traces show roughly parallel behavior (just as, in other TNT experiments, we saw NO and CO_2 evolution to have the same time history), but that the NO signal is larger than the NO_2 signal. It is interesting to note that in Fig. 3 the NO_2 disappears as soon as the heat is removed, while some NO is sampled for a longer decay time.

A surprising aspect of Fig. 3 is that NO and NO_2 concentrations are comparable, when other reports of thermal decomposition of TNT list NO, but not NO_2, as a significant product.[5-6] Low NO_2 levels are ascribed to rapid secondary reactions of NO_2 produced in initial bond-breaking steps, rather than to a lack of NO_2 formation from the earliest stages of TNT decomposition. In an effort to connect the two types of observations, we carried out a series of the heated tube experiments shown schematically in the inset of Fig. 2. Tube hot section wall temperatures and air flow rates (and hence residence times in the heated region of the tube) were varied. Integrals under curves like those of Fig. 3 were combined with flow rates to yield total moles of NO and NO_2 evolved, reported in Figs. 4 and 5 as ratios to the moles of TNT in the particle inserted into the heated region.

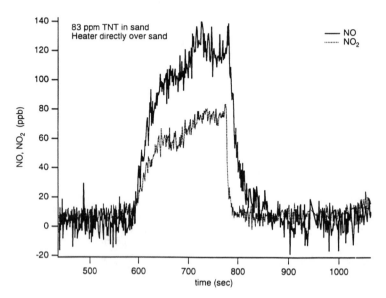

Fig. 3. Example of pyrolysis product concentrations measured in experimental configuration shown in Fig. 2. Signal rise times are determined by the rate of soil heating after heating element is applied, while NO_2 decay time shows time response of multipass cell, determined by gas flow rate.

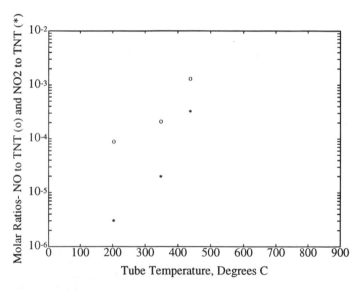

Fig. 4. Ratios of moles of product gases to moles of TNT in sample, as a function of tube wall temperature, for residence time in tube heated region of about 4 ms.

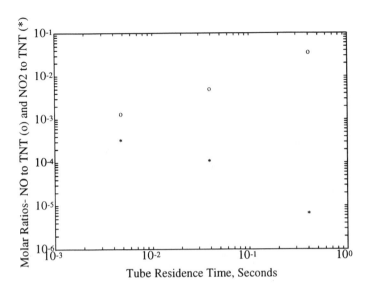

Fig. 5. Ratios of moles of product gases to moles of TNT in sample, as a function of residence time in tube heated region, for tube wall temperature in the 400-450 °C range.

DISCUSSION

The above experimental results coupled with some other observations allow us to put forward a conceptual model of TNT detection using a hot filament heating of contaminated soil followed by detection of gaseous products. We discuss NO (and NO_2) production first, then treat NO_2 loss.

TNT melts at about 80 °C, and simultaneously boils and decomposes at 240 °C. Its rate of thermal decomposition is characterized by a very strong temperature dependence, with activation energies for initial decomposition in the liquid phase measured in the range of 40-50 kcal/mole,[5] and with gas-phase activation energies higher than that, if related compounds are a guide. Three candidate mechanisms for NO production from TNT-contaminated soils using the pyrolysis probe shown in Figure 1 can be proposed. In the first, contaminated soil is rapidly raised to a high temperature, in the range of the roughly 900 °C filament temperature, and TNT is thermally decomposed in the soil matrix. In the second, contaminated soil in a low temperature regime (between the TNT melting and boiling points) evolves NO, NO_2 and other decomposition products. In the third, TNT is vaporized from the soil matrix and thermally decomposes at higher temperatures in the vicinity of the filament.

The results of our initial laboratory investigations suggest that the third mechanism is responsible for NO and NO_2 production from TNT-contaminated soils using a pyrolysis probe. Using embedded thermocouples, we found that filament heating of soil is unlikely to generate surface temperatures above 300 °C. A 900 °C filament placed 0.25 cm above the soil surface had a thermal penetration depth for temperatures above 200 °C of only the top few mm of soil. Soil/TNT dynamics therefore must be thought of in terms of a low temperature regime and the first candidate mechanism is considered unlikely. However, when TNT is heated to low temperatures (~180-240 °C), little NO evolves during TNT vaporization, as seen in the small fractions shown in Fig. 4. In further heated tube experiments, we found that when a 900 °C filament is placed in the flow of TNT vapor generated at low temperatures, a large increase in the NO signal is observed. These experiments indicate that contaminated soils heated to low temperatures will yield TNT vapor and little NO; however, the TNT vapor can be subsequently decomposed at high temperatures producing large amounts of NO. This is consistent with the third proposed mechanism for NO and

NO_2 production from TNT-contaminated soils using the pyrolysis probe, that of energetic material vaporization from the soil, followed by high temperature gas phase pyrolysis near the probe.

In the heated tube experiments reported in Figs. 4 and 5, it can be seen from the small ratios that most of the TNT evaporates and escapes the heated region before decomposing. Indeed, a coating of condensed TNT is observed on the cool section of the quartz tube between the heater and the multipass cell inlet. The fact that NO production increases both with increasing wall temperature and residence time in the hot region suggests that, in addition to NO evolved directly from the melted TNT particle (which must maintain a temperature of about 240 °C by evaporation), a substantial part of the evolved NO results from decomposition of TNT vapor which contacts the hot wall of the quartz tube. (Tube temperatures are low enough that we expect a negligible contribution from TNT vapor decomposition in the gas phase, as opposed to reaction on the walls.) The fact that NO_2 decreases with increasing residence time, on the other hand, points to reactions between NO_2 and TNT decomposition products (or even TNT itself) on the tube wall. Thus, in experiments where NO_2 is allowed significant contact time with molten TNT or condensed phase decomposition products, little or no NO_2 will be observed. On the other hand, in the spiral heater configuration of Fig. 2, there is little opportunity for such NO_2 destruction reactions, and substantial NO_2 is seen.

CONCLUSIONS

The initial goal of the research program described here is to attain an understanding of the mechanism of pyrolytic decomposition of energetic materials in soils. Quantitative diode laser detection of pyrolysis gases coupled with embedded thermocouple measurements of heated soil suggests that TNT is decomposed in the vapor phase near the pyrolyzer filament, not in the soil matrix. A significant difference between the decomposition products observed in these simulation experiments, and those reported in previous studies of TNT decomposition is the presence of significant NO_2 under conditions in which the decomposition gases are sampled before destruction reactions can occur.

ACKNOWLEDGMENTS

This work was supported by the U. S. Army's Environmental Restoration Research Program, managed by Dr. M. J. Cullinane, U. S. Army Engineer Waterways Experiment Station, under Contract DACA39-95-C-0036. S. Kallelis and J. Mulholland constructed and operated the experimental apparatus. M. Zahniser, D. D. Nelson, and J. B. McManus provided invaluable aid in setting up the laser system. Helpful discussions with Drs. E. R. Cespedes, W. M Davis, and Dr. J. Adams of Waterways Experiment Station are gratefully acknowledged.

REFERENCES

1. E. R. Cespedes, B. H. Miles, and S. H. Lieberman, in Optical Sensing for Environmental Monitoring, (Air and Waste Management Association SP-89, 1994), pp. 621-632.

2. E. R. Cespedes, S. S. Cooper, W. M. Davis, W. J. Buttner, and W. C. Vickers, Proc. SPIE **2367**, 33, (1994).

3. J. Wormhoudt, M. S. Zahniser, D. D. Nelson, J. B. McManus, R. C. Miake-Lye, and C. E. Kolb, Proc. SPIE **2546**, 552, (1995).

4. C. E. Kolb, J. C. Wormhoudt, and M. S. Zahniser, in Methods in Ecology: Trace Gases, edited by P. A. Matson and R. C. Harriss, (Blackwell Science Ltd., Oxford, England, 1995), pp. 258-290.

5. T. B. Brill and K. J. James, J. Phys. Chem. **97**, 8759 (1993).

6. T. B. Brill and K. James, Chem. Rev. **93**, 2667 (1993).

Part III

Combustion Mechanisms

EFFECT OF STRUCTURE OF ENERGETIC MATERIALS ON BURNING RATE

A.E.FOGELZANG, V.P.SINDITSKII, V.Y.EGORSHEV, V.V.SERUSHKIN
Department of Chemical Engineering, Mendeleev University of Chemical Technology,
9 Miusskaya Square, 125047, Moscow, Russia, svp@iht.mhti.msk.su

ABSTRACT

Data on the steady-state combustion in a constant-pressure bomb at 0.1-40 MPa are presented for energetic materials from the following classes: metal salts of organic explosive acids, salts of organic bases with inorganic oxidizing acids, explosive coordination compounds, and endothermic polynitrogen compounds.

For combustion of salts of organic bases with oxidizing acids it has been found that an increase in the oxidant redox potential, whose value serves as an estimate of the oxidizer reactivity, causes the burning rate value to increase. The same tendency has been disclosed for explosive coordination compounds which can formally be considered as metal-containing analogs of the salts of organic bases with oxidizing acids.

The introduction of a metal atom in an organic explosive acid has been shown to result generally in an enhancement of the burning rate, with the effectiveness of the metal as the combustion catalyst being dependent not only on the nature of the metal but on its position in the molecule as well.

Neither the nature of the metal, nor the nature and structure of the ligand really affects the combustion of coordination compounds of metal azides, whose combustion occurs at the expense of the heat produced in the decomposition process. All the coordination azides seem to have the same rate-limiting stage, namely, the decomposition of the intermediate HN_3 and differ from one another by their burning temperatures. The similar behavior is also characteristic of metalless analogs of the coordination azides: salts of HN_3 with amines.

INTRODUCTION

Research into burning of energetic materials has long been of considerable interest for specialists in the fields of combustion and explosion. Essentially all studies on combustion are restricted to compounds and formulations of practical significance, mainly to nitro compounds and compositions based on ammonium perchlorate. Little has been reported on combustion of individual explosives and energetic materials unrelated to solid propellant ingredients. A limitation on the objects under investigation allows the researchers to gain access to the root of the phenomenon but it may hinder disclosing the general regularities. While studying combustion behavior of substances from the same chemical class and of similar chemical natures it is difficult to obtain a correlation between the molecular structure of a substance and its burning rate. In this case flame temperature exerts primary influence on the burning rate.

The present work describes the results of the experimental study on the steady-state combustion of a wide series of energetic materials from the following classes: salts of organic bases with inorganic oxidizing acids, metal salts of explosive acids, explosive coordination compounds and endothermic polynitrogen compounds. The main goal of the work is to elucidate how we can affect the burning rate of energetic materials by altering the chemical structure and to reveal the reasons responsible for the burning rate level.

EXPERIMENTAL

Parameters of the steady-state combustion of all energetic materials under investigation were measured in a window constant-pressure vessel in the pressure range of 0.1-40 MPa using a photographic recording. An unique procedure of preparing high-density strands was devised for measuring the burning rates of fast-burning explosives [1].

A microthermocouple technique was extensively employed to measure temperature profiles during combustion using 5-μm-thick Π-shaped tungsten-rhenium tape thermocouples in order to recognize the combustion mechanism[2].

RESULTS

Energetic materials may be divided into two classes: substances capable of burning at the expense of the heat evolved during redox reactions and substances which can burn through the heat released at the destruction of endothermic groups.

One of the best-studied cases of combustion of energetic materials is that occurred through the heat of redox reactions. All nitro compounds fall basically in this category. It is well known that for combustion of usual C-nitro compounds, the flame temperature has the major influence on the burning rate[3]. Analogous regularities have long been found for combustion of double-base propellants[4]. Molecular structure of nitro compounds and chemical nature of functional groups have been revealed to exert lesser influence on the burning rate. This is not surprising because the rate-limiting stage for the most part is interaction between nitrogen oxides and fuel intermediates of close reactivity both produced at initial decomposition of nitro compounds. What will become of the burning rate, if we take compounds which includes alternative oxidizers in the molecule?

A change-over to the systems with other oxidizers reveals a wide range of possible burning rates, as for combustion of ammonium salts of various inorganic oxidizing acids (Table I). A correlation between the burning rate and flame temperature or thermal stability of the salt cannot be found. On the other hand, it is apparent that the burning rate is mainly determined by the oxidizer reactivity which may be estimated in terms of its standard redox potential (E_o).

Table I. Physicochemical and combustion characteristics of ammonium salts

Compound	Burning rate at 10 MPa $g{\cdot}cm^{-2}{\cdot}s^{-1}$	Ignition Temperature[a], °C	Flame Temperature, K	Electrode Couple	Redox Potential, E_o, V [5]
NH_4MnO_4	38	130	1690	MnO_2/MnO_4^-	1.70
NH_4BrO_3	24	70	1940	Br_2/BrO_3^-	1.52
NH_4IO_4	22	200	1005	IO_3^-/IO_4^-	1.50
NH_4ClO_3	8.6	130	1670	Cl_2/ClO_3^-	1.47
NH_4NO_2	5.6	90	2160	N_2/NO_2^-	1.45
NH_4ClO_4	2	315	1400	Cl_2/ClO_4^-	1.39
$(NH_4)_2Cr_3O_{10}$	0.76	190	945	Cr^{3+}/Cr^{6+}	1.33
$(NH_4)_2Cr_2O_7$	0.70	240	1285	Cr^{3+}/Cr^{6+}	1.33
NH_4IO_3	0.65	150	775	I_2/IO_3^-	1.19
NH_4NO_3	Does not burn	250	1250	NO/NO_3^-	0.96

[a] The heating rate is 20 degrees per minute

Analogous regularities are observed (Fig. 1) for ethylenediamine, $H_2NCH_2CH_2NH_2$, (En) salts of common formula of En·2HX (where HX = HNO_3, HNO_2, HIO_3, $HClO_4$, $HClO_3$, $HBrO_3$, HIO_4), as well as at combustion of salts of methylamine, benzylamine and guanidine [6]. The change in oxidizer reactivity may result in the burning rate variations as much as two orders of magnitude.

The burning rate of energetic materials is also affected by the reduction ability of the fuel. Combustion characteristics of perchlorate and nitrate salts of readily oxidizable ($E_o \sim 0.7$ V) carbohydrazide, $H_2NHNCONHNH_2$, (Chz) and less oxidizable ($E_o \sim 0.8$ V) ethylenediamine is shown in Fig. 2. The more easily fuel is oxidized, the higher the burning rate.

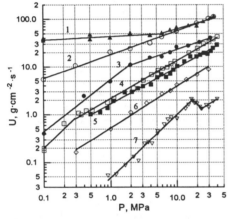 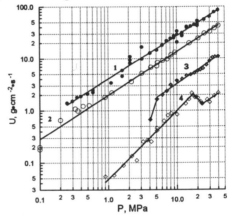

Fig. 1. Burning rate vs. pressure for ethylenediamine salts, En·2HX. HX = HIO_4 (1), $HBrO_3$ (2), $HClO_3$ (3), $HClO_4$ (4), HIO_3 (5), HNO_2 (6), HNO_3 (7).

Fig. 2. Burning rate vs. pressure for perchlorates(1), (2) and nitrates (3), (4) of carbohydrazide and ethylenediamine respectively.

The effect of oxidant and fuel activities is clearly traced at combustion of yet another class of energetic materials, namely, explosive coordination compounds of common formula of $[M(L)_n]X_2$, where M is metal, X is oxygen-containing anion and L is organic ligand. The molecular structure of such compounds allows anion-oxidizer, ligand-fuel and the central metal atom to be easily manipulated. As illustrated in Fig. 3, the complexes of BrO_3, ClO_3 and ClO_4 anion-oxidizers having the greatest redox potentials exhibit the highest burning rate values. Coordination compounds with easily oxidizible carbohydrazide as the ligand burn faster than those with ethylenediamine (Fig. 4). When changing the oxidant chemical nature the burning rate of the complexes may vary over 2 orders of magnitude, whereas in the case of different fuels it ranges five to ten times.

Having introduced the metal atom into the molecule of energetic materials, we face another important factor, i.e. catalysis. It is well known that metal-based compounds are widely used for catalyzing double-base propellants and composite formulations. In the present work the following metals were taken as the central metal atoms in the coordination compounds: Ni, Co, Cu, Cd, Pb, Zn, Mg, Ca, Ba, Sr. Carbohydrazide was used as the ligand. In order to recognize the catalytic effect of the metal during combustion, the salts of carbohydrazide with corresponding acid, Chz·HX, were employed as references.

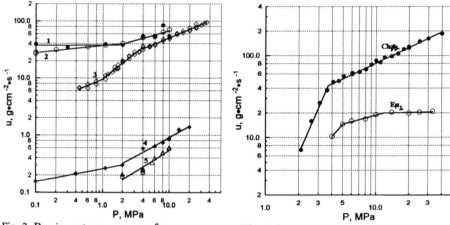

Fig. 3. Burning rate *vs.* pressure for [Cu(En)₂](X)₂. X = BrO₃(1), ClO₃(2), ClO₄(3), NO₃(4), NO₂(5).

Fig. 4. Burning rate *vs.* pressure for [Cd(Chz)₂](ClO₄)₂ (1) and [Cd(En)₂](ClO₄)₂ (2).

A comparison between burning rate of perchlorate complexes of carbohydrazide and Chz•HClO₄ as a control demonstrates (Fig. 5) that all compounds separated into two groups: the compounds which burn faster than the control (the complexes of Cu, Ni, Co, Cd, Pb) and the compounds which have lesser burn rate (the complexes of Zn, Mg, Ca, Ba, Sr).

A similar situation has been observed at combustion of coordination compounds with nitrate anion as an oxidizer (Fig. 6). In this case, however, the number of compounds which burn faster than the control nitrate carbohydrazide is well below than in the case of perchlorates (only com

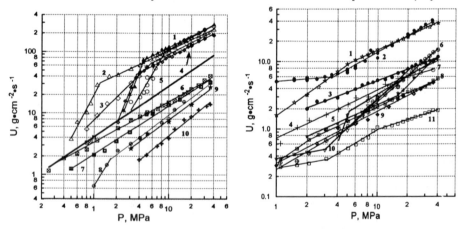

Fig. 5. Burning rate *vs.* pressure for [M(Chz)ₙ](ClO₄)₂, where M = Ni (1), Pb (2), Cu (3), Cd (4), Co (5), Zn (6), Ba (7), Ca (8), Mg (9), Sr (10). Solid line without points is Chz· HClO₄.

Fig. 6. Burning rate *vs.* pressure for [M(Chz)ₙ](NO₃)₂, where M = Cu (1), Pb (2), Cd (3), Mn (4), Ca (5), Co (6), Ba (7), Zn (8), Mg(9), Sr (10), Ni (11). Solid line without points is Chz•HNO₃.

plexes of Cu and Pb).

Somewhat different behavior of metal atoms is inherent in combustion of metal salts of nitro compounds (Fig. 7). In this case all studied metals (Pb, Li, Na, K, Cs, Rb, Ca, Ba, Ag, Zn, Fe) exhibit catalytic activity in some extent or another. Fastest of picric acid salts, lead picrate, burns 50 times as high as the parent acid. In passing to lead salt of trinitrobutyric acid, only 5-fold increase has been obtained (Fig. 8). Similar picture is observed if comparing salts of dinitrophenols with salts of dinitrobenzoic acids [7].

It is known that the lead atom introduced in nitrophenol salts is bound to oxygen atoms of not only hydroxyl group but nitro group as well thus producing aci-form [8].

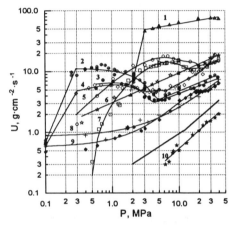

Fig. 7. Burning rate *vs.* pressure for picric acid (line without points) and its salts: Pb (1), K (2), Ag (3), Cs (4), Ba (5), Zn (6), Li (7), Mg (8), Fe (9), NH₄ (10).

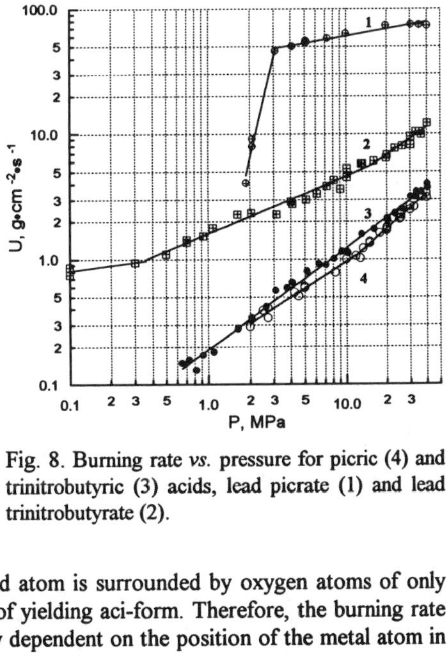

Fig. 8. Burning rate *vs.* pressure for picric (4) and trinitrobutyric (3) acids, lead picrate (1) and lead trinitrobutyrate (2).

Contrastingly, in lead trinitrobutyrate the lead atom is surrounded by oxygen atoms of only carboxylic groups and the molecule is incapable of yielding aci-form. Therefore, the burning rate of salts of nitro compounds turns out to be highly dependent on the position of the metal atom in the molecule.

When viewed as a whole, the burning rate of energetic materials, which combustion occurs through redox reactions, may be altered either by change in the reactivity of oxidizer and fuel or by introducing metal atoms into the molecule.

Let us consider now how the chemical structure of energetic materials capable of burning at the expense of heat evolved at the decomposition of endothermic groups may affect the burning rate. In studying combustion of 5-substituted tetrazoles the burning rate has been shown to correlate with thermal stability of tetrazole ring, which, in turn, is determined by electronic properties of the substituents (See Fig. 9.) [9]. Effect of entering metal atoms into endothermic energetic materials may be examined by use of coordination compounds with azide group as anion.   Fig. 10 presents the burning rate dependencies on pressure for complex compounds of Pb, Ni, Mn, Cd and Zn azides with carbohydrazide. It is notable that in this case too the lead compound burns faster than other complexes.

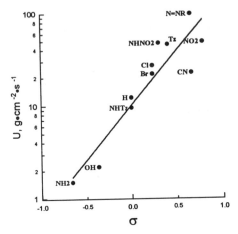

Fig. 9. Correlation between the mass burning rate of 5-substituted tetrazoles at 30MPa and Hammett's σ-values for substituents

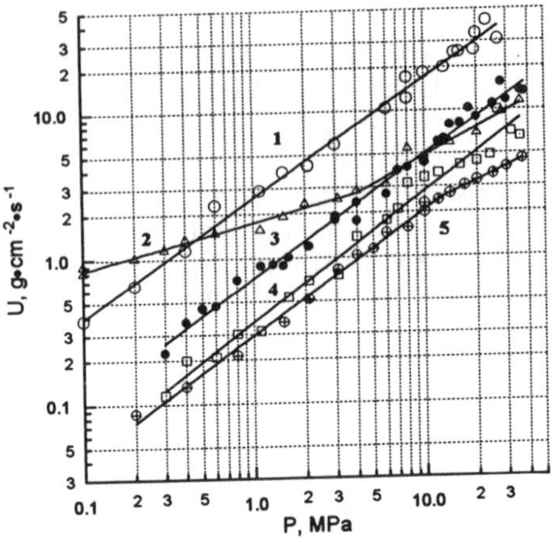

Fig. 10. Burning rate *vs.* pressure for [M(Chz)$_n$(N$_3$)$_2$,], M = Pb (1), Ni (2), Cd (3), Zn (4) and Mn (5).

DISCUSSION

A dramatic effect observed of both oxidizer and fuel on the burning rate of energetic materials is believed to indicate conclusively that the slowest stage of combustion involves redox reactions, whereas the rate of the initial decomposition reaction is subject to the rate of the subsequent oxidation stage. The reactivity of various reagents in redox reactions is often associated today with their redox potentials, the electrode reactions being considered as convenient models of the elementary electron-transfer step during redox reactions [10]. However, since the redox potentials are a measure of the free energy change in the thermodynamically reversible equilibrium, the usage thereof in the irreversible combustion process may seem to be incorrect. Actually this is not the case because it is conventional to consider elementary steps of redox reaction in the flame as being reversible in spite of the irreversible character of the combustion process as a whole [11].

Since redox potentials are a measure of the free energy change (ΔG^o) [12], when estimating the reactivity of oxidizers and fuels we simply correlate the burning rate of energetic materials with driving force of redox reactions:

$$\Delta G^o = - F (E_o^{ox} - E_o^{red}),$$

where F is the Faraday's number, E_o^{ox} and E_o^{red} are redox potentials of oxidant and reductant, respectively.

This correlation, of course, has merely a qualitative character due to the assumption that the change in the conditions from standard to high-temperature conditions in the flame leaves the ratio between the potentials substantially invariant. Besides, the free energy changes have been calculated for standard conditions rather than for flame ones. Nevertheless, in some instances, as in combustion of ethylenediamine salts (Fig. 11), there is a reasonable correlation between the burning rates and the free energy changes for ethylenediamine redox reactions with oxygen-containing acids.

156

In order to demonstrate the validity of a relationship between kinetic and thermodynamic combustion parameters an attempt has been made to describe kinetics of the leading chemical stage by the classic Marcus' electron transfer theory. It is widely used for analyzing and predicting the behaviors of various redox systems [12,13]. The Marcus' theory for electron transfer reactions relates the free energy of activation, $\Delta G^{\#}$, to the driving force of the reaction, ΔG^{o}.

$$\Delta G^{\#}= \gamma/4[1+(\Delta G^{o})/\gamma]^{2},$$

where γ is energy of change in bond lengths and angles of redox reagents during electron transfer. Calculated from Marcus' theory rate constants (k) of oxidation of organic bases (B) by oxygen-containing acids (HX),

$$k = Z\exp(-\Delta G^{\#} / RT_{g}),$$

where Z is collision frequency, have been used in calculating burning rates by Zeldovich's equation for the second-order gas-phase rate-controlling reaction [14, 15]. The calculations have been performed for the salts of B•nHX, where HX = HNO_3, HNO_2, HIO_3, $HClO_4$, $HClO_3$, $HBrO_3$, HIO_4; B = NH_3, CH_3NH_2, $PhNH_2$, p-H_2N-C_6H_4-$COOH$, o-ClC_6H_4-NH_2, o-CH_3-C_6H_4-NH_2, o-HO-C_6H_4-NH_2, and pyridine (n = 1); B = ethylenediamine, piperazine, o-H_2N-C_6H_4-NH_2, m-H_2N-C_6H_4-NH_2, p-H_2N-C_6H_4-NH_2, H_2N-C_6H_4-C_6H_4-NH_2 (n = 2).

A comparison of the observed burning rates with calculated ones is shown in Fig. 12. Despite many uncertainties involved, there is a certain correlation between calculated and experimental burning rates. This indicates that redox reactions play dominant role in combustion of energetic materials and obey the Marcus' rate-equilibrium relationship, which, in turn, is believed to be useful for understanding the combustion mechanism of different energetic materials.

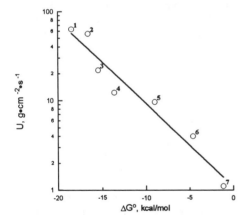

Fig. 11. Plot of the burning rate of ethylenedia-mine salts with inorganic acids at 10 MPa vs. free energy changes for oxidation of ethylenediamine by oxygen-containing acids. Point numbering corresponds to Fig. 1.

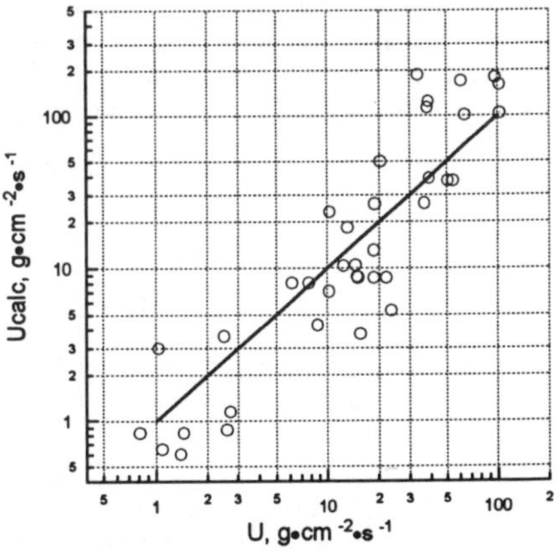

Fig. 12. Comparison of the experimental burning rates (U) of redox systems with calculated ones (Ucalc) from Zeldovich's and Marcus' equations.

The introduction of metal atoms into systems, in which the combustion process is determined by redox reactions, as for coordination compounds, results in the fact that certain of the compounds are beginning to burn faster than the control while the other compounds have a lower burning rate. The enhancement of the burning rate is attributable to the catalytic action of some metals during the redox process.

In our opinion the principle of catalysis implies that the metal atom M(II) capable of changing its valent state reacts with anion-oxidizer[1] in the combustion wave to yield an oxygen-containing specie, $[M(IV)=O]^{++}$ or $[M(III)-O^{\bullet}]^{++\,2}$, which includes the metal of a high-valent state. These species as the main active sites have been described in the works dealing with modeling ferment oxidation reactions [16,17]. The species have been shown by numerous studies on catalysis of oxidation reactions to be more active toward a reductant than the parent oxidizer itself [16, 17, 18]. The role of the metal consists substantially in facilitating the oxygen transfer from oxidant to fuel[3] [19],

$$[M(II)]^{++} + OX^- \rightarrow [M(IV)=O]^{++} \text{ (or } [M(III)-O^{\bullet}]^{++}) + X^-$$
$$L + [M(IV)=O]^{++} \{ \text{or } [M(III)-O^{\bullet}]^{++} \} \rightarrow L=O + [M(II)]^{++}.$$

According to the mechanism proposed, only metals capable of forming high-valent metal oxospecies, which are strong two-electron oxidants, can produce the catalytic effect. The stronger the oxidative properties of anion, the larger a body of metals capable of the catalytic activity. For example, among perchlorate complexes even Cd(II) (E^0 $Cd^{2+/3+} \sim 3$ V) exhibits the catalytic activity, whereas for nitrate complexes only lead (E^0 $Pb^{2+/4+} \sim 1.7$ V) can catalyze combustion (see Fig. 5-6).

When studying the combustion of ammonium perchlorate-based composite propellants doped with various metal-containing additives the catalytic effect was observed only for metals capable of transition to high-valent state [20]. This is likely to suggest a related mechanism of catalysis for the composites as well.

The catalytic effect like that observed at combustion of complex compounds of copper nitrate is characteristic of metals having oxidizing properties (E_o $Cu^{3+/2+} = 0.15$ V). Here, the metal activates fuel rather than oxidant. In the combustion wave, Cu(II) is able to oxidize ligand-fuel L to form the cation-radical $L^{+\bullet}$ which, in turn, splits out a proton to form $(L-H)^{\bullet}$ radical,

$$Cu(II) + L \rightarrow Cu(I) + L^{+\bullet}$$
$$L^{+\bullet} \rightarrow (L-H)^{\bullet} + H^+$$
$$H^+ + NH_3^- \rightarrow HNH_3$$
$$(L-H)^{\bullet} + HNH_3 \rightarrow \text{Oxidation products.}$$

Aminoalkyl radicals are known to be more strong reductants than parent amines [12], and, hence, the succeeding redox reaction between the radical and HNO_3 is favorable in terms of free energy changes and occurred faster than oxidation of the parent fuel by HNO_3.

Unlike explosive coordination compounds, metal atom introduced into the molecule of a nitro compound invariably causes an increase in the burning rate (see Fig. 7.). It may be suggested, that metal-based agents (oxides, carbonates, etc.) formed at the burning surface during combustion of the salts are serving as classic catalysts of oxidation. In this case the metal atom does not undergo changing in its oxidation state. Instead, the metal activates via prior coordination the oxidant or substrate, both formed during destruction of the nitro compound molecule, or arranges them in favorable geometry for oxidation to proceed and thus contributes to the redox reaction rate [21].

---

[1] It may be depicted as OX^-, where O is an oxygen atom.

[2] In brackets is a valent state of metal.

[3] L=O designates an oxidized form of ligand.

Since this kind of catalysis depends on the capacity of metal-based agents to coordinate the redox reaction reagents, the effect of catalysis is variable with temperature and pressure. It is well known for combustion of double-base propellants that the catalytic effect is practically lacking at pressures above 20 MPa [22]. For combustion of salts of nitro compounds, it decreases with pressure, but remains detectable even at 40 MPa.

In our opinion, this indicates that the role of the metal atom in combustion of salts of nitro compounds is not limited to the "classic" catalysis.

In recent years, specialists in the nitrocompound decomposition chemistry include more and more commonly the nitro-nitrite rearrangement in the reaction scheme [23],

$$R\text{-}NO_2 \;\rightarrow\; RONO \;\rightarrow\; RO\bullet + NO.$$

In combustion chemistry the mechanism involving the thermolytic cleavage of the C-N bond followed by interaction between NO_2 and fuel fragments is more accepted [24]. It is very likely, indeed, that in combustion of nitro compounds the nitro-nitrite rearrangement does not play a notable role due to its kinetic parameters. However, the situation is subject to variation if the reaction is catalyzed by metal atoms. Nitro - *aci*-nitro tautomerism may be yet another important rearrangement for thermal and redox reactions of nitro compounds. *aci*-Form is an intermediate in the Nef reaction which converts nitroalkanes to ketones. The assumption that catalysis of intramolecular rearrangements, together with classic catalysis, is at the basis of catalytic performance of metal atoms in combustion of salts of nitro compounds invites further investigations. Nevertheless, this assumption can explain well the catalytic activity observed for metals incapable of transforming to the high-valent state as well as the lower catalytic performance of metals which are introduced into the molecules incapable of nitro - *aci*-nitro tautomerism.

Combustion of some classes of energetic materials such as organic azides, salts of hydrazoic acid and tetrazole derivatives propagates through the heat released in decomposition of endothermic molecule fragments. Our investigations of combustion of these compounds have shown that, unlike above types of energetic materials, it is the decomposition reaction which determines the burning rate [9,25].

Coordination compounds of metal azides of common formula of $M(L)_n(N_3)_2$ belong to this class too, but include a metal atoms. The emergence of a ligand in the metal azide molecule capable initially of producing $N_3^\bullet$ radicals may result in that $N_3^\bullet$ can not only recombine but also react with the ligand to form HN_3. Further HN_3 multistage degradation to form nitrene moieties as well as associated progressive heat evolution are likely to be responsible for steady-state combustion of azide complexes which original metal azides fail to shows stable burning in themselves.

Burning rate data obtained for complex azides of Ni(II), Co(II), Mn(II), Cd, Zn, Pb(II) and Cu(II) with carbohydrazide, 4-amine-1,2,4-triazole and hydrazine appeared to fall satisfactory on a straight line in the lnU–1/T coordinates (Fig. 13). The activation energy (E_a) of the leading reaction at combustion of these materials proved to be close to E_a of the HN_3 decomposition reaction (168 kJ/mol) [26]. This may suggest that combustion of coordination azides too is governed by decomposition proc-

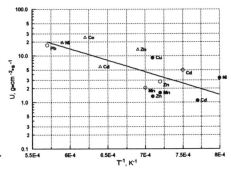

Fig. 13. Plot of the mass burning rate at 10 MPa *vs.* reciprocal flame temperature for azide complexes with carbohydrazide (O), 4-amino-1,2,4-triazole (●) and hydrazine (Δ).

ess. But, in contrast to redox systems, there is no association between the chemical nature of the metal atom and the burning velocity. The increased burning rate of the lead compound is due to higher combustion temperature rather than catalytic activity of the lead.

CONCLUSION

Burning rate characteristics of energetic materials from different chemical classes have been studied. Redox process is proved to be the rate-limiting stage in combustion of substances composed of fuel and oxidative fragments. The burning rate of such systems is subjected to variation by either changing in the reactivity of oxidizer and fuel or catalyzing the redox process with metal atoms as the molecule constituents.

For energetic systems capable of combustion at the expense of heat produced in decomposition of endothermic fragments of the molecule, it is the destruction process that controls the rate of burning. In this case the burning velocity is variable with chemical structure which, in turn, determines the rate of the decomposition.

The authors gratefully acknowledge support from the International Science Foundation and Russian Government through Grants M6Y000 and M6Y300.

REFERENCES

1. A.E.Fogelzang, V.Yu.Egorshev, V.P.Sinditskii, and M.D.Dutov, Combust. Flame **87**, 123 (1991).

2. N.N.Bakhman, Y.S.Kichin, S.M.Kolyasov, and A.E.Fogelzang, Combust. Flame **26**, 235 (1976).

3. B.N.Kondrikov, V.M.Raikova and B.S.Samsonov, Fizika Gorenia i Vzryva **1**, 84 (1973) (in Russian).

4. R.L.Simmions, Proc. 26th Int. ICT-Conference, Karlsruhe, Federal Republic of Germany, 4-7 July, 1995, p. 20-1.

5. V.M.Latimer, The Oxidation State of the Elements and Their Potentials in Aqueous Solution (William & Wilkins, N.Y., 1952).

6. A.E.Fogelzang, V.Ja.Adzhemian, and B.S.Svetlov, in Combustion and Explosion (Proc. III All-union Symp. on Comb. and Explosion, Nayka, Moscow, 1972), p. 63

7. A.E.Fogelzang, V.Ja.Adzhemian, and B.S.Svetlov, Dokl. Akad. Nauk SSSR, **236**, 688 (1977) (in Russian).

8. M.A.Pierce-Butler, Acta Cryst., C **40**, 63 (1984).

9. A.E.Fogelzang, V.Yu.Egorshev, V.P.Sinditskii, and M.D Dutov, Proc. 17th Int. Pyrotech. Seminar Combined with 2nd Beijing Int. Symp. on Pyrotech. & Expl., Beijing, 1991, Vol 2, p. 618.

10. R.A.Marcus, N. Sutin, Biochem. Biophys. Acta **811**, 265 (1985).

11. C.F.Melius, in Chemistry and Physics of Energetic Materials, edited by S.N.Bulusu (Kluwer Academic Publisher, the Netherlands, 1990), p. 51.

12. L.Eberson, Electron Transfer Reactions in Organic Chemistry (Springer-Verlag, Berlin, 1987).

13. R.A.Marcus, Ann. Rev. Phys. Chem. 15, 155 (1964).

14. Ya.B.Zel'dovich, Zhurnal Eksperimental'noi i Teoreticheskoi Fiziki 12, 498. (1942) (in Russian).

15. M.S.Miller, Combust. Flame 46, 51 (1982).

16. J.P. Collman, J.I.Brauman, and B.Meunier, J. Amer. Chem. Soc. 107, 2000 (1985).

17. J.T.Groves, R.C.Hausshalter, M.Nakamura, T.E.Nemo, and B.J.Evans, J. Amer. Chem Soc. 101, 1032 (1979).

18. J.Muzart, Chem.Rev. 92, 113 (1992).

19. V.P.Sinditskii, A.E.Fogelzang, V.Yu.Egorshev, and V.V.Serushkin, Proc. 21th Int. Pyrotech. Seminar, Moscow, 1995, p. 747.

20. C.W.Fong and B.L.Hamshere, Combust. Flame 65, 71 (1986).

21. R.A. Sheldon and J.K. Kochi, Metal-Catalysed Oxidation of Organic Compounds, (Acad.Press, New York, 1981).

22. G.Legelle, A.Bizot, J.Duterque, and J.F.Trubert, in Fundamentals of Solid-Propellant Combustion, edited by M.Summerfield (Progress in Astronautics and Aeronautics, 90, the American Institute of Astronautics and Aeronautics,1984), p. 301.

23. M.J.S.Dewar, J.P.Ritchie, and J. Alster. J. Org. Chem. 50, 1031 (1985).

24. C.F.Melius, Phil. Trans. R. Soc. Lond. A, 339, 365 (1992).

25. A.E.Fogelzang, V.Yu.Egorshev, V.P.Sinditskii, and B.I.Kolesov, Combust. Flame 90, 289 (1992).

26. O.Kajimoto, T.Yamamoto, and T.Fueno, J. Chem. Phys. 83, 429 (1979).

TOWARD QUANTITATIVE THERMODYNAMICS AND KINETICS OF PYROLYSIS OF BULK MATERIALS AT HIGH TEMPERATURE AND PRESSURE

G. K. WILLIAMS and T. B. BRILL
Department of Chemistry and Biochemistry, University of Delaware, Newark, DE 19716

ABSTRACT

An evaluation was made of whether T-jump/FTIR spectroscopy could be used to determine the decomposition kinetics (E_a and ln A) and thermochemical (ΔH_d) constants of an energetic material at high temperature and high heating rate. Polystyrene peroxide was chosen because of its known, simple, decomposition process. The kinetic constants are reasonable for O-O bond homolysis as the rate determining step: E_a = 39 kcal/mol, ln (A, s^{-1}) = 21.5. Significant uncertainty exists, however, in the estimation of ΔH_d.

INTRODUCTION

The experimental heat of reaction and kinetic parameters associated with the thermal decomposition of almost all energetic materials in the bulk state are difficult to obtain. DSC and TGA are common methods to obtain these global parameters. When applied to energetic materials, these techniques suffer several shortcomings. First, in DSC the heat release during thermal decomposition of most energetic materials is so rapid a time lag exists between the reaction and the recording of the response. Second, the decomposition chemistry can be affected by changes in the heating rate, pressure, and average temperature. DSC and TGA apply to relatively slow decomposition which may not be representative of combustion. Of greatest concern to us has been the kinetics and thermodynamics of fast decomposition which is representative of the surface reaction zone during combustion of an energetic material [1,2]. One approach to gain such data is to measure the time to ignition or explosion [3-5]. The desire to learn about species and rates simultaneously prompted development of T-jump/FTIR spectroscopy [6], which simultaneously records thermal events and gaseous decomposition products of a flash-heated material in near real-time. DSC and T-jump/FTIR spectroscopy are similar in that both record electrical requirements for maintaining a programmed temperature, but do so at very different heating rates. The voltage is extracted from T-jump/FTIR whereas DSC records power. This precludes direct calculation of the heat of reaction as in DSC. Therefore, an energetic material was sought with a "one-step," exothermic decomposition mechanism and known heat of reaction. The integrated voltage change in the T-jump/FTIR experiment might be used as a calibration standard to gain thermal data on other energetic materials. Kinetic data might be extracted either from the rate of the voltage change or time-to-exotherm. It would be desirable if the material could be cast as a thin film to maximize sample-to-filament contact for uniform heating.

Polystyrene peroxide (PSP) was chosen because it decomposes exothermally to an equimolar ratio of benzaldehyde and formaldehyde which account for more than 95% of the products. Kishore, et al., [7-11] have studied many aspects of PSP. Pressure or heating rate differences do not appear to change the decomposition mechanism. PSP is easily cast as a film from benzene. The unusually simple decomposition chemistry could also enable comparisons to be made among the kinetic methods of DSC, TGA, and T-jump/FTIR spectroscopy.

EXPERIMENTAL AND RESULTS

PSP was prepared by the "equimolar copolymer" procedure [12] and stored at 253 K in the dark. The resulting polymer had a monomodal distribution with weight-average (M_w = 19900) and number-average (M_n = 12400) molecular weights determined by GPC using polystyrene standards and THF solvent. Proton NMR and solid state IR spectra closely matched previous reports [13, 14]. The equimolar ratio of gaseous benzaldehyde and formaldehyde was confirmed in the IR spectrum by integrating v_8 of benzaldehyde [15] and v_3 of formaldehyde [16] and establishing the concentrations by the absolute intensity of these modes.

Kinetic Constants

In a typical nonisothermal TGA (DuPont Instruments Model 951) experiment, 16-20 mg of PSP was placed in a open Al pan situated on the hanger of the thermobalance under a flow of Ar, and was heated from ambient temperature to 438 K at 1 K/min. The first derivative of the TGA mass loss curve revealed a single mass loss step. The method of Coates and Redfern [17] was utilized to determine the kinetic parameters for this process. Table 1 gives the Arrhenius activation energy (E_a) in the 345-400 K range.

Nonisothermal DSC experiments were conducted on a DuPont Instruments Model 910 calibrated with an In metal standard. Samples of 5-15 mg were heated at rates of 1-10 K/min in open or hermetically sealed Al pans under a flow of Ar. The method of Barrett [18] was used to calculate the kinetic parameters. The heat of degradation (ΔH_d) of PSP (217±3 cal/g) corresponds to the area, W, of the exotherm. Any fraction, w, of W corresponds to the heat released in a particular time, t. Assuming that the amount of heat evolved is proportional to the number of moles of reacted, then $k = (dH/dt)/(W-w)$ for a first-order reaction. Since the temperature was continually rising, several rate constants are obtained from a single experiment. A plot of ln k vs $1/T$ yields Arrhenius data given in Table 1.

At high heating rates, kinetics were determined by T-jump/FTIR spectroscopy [6]. A Pt ribbon filament was situated inside a gas cell in an Ar atmosphere having a chosen, constant pressure. Detailed heat transfer models of this device are available [19, 20]. To maximize the heat transfer between the Pt ribbon filament and the sample, approximately 200 µg of PSP was cast onto the filament from benzene solution. The filament was placed under 0.10 torr for 10 minutes to remove the benzene. However, no combination of reduced pressure and mild heating could remove all of the benzene as evidenced by the IR spectra collected during thermolysis. By a separate TGA experiment, approximately 10% of the mass deposited on the filament remained as benzene after pumping. Because of the chemical inertness of C_6H_6, none of the kinetic features of the exotherms discussed below can be attributed to C_6H_6. Since pressure differences of 1-60 atm had no apparent effect on the decomposition of PSP, all T-jump/FTIR experiments were conducted at 1 atm Ar. A high-gain, fast-response power supply heated the Pt filament at 2000 K/s to a predetermined temperature. The power supply rapidly adjusted the power requirements to maintain the final temperature of the filament at a constant value during the experiment. An exothermic event required a decrease (negative deflection) in the voltage (V) necessary to maintain the filament at the set temperature. This control voltage was stored at 50 points/s during the experiment. As shown in Figure 1, there is a generally decreasing value of the control voltage throughout the experiment because of convective heat loss to the surrounding Ar atmosphere. Superimposed on this profile are the thermal events of PSP.

Table 1. Arrhenius Constants for PSP.

Method	T, K	E_a	ln (A, s^{-1})
TGA	345-400	21.5	----
DSC	362-396	27.5	32.4
DSC^b	405-415	32.5^c	----
T-jump	435-451	38.9	45.9
TTX	430-440	19.5	21.5

[a] kcal/mol   [b] Data from reference 8
[c] Reaction order changes from 2 at
   405 K to 1.2 at 415 K.

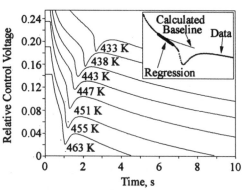

Figure 1. Difference control voltage traces
for PSP from T-jump/FTIR.

Figure 1 reveals a gradual release of heat prior to the main exothermic event. A similar, possibly related, sequence was observed in several of the DSC thermograms (Figure 2), although only a single mass loss step was observed by TGA. By IR spectroscopy, formaldehyde and benzaldehyde appear in the gas phase concurrently with the onset of heat release, and they rise rapidly in concentration at the strongly exothermic point.

The activation energy of the initial step of heat release was determined in the T-jump experiment by adapting Rogers' method [21] of using DSC data to obtain rate data. PEAKFIT software was used to calculate the baseline (a double exponential) and to fit the initial stage of the exotherm (an asymmetric double sigmoidal). At all temperatures $R^2 > 0.99$. The difference between the sample trace and the calculated baseline was measured. Since DSC and T-jump data reflect the rate of heat release directly, the deflection from the baseline, b, is directly proportional to the rate of reaction, $d\alpha/dt$, where α is the degree of conversion. Thus,

$$\beta b = d\alpha/dt = k(1-\alpha) \qquad (1)$$

where β is a proportionality constant and k is the rate constant. The relationships in equation (1) are valid for small values of α. Hence, $\ln b = \ln k/\beta + \ln (1-\alpha)$. For a first order reaction $\ln (1-\alpha) = -kt + c$, where c is a constant. By substituting and combining constants, $\ln b = C - kt$ is obtained, and rate constants for a first order process can be obtained directly from a plot of $\ln b$ vs t. Figure 3 shows the rate data and Table 1 contains the Arrhenius constants. A similar procedure was attempted for the initial exotherm in the DSC thermogram of Figure 2, but was inconclusive because of large uncertainty in the integration of this small exotherm.

Finally, zeroth-order induction time kinetics based on the Semenov model [4, 5] of the time-to-exotherm (TTX) [22] was explored. Here the time-to-exotherm as a function of temperature was obtained from the data in Figure 1. Equation (2) relates E_a and $\ln A$ to these data, where B is the intercept and C_p is the specific heat.

$$\ln TTX = \frac{E_a}{RT} + \frac{C_p RT^2}{\Delta H_d B E_a} \qquad (2)$$

A plot of $\ln(1/TTX, s^{-1})$ vs $1/T$ and equation (2) yields the apparent Arrhenius data in Table 1.

Thermochemistry

The integrated area of the control voltage trace should be directly proportional to the heat generated by PSP provided that the heat transfer coefficient between the filament and film of PSP is very large and little heat is transferred to any part of the system (e.g. the surrounding atmosphere) other than the filament. Of course, neither of the presumptions is strictly true, making the amount of heat sensed by the filament always less than the absolute amount produced by the sample. The fall-back position is to take the amount of heat generated from the decomposition of PSP (ΔH_d) which was determined above by DSC, and to correlate this value with a specific area defined by the control voltage deflection from the baseline. Figure 4 shows that the area from the average of five runs with a series of masses of PSP is approximately linearly related to the area of the control voltage trace, ΔV, but that the range of uncertainty is large. Equation (3) on Figure 4 gives the mass-area relationship. If equation (3) is used for calibration purposes, then equation (4) relates the heat generated by a compound x, ΔH_d (x), to the area ΔV, for a specific mass of x. For 200 µg of PSP, $\gamma = 13.1$ cal/volt-sec.

$$\Delta H_d(x) = \frac{\Delta H_d(PSP) \bullet \Delta V(x)}{\Delta V(PSP)} = \Delta V(x) \bullet \gamma \qquad (4)$$

DISCUSSION

The Arrhenius constants (Table 1) that are obtained for PSP from the deviation of the voltage trace from the baseline trace by T-jump/FTIR spectroscopy reflect the initial reaction stage before the process becomes strongly autothermal. The resultant activation energy of $E_a =$ 39 kcal/mol lies in the range of values for thermal homolysis of dialkyl peroxides of 30-39 kcal/mol [23]. However, ln $(A, s^{-1}) = 45.9$ from the T-jump/FTIR data on solid PSP is larger than the highest value reported [23] for dialkyl peroxides in the gas phase [ln($A, s^{-1} = 16.5$)]. The meaning of ln A for solid phase decomposition processes is much less certain than that for the gas phase. The larger A factor for the solid phase process could mean that the heat released during decomposition accelerates the rate to some extent compared to the reaction of more isolated molecules in the gas phase. The value of $E_a = 39$ kcal/mol resembles the O-O single bond strength of 38 kcal/mol for a dialkyl peroxide [24]. These facts suggest that the control voltage trace during flash thermolysis of PSP is capturing the homolysis rate of the O-O bond of PSP, although probably with an extraneous contribution from the autothermal nature of the process.

The value of E_a obtained in this work by DSC is lower than the value reported by Kishore [8], which was based on three data points in the 405-415 K range. No A factor was reported. Since ln A was determined from the DSC data in Figure 2, a comparison of rates by DSC, T-jump/FTIR spectroscopy and TTX can be made in Figure 5. Despite the apparent differences in E_a in Table 1, the compensation effect between E_a and A produces approximately the same rate of the overall process in the 400±20 K range. This suggests that all of the measurements are sensing essentially the same process, namely O-O homolysis. Extrapolation of the rate measured by any of these techniques into another temperature range yields a rather different prediction.

Unlike the kinetic analysis by T-jump/FTIR spectroscopy, the determination of ΔH_d for a bulk material at high temperature is subject to a rather large uncertainty. However the uncertainty appears to be least when approximately 200 µg sample are used. An uncertainty in ΔH_d of ±20% would be likely.

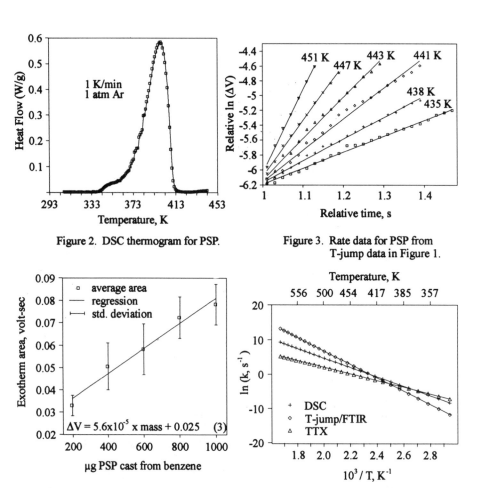

Figure 2. DSC thermogram for PSP.

Figure 3. Rate data for PSP from T-jump data in Figure 1.

$$\Delta V = 5.6 \times 10^{-5} \times \text{mass} + 0.025 \quad (3)$$

Figure 4. Relationship between the area of the thermal trace in Figure 1 and the sample mass.

Figure 5. Comparison of the Arrhenius plots from data in Table 1.

CONCLUSIONS

In principle, the rates of decomposition and the heat released from energetic materials heated at high rates to high temperatures can be obtained by T-jump/FTIR spectroscopy. The degree of success on the kinetics of PSP is good, but the thermochemistry is only marginally acceptable. From experience with the thermal traces of other energetic materials, we know that the successful analysis of the kinetics for PSP is partly attributable to the gradual slope of the exotherm before the violent release of heat occurs. Other energetic materials with this behavior can, therefore, be analyzed. On the other hand, the kinetics of decomposition of compounds which release a great deal of heat very early in the decomposition scheme (e.g. azide-containing polymers [25]) probably could not be characterized in this manner.

ACKNOWLEDGMENT

We are grateful to the Air Force Office of Scientific Research (NA) for support on F49620-94-1-0053.

REFERENCES

1. T. B. Brill, H. Arisawa, P. J. Brush, P. E. Gongwer and G. K. Williams, J. Phys. Chem. **99**, p. 1384 (1995).
2. T. B. Brill, J. Propul. Power **11**, p. 740 (1995).
3. J. Wenograd, Trans. Farad. Soc. **57**, p. 1612 (1961).
4. A. G. Merzhanov, Combust. Flame **11**, p. 201 (1967).
5. N. N. Semenov, Z. Phys. **48**, p. 571 (1928).
6. T. B. Brill, P. J. Brush, K. J. James, J. E. Shepherd and K. J. Pfeiffer, Appl. Spectrosc. **46**, p. 900 (1992).
7. R. P. Rastogi, K. Kishore and B. K. Chaturvedi, AIAA Journal **12**, p. 1187 (1974).
8. K. Kishore, J. Therm. Anal. **21**, p. 15 (1981).
9. K. Kishore and K. Ravindran, Macromolecules **15**, p.1638 (1982).
10. K. Kishore, V. Gayathri and K. Ravindran, J. Macromol. Sci. Chem. **A19**, p. 943 (1983).
11. K. Kishore and T. Mukundan, Nature **324**, p. 130 (1986).
12. R. E. Cais and F. A. Bovey, Macromolecules **10**, p. 169 (1977).
13. K. Kishore, J. Chem. Eng. Data **25**, p. 92 (1980).
14. K. Kishore and K. Ravindran, J. Anal. Appl. Pyrolysis **5**, p. 363 (1983).
15. R. T. C. Brownlee, D. G. Cameron, R. D. Topsom, A. R. Katritzky and A. J. Sparrow, J. Mol. Struct. **16**, p. 365 (1973).
16. T. Nakanaga, S. Kondo and S. Saëki, J. Chem. Phys. **76**, p. 3860 (1982).
17. A. W. Coates and J. P. Redfern, Nature **201**, p. 68 (1964).
18. K. E. J. Barrett, J. Appl. Polym. Sci. **11**, p. 1617 (1967).
19. J. E. Shepherd and T. B. Brill, 10th International Detonation Symposium, Office of Naval Research, Arlington, VA 1993, pp. 849-855.
20. S. T. Thynell, P. E. Gongwer and T. B. Brill, J. Propul. Power, submitted.
21. R. N. Rogers, Anal. Chem. **44**, p. 1336 (1972).
22. G. K. Williams and T. B. Brill, Combust. Flame **102**, p. 481 (1995).
23. R. Hiatt in Organic Peroxides, Vol. III, D. Swern, Ed. (John Wiley & Sons, Inc., New York, 1972), pp. 1-66.
24. S. W. Benson and R. Shaw in Organic Peroxides, Vol I, D. Swern, Ed. (John Wiley & Sons, Inc., New York, 1970), pp. 106-140.
25. T. B. Brill, P. J. Brush, D. G. Patil and J. K. Chen, 24th Symposium (International) on Combustion, The Combustion Institute, Pittsburgh, PA 1992, pp. 1907-1914.

THREE-PHASE COMBUSTION MODELLING:
FROZEN OZONE, A PROTOTYPE SYSTEM

MARTIN S. MILLER
Army Research Laboratory, Aberdeen Proving Ground, MD 21005

ABSTRACT

A number of efforts are currently underway in the U.S. to model the self-sustained combustion of solid energetic materials at the level of fundamental physical processes and elementary chemical reactions. Since many of these materials burn, at least at some pressures, with a liquid surface layer, these models must address phenomena in three physical phases. Most of these modeling efforts have focussed on pure RDX as a prototype. From a systems viewpoint this was a natural choice since it, or materials like it, is one of the components of modern propellants. It is argued here that, from a scientific viewpoint, RDX is already too complex with too many uncertain mechanisms and unavailable supporting data to serve this role effectively. We discuss here a number of issues related to the condensed phase and interfacial phenomena which have not been previously identified and which warrant more detailed research. In this paper frozen ozone is adopted as a more suitable prototypical 3-phase system. In this paper progress toward addressing this system is presented. The first new process to be modelled was a gas/surface reaction A detailed analysis was performed on the reaction mechanisms at play in the ozone flame and how they are affected by the heterogeneous reaction. Other new mechanisms are associated with the multicomponent nature of the liquid surface layer and will be addressed in future work.

INTRODUCTION

Over the last decade there has been an increasing interest and optimism that one might be able to describe the fundamental chemical and physical processes of energetic-material combustion in an *a priori* mathematical model. Nitramine combustion has been the subject of the bulk of this activity. One of the first of these models was proposed by Ben-Reuven, et al.[1,2,3] who, though treating the gas-phase chemistry as two global steps and the liquid chemistry as one global step, nevertheless considered all three phases, coupling the mass, species, and energy conservation equations using appropriate boundary conditions at the phase interfaces (including phase-transition enthalpies). Little decomposition of either RDX or HMX was found to occur in the condensed phase so that the dominant surface regression mechanism was nitramine evaporation at the surface. Later, Williams and coworkers[4,5], developed models for RDX and HMX combustion which considered a single global reaction in each of the liquid and gas phases but also allowed for nitramine vaporization at the surface. While agreeing with the Ben-Reuven, et al. that little decomposition occurs in the liquid phase for RDX, they found that 30% or more of HMX decomposed in the condensed phase. They did not treat the liquid layer explicitly and but did consider the effects of two-phase flow in the condensed-phase. Next, Melius[6] ratcheted up the complexity in his RDX combustion model by considering 38 gas-phase species related by 158 reactions, one global condensed-phase reaction, and evaporation of RDX at the surface. He found that little RDX decomposed in the condensed phase. Most recently, Liau and Yang[7] modeled RDX combustion, utilizing both the gas-phase reaction schemes of Melius and that of Yetter, et al.[8], which consists of 38 species and 178 reactions. Three condensed-phase reactions were considered along with evaporation of RDX. Bubbles in the liquid layer were treated using

169

a variable porosity concept which allowed retention of the one-dimensional formulation. This model explicitly considers solid, liquid, and gas phases phase. It was found that up to half of the RDX decomposes in the subsurface two-phase zone and that the gas bubbles occupy 45% of the volume at the surface.

For those models which find little nitramine decomposition in the condensed phase, the mechanism of surface regression is primarily evaporation. For RDX this is true in three of the four models mentioned above. This evaporation process, therefore, warrants close scrutiny. In the Ben-Reuven, et al. model, for example, the products of RDX decomposition in the condensed phase (c-phase) are assumed to be dissolved in the remaining liquid RDX and simply entrained in this liquid as it is convected to the surface. These c-phase products then desorb at a rate which is linearly proportional to the RDX evaporation rate. With the possible exception of the Melius model (evaporation rate computation not described), all of these models make use of the equilibrium vapor pressure of pure RDX as a function of temperature. In every case they assume that the heat of vaporization of RDX is not changed by the presence of the c-phase decomposition products. Such an assumption may ultimately prove to be a good approximation where little RDX has decomposed in the c-phase, however, in general it is not. The above models also assume that the sticking coefficient, i.e., the fraction of molecules impinging on the surface that actually are absorbed into the surface, for vapor-phase RDX is unity. This value of unity is in fact quite a good assumption for thermal energy vapors; however, in all of the above treatments gas-phase species other than RDX are implicitly assumed to have zero sticking coefficients, i.e., only RDX molecules may reenter the surface, despite the fact that the randomly directed thermal velocities are orders of magnitude larger than their convection velocity away from the surface. If one were to allow the non-RDX molecules to reenter the c-phase, it would be necessary to consider molecular diffusion in the liquid layer since their gradient would be steep at the surface and therefore diffusion could well affect the surface concentration. None of the models to date have considered it important to include liquid-phase diffusion.

The subtleties described above could be important since they affect the very mechanism by which the surface regresses. Yet the models discussed above for nitramine combustion are already formidably complex. Drawing firm conclusions about the importance of certain mechanisms using these models is greatly hampered by the considerable uncertainties associated with both the supportive data and physical/chemical mechanisms not being tested. Furthermore, sensitivity analysis cannot entirely relieve the dilemma because it is based on the assumption of small perturbations, and our understanding of many facets of the problem is still rudimentary at this stage. What is needed is a prototypical system which possesses key features of the nitramine problem but with less pervasive uncertainties in supporting data and a less distracting degree of chemical complexity. In our opinion, a study of the steady deflagration of frozen and liquid ozone meets these criteria, and this paper describes our current state of progress toward analyzing that problem. It should be admitted, however, that while the frozen ozone system is clearly the best prototypical system from a heuristic viewpoint, it is probably not the most convenient system to adopt from an experimental viewpoint. The sensitivity of the condensed phases of ozone to detonation is legendary and one must work at temperatures below 100 K. After exploiting the ozone case conceptually, it would be wise to seek another prototype of intermediate complexity enabling more intensive comparisons of model to experiment.

OZONE PROPERTIES AND DECOMPOSITION CHEMISTRY

In the 1950s liquid ozone was considered as a rocket-propellant ingredient and, consequently, much of the work to characterize its properties, production, and behavior was done during those years. At 1 atm ozone freezes at about 80 K and boils at 161.3 K. The color of gaseous ozone is light blue, liquid ozone indigo blue, and solid ozone deep blue-violet[9]. In any phase O_3 will readily detonate if subjected to heat, spark, flame, or shock. Liquid ozone is particularly sensitive to detonation as rapid heating or even cooling can cause initiation. A common cause of hyper-sensitivity was found to be certain organic impurities, and eventually it was learned how to avoid such contamination with a consequent increase in safety as well as reproducibility of measurements[10]. For example, it was found that suitably purified 100% gaseous ozone detonates reproducibly when its temperature reaches 105°C, whereas contaminated samples had detonated at much lower and unpredictable temperatures. The rate of thermal decomposition of liquid ozone appears not to have been measured, but one can infer that it must be quite slow at temperatures up to 243 K as it was determined not to have affected the measurement of vapor pressure up to that temperature[11]. Thus in a model of frozen or liquid ozone combustion, one can safely assume that the effects of any c-phase decomposition are negligible on the time scale pertinent to a steady deflagration wave. This eliminates a large source of uncertainty in the ozone case relative to the nitramine case. In addition, unlike RDX, the specific heat[12], the thermal conductivity[13], and the mass density[9] have all been measured for the liquid phase of ozone. The vapor pressure of ozone has been measured by a number of researchers[11,14,15] with consistent results.[16] The critical temperature for ozone was measured as 261 K corresponding to a deduced critical pressure of 54.6 atm.[9] Thermodynamic data and transport data used in the calculations reported in this paper are given in detail elsewhere.[16]

The reaction mechanism for the decomposition of gaseous ozone has been determined with considerable confidence[17] and consists of the following three reversible elementary reactions.

$$O_3 + M \rightleftharpoons O_2 + O + M \qquad \Delta H^0_{298.15K} = +25.65 \ kcal/mol \qquad \text{(I)}$$

$$O_3 + O \rightleftharpoons O_2 + O_2 \qquad \Delta H^0_{298.15K} = -93.41 \ kcal/mol \qquad \text{(II)}$$

$$O + O + M \rightleftharpoons O_2 + M \qquad \Delta H^0_{298.15K} = -119.06 \ kcal/mol \qquad \text{(III)}$$

The rate coefficients for the forward directions are taken from Heimerl and Coffee[18] who chose ozone as their prototype of an unbounded gas-phase flame before undertaking flames with more complex chemistry. Rate coefficients for the reverse reactions are computed from the equilibrium constants. These authors did a thorough review[19] of the literature on Reaction I and developed a new expression for the rate coefficient which best represents the experimental rate data over the temperature range 300-3000 K. Because we intended to use this rate at temperatures down to about 160 K, we compared their rate coefficient to the available cold temperature data[17]. The fit is quite good in the very low temperature range as well.[16] This single two-parameter Arrhenius rate coefficient accurately represents the rate data over 25 orders of magnitude! Since we intended to do calculations at pressures considerably above 1 atm, we modified the rate coefficient of Ref. 18 for Reaction I to account for the transition of the three-body reaction to a two-body reaction at high pressures. In the absence of more detailed information we used a Lindemann functionality to describe this pressure saturation effect. The limiting two-body rate was taken from Popovich, et al.[20]. The rates as used in the calculations reported here in the form required by the PREMIX code are given in Ref. 16.

In addition to the three homogeneous gas-phase reactions discussed above we included for study the following gas/surface reaction which seems probable.

$$O_3(l) + O(g) \rightleftharpoons O_2(g) + O_2(g) \qquad\qquad \Delta H_{298.15K} = -92.67 \ kcal/mol \qquad \text{(IV)}$$

This equation represents a gas-phase oxygen atom recombining with a liquid-phase ozone molecule on the surface to produce two gas-phase oxygen molecules. This heterogeneous reaction has a large exothermic heat of reaction which is deposited directly on the surface. Reactions II and III are also highly exothermic but heat from those reactions must be conducted from their distributed sites of deposition upstream to the surface. Since the O atom concentration in the gas at the surface is computed in the code, we use the flux of these atoms at the surface temperature multiplied by a reaction probabilty to compute the reactive flux associated with Reaction IV. Implementation of this new type of reaction requires a modification of the usual the boundary conditions in the gas phase at the surface, and will be discussed in the next section. Data on this reaction is not yet available but its effects can be bracketed by using the limiting values of 0 and 1 for the reaction probability. This appears to be the first time that the effects of a heterogeneous reaction on the burning rate of an energetic material has been calculated using a detailed chemistry code.

DESCRIPTION OF THE CODE

The model developed here considers each phase as a separate mathematical domain in which the mass, energy, and species conservation equations are solved subject to the appropriate boundary conditions conserving the various fluxes across each interface. Pressure changes are considered negligible in this and all of the models of this type, as is molecular diffusion in the liquid phase. The boundary conditions for the case without heterogeneous reaction are fairly standard and have been discussed elsewhere[3]. The presence of Reaction IV affects the boundary conditions on the fluxes of species and energy at the gas/liquid interface and the evaporative surface regression mechanism discussed below. The modified species boundary conditions are expressed as follows, P_{hrxn} being the heterogeneous-reaction probability (in this work taken to be either 0 or 1):

$$Y_O^{-0}\dot{m} = Y_O^{+0}\dot{m} + \rho(YV)_O^{+0} + \frac{1}{4}P_{hrxn}(\rho_O\bar{v}_O)^{+0} \qquad\qquad (5)$$

$$Y_{O_2}^{-0}\dot{m} = Y_{O_2}^{+0}\dot{m} + \rho(YV)_{O_2}^{+0} - \frac{1}{2}\frac{W_{O_2}}{W_O}P_{hrxn}(\rho_O\bar{v}_O)^{+0} \qquad\qquad (6)$$

$$Y_{O_3}^{-0}\dot{m} = Y_{O_3}^{+0}\dot{m} + \rho(YV)_{O_3}^{+0} + \frac{1}{4}\frac{W_{O_3}}{W_O}P_{hrxn}(\rho_O\bar{v}_O)^{+0} \qquad\qquad (7)$$

where $\dot{m}$ is the total mass flux, Y_i^{-0} and Y_i^{+0} are the mass fractions of species i on the liquid and gas sides of the surface, respectively, $(YV)_i^{+0}$ is the diffusion flux of the ith species on the gas side of the surface, $(\rho_O \, \bar{v}_O)^{+0}$ is the product of the mass density of atomic oxygen and the average molecular speed of atomic oxygen evaluated at the surface.

The boundary condition on the energy flux at the gas/liquid boundary is modified by the heterogeneous reaction as follows:

$$\lambda \left(\frac{dT}{dx} \right)^{+0} = \dot{m} \left(h_{O_3}^{liq}(T_s) - h_{O_3}^{c-phase}(T_0) \right) + \left(\dot{m} - \frac{1}{4} P_{hrxn}(\rho_O \bar{v}_O)^{+0} \right) \Delta h_{O_3}^{vap}(T_s)$$

$$+ \frac{1}{4 W_O} P_{hrxn}(\rho_O \bar{v}_O)^{+0} \Delta H_{hrxn}(T_s) \tag{8}$$

where the specific enthapy of vaporization at T_s is

$$\Delta h_{O_3}^{vap}(T_s) = h_{O_3}^{gas}(T_s) - h_{O_3}^{liq}(T_s) \tag{9}$$

and the molar reaction enthalpy at T_s for the heterogeneous reaction (Reaction IV above) is given by

$$\Delta H_{hrxn}(T_s) = 2 H_{O_2}^{gas}(T_s) - H_{O_3}^{liq}(T_s) - H_O^{gas}(T_s) \tag{10}$$

In addition to the heterogeneous reaction at the gas/liquid interface, the model considers evaporation of the energetic material. If there were no reactions, liquid ozone would exist at the surface in equilibrium with its vapor, and the gross evaporation flux would just equal the gross condensing flux. Gas-phase reactions serve to deplete the concentration of vapor molecules and therefore diminish the condensing flux. The evaporating flux is undiminished so that under these conditions a net evaporation flux leads to a regressing surface. One can infer the magnitude of the gross evaporation flux by using the equilibrium value of the condensing flux, which can be calculated from the equilibrium vapor pressure at the surface temperature and the kinetic theory result that the molecular flux crossing a plane in one direction is simply one fourth the product of the number density and the average molecular speed. Under reactive conditions one can use the computed gas-phase density of ozone at the surface to determine the condensing flux. Thus the expression for the net evaporating mass flux of O_3 is

$$\dot{m}_{O_3} = \frac{\alpha W_{O_3}}{4 R T_s} \left(\frac{8 R T_s}{\pi W_{O_3}} \right)^{\frac{1}{2}} \left(X_{O_3}^{-0} \, p_{O_3}^e - X_{O_3}^{+0} \, P_{total} \right) + \frac{W_{O_3}}{4 W_O} P_{hrxn}(\rho_O \bar{v}_O)^{+0} \tag{11}$$

where α is the sticking coefficient of the condensing molecules, W_{O3} is the molecular weight of ozone, T_s is the surface temperature p_{O3}^e the equilibrium vapor pressure at T_s, X^{-0} is the mole fraction of O_3 on the liquid side of the surface, X^{+0} is the computed mole fraction of gas-phase ozone at the surface, and p_{total} is the total pressure. The sticking coefficient is expected to be unity for thermalized vapors and this value is used exclusively in the calculations reported here. The equilibrium vapor-pressure expression of Jenkins and Birdsall[11] is used in all calculations.

If there are liquid components other than O_3 , i.e., $X^0 < 1$, then the total mass flux can be determined by the following artifice used by Ben-Reuven, et al.[3]

$$\dot{m} = \dot{m}\left(1 - Y_{O_3}^{-0}\right) + \dot{m}_{O_3} \tag{12}$$

$$\dot{m} = \frac{\dot{m}_{O_3}}{Y_{O_3}^{-0}} \tag{13}$$

where Y^{-0} is the mass fraction on the liquid side of the surface. These relations assume that the components other than O_3 escape the surface according to the same dynamics as O_3 , i.e., these other components are simply entrained by the O_3 .

The solutions for each phase are coupled through these boundary conditions and an iterative approach is used here solving the equations in each phase for a trial set of boundary conditions which are then adjusted and new solutions obtained in each domain until convergence is judged to have been achieved. The mass flux $\dot{m}$, surface temperature T_s , and the liquid-layer thickness X_{liq} are the eigenvalues of the problem and values are sought which satisfy the heat flux boundary conditions at the solid/liquid and liquid/gas interfaces and the above evaporation mechanism constraint. In the gas phase the solution is obtained using the PREMIX code[21]. Though for the ozone case we consider no chemical reactions in the liquid phase, in anticipation of applying the code to other energetic materials we have created a special version of PREMIX to treat the liquid phase. Therefore our code is capable of treating an arbitrary number of gas and liquid reactions. Although the same strategy could be followed for the solid phase, at present there is no capability for reactions there, however, a numerical integration over the solid phase is utilized permiting temperature-dependent properties there. The calculations reported here include the effects of thermal diffusion, though the burning rate changes by only a few percent if it is not considered. Also, the multicomponent transport properties were evaluated and used in the gas phase. Numerical accuracy of the burning rate and heat feedback was examined using up to 15,000 grid points in the gas phase. Results presented here used about 4500 grid points in the gas phase and should insure accuracy of a few tenths of a percent in the computed burning rates and heat feedback.

BURNING RATES AND THERMAL STRUCTURE

The computed burning rate for frozen ozone at an initial temperature of 40 K as a function of pressure is given in Fig. 1 and Tables 1 and 2. The first thing to notice is the relatively high value of the burning rate, about 2 cm/s at 10 atm. At that pressure and an initial temperature of 300 K, RDX burns an order of magnitude more slowly! The pressure dependence is also well approximated by the power-law dependence characteristic of many energetic materials; the pressure exponent is slightly less than 1, similar to RDX.

Fig. 2 shows the temperature and species profiles in each phase at 1 atm. On the scale of this figure the results with maximum heterogeneous reaction and none are indistinguishable, and the major difference in results between the two assumptions is in the O-atom concentration at the surface, being about 30 times smaller for the maximized effect of Reaction IV. Tables 1 and 2 summarize the numerical results.

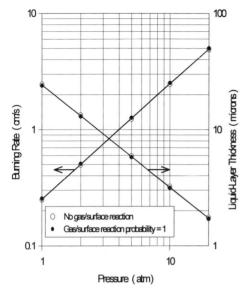

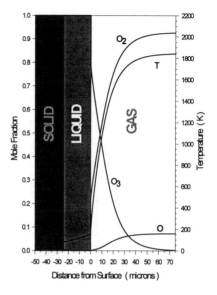

Fig. 1. Computed burning rates and liquid-layer thicknesses vs. pressure for frozen ozone at 40 K assuming maximum and minimum effects of gas/surface reaction.

Fig. 2. Computed temperature and species profiles in three phases for deflagrating frozen ozone at 1 atm and an initial temperature of 40 K.

TABLE 1. Summary of Results for 100% Solid Ozone without Gas/Surface Reaction at 40 K Initial Temperature. T_{bp} is the boiling point temperature.

p (atm)	r (cm/s)	T_s (K)	X_{liq} (microns)	$\lambda(dT/dx)^{+0}$ (cal/cm²-s)	T_{bp} (K)
1	0.2497	157.9	24.32	59.77	161.3
2	0.4985	168.4	13.22	120.8	172.3
5	1.242	184.8	5.89	307.0	189.6
10	2.470	199.6	3.20	621.7	205.5
20	4.906	217.4	1.73	1262.	224.7

TABLE 2. Summary of Results for 100% Solid Ozone with Unit Probability for Gas/Surface Reaction at 40 K. Φ_{hrxn} is the last term in Eqn. (8).

p (atm)	r (cm/s)	Ts (K)	X_{liq} (microns)	$\lambda(dT/dx)^{+0}$ (cal/cm^2-s)	Φ_{hrxn} (cal/cm^2-s)
1	0.2526	158.1	24.07	51.39	8.73
2	0.5052	168.5	13.06	104.2	17.6
5	1.257	184.9	5.83	264.8	44.4
10	2.505	199.8	3.16	538.5	89.2
20	4.953	217.6	1.72	1090.	179.

We were able to find only one experimental value of the burning rate of condensed-phase ozone. Streng[22] measured a value of 0.4 cm/s for a liquid mixture of 90% ozone with 10% molecular oxygen at an initial temperature of 90 K. Our calculations predict 0.27 cm/s at these conditions. While the computed value is in rough agreement with the experimental value, we believe that the quality of our input data warrants closer agreement. Of course, the experimental value may be inaccurate; the liquid was contained in a 9 mm Pyrex tube cooled outside by liquid oxygen and therefore the flame could have led to preheating of the liquid by conduction along the tube walls. However, our calculations show that an initial temperature of even 160 K is not enough to boost the rate to 0.4 cm/s. There is also the possibility of hydrodynamic instabilities in the liquid surface; this also would give a falsely high value to the measured rate. No visual observations of this sort were noted by the authors, who generally reported their experiments with considerable care. On the other hand, the boiling point of O_2 is very much less than that of O_3 (90 K compared to 161 K); therefore, the O_2 component should lower the boiling point of O_3 in the mixture and decrease its heat of vaporization. We may be seeing a case here where the heat of vaporization is significantly affected by the presence of as little as 10% other molecules. This issue may have direct relevance to the RDX case. (See discussion in FUTURE WORK section below.)

MECHANISM ANALYSIS

Fig. 1 shows that the maximum effect of the gas/surface reaction is slight, enhancing the rate by about 1% at 1 atm and 20 atm. Calculations have not been carried out at pressures greater than 20 atm because the model would have to be modified to account for the approach to the thermodynamic critical point at about 55 atm. There are two mechanisms by which the heterogeneous reaction might bolster the rate. Its heat of reaction is deposited directly at the surface whereas the heat from gas-phase reactions must be conducted upstream, therefore the heterogeneous reaction is more efficient in having its heat affect the burning rate. The other mechanism is the destruction of liquid-phase ozone by oxygen atoms, a process which contributes directly to the regression rate. Analysis of the calculations show that most of the heterogeneous-reaction contribution to the burn rate is by this latter mechanism. To understand how the heterogeneous reaction affects the gas-phase reaction mechanism, it will be helpful to examine the role of the various reactions in different parts of the flame.

Fig. 3 shows how the net rate of production of each species is spatially distributed through the flame. Four spatial zones are identified in the figure. In Zone A, adjacent to the surface, O_3 is being produced at the expense of O_2 and O. The region of most intense reaction activity is divided into two zones, B and C. In Zone B O_3 is now being destroyed, O_2 is now being produced, and O is being destroyed more vigorously than at the surface. In Zone C, O_3 continues to be destroyed and O_2 continues to be formed, but now O is being produced. Zone C is the site of most of the O_3 destruction. Finally, by Zone D, O_3 has been consumed and, though the scale of this figure does not show it, O_2 is being produced slowly and O is being destroyed slowly.

The mechanism is more completely revealed by examining the contribution to each species production rate by each reaction. An examination of these rates shows that immediately adjacent to the surface in Zone A, Reaction I is dominating and it is running in reverse! This is enabled by the diffusion of O atoms upstream from where they are produced in Zone C. Note that this Zone-A reaction is exothermic. Reaction III, in the forward direction, also contributes here. In Zone B Reaction II dominates, in the forward direction, producing O_2 at the expense of O_3 and O. Zone C is dominated by Reaction II and Reaction I, this time in the forward direction, producing O_2 and O at the expense of O_3. Finally, the recombination reaction III dominates in the post-flame region, Zone D.

An unexpected outcome of this mechanism analysis is that **the first chemical step (spatially) in the steady deflagration of frozen ozone is the production of ozone!** This unusual result had never been noted previously in studies of ozone gas flames[18], so we did an unbounded flame calculation for gas-phase ozone at 161 K. A small but distinct peak in the total heat-release profile due to Reaction I running in reverse could be seen in the leading edge of the flame consistent with the 3-phase calculations. When the free flame was given an initial temperature of 300 K (where previous studies had been conducted), this stage disappeared completely.

Returning to the 3-phase case, the effect of the reactions on the heat release profiles at 1 atm is shown in Fig. 4. It can be seen that the "simple" but real ozone chemistry manifests three distinct gas-phase stages of heat release. The stage closest to the surface results from the recombination of O and O_2 to form O_3, Reaction I running in reverse and the forward direction of Reaction III. This first stage coincides with Zone A. The next stage comprises Zones B and C. As can be seen in the figures, only Zones A and B are affected by the heterogeneous reaction which robs both zones of its O reactant. The final heat release stage is Zone D where O atoms slowly recombine to form the final product O_2.

The reason the heterogeneous reaction has such a slight effect on the burning rate can now be understood. First, note that the heat feedback to the surface can be expressed in terms of the volumetric heat release from the gas-phase reactions, $q(x)$, by the following simple expression (valid for constant c_p and λ)

$$\lambda \left(\frac{dT}{dx} \right)^{+0} = \int_0^\infty q(x) \, e^{-\frac{\dot{m} c_p x}{\lambda}} \, dx \qquad (14)$$

This expression shows that the heat released within a distance of about $\lambda/\dot{m}c_p$ is returned to the surface with good efficiency. At 1 atm this characteristic distance is of the order of 10 microns. By Fig. 4 one can see that the heterogeneous reaction decreases the gas-phase heat release within this distance. Thus heat gained at the surface from the heterogeneous reaction would have come back to the surface conductively from the gas phase anyway. The net effect is therefore almost neutral.

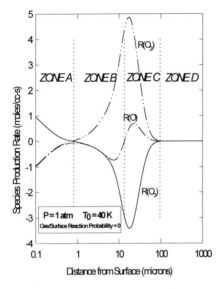

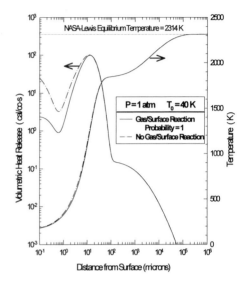

Fig. 3. Net species production rates in the gas phase of deflagrating frozen ozone at 1 atm and 40 K without heterogeneous reaction.

Fig. 4. Gas-phase heat release and temperature profiles showing effect of heterogeneous reaction at 1 atm. Both Zones A and B of Fig. 3 are affected.

FUTURE WORK

A number of questions remain on the ozone problem which have an important bearing on the modeling of other energetic materials. As alluded to in the Introduction, there is an inconsistency in all of the existing models (including the one in this paper) in the fact that the vapor-phase energetic material (ozone, RDX, HMX, etc.) is assumed to have unit sticking coefficient whereas the other gas-phase species are assumed to have a zero value. Relaxing this unphysical assumption, i.e., the zero values, will require consideration of molecular diffusion in the liquid layer. While it is true that diffusion in the liquid phase is generally much slower than in the gas phase, the importance of the process depends on the magnitude of the concentration gradients and these are expected to be very large close to the surface.

Secondly, the evaporation process has been highly idealized in existing models (again including the present one). Both the heat of vaporization and the rate of escape from the liquid surface has been assumed not to depend on the presence of other liquid-phase molecules. These other liquid phase molecules could be either decompositon products of liquid-phase reactions or gases adsorbed due to a nonzero sticking coefficient. What is needed to explore these effects is a broadly applicable theory for predicting heats of desorption from multicomponent liquids and a kinetic theory of evaporation from multicomponenent liquids. The current artifice of using the measured vapor pressure to estimate the gross rate of evaporation should be valid for a pure substance, but of little value for a multicomponent liquid, where both the heat of desorption and the rate of escape from the surface will be affected by the presence of other molecules.

Thirdly, extending model validity to pressures through the thermodynamic critical point may have considerable importance to the combustion of liquid energetic materials. This extension is related to the ability to predict heats of desorption from multicomponent liquids, as such a theory must account for the heat of desorption (or vaporization for a pure substance) going to zero at the critical point. Although beyond the scope of this paper, these areas are now being addressed and progress will be described in a future report.

We hope that through this paper and the outline given of future work, that a compelling case has been made for the value of modeling prototypical systems. While frozen or liquid ozone is the obvious choice from the standpoint of theoretical convenience, from an experimental viewpoint it is an extremely hazardous and inconvenient material to work with due to its detonation sensitivity and low melting/boiling points. Hydrazine (N_2H_4), with melting and boiling points near those of water, may be a natural compromise for a combined experimental and theoretical effort. We plan to explore further the feasibility of studying that system.

REFERENCES

1. M. Ben-Reuven and L.H. Caveny, AIAA Journal 19, 1276 (1981).

2. M. Ben-Reuven, L.H. Caveny, R. Vichnevetsky, and M. Summerfield, Proceedings of the 16th Symposium (International) on Combustion (The Combustion Institute, Pittsburgh, PA, 1976), p. 1223.

3. M. Ben-Reuven and L.H. Caveny, MAE Report 1455, Princeton University, Princeton, NJ, Jan. 1980.

4. T. Mitani and F.A. Williams, Twenty-first Symposium (International) on Combustion (The Combustion Institute, 1986), p.1965.

5. S.C. Li, F.A. Williams, and S.B Margolis, Combustion and Flame 80, 329 (1990).

6. C.F. Melius, Chemistry and Physics of Energetic Materials, edited by S.N. Bulusu (Kluwer Academic, The Netherlands, 1990), p.51.

7. Y.C. Liau and V. Yang, J. of Propulsion and Power 11, 729 (1995).

8. R.A. Yetter, F.L. Dryer, M.T. Allen, and J.L. Gatto, J. of Propulsion and Power 11, 683 (1995).

9. A.G. Streng, J. of Chemical and Engineering Data 6, 431 (1961).

10. C.E. Thorp, Bibligraphy of Ozone Technology, Vol. 2, Physical and Pharmacological Properties (John S. Swift Co., Inc., Chicago, 1955), p. 30.

11. A.C. Jenkins and C.M. Birdsall, J. Chemical Physics 20, 1158 (1952).

12. R.I. Brabets and T.E. Waterman, J. of Chemical Physics 28, 1212 (1958).

13. T.E. Waterman, D.P. Kirsh, and R.I. Brabets, J. of Chemical Physics 29, 905 (1958).

14. D. Hanson and K. Mauersberger, J. Chemical Physics 85, 4669 (1986).

15. R.C. Reid, J.M. Prausnitz, and T.K. Sherwood, The Properties of Gases and Liquids (McGraw-Hill, New York, 1977), p 629.

16. M.S. Miller, Proceedings of the 32nd JANNAF Combustion Meeting, Huntsville, AL, October, 1995.

17. H.S. Johnston, Report NSRDS-NBS 20, National Bureau of Standards, Sept. 1968.

18. J.M. Heimerl and T.P. Coffee, Combustion and Flame **39**, 301 (1980).

19. J.M. Heimerl and T.P. Coffee, Combustion and Flame **35**, 117 (1979).

20. M.O. Popovich, G.V. Egorova, and Yu. V. Filippov, Russ. J. Phys. Chem. **59**, 273 (1985).

21. R.J. Kee, J.F. Grcar, M.D. Smooke, and J.A. Miller, Report SAND85-8240, Sandia National Laboratories, March, 1991.

22. A.G. Streng, Explosivstoffe **10**, 218 (1960).

A Theoretical Study of BF + OH and BO + HF Reactions

M. R. SOTO Laboratory for Computational Physics and Fluid Dynamics, Code 6410, Naval Research Laboratory, Washington, DC 20375-5344, soto@lcp.nrl.navy.mil

ABSTRACT

The reactions of BF + OH and BO + HF are critical reactions in the reaction mechanism of fluorine-enriched boron combustion.[1] In this study, ab initio multiconfigurational methods have been used to calculate energies, optimized geometries, harmonic vibrational frequencies and zero-point energies for reactants, products and intermediates of these reactions. Results for the following reactive pathways will be discussed:

$$BF + OH \longrightarrow FBO + H$$
$$\longrightarrow trans{-}FBOH \longrightarrow FBO + H$$
$$BO + HF \longrightarrow H(F)BO \longrightarrow FBO + H$$
$$\longrightarrow cis{-}FBOH \longrightarrow FBO + H$$

INTRODUCTION

Reactions leading to the production of FBO have been identified as critical elementary steps in the reaction mechanism of fluorine-enriched boron combustion.[1] However, very little kinetic or mechanistic information is available on these reactions. Two of these reactions are:

$$HBO + F \longrightarrow FBO + H \qquad (1)$$

$$BF + OH \longrightarrow FBO + H \qquad (2)$$

A previous theoretical study[2] of different reactive pathways stemming from HBO + F proposed that reaction 1 proceeds via the formation of a H(F)BO complex. That is:

$$HBO + F \longrightarrow H(F)BO \longrightarrow FBO + H \qquad (1a,b)$$

In this study it was reported that reaction 1a is exothermic by 56.5 kcal/mol and proceeds without a barrier, whereas reaction 1b is endothermic by 9.9 kcal/mol and has a barrier of 13.0 kcal/mol. Furthermore, rate constant calculations done using transition state theory (TST) and canonical variational transtion state theory (CVT) showed that the complex formation dominates the HBO + F reaction at low temperatures, but at temperatures exceeding 1500 K it competes with the abstraction pathway, i.e.:

$$HBO + F \longrightarrow HF + BO$$

In recent experimental work by S. Anderson and his group[3] they found that the elimination of an FBOH complex is the dominant surface reaction when HF reacts with boron oxide clusters. They found that complex formation was more likely than the direct formation of FBO.

From both of these studies it appears that the formation of a FBOH complex or complexes plays a role in the combustion of B/O/H/C/F systems that is not fully understood. In an effort to understand the role of complex formation in this process and in order to obtain essential thermochemical and kinetic parameters for the model, ab initio multiconfigurational methods were used to study reaction 2 and reactive pathways stemming from BO + HF. More specifically, the reactions studied were:

$$BF + OH \longrightarrow FBO + H \qquad (2)$$
$$\longrightarrow trans-FBOH \longrightarrow FBO + H \qquad (3a, b)$$
$$BO + HF \longrightarrow H(F)BO \longrightarrow FBO + H \qquad (4a, b)$$
$$\longrightarrow cis-FBOH \longrightarrow FBO + H \qquad (5a, b)$$

THEORETICAL METHOD

Electronic structure calculations of the reactants, products and transition states of reactions 2-5 were done using a 17-electron-in-13-orbital (17/13) complete active space self-consistent field (CASSCF) wavefunction.[4] This wavefunction includes all single, double, triple, etc. excitations that are possible by exciting 17 electrons among the 13 orbitals. This CAS includes all of the bonding electrons, the lone pairs on O and the lone pairs on F. For the species discussed herein the optimized structures, harmonic vibrational frequencies and zero-point energies were obtained at the 17/13 CASSCF level. The calculations for BF + OH, HF + BO and FBO + H were done as supermolecules where the molecular fragments were placed 10 $\mathring{A}$ apart.

In many cases it is desirable to obtain energy differences at a high level of theory such as configuration interaction (CI) or multireference configuration interaction (MRCI). However, the cost of optimizing geometries at these levels can be prohibitive. In this event, it is helpful to do calculations at a higher level of theory using the geometries optimized at a lower level of theory.[5] In this study, single-point energy calculations were done using MRCI. These calculations include all doublet configurations that result from single and double substitutions from the configurations in a defined CASSCF reference space. These calculations exclude excitations out of the 1s core orbitals of oxygen, boron and fluorine and into the corresponding virtual orbitals. The MRCI calculations done here used a 5/5 CASSCF reference space and are referred to as 5/5 MRCI. The MRCI wavefunction consisted of 9.1 million configurations.

The basis set used was the augmented cc-pVDZ (aug.cc-pVDZ) developed by Dunning and his coworkers.[6,7] The aug.cc-pVDZ is the cc-pVDZ basis set augmented by (1s1p1d). The cc-pVDZ basis consists of a (9s5p1d) primitive set contracted to [3s2p1d] for first-row atoms and a (4s) primitive contracted to [2s] and augmented with a (1p) polarization function for hydrogen.

All calculations were done using MESA[8] on several supercomputer centers. These were the C90 and Cray YMP at CEWES in Vicksburg, Mississippi; the Cray YMP at the Arctic Region Supercomputer Center (ARSC); and the Cray YMP-EL at the Naval Research Laboratory (NRL).

RESULTS AND DISCUSSION

Figure 1 is the energy diagram of the stationary points for reactions 1-5.

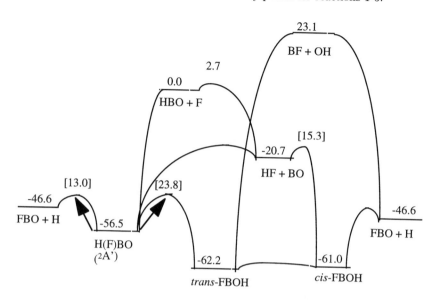

Figure 1. Energy Diagram for Stationary Points on the PES of FBO+H

All of the values shown in this figure are in kcal/mol. The numbers in square brackets are the barriers for the reaction in the direction shown by the arrows. The heats of reaction, and the barriers wherever available, are summarized in Table I. The values for reactions 1a and 1b were obtained from reference 2 and were not obtained using a 17/13 CASSCF. These were obtained using optimized geometries from a 11/10 CASSCF and single-point energy calculations at a 3/3 MRCI level.

Table I. E_A's and ΔH_R's in kcal/mol (including zero-point energy corrections) obtained at the 5/5 MRCI Level Using Optimized Structures from the 17/13 CASSCF Level

Reaction	E_A	ΔH_R
Reaction 1a	-7.4[2]	-56.5[2]
Reaction 1b	13.0[2]	9.9[2]
Reaction 2		-67.9
Reaction 3a		-85.3

Reaction 4a		-34.1
Reaction 4b	13.0	10.1
Reaction 5a	15.3	-40.2
Reaction 5b		16.2
$cis-FBOH \longrightarrow trans-FBOH$		-1.2
$H(F)BO \longrightarrow trans-FBOH$	23.8	-7.3

The complexes H(F)BO, *trans*-FBOH and *cis*-FBOH are shown below in Figure 2.

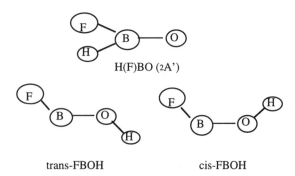

H(F)BO (2A')

trans-FBOH   cis-FBOH

Figure 2. Complexes of FBOH

The optimized geometrical parameters and harmonic vibrational frequencies obtained at the 17/13 CASSCF level are displayed below in Table II.

Table II. Optimized Structural Parameters and Harmonic Vibrational Frequencies for FBOH Complexes using a 17/13 CASSCF/aug-ccpVDZ

	H(F)BO	*trans*-FBOH	*cis*-FBOH
$R_{BO}{}^a$	1.366	1.361	1.358
R_{BF}	1.346	1.335	1.344
R_{BH}	1.189	—	—
R_{OH}	—	0.969	0.975
$\angle FBO^b$	117.9	118.9	122.0
$\angle HBO$	119.4	—	—
$\angle HOB$	—	110.5	110.0
$\nu_1{}^c$	2775.2	3824.7	3721.0
ν_2	1392.7	1384.5	1364.9
ν_3	1108.9	1177.4	1203.9
ν_4	1017.2	990.6	992.0
ν_5	889.3	583.2	610.7

| ν_6 | 439.5 | 484.8 | 467.8 |
| zpe^d | 10.91 | 12.08 | 11.96 |

[a] All bond distances are in Å. [b] All bond angles are in degrees. [c] All frequencies are in cm^{-1}. [d] All zero-point energies are in kcal/mol.

The lowest energy complex is the planar *trans*-FBOH, followed by the *cis*-FBOH and the trigonal planar H(F)BO. The calculation for the barrier for the *cis-trans* isomerization is in progress. The barrier for interconversion from the H(F)BO isomer to the *trans*-FBOH isomer is very large at 23.8 kcal/mol.

For BF + OH the exothermicity for complex formation is larger than the pathway leading to direct FBO formation. For reaction 2 (direct FBO formation) the exothermicity is 67.9 kcal/mol versus 85.3 kcal/mol for reaction 3a (complex formation). At this time it is not possible to comment on the barrier for these reactions or the conditions under which they may compete. Calculations are underway to characterize the transition states. Nonetheless, due to the large exothermicites predicted for both of these reactions it is expected that they both proceed with a small or zero barrier.

BO + HF are the products of the hydrogen abstraction pathway from HBO + F. As mentioned previously this reaction competes with reaction 1a at temperatures exceeding 1500 K. That is, BO + HF are possible products in the temperature regime of combustion. Therefore, it was important to explore the reactive pathways stemming from these products especially pathways leading to FBO. From this study it appears that these products will proceed to form one of the complexes H(F)BO or *cis*-FBOH. Both of these reactions are highly exothermic at -34.1 and -40.2 kcal/mol, respectively. The calculations of the barrier to reaction 4a has not been completed at this time. The barrier to reaction 5a is predicted to be 15.3 kcal/mol. Until the barrier to reaction 4a is obtained it is difficult to predict the conditions under which these two pathways will compete. Nonetheless, these preliminary calculations show that in the event that complex formation is bypassed in the reaction of HBO + F to form BO + HF, it is very likely that these products will proceed to form one of the two complexes mentioned. Furthermore, these complexes may proceed to form FBO + H.

In summary, these calculations show that pathways leading to a complex of FBOH from the reactants HBO + F and BF + OH are highly exothermic and proceed with little or no barrier. Also, there are exothermic pathways stemming from BO + HF that lead to FBOH complexes. The results of this study show that complex formation is the most likely mechanism by which FBO is formed from these reactants in fluorine-enriched boron combustion. The reactive pathways studied here along with the thermochemical and kinetic parameters calculated will be incorporated into the reaction mechanism developed by Brown, Kolb, Yetter, Dryer and Rabitz; nonetheless, experimental validation of these findings is needed.

ACKNOWLEDGMENTS

The author wishes to acknowledge fruitful discussions with Dr. Michael Page of North Dakota State University. This work was supported by the Office of Naval Research (ONR) under contract 95WX20198. The support of Dr. Richard Miller at ONR is gratefully acknowledged. Also, the computational work was done with grants for supercomputing time through the Department of Defense's High Performance Computing Modernization Plan (HPC-MP).

References

1. R. C. Brown, C. E. Kolb, R. A. Yetter, F. L. Dryer, H. Rabitz, Combustion and Flame, **101**, 221 (1995).
2. M. R. Soto, J. of Phys. Chem., **99**, 6540 (1995).
3. S. Anderson, private communication.
4. B. O. Roos, P. R. Taylor and P. E. M. Siegbahn, Chem. Phys., **48**, 152 (1980).
5. W. J. Hehre, W. J., L. Radom, P. v. R. Schleyer, J. A. Pople, Ab Initio Molecular Orbital Theory, John Wiley & Sons, New York, 1986, p. 95.
6. T. H. Dunning, Jr., J. Chem. Phys., **90**, 1007 (1989).
7. R. A. Kendall, T. H. Dunning, Jr. and R. J. Harrison, J. Chem. Phys., **96**, 6796 (1992).
8. P. Saxe, R. Martin, M. Page and B. H. Lengsfield, MESA (Molecular Electronic Structure Applications).

HIGH PRESSURE IGNITION OF BORON IN REDUCED OXYGEN ATMOSPHERES

R.O. FOELSCHE*, M.J. SPALDING**, R.L. BURTON*, H. KRIER**
University of Illinois, Urbana, IL 61801
  *Dept. of Aeronautical and Astronautical Engineering
  **Dept. of Mechanical and Industrial Engineering

ABSTRACT

Boron ignition delay times for 24 μm diameter particles have been measured behind the reflected shock at a shock tube endwall in reduced oxygen atmospheres and in a combustion bomb at higher pressures in the products of a hydrogen/oxygen/nitrogen reaction. The shock tube study independently varies temperature (1400 - 3200 K), pressure (8.5, 34 atm), and ignition-enhancer additives (water vapor, fluorine compounds). A combustion chamber is used at a peak pressure of 157 atm and temperature in excess of 2800 K to study ignition delays at higher pressures than are possible in the shock tube.

INTRODUCTION

Boron has been considered for many years as a prime candidate for high enthalpy fuel formulations, as an additive to solid propellant formulations, and as an additive for tailoring the energy release within explosive grains. Theoretically, boron shows great promise based on the high potential energy release both on a volumetric and gravimetric basis. The high enthalpy of combustion, high combustion temperature, and low molecular weight products also suggests a promising application in rocket propellants.

In practice the energy potential of boron additives in propellant formulations has been difficult to harness. Ignition is hindered by a protective oxide coating which is continually replenished by exothermic reactions until it can be removed from the particle surface. Boron combustion is made difficult by the high vaporization temperature (T_v = 4100 K at 1 atm) of boron which limits oxidation to slower heterogeneous surface reactions. Condensation of the product species (B_2O_3) is also a slow process which must occur at relatively low temperatures, and in the presence of hydrogenated species the product routes change in favor of gas phase species (HBO_2), reducing additional energy release from condensation.

Several theoretical studies [1,2] and more recent experimental findings [3-6] have warranted the examination of oxide layer removal-enhancing agents. Hydrogenated and fluorinated compounds show promise and many of these compounds already exist in typical propellant formulations. Experimental results from Macek [7] and Krier, et al. [3] indicate that operation at elevated pressures also improves boron particle ignition.

Macek and Semple [8] experimentally found reduced boron ignition times for wet flat-flames compared to dry flames over a range of temperatures below 2900 K at 1 atm pressure. More recently, Krier, et al. [3] have provided further evidence of the ignition-enhancing effects of water vapor and fluorine at higher pressures. Results from shock-initiated combustion of 24 μm boron particles indicate that water vapor (30%) in oxygen marginally reduces delays at ambient temperatures above 2600 K at 8.5 atm, whereas traces of fluorine (obtained by dissociating sulfur hexafluoride, SF_6, 1 and 3%) in oxygen significantly reduce ignition delay times between 1400 and 2700 K at 8.5 and 34 atm, when compared to pure oxygen alone. Hydrogen fluoride (HF, 6 and 12%) added to oxygen did not produce any beneficial decrease in ignition delay times between 2500 and 3100 K at 8.5 atm, suggesting that free fluorine atoms are likely the active agent, and not necessarily fluorine compounds as suggested earlier [9].

In this paper we introduce newer measurements of boron ignition delays in reduced oxygen atmospheres. Particle data are obtained behind reflected shock waves at 34 atm and at higher pressures in nitrogen diluted hydrogen/oxygen mixtures in a combustion bomb. We compare these newer data to previous determinations at lower pressures.

EXPERIMENTS

Shock-initiated particle combustion experiments are conducted behind the reflected shock wave in the endwall region of a 12 meter long, double diaphragm, cylindrical cross-section stainless steel shock tube, fully described elsewhere [3,10]. The driven section is 8.4 m long by 8.9 cm in diameter and is coupled to a 3.3 m long, 16.5 cm diameter driver section by a converging nozzle and diaphragm section. Optical access to the near-endwall region is obtained axially through the endwall by focusing optics. Test temperature, pressure, and composition of the oxidizing environment are independently varied to cover a wide range of operating conditions. Particles are mounted on a hobby knife blade and are swept off by the incident shock wave to be ignited in the hot, stagnated gases behind the reflecting shock [10].

The high pressure combustion chamber is a megajoule-class combustion bomb designed for transient pressures to 3400 atm and transient temperatures to 4000 K. The 66 cm high by 51 cm diameter vessel is fabricated from heat-treated 4340 carbon-steel. The vessel is capped by a 25 cm diameter plug which closes the central 2.0 liter cylindrical chamber cavity. For these experiments, a chamber insert has been used to given an effective 1.0 liter volume. All diagnostics are mounted in the plug, consisting of both static and dynamic pressure measurements, inlet gas temperature, and centerline-viewing optical emission detection through a high pressure optical window. Ignition of the gas mixture is achieved by an electrical discharge through a foil-type igniter which produces an intense spark at the anode/foil point of contact, even at high initial pressure. Particles are mounted in a recess on the tip of a cone situated 2 cm below the optical window, and are dispersed by the advancing combustion front. Calculations show that the particles do not reside in the cone's thermal layer nor are they dispersed beyond the optical field of view.

Several important characteristic timescales for both experiments are listed and compared in Table I. The fast-rising pressure and temperature profile of the reflected shock wave is ideal for particle ignition studies, although tests in our shock tube are limited to 34 atm and 3200 K by driver-section design considerations. The combustion chamber employs hydrogen/oxygen mixtures diluted with considerable nitrogen to avoid detonation. The rise time to ambient conditions is limited by flame front propagation through the chamber, and the pressure rise time is even longer, since the gases are free to expand in front of the advancing flame front. The temperature wave is localized, and

Characteristic Times, [μsec]	Shock Tube	Combustion Chamber
Pressure Rise	< 1 [a]	640 [b]
Temperature Rise	< 1 [a]	10 [c]
Total Test	500-1000 [d]	3000 [d]
Heat Loss	5000 [e]	5000 [e]
Wall Thermal Layer Growth	50 [e]	50 [e]

Table I : Characteristic times for shock tube and combustion chamber experiments.

[a] across reflected shock.
[b] of entire chamber.
[c] across deflagration front.
[d] within 10% of peak.
[e] predominantly radiation.

Fig. 1 : Photomicrograph of 24 μm crystalline boron particles with attached sub-micron parasitic particles, at 300x magnification. White micron bar length is 100 μm.

peak temperature is achieved immediately behind the flame front. Constant pressure conditions are achieved for several milliseconds after peak pressure is established.

Crystalline boron particles were obtained by sieving raw samples from Aldrich Chemical Co. (purity 99.0%) and Goodfellow, Corp. (purity 99.6%). The separate batches showed similar particle sizing and ignition characteristics. Statistical sizing analysis gives an effective mean diameter of 23.9±3.9 μm, determined by sedimentation technique [11]. A sample photomicrograph of the powder (Fig. 1) shows sub-μm sized parasitic particles clinging to the surface after sieving. Higher magnification exposures suggest that many of the smaller particles may be fused to the large particle surface. Smaller ~1 μm crystalline boron and unsieved amorphous boron oxide (B₂O₃) were obtained respectively from the Naval Research Laboratory [12] (purity 99.5%) and Aldrich Chemical Co. (purity 99.98%). The unsieved boron oxide particles range in size from sub-μm to greater than 100 μm.

SHOCK TUBE AND COMBUSTION CHAMBER RESULTS

Particle combustion time histories are measured indirectly by monitoring combustion emissions in the broadband and at selected wavelengths. Optical bandpass filters are selectively employed to monitor two reactive intermediary gas-phase species, the BO₂ molecule about the 545 nm band and BO about its 436 nm band [13]. These species are expected to form throughout the combustion event. This choice is supported by the experience of others [4,6] and our own spectral analysis obtained with a gateable optical multichannel analyzer (OMA). A sample shock tube spectrum obtained over a 200 μsec exposure initiated at the time of shock reflection from the endwall at 8.5 atm in pure oxygen burning ~1 μm crystalline boron is shown in Fig. 2.

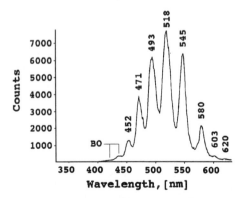

Fig. 2 : OMA spectrum of ~1 μm crystalline boron particles in pure oxygen showing characteristic BO₂ bands. BO bands at 422.8 and 436.3 nm are also detectable. Data obtained in shock tube at 8.5 atm, 3100 K with 200 μsec exposure.

Bandpass-filtered photodiode output as shown in Figs. 3 and 4 is analyzed to determine particle ignition

189

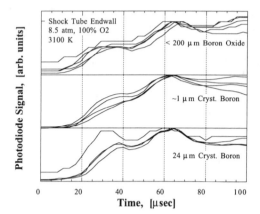

Fig. 3 : Normalized emission of amorphous boron oxide (B_2O_3) and crystalline boron particles detected by photodiode filtered at 546.1±5 nm.

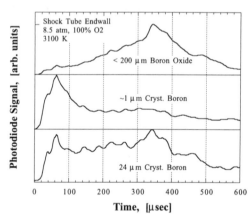

Fig. 4 : Amorphous boron oxide (B_2O_3) and crystalline boron particle emission detected by photodiode filtered at 546.1±5 nm.

and burn times for the shock tube. Photodiode traces in the combustion chamber have similar characteristics to the shock tube, but display high background gas emission at long times. The net combusting boron particle signal is obtained by subtracting out the background emission signal, which then shows the net boron signal returning to zero after ~ 2 msec (see Fig. 5).

Ignition is assumed to occur at the half-maximum point in the filtered photodiode output signal. Ignition delay time is defined in practice as the time from the temperature stimulus (reflected shock, combustion front) to the point of ignition, and the combustion (burn) time as the time from ignition to the time when the filtered photodiode output signal has fallen back to half-maximum [3,10]. For combustion chamber experiments, ignition and burn times are obtained from the net boron signal. Work to be reported [14] will include detailed data for the burn time of boron particles at these conditions.

Previous results in the shock tube for boron particle ignition [3] in pure oxygen reveal ignition delays (t_{ign}) which are a strong function of temperature below 2400 K for pressures of 8.5 and 34 atm and which asymptote to ~300 μsec at higher temperature for both pressures. The high pressure data can be expressed as a function of temperature by an equation of the form:

$$t_{ign} = A_0 \exp(E_0/T) \qquad (1)$$

where A_0 and E_0 are empirically fit constants. In pure oxygen at 34 atm $A_0 = 80$ μsec and $E_0 = 4100$ K between 1850 and 3000 K. With 1% SF_6 additive to oxygen at 34 atm, $A_0 = 44$ μsec and $E_0 = 4300$ K between 1450 and 2650 K. For 30% H_2O in oxygen at 8.5 atm, $A_0 = 8.3$ μsec and $E_0 = 9800$ K between 2150 and 2700 K. Ignition delays in pure oxygen are slightly reduced below 2400 K as pressure is increased from 8.5 to 34 atm, while delays with traces of fluorine are significantly reduced as pressure is increased.

Recent experiments in oxidizing mixtures of O_2 in Ar and O_2 + 1% SF_6 in Ar were performed at 34.3±1.9 atm pressure and 2950±87 K. Oxygen mole fractions were varied from 100% to 5%. The ignition delay times follow a relationship of the form:

$$t_{ign} = C_0 \exp(-D_0 X_{O2}) \quad (2)$$

where C_0 and D_0 are empirical constants and X_{O2} is the mole fraction of oxygen. For O_2 in Ar, the constant $C_0 = 840$ μsec and $D_0 = 1.0$. For O_2 + 1% SF_6 in Ar, $C_0 = 510$ μsec, $D_0 = 0.5$. Thus ignition delays roughly increase a factor of 2.5 for pure oxygen and a factor of 1.5 with fluorine present, when oxygen mole fractions are reduced to 5%.

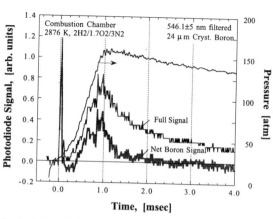

Fig. 5 : Filtered photodiode signal (546.1±5 nm) measurement showing full and net emission signals from 24 μm crystalline boron particles igniting in oxygen/water vapor/nitrogen mixture, in combustion chamber.

Ignition delay times are measured for 24 μm crystalline boron particles igniting in the reaction products of a 2.0 mole hydrogen/1.7 mole oxygen/3.0 mole nitrogen mixture. Data (Fig. 6) indicate the ignition delay time is 390±100 μsec. Peak combustion pressure is 157 atm, but particle ignition occurs while pressures are still rising in the chamber, typically at one-third of peak pressure. Equilibrium calculations predict the reaction products are composed of 11% O_2, 34% H_2O, 2% OH, 2% NO, and 51% N_2 and attain 2876 K with negligible decrease during the test time due to heat loss (Table I).

DISCUSSION

Measured ignition delays in the combustion chamber at ~52 atm is similar to reduced-oxygen shock tube data in Fig. 6 but is longer than shock tube data described by Eq. (1) at 8.5 atm when water vapor is present. Figure 6 also shows atmospheric pressure ignition delay data from Macek and Semple [8] using 34.6 μm particles. The Macek data show an order-of-magnitude longer ignition delay, which we

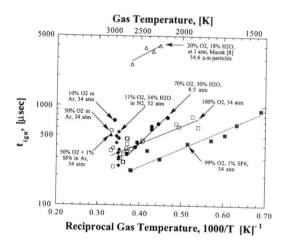

Fig. 6 : Ignition delay time for 24 μm crystalline boron particles at high pressure and various oxidizers. Flat-flame burner data [8] used 34.6 μm particles at 1 atm.

ascribe to the effect of pressure.

Our analysis of experimental data [11] on pressure effects indicates a small decrease in ignition delay in pure oxygen for a pressure increase from 8.5 to 34 atm, between 1900 and 2400 K, and no decrease above 2400 K. The recent modeling work of Yeh and Kuo [15] at 1800 K predicts a 3-fold decrease in ignition delay in a reduced oxygen/water vapor mixture when pressure is increased from 1 to 8.5 atm total pressure. Their model predicts an additional 25% decrease in delay when pressure is increased from 8.5 to 34 atm, showing a similar trend to our pure oxygen experiments. Additional delays may also be caused by finite particle heating times experienced in the Macek experiments. The difference in particle size is negligible based on numerous experimental determinations [4,7,11].

We observe a multiple-stage ignition and combustion process for crystalline boron. Bandpass-filtered emission signals (546.1±5 nm) for three distinct types of particles are compared in Figs. 3 and 4 from identical shock tube experiments at 8.5 atm and 3100 K. In Fig. 3 we compare the first 100 μsec of the normalized photodiode output for 24 μm crystalline boron, ~1 μm crystalline boron, and unsieved amorphous boron oxide (B_2O_3) powder. Figure 4 shows the entire emission signal for these same experiments out to 600 μsec.

Three local maxima are observed in the photodiode outputs for each of the three types of powders. The first two local maxima occur at 35 μsec and 65 μsec, displaying the same signature irrespective of particle size and composition. Since all three powders exhibit similar time histories up to 100 μsec, we believe that the thin (10-100Å) boron oxide layer is being removed in all three cases, leading to gaseous BO_2. The oxide layer thickness is postulated from boron combustion models [1,4,15].

We identify the first two peaks as the first combustion stage, defined in the literature as the oxide layer removal stage. Figure 4 shows that the oxide layer is substantially removed over the first 100 μsec. This suggests a lower limit for ignition delay time of 100 μsec in pure oxygen at 3100 K, 8.5 atm and indicates that the use of boron particles of any size requires significant ignition-enhancement to achieve a sub-100 μsec ignition delay. In Fig. 4 a third local maximum at ~350 μsec is shown for B_2O_3 and 24 μm boron particles. We attribute the third maximum for 24 μm boron to full-fledged particle combustion; i.e., the second stage combustion of the oxide-free boron particle. The third peak for B_2O_3 is presumable due to the emission signal of dissociating oxide from the larger (< 200 μm) boron oxide particles. For ~1 μm boron particles the first and second combustion stages are indistinguishable.

SUMMARY AND CONCLUSIONS

In our experiments the boron particles are exposed to high temperature and high pressure which quickly melt the particle and its oxide shell. The oxide removal is then controlled by vaporization (T_v = 2450 K) and surface chemistry processes [1,15]. During ignition, and later during combustion, the sub-oxides BO, BO_2, and HBO in hydrogenated mixtures, are liberated at the liquid-gas interface.

Ignition-enhancing agents which reduce particle ignition delays are believed to assist actively in the gasification of the oxide layer. Reactive species including H_2O, OH, H, F, and boron sub-oxides (BO, BO_2) are absorbed at the liquid oxide-layer surface and chemidesorb gas phase species after causing chemical bond rearrangement. Cluster beam experiments [5] indicate that H_2O and HF have greater reaction probabilities than oxygen. Our observations of ignition delays indicate that enhancement occurs marginally with H_2O above 2600 K, and does not occur significantly with HF.

In summary, the data indicate that boron particles at high pressure experience an oxide removal stage lasting ~100 μsec at 3100 K, that the high pressure (8.5-52 atm) reduces ignition delay by an order-of-magnitude compared to 1 atm, that the reduction in O_2 mole fraction to 5% increases ignition delay by a factor of ~2 near 3000 K, and that SF_6 exhibits significant ignition enhancement.

ACKNOWLEDGMENTS

The authors would like to thank the many contributions of S. Pirman, R. McMullen, C. Meyers, and D. Schneider for help in execution of the experiments, and the expert work of P. Hetman, D. Roberts, and R. Coverdill during fabrication of the combustion chamber facility and their continued expert advice. We are grateful for the many helpful comments of J. Black during the design of the combustion chamber. The authors acknowledge the sponsorship of the Office of Naval Research, Grant No. N00014-93-1-0654. Dr. R. Miller is the Project Director.

REFERENCES

1. R. C. Brown, et al., *International Journal of Chemical Kinetics* **26**, pp. 319-332 (1994).
2. R. A. Yetter, H. Rabitz, F. L. Dryer, R. C. Brown, C. E. Kolb, *Combustion and Flame* **83**, pp. 43-62 (1991).
3. H. Krier, R. L. Burton, S. R. Pirman, M. J. Spalding, *Journal of Propulsion and Power*, in press (1996).
4. S. C. Li, F. A. Williams, *Twenty-Third Symposium (International) on Combustion*, The Combustion Institute, Pittsburgh, 1990, pp. 1147-1154.
5. J. Smolanoff, et al., submitted to *Combustion and Flame* (1995).
6. S. Yuasa, H. Isoda, *Combustion and Flame* **86**, pp. 216-222 (1991).
7. A. Macek, *Fourteenth Symposium (International) on Combustion*, The Combustion Institute, Pittsburgh, 1972, pp. 1401-1411.
8. A. Macek, J. M. Semple, *Combustion Science and Technology* **1**, pp. 181-191 (1969).
9. M. K. King, *Journal of Spacecraft and Rockets* **19**, pp. 294-306 (1982).
10. T. A. Roberts, R. L. Burton, H. Krier, *Combustion and Flame* **92**, pp. 125-143 (1993).
11. S. R. Pirman, Masters Thesis, Department of Mechanical and Industrial Engineering, University of Illinois at Urbana/ Champaign (1994).
12. Crystalline boron sample (~1 μm) courtesy of Dr. T. Russell, Naval Research Lab.
13. R. W. B. Pearse, A. G. Gaydon, *The Identification of Molecular Spectra*, Chapman and Hall, London, 1976, pp. 57-59.
14. R. O. Foelsche, Ph.D. Thesis, Department of Aeronautical and Astronautical Engineering, University of Illinois at Urbana/ Champaign (1996).
15. C. L. Yeh, K. K. Kuo, "Theoretical Model Development and Verification of Diffusion/Reaction Mechanisms of Boron Particle Combustion," The Eighth International Symposium on Transport Phenomena (ISTP-8) in Combustion (1995).

HIGH-TEMPERATURE OXYGEN DISSOLUTION IN LIQUID ZIRCONIUM

I. E. MOLODETSKY [*], E. L. DREIZIN [**]
[*]   Princeton University, GEO Department, Princeton, NJ 08544, irina@spark.edu
[**]   AeroChem Research Laboratories, Inc., P.O. Box 12, Princeton, NJ 08542, eld@aerochem.com

ABSTRACT

The enthalpy of oxygen dissolution in liquid zirconium occurring during the combustion of Zr particles is estimated. The analysis presented uses direct experimental measurements of the temperature, size, and composition histories of burning Zr particles. The dissolution enthalpy is limited by the range of 700-830 kJ/mol and is somewhat less than that of Zr oxidation. This enthalpy determines the rate of the heat release during Zr combustion until stoichiometric ZrO_2 forms out of the supersaturated Zr/O solution. Stoichiometric ZrO_2 is formed near the end of combustion and the additional energy released at that time causes a rapid temperature increase which can trigger a particle explosion.

INTRODUCTION

Metals and metal additives are widely used in propellants and high energy fuels because of their high oxidation enthalpies. However, the theoretically expected energy release is often not achieved in practice due to incomplete reaction. The reaction pathways and the corresponding thermodynamic parameters for metal/oxygen interactions at high metal combustion temperatures need to be determined to analyze and enhance metal fuel performance. It was recently shown [1-6] that metal/oxygen liquid solutions (different from stoichiometric oxides), initially form upon high temperature metal/oxygen heterogeneous interaction. Oxygen dissolution is followed by the formation of stoichiometric oxide after the oxygen solubility limit is attained. These processes are particularly important in combustion of transition metals, i.e., Ti and Zr, which are able to dissolve significant amounts of oxygen. Recently, the role of oxygen dissolution during zirconium particle combustion in air was investigated experimentally [4,5]. Uniform spherical zirconium particles were produced and ignited using a novel pulsed micro-arc technique [7]. Real time particle combustion temperatures were measured using a three-wavelength optical pyrometer. Particles were rapidly quenched at prescribed times, embedded in epoxy, and cross-sectioned. Their internal compositions were determined using a wavelength dispersive spectroscopy scan (an SEM based technique). The experimental technique and results are summarized in Ref. [4,5]. The measurements have shown that a uniform Zr/O/N solution (rather than a layer or inclusions of stoichiometric oxide or nitride) forms in burning Zr droplets. The concentration of dissolved oxygen was observed to increase during combustion and, eventually, it exceeded the equilibrium solubility limit. By the end of a single particle combustion event, a supersaturated Zr/O solution was observed to transform into stoichiometric ZrO_2 oxide, which was accompanied by particle explosions. The formation of a Zr/O/N solution observed during Zr particle combustion implied that the enthalpies of oxygen and nitrogen dissolution, which could be different from the oxidation or nitridation enthalpies, support particle combustion until stoichiometric ZrO_2 starts to form. Therefore, the energy release during particle combustion can change throughout a single particle combustion event, i.e., from oxygen dissolution enthalpy to oxidation enthalpy. The energy balance was analyzed for a burning particle to uncover these changes and suggest when the reaction mechanism is modified. This analysis based on our experimental data [4,5] and its interpretation are presented below.

TECHNICAL APPROACH

An energy balance equation was constructed in which the particle heat content $\rho VC(dT/dt)$, where ρ is particle density, V is particle volume, C is its heat capacity, and dT/dt is the rate of particle temperature change, was balanced by the heat losses Q_{loss} and by an unknown term for the reaction heat release, Q_{react}:

$$\rho V C \cdot \frac{dT}{dt} = Q_{react} - Q_{loss} \qquad (1)$$

Both particle heat content and heat loss terms were determined using the experimentally measured temperature, size, and composition histories of burning zirconium particles of 240 μm initial diameter. The heat loss is comprised of radiative, Q_r, and convective, Q_c, terms. The radiative loss term for a spherical particle Q_r was computed as:

$$Q_r = \pi d^2 \sigma k \, (T^4 - T_\infty^4) \qquad (2)$$

where T is particle temperature, T_∞ is ambient temperature, d is particle diameter, σ is the Stefan-Boltzman constant, and k is the emissivity coefficient (the recommended value is 0.30 [8]). The convection loss term Q_c was determined as:

$$Q_c = \pi d \lambda_g Nu \, (T - T_\infty) \qquad (3)$$

where λ_g is gas heat conductivity and Nu is the dimensionless Nusselt number. For a sphere in a laminar flow [9]:

$$Nu = 2 + 0.6(dv/\eta_g)^{1/2}(\eta_g C_g/\rho_g \lambda_g)^{1/3} \qquad (4)$$

where v is the particle velocity, and η_g, C_g and ρ_g are the gas kinematic viscosity, heat capacity, and density, respectively.

Preliminary estimates showed that particle heat losses due to evaporation were negligible compared to the radiative and convective heat losses. These estimates used reference data on Zr and ZrO_2 evaporation, which are characterized by very low rates at combustion temperatures. The insignificant evaporation is consistent with our experimental results [4,5] as well as with earlier experiments [10-12] indicating that the change in the number of zirconium moles in a burning particle during combustion is negligible, and, therefore, the contribution of evaporation in heat and mass transfer processes could be disregarded.

The density ρ and heat capacity C of burning droplets consisting of a metal/gas solution are needed to determine the reaction enthalpy from Eq. (1) and are not known. However, the density and heat capacity of a Zr/O/N solution are expected to vary within the range covered by the densities and heat capacities of: a) zirconium; b) zirconium nitride; and c) zirconium oxide. To estimate ρ and C for the solution, we assumed that they were equal to those for Zr droplets with internal oxygen and nitrogen concentrations equal to the experimental values, but in which internal oxygen and nitrogen formed corresponding amounts of stoichiometric oxide and nitride.

RESULTS

The experimental droplet temperature history and the temporal changes of dissolved oxygen and nitrogen concentrations used as input data for the computations are shown in Figs. 1 and 2 (for a 240 μm diameter burning zirconium droplet). Particle enthalpy and heat losses were computed, and the unknown term for heat release was determined as a function of burning time using Eqs.(1)-(4). The computed heat release is shown in Fig. 3. In addition, the heat release that would be expected from zirconium oxidation and nitridation (i.e., the formation of stoichiometric products) with rates equal to the experimental oxygen and nitrogen uptake rates, respectively, was computed and is shown in Fig. 3.

The accuracy of the estimate made in this work depends on the accuracy of the experimental results [4,5] and on possible errors incurred because of the assumptions used. A possible systematic error of ±150°C (±6%) was associated with the temperature measurements, and random errors of droplet size and composition measurements were ±10% and ±5%, respectively. In addition, the possible inaccuracy caused by the unknown droplet thermophysical properties (density and heat capacity) used in the calculations was estimated by varying the density and heat capacity from those of pure zirconium to those of pure oxide. The bands in which the input parameters and the

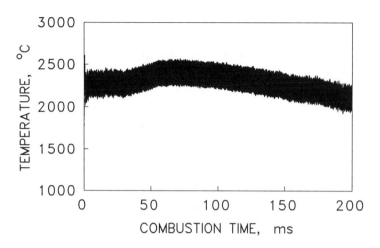

Figure 1. Experimental temperature history of a 240 μm diameter zirconium particle burning in air

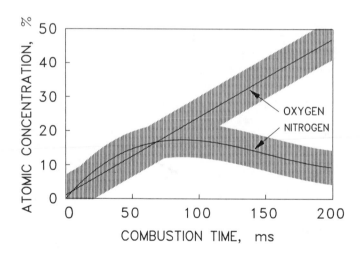

Figure 2. Experimental changes in concentrations of oxygen and nitrogen dissolved in a 240 μm diameter zirconium particle burning in air

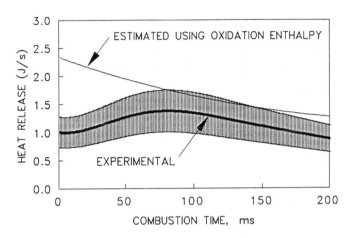

Figure 3. Estimated and experimental heat release of a 240 μm diameter zirconium particle burning in air

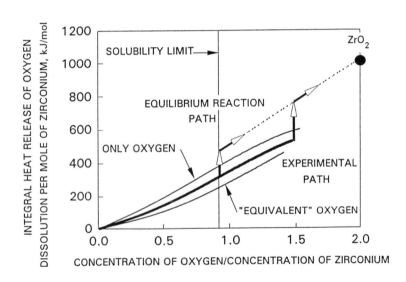

Figure 4. Integral heat release of oxygen dissolution in liquid zirconium and pathways of reactions occurring during zirconium particle combustion

enthalpies could be varied after accounting for the combined error are shown in Figs. 1-3. It can be seen that although possible errors are significant, the particle combustion heat release is lower than that inferred from the oxidation enthalpy. This effect should be considered while developing energetic materials or designing practical devices in which zirconium combustion is utilized.

The determined heat release was used to evaluate the enthalpy of oxygen dissolution during zirconium particle combustion. Possible variations in the dissolution enthalpy which could be caused by the temperature changes were disregarded. The enthalpy per mole of zirconium was estimated using a plot of the heat release integrated over the combustion time versus the concentration of dissolved oxygen (see Fig. 4 and the following discussion). A linear fit was found for the experimental curve, and its slope gave the dissolution enthalpy, which thus was assumed to be constant over the range of dissolved oxygen concentrations. The presence of dissolved nitrogen, in addition to oxygen, could somewhat alter the oxygen dissolution enthalpy, and was considered while making estimates, as discussed below.

The lower bound for the oxygen dissolution enthalpy (the curve marked "EQUIVALENT" OXYGEN in Fig. 4) was estimated assuming that both dissolved nitrogen and oxygen can be considered as an "equivalent dissolved gas" or "equivalent dissolved oxygen" [5]. The reason for such an assumption is that binary Zr/O and Zr/N systems are quite similar [13] in the temperature range in which zirconium combustion occurs. However, the Zr-N bonds in a Zr/O/N solution in which oxygen is the dominant solute are not as strong as the Zr-O bonds, i.e., the heat release would be greater if nitrogen atoms were substituted with oxygen atoms. Therefore, the dissolution enthalpy of 700 kJ/mol estimated for "equivalent" oxygen gives a value which is somewhat underestimated compared to the actual oxygen dissolution enthalpy.

An alternative treatment of the dissolved nitrogen was used to estimate the upper bound of the oxygen dissolution enthalpy (the curve marked ONLY OXYGEN in Fig. 4). In this treatment both the dissolved nitrogen atoms and the same number of zirconium atoms considered to be bonded to nitrogen, as in stoichiometric ZrN, were excluded from the Zr/O/N solution. Accordingly, the energy release determined from Eq. (1) and used to compute the enthalpy of oxygen dissolution was reduced by the enthalpy of ZrN formation. Because in the actual solution each of the Zr atoms was bonded to more than one atom of dissolved gases, ultimately approaching the pattern which would form in stoichiometric ZrO_2, the concentration of Zr atoms available in the considered "pure" Zr/O solution was somewhat reduced. Therefore, the dissolution enthalpy per mol of zirconium was somewhat overestimated. Another reason for some overestimation of the oxygen dissolution enthalpy in this case was that the experimental heat release was corrected using the enthalpy of formation of stoichiometric ZrN, which is less than that of an oxygen-rich Zr/O/N solution. The upper bound for oxygen dissolution energy obtained using the described approach was found to be equal to 830 kJ/mol.

DISCUSSION

The determined range of 700-830 kJ/mol, for which the enthalpy of oxygen dissolution in liquid zirconium can vary, is considerably different from the value of 1023 kJ/mol [14] for zirconium oxidation. Interestingly, no such differences were reported in earlier works [15,16] in which the enthalpy of formation of solid Zr/O solutions was measured. However, since zirconium particles melt during combustion, the difference which is observed to exist between the oxygen dissolution and oxidation enthalpies for liquid zirconium implies important corrections in the zirconium combustion thermochemistry. Indeed, until the stoichiometric oxide starts to form, the energy release of zirconium combustion is defined by oxygen dissolution enthalpy, which is some 19-32% lower than that of zirconium oxidation. On the other hand, the final product of zirconium combustion still is stoichiometric ZrO_2. For the reaction path at thermodynamic equilibrium, stoichiometric ZrO_2 should precipitate out of the saturated Zr/O solution. This path is indicated in Fig. 4 by the arrows marked EQUILIBRIUM REACTION PATH (the solubility limit is shown at the experimental Zr combustion temperature of ~2330°C). An equilibrium phase transition should occur at a constant temperature which is maintained by the latent heat of transition, as indicated by a vertical arrow. A further increase of the oxygen content in a binary Zr/O solution should result in an increased amount of stoichiometric ZrO_2 as shown by the second, oblique arrow. However, analyses of the cross-sections of rapidly quenched particles showed that a supersaturated Zr/O solution was formed under experimental conditions [4,5]. This is schematically represented by the second arrow set marked EXPERIMENTAL PATH in Fig. 4. In this case, the precipitation of stoichiometric oxide out of a metastable solution can result in a sudden temperature increase because of the release of the latent heat of transition; this can cause a temperature jump and/or particle explosion. This heat of transition is defined by the difference between the oxygen dissolution and oxidation enthalpies, and the range in which the heat of transition can vary is 193-323 kJ/mol. For a 240 μm diameter zirconium particle this is equivalent to a heat release of ~100-167 mJ. For reference, the energy in a micro-arc pulse which melts and heats up such a particle from a room temperature to the ignition temperature of ~2100°C is about 500 mJ. Therefore, it appears reasonable that such noticeable features as temperature jumps and/or particle explosions [4,5,10-12] can be caused by the precipitation of stoichiometric oxide out of supersaturated solution during zirconium particle combustion. These processes characterizing liquid zirconium/oxygen interaction should be considered when developing energetic materials utilizing zirconium combustion.

CONCLUSIONS

The enthalpy of high temperature oxygen dissolution in liquid zirconium has been determined in this work. This allows one to model zirconium combustion processes more accurately and predict the rates of heat release in practical combustion and propulsion systems in which zirconium energetic fuels are used.

ACKNOWLEDGMENTS

This work was supported by NASA Lewis Research Center under Contract No. NAS3-27259 and by the US Office of Naval Research under Contract No. N00014-94-1-0613.

REFERENCES

1. Dreizin, E.L., Suslov, A.V., and Trunov, M.A., *Combust. Sci. Tech.* **87**, pp. 45-48 (1992).
2. Dreizin, E.L., Suslov, A.V., and Trunov, M.A., *Combust. Sci. Tech.* **90**, pp. 79-99 (1993).
3. Dreizin, E.L. and Trunov, M.A., *Combustion and Flame*, **101**, pp. 378-382 (1995).
4. Dreizin, E.L., Molodetsky, I.E., and Law, C.K., Third International Microgravity Combustion Workshop, NASA Conference Publication 10174, 1995, pp. 129-134.
5. Molodetsky, I.E., Law, C.K., and Dreizin, E.L., (paper in preparation).
6. Dreizin, E.L., *Combustion and Flame*, accepted for publication, 1995.
7. Suslov, A.V., Dreizin, E.L., and Trunov, M.A., *Powder Technology* 74, pp. 23-30 (1993).
8. Lide, D.R. (ed.), CRC Handbook of Chemistry and Physics, 71st Edition, CRC Press, Boca Raton, 1991.
9. Bird, R.B., Stewart, W.E., and Lightfoot, E.N., Transport Phenomena, John Wiley & Sons, New York, 1960, pp. 408-410.
10. Nelson, L.S., Eleventh Symposium (Interantional) on Combustion, The Combustion Institute, 1967, pp. 409-416.
11. Meyer, R.T. and Breiland, W.G., *High Temperature Science*, **4**, pp. 255-271 (1972).
12. Nelson, L.S., Rosner, D.E., Kurzius, S.C., and Levine, H.S., Twelfth Symposium (Interantional) on Combustion, The Combustion Institute, 1968, pp. 59-70.
13. Massalski, T.B., Okamoto, H., Subramanian, P.R., and Kacprzak, L., (eds.) Binary Alloy Phase Diagrams, Second edition, ASM Publ., 1990.
14. "JANAF Thermochemical Tables," *J. Phys. Chem. Ref. Data*, **14**, Suppl. 1 (1985).
15. Mah, A.D. and Kelley, K.K., Report of Investigations 5316, Bureau of Mines, Washington, 1957.
16. Boureau, G. and Gerdanian, P., *High Temperatures - High Pressures*, **2**, pp. 681-693 (1970). (in French).

SELF-AFFINITY OF COMBUSTION-GENERATED AGGREGATES

A. V. NEIMARK, Ü. Ö. KÖYLÜ, D. E. ROSNER
Department of Chemical Engineering, Yale University, New Haven CT 06520-8286

ABSTRACT

A large population of combustion-generated soot aggregates (more than 3,000 samples) was thermophoretically extracted from a variety of laminar and turbulent flames and analyzed using transmission electron microscopy (TEM). It was shown that the scaling structural properties of these fractal aggregates cannot be exclusively characterized by a single mass fractal dimension. *Asymmetric* properties of the aggregates were considered here by first assuming and then demonstrating their *self-affinity via.* an affinity exponent reflecting scaling with respect to the length and width of the aggregate projections. In addition to the conventional fractal dimension, D_f, determined by using the geometrical mean of the longitudinal and transverse sizes as the characteristic length, the affinity exponent, H, and two complementary fractal dimensions, one longitudinal, $D_L = [(1+H)/2]D_f$, and one transverse, $D_W = [(1+H)/2H]D_f$, were introduced. By fitting TEM data for the entire population of aggregates, the values $D_f = 1.75$ and $H = 0.91$ were obtained. This more complete description of aggregate morphologies in terms of the self-affine scaling is expected to lead to a better understanding of the transport properties and restructuring kinetics of flame-generated aggregates.

INTRODUCTION

Fractal aggregates and patterns are produced in many engineering and natural environments. Although their morphology may look very complex, they are usually characterized by a single structural parameter, namely the mass fractal dimension, D_f[1-3]. Moreover, the products of aggregation are divided into a few "universality" classes, each characterized by a certain value of D_f. The most practically important universality classes are Diffusion Limited Aggregation (DLA), Diffusion Limited Cluster-Cluster Aggregation (DLCCA), Ballistic Cluster-Cluster Aggregation (BCCA), Reaction Limited Cluster-Cluster Aggregation (RCCLA), and Percolation Clusters (PC). It has been supposed that the fractal dimension, D_f, reflects the main features of the pattern formation procedure. For example, the products of DLCCA constitute a universality class, characterized by fractal dimensions *ca.* 1.75 in 3D. Typical examples of DLCC aggregates are liquid phase products like colloidal gold and gas phase products like aggregates of metal particles in a dense vapor, or flame-generated "soot" aggregates. The latter are the main focus of the experimental part of this paper.

Mass fractal dimension D_f is defined as the dimensionless exponent in a scaling relationship between the total number (or mass) of primary particles, N, and linear size of aggregates, R, 'reduced' by the primary particle radius, a_p:

$$N = \alpha \ (R/a_p)^{D_f} \tag{1}$$

The coefficient α, denoted by Mandelbrot [1] as a first order *lacunarity,* is a complementary structural parameter characterizing the aggregate density and cutoff of fractality. It is usually assumed that Eq. (1) is valid for any characteristic linear size of the aggregate, and, in contrast to the lacunarity, the choice of this size does not affect the inferred fractal dimension. This is one consequence of the hypothesis of *self-similarity* of aggregates. Theoretically, the most straightforward choice is to use the radii of gyration of aggregates. However, for 3D aggregates with $D_f < 2$, comprised of small spherical particles, while projected lengths and primary particle radii are easily obtained from TEM data by analyzing 2D projections of the aggregates, the gyration radius can not be measured directly. Instead, following the work of Weitz and Huang [4], the geometric mean $(LW)^{1/2}$ of the length of the longest axis, L, and the width, W, of the aggregate projection measured perpendicular to this axis is commonly employed as the characteristic linear size to be used in the scaling relationship of Eq. (1) for defining fractal dimensions, *i.e.* :

$$\langle N \rangle = \alpha \left((LW)^{1/2} \right)^{D_f} \tag{2}$$

Here $\langle ... \rangle$ denotes averaging over aggregates with given length L and width W of the plane projection. Note that here the length L and width W are dimensionless, *i.e.*, they are measured in multiples of the mean primary particle diameter. The structural parameters, lacunarity α and "geometric mean" fractal dimension, D_f, are then obtained by using least-squares fitting for a given population of fractal aggregates, or, more precisely, for a given set of TEM images.

In previous papers of Botet and Jullien [5], and Lindsay et al. [6], the anisotropic properties of cluster-cluster aggregates were discussed based on the results of numerical simulations. The simulated aggregates were essentially found to be anisotropic, however the anisotropic structural parameters were scale invariant and could be expressed in terms of self-similarity. The goal of this paper is to consider the aggregates found in real engineering environments. Based on a statistical analysis of a large population of combustion-generated carbonaceous soot aggregates (more than 3,000 samples), we critically examine the hypothesis of self-similarity with respect to the samples thermophoretically extracted from a wide variety of flames. We show that these aggregates exhibit anisotropic scaling which is size-dependent, and therefore their morphology can be better explained in terms of *self-affinity* rather than self-similarity.

EXPERIMENTAL METHODS

Since the present study deals mainly with the detailed characterization of combustion-generated aggregates, the following description of the experimental procedures is brief, and additional details can be found in Köylü and Faeth [7,8], Sunderland *et al.* [9].

Thermophoretic sampling procedures to extract representative aggregates from flame environments were based on the experimental and theoretical methods established by Dobbins and Megaridis [10], and Rosner *et al.* [11], respectively. The sampling surfaces were carbon-supported copper grids used to hold TEM specimens, aligned parallel to the mean flow direction. The probes were stored outside the flames and inserted briefly into the flame environment using a double-acting pneumatic cylinder. Sampling times were 30-100 ms in order that the collected soot aggregates cover no more than 10% of the TEM grid.

The samples were observed using a JEOL 2000FX analytical electron microscope system with a 1 nm edge-to-edge resolution. The images were processed on a computer to find aggregate areas and geometric dimensions, as well as the sizes of primary particles. A typical carbonaceous soot aggregate sampled from a turbulent acetylene flame is shown in Fig. 1. Primary particle sizes

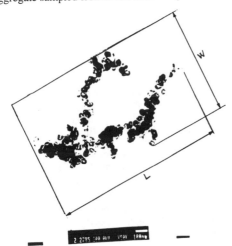

Figure 1. TEM photograph of a typical combustion-generated carbonaceous soot aggregate thermophoretically sampled from a turbulent acetylene/air diffusion flame.

in a set of aggregates sampled from a particular flame condition can usually be represented by a single mean radius, a_p, due to narrow size distributions. Since $Df < 2$, the number of primary particles, N, in an aggregate was estimated from aggregate projected area [7,12].

The TEM measurements involved thermophoretic sampling of carbonaceous soot aggregates from a variety positions in different diffusion flame environments, including the non-luminous (overfire) region of buoyant turbulent flames fueled with various gases and liquids [7], luminous (fuel-rich) region of buoyant laminar flames fueled with acetylene or ethylene [8], and of a weakly-buoyant acetylene flame operating at 1/4 atmosphere [9]. These various experimental conditions resulted in an aggregate population of 3080 images, which were analyzed following the specific methods of Köylü et al. [12]. Experimental uncertainties (95% confidence interval) of aggregate morphological parameters obtained from TEM images were estimated to be generally less than 1-2% due to our large population of 3080 aggregates.

RESULTS AND DISCUSSION

Self-Affinity of Fractal Aggregates

The scaling relationship (2) is valid for the combustion-generated aggregates under consideration. Figure 2 illustrates number of primary particles in aggregates, N, as a function of characteristic linear size, $(LW)^{1/2}$, with the least squares fit to data yielding the fractal dimension $D_f = 1.75 \pm 0.01^*$ that is typical for DLCCA. TEM images (see Fig. 1) show that the aggregate shapes are mostly asymmetric, and the aggregate projections are almost always elongated, with the length and width being considerably different. The aspect ratio L/W is found to vary in random fashion between 1 and 4, tending to increase with increased aggregate size. This observation suggests that the aggregates may be *self-affine* rather that self-similar. Supposition of statistical self-affinity implies that the aggregate shape is invariant (in a statistical sense) with respect to anisotropic scaling transformations (see, for example, [13-15]), i.e., with respect to shrinking or stretching with different scaling factors, n_L and n_W, in the longitudinal and transverse directions:

$$L \to n_L L \quad ; \quad W \to n_W W \tag{3}$$

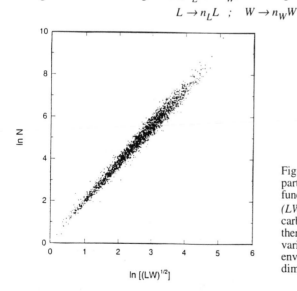

Figure 2. Number of primary particles, N, in aggregates as a function of the characteristic length, $(LW)^{1/2}$ for a population of 3080 carbonaceous soot aggregates thermophoretically sampled from various diffusion flame environments, yielding the fractal dimension, D_f.

---

*Here and below, the margin of error is calculated using the standard method of random uncertainty with a confidence level of 95%.

Thus, the length L and width W of self-affine aggregates are related through the so-called affinity exponent, H, defined as a ratio of the logarithms of longitudinal and transverse scaling factors (32):

$$H \equiv ln(n_W) / ln(n_L) \qquad (4)$$

Accordingly, an increase in length by the factor n is associated with an increase in width by the factor n^H, so that, on the average,

$$W = W_0 (L / L_0)^H \qquad (5)$$

Here L_0 and W_0 are the longitudinal and transverse dimensions of some reference aggregate.

For self-affine systems the affinity exponent is less than unity, $H < 1$, and it follows from Eq. (5) that with an increase in size the aggregates tend to become more and more elongated. For self-similar systems, $H = 1$ and the aggregate shape is invariant with respect to isotropic scaling transformations. In other words, length and width are directly proportional only for self-similar aggregates.

Self-affinity implies more than geometric allometry. It implies an hierarchical structure of given objects. Self-affine objects can be decomposed into non-overlapping parts, each of which is similar to the whole with respect to the anisotropic scaling transformation [3]. Thus, to check self-affinity of the aggregates one must prove anisotropic invariance for the constituent parts of the aggregates. Because experimentally this is hardly feasible, the only practical approach is to test the scaling properties of the entire aggregates themselves rather then their parts. It is well known [14] that contrary to a self-similar system, whose scaling properties are characterized by a single parameter called fractal dimension, for a self-affine system a multitude of different fractal dimensions can be introduced depending on the definition of the characteristic size of aggregates. Considering scaling relationships with respect to the aggregate length and width taken separately, we introduce the following definitions, which represent the lengthwise and widthwise scaling:

$$<N> = \alpha_L L^{D_L} \qquad (6)$$

where the "longitudinal" fractal dimension D_L is introduced together with its associated "longitudinal" lacunarity α_L, and

$$<N> = \alpha_W W^{D_W} \qquad (7)$$

where the "transverse" fractal dimension D_W is introduced together with its associated "transverse" lacunarity α_W.

In general, any combination of length and width, $L^\gamma W^{1-\gamma}$, can be employed as characteristic aggregate size, and the corresponding fractal dimension, D_γ, with its associated lacunarity, α_γ, can be defined by:

$$<N> = \alpha_\gamma (L^\gamma W^{1-\gamma})^{D_\gamma} \qquad (8)$$

The fractal dimensions, D_f, D_L, and D_W, defined above are particular cases of D_γ at $\gamma = 1/2$, 1, and 0 respectively. By combining Eqs. (5) and (8), the following relationship is obtained:

$$<N> = \alpha_\gamma (L^\gamma W^{1-\gamma})^{D_\gamma} = \alpha ((LW)^{1/2})^{[(\gamma+H(1-\gamma)/((1+H)/2)]D_\gamma} \qquad (9)$$

Thus, the conventional "geometric mean" or "gap" fractal dimension D_f defined in Eq. (2) is related to the fractal dimension D_γ defined in Eq. [6]:

$$D_f = [(\gamma+H(1-\gamma)/((1+H)/2)]D_\gamma \qquad (10)$$

By taking D_f as a reference, the longitudinal D_L and D_W transverse fractal dimensions can be expressed as follows:

$$D_L = [(1+H)/2] D_f \quad ; \quad D_W = [(1+H)/2H] D_f \qquad (11)$$

These general scaling relationships for self-affine systems described above were verified for our population of combustion-generated thermophoretically sampled aggregates. The least-squares fitting of Eq. (5) with regard to the whole population of 3080 soot aggregates yielded the affinity exponent $H = 0.91\pm0.01$ (see Fig. 3). Thus, using this H value and $D_f = 1.75$, Eq. (11) predicts noticeably different values for the "longitudinal" and "transverse" fractal dimensions: $D_L = 1.67$, and $D_W = 1.83$. On the other hand, the corresponding experimental values, obtained formally by the least-squares fittings of Eqs. (6) and (7) with regard to the whole population of soot aggregates, are: $D_L = 1.68 \pm 0.01$, and $D_W = 1.72 \pm 0.01$. The experimental value of D_L is in perfect agreement with the predicted one, whereas the experimental value of D_W is about 7% smaller than the theoretical prediction. This latter deviation may be attributable partially to the experimental definition of the width, W, of aggregate projections which leads to substantially larger uncertainty in W than in L. Thus, it is conjectured that *combustion-generated soot aggregates should be regarded as self-affine rather than self-similar*. Possible implications of this intrinsic 'asymmetry' will be considered briefly below.

Practical Implications

The anisotropic structural properties of combustion-generated aggregates should also be reflected in some of their transport properties, especially prior to orientation-averaging. To date, several important transport properties of large fractal aggregates have been estimated *via*. quasi-continuum methods exploiting the assumption of a quasispherical shape [16,17], in part based on the notion that this might be an adequate approximation for *orientation-averaged* properties. However, it is well known from both experimental and theoretical studies that the effect of nonsphericity should be taken into account in predicting viscous drag, dynamic mobility, sedimentation, especially for applications in which all orientations are *not* equally probable (see, for example, [11,18,19] and references cited therein). The self-affinity established in this study can be used to modify some of these relationships, derived previously for symmetric particles, to apply more accurately to large 'asymmetric' fractal aggregates.

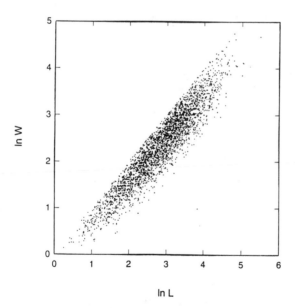

Figure 3. Width, W, perpendicular to the maximum projected length as a function of this longest axis, L, of aggregates for carbonaceous soot from various diffusion flame environments, yielding the affinity exponent, H (Eq. (4)).

SUMMARY AND CONCLUSIONS

Using the prominent example of thermophoretically sampled combustion-generated aggregates, it was shown that the scaling structural properties of DLCCA products cannot be characterized solely by using the now familiar mass fractal dimension, D_f. The asymmetric properties of aggregates were characterized here by first assuming and then verifying their *self-affinity*, introducing different exponents reflecting a scaling invariance with respect to the length and width of the aggregate projections. The affinity exponent, H, and two complementary fractal dimensions, one longitudinal, $D_L = [(1+H)/2]D_f$, and one transverse, $D_W = [(1+H)/2H]D_f$, were introduced. By fitting the TEM experimental data for the entire population of 3080 carbonaceous soot aggregates, the values of $D_f = 1.75$ and $H = 0.91$ were obtained. Therefore, it is concluded that the affinity exponent gives a more detailed description of the aggregates than the mass fractal dimension taken alone, and is expected to be useful for practical applications, in particular for more accurately predicting transport properties (*e.g.*, Brownian diffusivity and coagulation rate constants) and the kinetics of morphological changes (see below). Additionally, while we have explicitly demonstrated that these morphological regularities for a large population of carbonaceous soot aggregates, it is expected that these morphological characteristics carry over to many other inorganic flame-generated aggregates, *e.g.*, Al_2O_3 *via*. $Al(CH_3)_3$-seeded CH_4/N_2 flames because of similar fractal dimensions and first order lacunarities [12]. Our premise remains that a more precise description of aggregate morphologies in terms of the self-affine scaling should lead to a better understanding of the anisotropic transport properties and restructuring kinetics of combustion-generated aggregates and other disordered 'asymmetric' agglomerates.

Acknowledgments. This research was sponsored in part by AFOSR Grant No. 94-1-0143 with J. M. Tishkoff serving as Technical Monitor, the Yale *HTCRE Laboratory Industrial Affiliates*: DuPont and ALCOA, and the Mandelbrot Foundation for Fractals. The authors also would like to acknowledge valuable discussions with Dr. Benoit B. Mandelbrot.

REFERENCES

1. B. B. Mandelbrot, The Fractal Geometry of Nature, Freeman & Co., New York, 1983.
2. R. Jullien and R. Botet, Aggregation and Fractal Aggregates, World Scientific, Singapore, 1987.
3. T. Vicsek, Fractal Growth Phenomena, World Scientific, Singapore, 1992.
4. D. A. Weitz and J. S. Huang, in Kinetics of Aggregation and Gelation, edited by F. Family and D. P. Landau, p. 19, Elsevier Science, Amsterdam, 1984.
5. R. Botet and R. Jullien, *J. Phys. A: Math. Gen.* **19**, L907 (1986).
6. H. M. Lindsay, R. Klein, D. A. Weitz, M. Y. Lin and P. Meakin, *Phys. Rev. A* **39**, 3112 (1989).
7. Ü. Ö. Köylü and G. M. Faeth, *Combust. Flame* **89**, 140 (1992).
8. Ü. Ö. Köylü and G. M. Faeth, *ASME J. Heat Trans.* **116**, 971 (1994).
9. P. B. Sunderland, Ü. Ö. Köylü and G. M. Faeth, *Combust. Flame* **100**, 310 (1995).
10. R. A. Dobbins and C. M. Megaridis, *Langmuir* **3**, 254 (1987).
11. D. E. Rosner, D. W. Mackowski and P. Garcia-Ybarra, *Combust. Sci. Tech.* **80**, 87 (1991).
12. Ü. Ö. Köylü, Y. Xing and D. E. Rosner, *Langmuir* , in press (1995).
13. B. B. Mandelbrot, *Phys. Scr.* **32**, 257 (1985).
14. B. B. Mandelbrot, in Fractals in Physics, edited by L. Pietronero and E. Tosatti, Elsevier, Amsterdam, 1986.
15. A. V. Neimark, *Phys. Rev. B* **50**, 15435 (1994).
16. D. E. Rosner and P. Tandon, *AIChE J.* **40**, 1167 (1994).
17. P. Tandon and D. E. Rosner, *I&EC Research* **34**, 3265 (1995).
18. P. Garcia-Ybarra and D. E. Rosner, *AIChE J.* **35**, 139 (1989).
19. P. Tandon and D. E. Rosner, *Chem. Eng. Comm.*, in press (1995).

SOLID PROPELLANT FLAME STRUCTURE

T. P. Parr and D. M. Hanson-Parr

Combustion Diagnostics Laboratory, Research Division, Code 474320D, Naval Air Warfare Center, China Lake, CA 93555-6001, (619) 939-3367 939-6569 (FAX), tim@suns.chinalake.navy.mil

ABSTRACT

Planar Laser Induced Fluorescence (PLIF), UV/Vis Absorption, and thermocouple measurements were done for HNF, RDX, HMX, and XM39 deflagration with and without CO_2 laser-support. RDX and especially HNF have very short self-deflagration flame length scales. HMX and XM39 have taller self-deflagration flames. XM39 has a marked dark zone with plateau temperature about 1400K. RDX's dark zone, present under laser supported deflagration, collapses when the external laser flux is removed. PLIF was used to measure the 2D NH, OH, and CN species profiles for these materials and OH temperature profile for RDX and HNF under non-laser supported conditions. The best spatial resolution for the RDX PLIF was about 4μm. Sandwiches of HNF and various binders were studied with PLIF and while obvious diffusion flames were present at low pressure, they are weak and are not expected to be burn rate controlling.

INTRODUCTION

A solid propellant flame is a complex system, and it is necessary to understand the decomposition processes and the kinetic mechanisms of the diffusionally mixed products of the materials which comprise the propellant in order to understand the system as a whole. Once understood, the system can be modeled, and the models used for a-priori prediction of global ballistic properties.

The effects of many parameters have to be studied. For example, how does the pressure influence the burning rate, species concentration and temperature profiles? What effects do particle sizes have on the flame structure and rocket motor combustion instability? What material can be added to enhance or reduce the propellant burning rate over a given pressure range? In order to answer these questions techniques with adequate spatial and temporal resolution must be employed so that species and temperature profiles can be measured and mechanisms deduced. This is especially true the higher the pressure.

Substantial work has been done in our laboratory and others to characterize the flame zone of neat RDX laser-supported and self- deflagration.[1-9] XM39, an RDX-containing LOVA gun propellant, has also been studied [10-14] at elevated pressures. It has been found for RDX that the support laser flux has an effect on certain species concentrations[1,3,4] specifically that the amount of NO_2, and therefore NO, as well as CN increased with increasing laser flux, but how the temperature profile is affected has not been studied experimentally. Modelers[15] have theorized that with added flux the flame temperature should increase. There has also been some controversy about whether or not the temperature plateau seen for the added flux case[1] is present in the self-deflagration (no added flux) case.

HMX and RDX are both non-aromatic ring nitramines, with HMX being an 8-membered ring and RDX a 6-membered. They are called "monopropellants" since they can sustain a flame without added oxygen and are used as oxidizers in some propellants (although they are really slightly fuel rich). The vapor pressure for RDX is higher than for HMX[16] and it has been postulated that solid RDX may for the most part vaporize from the surface during deflagration whereas the products from the surface of HMX may already be those from partial decomposition.

From a recent workshop[17], it was decided that data on non-laser-supported deflagration of RDX was needed and that as a check against other techniques, UV/Vis absorption measurements should be done on XM39 propellant burning at 1 atm.

Approved for Public Release, Distribution Unlimited. Work funded by ONR, Scientific Officer Dr. Richard Miller

Mat. Res. Soc. Symp. Proc. Vol. 418 © 1996 Materials Research Society

Therefore, a review of recent work done in our laboratory using Planar Laser Induced Fluorescence (PLIF), UV/Vis Absorption, and thermocouple measurements for RDX, HMX, and XM39 self-deflagration with and without CO_2 laser-support will be discussed.

Binders are a vital part of a solid rocket propellant. As the name suggests, these pliable materials are used so that propellants can be molded into various shapes for practical uses. It can be a difficult task to choose the best binder (and other ingredients) to go with a given oxidizer so as not to significantly degrade performance but also to keep the propellant from falling apart upon aging. Therefore, it is important to also do studies of oxidizer with binder to reveal the interactions between them in diffusional flame experiments.

It is a continuing effort of this laboratory to also study *new* oxidizers such as ADN (ammonium dinitramide). Presently work has been done with neat HNF (hydrazinium nitroformate, or $N_2H_5^+C(NO_2)_3^-$ and "sandwiches" of HNF with energetic binders and other oxidizers such as HMX and RDX. These and other recent 2D diffusion flame structure results from our laboratory will be discussed.

EXPERIMENTAL

The sample cylinders used in the experiments were pressed neat at about 80,000 PSI to 96%-97% TMD. The XM39 were 0.25" diameter extruded cylinders. For self-deflagration tests, the ignition source (a CO_2 laser) was turned off shortly after the sample was ignited, and the data used were from the time frame of steady state self-deflagration without CO_2 laser support. In the laser-supported work, the CO_2 laser flux was continued during the entire experiment.

The flame structure was measured using Planar Laser Induced Fluorescence (PLIF) imaging[2,18-22], and UV-Visible absorption[1,2]. The apparatus diagrams for the PLIF and absorption spectroscopy setups are shown in Figs. 1A and 1B, respectively.

The laser used to excite PLIF was a Nd-Yag pumped tunable dye laser with non-linear crystal mixing and doubling into the ultraviolet. The ultraviolet beam is expanded and formed into a sheet and passed through the center of the flame. The camera used to image PLIF for recent tests was a gated image intensified CCD with 752 x 480 pixels. For PLIF work the camera is gated on for about 80 nsec only during or just after the diagnostic laser pulse; this discriminates against natural flame emission and monitors only the laser induced fluorescence.

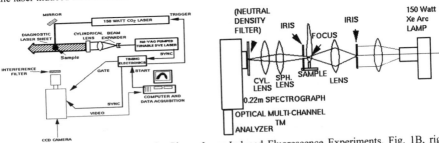

Fig. 1A, left: Apparatus Diagram for Planar Laser-Induced Fluorescence Experiments. Fig. 1B, right: Apparatus Diagram for UV-Visible Absorption Experiments.

PLIF imaging monitors the ground electronic state of flame species while chemiluminescence comes only from excited electronic states formed in low probability energetic reactions. Thus it is often not a good idea to use chemiluminescence to monitor flame structure as it tracks low probability secondary reactions rather than the main flame chemistry pathways. For OH radicals, for example, the chemiluminescence does not track ground state OH concentrations at all.[23] This is why we chose to use PLIF. In addition, PLIF provides a narrow planar slice through the flame while emission imaging is a line of sight average technique. The thickness of the laser sheet was less than 150 μm and the resolution of the camera system was from 4 to 9 microns per pixel. This fine a resolution was required due to the very short RDX and especially HNF flames.

CN radicals were monitored using PLIF by exciting the (0,0) B-X bandhead at 388.34 nm (all wavelengths reported here are in air, not vacuum) while imaging (0,1) emission using a 420 nm interference filter. The laser power used saturated the CN transition thereby partially limiting problems due to quenching. The CN PLIF image profiles are therefore expected to be a good relative image of CN radical concentration profiles. NH PLIF was obtained by tuning the laser to the Q_3 (1,0) (A-X) bandhead near 305 nm and monitoring LIF from the $\Delta v = 0$ spectral region around 336 nm using a combination of WG335 and UG11 filters. OH PLIF was done by tuning the laser to a given R_1 line of the (1,0)(A-X) transition and monitoring the LIF using a combination of two WG305 with one UG11 filters..

The absorption measurements have been described previously.[1,2] Basically, a light beam, focused at the center of the sample to avoid beam steering, passes through the flame of a deflagrating sample and the transmitted light is measured spectrally using an intensified linear diode array detector. The absorption spectrum is obtained by ratioing the spectrum obtained without flame to that with flame. Species concentration and temperature (if the spectroscopy of the species is known) are then obtained by comparing with calculated spectra, or (if the spectroscopy is not known) with spectra from the literature.

RESULTS

Laser Supported Deflagration of RDX and HMX at 1 atm

Work has been done in our laboratory to image the flame structure of HMX, RDX, HNF, ADN, and other advanced energetic materials at pressures between 1 atm and 18 atm during the ignition and steady-state deflagration stages using a CO_2 laser as an ignition source[1,2,18-22]. The added CO_2 laser heating was used to enhance the low pressure burning rate of these neat materials and to more closely simulate the surface heat flux conditions seen by samples self deflagrating at rocket motor-like pressures. In addition, the external laser flux stretches out the flame zone, making it easier to use diagnostics to resolve the species and temperature profiles. The evolution of the flame during the ignition stage could be studied with PLIF by varying the time between the ignition laser and probe laser. Previous results for HMX have been reviewed earlier.[2] Results for RDX laser-supported deflagration at 1 atm are shown in Fig. 2 (all raw data for any plot in this paper is available to anyone via e-mail by requesting from tim@suns.chinalake.navy.mil). These profiles were obtained from the PLIF images via selection of a centerline slice and from the UV-Visible absorption traces via deconvolution of path lengths measured from PLIF and emission imaging. RDX shows considerably shorter flame heights than HMX: under the same conditions of external laser flux the height to the peak in the CN flamesheet is 2.5 mm for RDX and 4 mm for HMX.

The temperature trace of Fig. 2 is a combination of data from three sources: within the condensed phase and just above it was obtained from 5 micron thick thermocouples, the temperature profile in the dark zone region was obtained from fitting of NO UV absorption spectra, and the temperature profile from the dark zone through the secondary flame and beyond was obtained from OH rotational temperature (OH T_r) PLIF. Several R_1 transitions of the (1,0) A-X band were excited and Boltzmann populations used to obtain rotational temperature. The models of RDX deflagration predict that the temperature should rise above the adiabatic flame temperature in laser assisted experiments, due to the heat input from the laser, yet our prior OH T_r measurements peaked at only about 2600K.[19,21] This prompted a reexamination of the analysis procedure for OH rotational

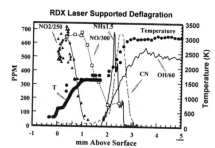

Fig. 2. Temperature and species profile for laser-supported deflagration of RDX at 1 atm. Laser flux of 400W/cm².

209

temperature. It was found that high K transitions, $R_1(18)$ and $R_1(20)$, that were added to improve the T_r measurements, actually do not fit the Boltzmann distribution very well. When these two transitions were left out of the analysis the remaining six transitions ($R_1$3, 5, 8, 10, 12, and 14) gave much more linear Boltzmann plots, much higher correlation coefficients, generally 0.99 or better, and excellent agreement with the model T profile. Figure 3 shows an example Boltzmann plot for self deflagration of RDX at 1 atm (here only three rotational lines, $R_1$3, $R_1$10, and $R_1$14, were used, as they covered the range evenly and gave good correlation coefficients).

These laser-supported flames were 3D (three-dimensional) rather than 1D, due both to the bell-shaped energy profile of the ignition laser and the heat losses from the edges of the flame. Using a flatter energy profile, at lower average flux, improved the flame structure, but the flame was still far from mathematically 1D. Modeling these flames in full 3D would obviously be computationally prohibitive, so we used PIV (particle imaging velocimetry) to measure the flow streamlines and therefore the flame area ratio function. (In our geometry the velocity is nearly constant through the dark zone and secondary flame sheet so that the energy release of the secondary flame appears to mostly spread out the flame rather than speed up the flow.) The area ratio curve was given to modelers (such as Prof. Mitch Smooke and Dr. Kuldeep Prasad of Yale) to include in their 1D model as a correction factor. When this was taken into account there was quite good

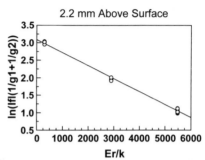

Fig. 3. Plot of $\ln(I_{fl}(1/g_1 + 1/g_2))$ vs. E_r/k for RDX self-deflagration at 1 atmosphere in air. The height above the surface was 2.2 mm. E_r is the initial rotational energy of the pumped transition, k is Boltzmann's constant, I_{fl} is the measure LIF signal, and g_1 and g_2 are the degeneracies of the two levels involved in the transition and the transitions are saturated.

agreement between the model[24] and our species and temperature profiles as well as China Lake's (Atwood and Curran[25]) burn rate, pressure exponent and temperature sensitivity measurements. This is the first time that an a-priori model with detailed kinetics has been able to predict the ballistic properties of solid propellant combustion, as well as match the species and temperature profiles. Experimental measurements were important inputs in developing and validating this model.

Nevertheless, modelers wanted some results from a truly 1D flame without added laser flux as a validation. It was also suggested that using added energy from a laser may produce "artifacts" in the flame, such as liquid spray, which may not be present without the laser. Laser Mie scattering experiments showed that there is indeed a liquid spray above the surface with laser supported deflagration of RDX or HMX, but also that this spray is present, at a lower number density, for self deflagration of HMX (the dark zone height of self deflagrating RDX is so small, even at 1 atm, that few liquid droplets survive to be seen).

Self deflagration of RDX and HMX at 1 atm

Because over the years our detection system for PLIF improved to the point where 4 micron resolution was obtainable and 50 micron resolution was obtainable for the absorption measurements, it was decided that deflagration studies of RDX without added laser flux were possible. At 1 atm pressure the RDX flame height (to the peak in the CN flame sheet) is only about 0.47 mm. Although modelers would like data from an infinite plane sample, use of a pressed pellet width of 3/8" (vs. 1/4" previously used) produced a very flat flame (Fig. 4).

Despite the much shorter flame standoff distance, measurements of non-laser supported deflagration, i.e. self deflagration, of RDX at 1 atm were made using UV-Visible absorption and PLIF (at down to 4 micron spatial resolution). Despite the much smaller length scales temperature and many species profiles were measured. Absorption quantified NO, H_2CO, HONO, OH, CN, and NH. The flame was too short to resolve NO_2 and the "365 nm unknown"[1] profiles, which die quite close to the surface even for laser supported combustion. Peak concentrations were: 310 ppm for CN and 260 ppm for NH, and 3.6% (mole percent) for OH, 17% for NO and <1% for H_2CO and HONO. PLIF was used to measure profiles for CN, NH, and, OH, and OH rotational temperature. Species concentration and temperatures profiles are shown in Fig. 5. For OH, the $R_1(10)$ line was used and was corrected for the effects of temperature, but it was found that this correction was minor.

The flame standoff for self deflagrating HMX is much higher than for RDX but without the external laser flux to support the flame the burn rate and flame height are variable and this makes comparison of species profiles, which are measured for separate tests, difficult and temperature ratios impossible.

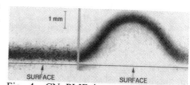

Fig. 4   CN PLIF images using 10 mm diameter RDX pellets at 1 atm for (left) self deflagration showing an extremely flat flame, compared with (right) laser-supported deflagration showing a decidedly nonflat flame.

Effect of Pressure on HMX Flame Structure

The 4 micron resolution obtainable in the PLIF work prompted a restudy of the change in HMX laser supported flame structure with pressure. It was found that the previous results[2,20] for flame thickness had been in error due to the relatively poor MTF (modulation transfer function) of the lens used with the diode array camera. Fig. 6 shows the revised findings: the secondary flamesheet thickness, as monitored with CN radicals, decreases with pressure as $1/P^{0.98}$ and the flame height decreases about inversely with pressure as well. The range of pressures covered was 1 to 18 atm. Other PLIF measurements have been done up to 25 atm,

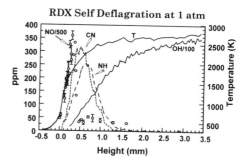

Fig. 5   PLIF measured and absorption calibrated species and temperature profiles for *self* deflagration of RDX at 1 atm (no external laser heating flux). Temperatures below 0.3 mm come from 5 micron thermocouple. Without substantial external laser flux the dark zone temperature plateau disappears due to the diffusive smearing caused by the close proximity of the radical profiles of the secondary flame.

which is getting near the actual rocket motor operating pressures. Loss of signal due to quenching does not seem to be a problem, but Fig. 6 shows that for nitramines the flame structure length scales become exceedingly small at realistic rocket motor pressures. Extrapolation of Fig. 6 to 51 atm, a realistic rocket motor operating pressure, results in a flame height of only 24 microns. Virtually no diagnostic technique can make profile measurements within that length scale. This is why work is done at "low" pressures, i.e. 1 to 10 atm.

HMX self deflagration at 1 atm

We made measurements of HMX deflagration flame structure *without* laser flux support, but found that the flame was unstable at 1 atm and a "fuel spray" was present which interfered especially with the NO measurements. We saw a maximum mole fraction of about 6.5% for both H_2CO and NO_2, about 210 ppm of CN and 3.6% OH. The "365 nm unknown" was not observed.

Some energetic materials have extremely short flame length scales even at 1 atm. Emission images of deflagrating HNF (hydrazinium nitroformate, $N_2H_5^+C(NO_2)_3^-$) show an extremely short dark zone (ca 40 microns) with a complex flame structure. The first visible flame, starting at 40 microns, centered at 180 microns with a FWHM thickness of 180, shows whitish on the video and has the majority of its spectral emission to the red of 515 nm. A *preliminary* assignment is the NH_2 radical. Within this flamesheet, and extending beyond, are emissions from CN, NH, and CH. Emission spectroscopy showed that the CH was much stronger than for nitramines (surprising as HNF has a single C atom and HMX has 4). The CN and NH emission was much weaker than for nitramines, and NH was greater than CN, opposite to nitramines. The NH profile peak appeared to occur closer to the surface. PLIF showed that the CN profile peaked at 360 microns and the NH at 300 microns. Above the bluish flame is broadband orange brown emission (probably $NO+O = NO_2^*$ recombination chemiluminescence) which extends far into the burnt gas region. The OH PLIF profile rose rapidly through and outside the CN flamesheet and then transitioned to a more gradual rise in the burnt gas region. The OH T_r also rose sharply at the CN flamesheet but did not reach adiabatic at this point. Instead, it continued to rise at a slower slope. The species and temperature profiles from both emission and PLIF imaging are shown in Fig. 7.

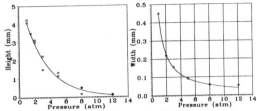

Fig. 6 HMX secondary flame standoff (left) and thickness (right) as function of pressure.

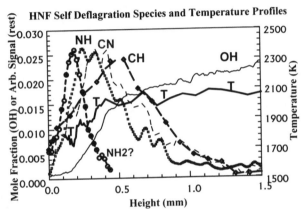

Fig 7. PLIF measured species and temperature profiles for self deflagration of HNF at 1 atm. Also included are selected emission profiles (NH_2 and CH). The NH_2 (postulated) is the Gaussian curve peaking closest to 0 mm (circles), NH is the next Gaussian curve out (squares), CN next (short dash), and CH is the furthest Gaussian curve out (long dash with diamonds). Anyone interested in the original data can get it via e-mail by contacting tim@suns.chinalake.navy.mil.

XM39 self and laser supported deflagration at 1 atm

The next step up in complexity from neat energetic materials such as HMX, RDX, and HNF is homogeneous composite propellants based on these ingredients. Here the intent is to a) stretch out the flame structure of nitramines by adding another ingredient and b) supply data to modelers who can add the binder chemistry to the model without adding the dimensional complexity of actual diffusion flames. Several additives were tried. Paraformaldehyde merely slowed the burn rate while slightly *shortening* the flame standoff. UV-Visible absorption measurements confirmed that the NO_2 concentration was less with paraformaldehyde added, indicating that the exothermic $H_2CO + NO_2$ reaction may have actually been accelerating the dark zone chemistry. Cyanuric acid (makes HNCO in flame) also slowed the burn rate but did not increase standoff. CAB (cellulose acetate butyrate) greatly slowed the burn rate, and slightly increased the standoff, but the burn was variable due to melting of the entire sample and char formed on the surface.

XM39 is an RDX based LOVA propellant that naturally has a tall flame standoff distance, even for self deflagration at 1 atm, and there is significant work in the literature from other groups for comparison so tests were undertaken with transient UV-visible absorption and PLIF imaging to quantify the flame structure of ARL supplied XM39 as a kind of round-robin standard. The CN flamesheet height at zero laser flux varied from about 8 mm initially to about 4 mm toward the end of the burn, as the sample became hemispherical, similar to what happened to the RDX/CAB samples mentioned above. The flame height increased to 11 mm at 70 W/cm^2 and more than 16 mm at 350 W/cm^2. Fitting the NO absorption spectrum in the dark zone gave a plateau temperature of 1400K and a peak NO concentration of about 0.073 for self deflagration. Increasing the laser flux lead to higher NO concentrations (up to 0.13). Thermocouple results (5 to 10 micron) showed a surface temperature of 610K and matched the dark zone plateau temperature measured with NO (1400K). The primary stage flame was about 1 mm thick as measured by the thermocouple/NO T profile. Figure 8 shows the results of the thermocouple traces with the NO rotational temperatures determined for XM39 self-deflagration. Figure 9 shows the [NO] mole fraction as a function of height off the surface and as a function of laser flux.

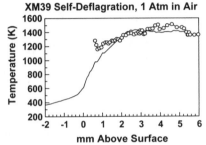

XM39 Self-Deflagration, 1 Atm in Air

Figure 8. Temperature profile for self-deflagration of XM39 at 1 atm in air. Shown are results of thermocouple measurements (solid line) and NO rotational temperature (circles).

The NO_2 profile was extremely narrow for self deflagration (decays within 1 mm of the surface) with a peak mole fraction of 0.024. The peak NO_2 mole fraction increased with increasing laser flux (0.08 at 350 W/cm^2) and the profile reached further from the surface as seen in Figure 10. Litzinger's[4] original mass spectroscopy results for neat RDX showed considerable concentrations of H_2CO at reasonable distances away from the surface, but China Lake absorption and PLIF experiments placed the H_2CO mole fraction below 1-2% for laser supported deflagration of pure RDX. Subsequent triple quadruple measurements by Litzinger's group has brought the two groups into better agreement as Litzinger now sees only low concentrations of H_2CO in the gas phase of pure RDX flames. With XM39 however, both Litzinger[13,14] and China Lake see H_2CO formation and the peak concentrations agree fairly well. The binder components of the XM39

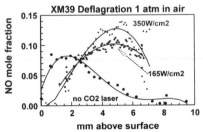

XM39 Deflagration 1 atm in air

Figure 9. NO mole fraction profiles for XM39 deflagration at 1 atm in air as a function of CO_2 laser flux.

slow the burn rate and apparently allow more branching via the $H_2CO + N_2O$ channel vs. HCN + NO_2/NO channel. The China Lake H_2CO profile for nearly self deflagration is, like the NO_2 profile, extremely narrow with a peak mole fraction of 0.075 (Litzinger got 0.12 at a somewhat higher flux). Figure 11 shows the formaldehyde results for XM39.

The secondary flame sheet is not as intense with XM39 as pure RDX: the CN peak concentration was found to be only 34 ppm vs. 310 ppm for self deflagration of RDX. PLIF results showed that the NH concentration was also substantially lower than for pure RDX. The peak [OH] was 0.0112 at 2400K measured NO T_r. The CN profile peaked at 6.9 mm with a FWHM of 2.06 mm, much larger than for HMX or RDX.

OH temperature was not attempted for XM39 due to the aforementioned fluctuating flame height observed during burning. The OH $R_1(10)$ line profile for self-deflagration is shown in Fig. 12.

Other researchers[10,13] have obtained [NO] and darkzone temperature for XM39 self-deflagration at elevated pressure and [NO] and [NO2] profiles for laser-supported regression (no visible flame) at elevated pressure, both in an inert atmosphere. In general, they have found the dark zone temperature to be somewhat lower than that shown in Fig. 8 (about 1250K vs. 1400K) and higher [NO] of either 15% (18 atm)[10], or 20% (laser assisted, 3 atm, 100W/cm2)[13], compared with 10% from the present results (165W/cm2). More recent results (FTIR absorption and thermocouple)[26], however, tend to agree much better with those presented here. The [NO2] peaked at about 5% close to the surface at 3 atm[13] (100W/cm2), compared with 5.5% from the present results (70W/cm2).

The "365 nm unknown" seen in RDX laser-supported deflagration[1] was also seen in XM39 laser-supported deflagration (Fig. 13) and for self-deflagration. The unknown was only seen for self-deflagration, however, after a large bubble was seen to pop on the surface: it's concentration rapidly decayed after initial appearance, which indicates that the unknown is associated with the surface decomposition products.

Diffusion Flame Studies via Sandwiches

Another method to study the interaction between ingredients is to use "sandwiches"[2,18,27-31] consisting of planes or slabs of two ingredients in close contact. This allows imaging of the diffusive flame structure from known geometries. The approach in this laboratory was to use PLIF to get a 2D image of a particular species in the flame of a sandwich and study the effects of lamina thickness, lamina type, and pressure on the flame structure.

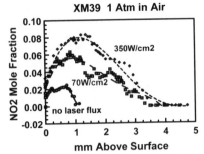

Figure 10. NO2 mole fraction profiles for XM39 deflagration at 1 atm in air as a function of CO2 laser flux.

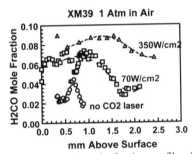

Figure 11. H2CO mole fraction profiles for XM39 deflagration at 1 atm in air as a function of CO2 laser flux.

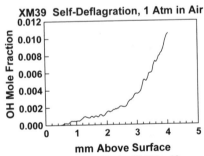

Fig. 12. OH PLIF Profile for XM39 self-deflagration at 1 atm in air scaled to absorption measurements.

How much kinetic delay there is to significant heat release can be seen by the length of the dark zone in these "model propellants". If the decomposition products of the oxidizer and fuel mix and react with fast kinetics, then the flame standoff seen will be short and if the kinetics is slow, the standoff will be tall. By comparing these standoffs with

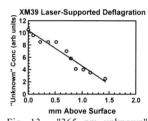

Fig. 13. "365 nm unknown" species relative concentration in an XM39 flame as a function of distance from the surface. The CN flame sheet was at about 10 mm above the surface.

the neat material one can judge if there is any positive diffusion flame interaction between the ingredients which might lead to a particle size control of ballistics.

Extensive measurements in our laboratory[2,18,31] have been made of diffusion flame structure of nitramines (HMX, RDX, TNAZ), ammonium dinitramide (ADN), and AP composite sandwich solid propellants. Recent work[18] done with AP has shown clear diffusion flames that have persisted at elevated pressures. The combination of HMX or TNAZ with either AP or ADN lead to obvious flame interactions[18]: the AP or ADN decomposition products accelerated the nitramine dark zone kinetics. From previous sandwich diagnostics measurements made on ADN/binder sandwiches, predictions were made that ADN-based formulations would have weak or no diffusion flame interaction between oxidizer and binder and therefore little or no particle size control over ballistic properties. This finding was recently corroborated through discussions with Russian scientists who have ADN-based propellants. Work done in our laboratory[18] showed that HMX clearly has no diffusion flame with non-energetic binders. Figure 14 shows an example: even with an energetic binder such as N5, the HMX secondary flame sheet is lifted in the diffusive mixing layer, not shortened.

Understanding the interaction between ingredients and diffusion flame structure is important to understanding the effects of particle size on propellant ballistic properties. For example, most AP based propellants are thought to have diffusion flames between the AP and fuel decomposition products and such propellants show particle size effects on ballistic properties[32,33] Changing the AP particle size in AP composite propellants can lead to nearly a factor of ten change in burn rate. The smallest particles lead to the highest burn rate as the hot primary diffusion flame between AP and fuel dominates the heat feedback to the AP and leads to a high regression rate. Large AP particles see mostly the cooler AP monopropellant flame leading to a lower burning rate[33]. Much work has already been done in understanding the nature of propellant diffusion flame structure of AP based formulations.[27-29,31]

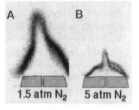

Fig. 14 CN PLIF image of HMX/N5/HMX sandwich flame.

<u>HNF Composite Flame Microstructure</u>

PLIF measurements were made on the diffusion flame structure between HNF and various energetic and non energetic binders including HTPB, wax, polyurethane (PU), GAP, and BAMO/NMMO in sandwich configurations. Unlike ADN, HNF entirely consumes even non energetic binders for low flux laser supported deflagration at 1 atm. ADN did not even consume energetic binders, at 10 atm, for a laser heating flux nearly nine times as large as for the HNF tests. Diffusion flames between the HNF and binder were clearly evident in the emission and PLIF imaging experiments and these flames had much lower kinetic standoff distances than those seen with ADN sandwiches (Fig. 15).

The HNF and binder regression rate was highest in the interface region where the diffusion flame heat and chemistry are located. This behavior is understandable: neat HNF has a relatively hot reactive flame extremely close to the surface while ADN has no visible flame. PLIF measurements of CN, NH, and OH were undertaken for the various binder combinations and binder length scales. These experiments were then extended to elevated pressure to determine if the diffusion flames seen are also important at higher pressures.

It was found that as with ADN, as the pressure increases the burn rate of the HNF increases and it starts to leave non-energetic binders behind. Even at 1 atm the HTPB lamina sticks up above the HNF.

At 10 atm HTPB lamina of 200 microns were recovered only charred. The PLIF images showed weaker diffusion flames at elevated pressures and HTPB decomposition products seemed to pass through the diffusion flame (Fig. 16).

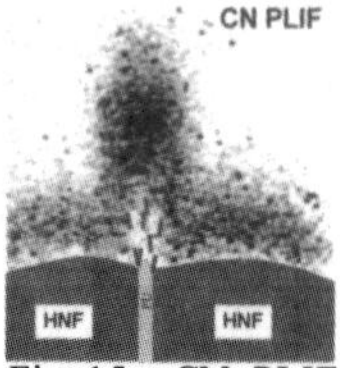

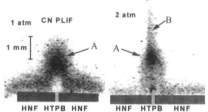

Fig 15 CN PLIF image of diffusion flame at 1 atm between HNF (outer lamina) and polyurethane binder (center). The polyurethane is 190 microns thick.

When HNF was used with energetic binders, such as BAMO/NMMO the binder was able to keep up with the increasing HNF regression rate at higher pressures and obvious diffusion flames were seen even at elevated pressures. Figure 17 compares the flame structure of HNF/BAMO sandwiches at 1 and 5 atm. Notice that as per diffusion flame theory the diffusion flames height is approximately independent of pressure while the HNF flame height decreases dramatically. The diffusion flame height is much taller than the neat HNF flame so even in this case it is probably not burn rate controlling and any particle size controllability of HNF propellant ballistics will likely be much weaker than for AP.

HNF with either RDX or HMX as the outer laminae were also studied (Fig. 18). Past work[10] showed that ADN somewhat accelerates the HMX dark zone chemistry and shortens the HMX secondary flame while AP greatly accelerates it. HNF has the same effect on both the RDX and HMX secondary flames and appears closer to AP's effectiveness than ADN's.

SUMMARY

Planar Laser Induced Fluorescence (PLIF), UV/Vis Absorption, and thermocouple measurements were done for HNF, RDX, HMX, and XM39 deflagration with and without CO_2 laser-support. RDX and especially HNF have very short self-deflagration flame length scales. HMX and XM39 have taller self-deflagration flames. XM39 has a marked dark zone with plateau temperature about 1400K. RDX's dark zone present under laser supported deflagration collapses when the external laser flux is removed. The peak [NO] seen for XM39 was about 7.5% and about 14% for RDX under the same experimental conditions. Due to a significant amount of liquid present in the

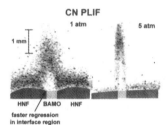

Fig. 16 HNF/HTPB sandwich combustion at 1 atm (left) and 2 atm(right). Arrows marked A point to CN PLIF signal while B points to HTPB binder decomposition products. Note that the HTPB sticks up above the HNF in the 2 atm image.

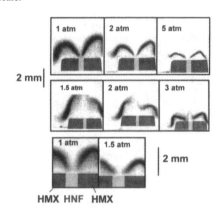

Fig 17. CN PLIF from HNF/BAMO sandwiches at 1 atm (left) and 5 atm to the same scale.

Fig. 18 CN PLIF images for HMX/X/HMX sandwiches at various pressures where X = AP in the top three frames, X = ADN in the middle, and X = HNF in the bottom.

HMX flame almost up to the CN flame sheet, it was not possible to obtain an [NO] profile very close to the surface. Whereas only about 1% formaldehyde mole fraction was seen for RDX, HMX produced about 8% and XM39 also produced about 8%, but the peak NO_2 concentration was only 2.5% for XM39 compared with 15% for HMX self-deflagration and 15% for RDX (laser flux of 160 W/cm^2). Surface temperatures for RDX and XM39 appear to be the same (about 600 - 610K) compared with about 660K for HMX. PLIF was used to measure the 2D NH, OH, and CN species profiles for these materials and OH temperature profile for RDX under non-laser supported conditions. The best spatial resolution for the RDX PLIF was about 4μm. Sandwiches of HNF and various binders were studied with PLIF and while obvious diffusion flames were present at low pressure, they are weak and are not expected to be burn rate controlling.

ACKNOWLEDGMENTS

The authors gratefully thank Dr. Martin Miller or ARL for supplying the XM39, Dr. May Chan, Jerry Finlinson, and Vicki Brady of NAWC and Dr. Jerry Manser of Aerojet for supplying us with binders, Dr. David Vanderah for supplying us with ADN, Dr. Jay Levine of Edwards for the HNF, and especially Dr. Richard Miller of ONR for funding this work.

REFERENCES

1. "RDX Laser Assisted Flame Structure", Donna Hanson-Parr and Tim Parr, Proceedings of the 31st JANNAF Combustion Subcommittee Meeting, Vol. II, page 407 (1994), CPIA publication 620.

2. "Solid Propellant Flame Structure", Tim Parr and Donna Hanson-Parr, in **Non-Intrusive Combustion Diagnostics,** Kenneth K. Kuo and Tim Parr editors, Begell House, Inc. New York (1994), pp. 571-599.

3. "A Study of the Gas-Phase Processes of RDX Combustion Using a Triple Quadrupole Mass Spectrometer", Y.J. Lee, C.J. Tang, and T.A. Litzinger, Proceedings of the 31st JANNAF Combustion Subcommittee Meeting, Vol. II, page 425-430 (1994), CPIA publication 620.

4. "Chemical Structure of the Gas Phase above Deflagrating RDX: Comparison of Experimental Measurements and Model Predictions", B.L. Fetherolf, and T.A. Litzinger, Proceedings of the 30[h] JANNAF Combustion Committee Meeting, CPIA pub. 606, Vol. II, p 15, (1993).

5. "Dynamic Flame Probe Mass Spectrometry and Condensed-System Decomposition ", O.P. Korobeinichev, Comb. Expl. and Shock Waves **23** 565 - 576 March 1988.

6. "Processes in Hexogen Flames", N.E. Ermolin, O.P. Korobeinichev, L.V. Kuibida and V.M. Fomin, Comb. Expl. and Shock Waves **24** 400 - 407 Jan. 1989.

7. "CARS Probe of RDX Decomposition", Kenneth Aron and L.E. Harris, Chem. Phys. Lett. **103(5)**, 413-417 (1983).

8. "CARS Diagnostics of Solid Propellant Combustion at Elevated Pressures", John H. Stufflebeam and Alan C. Eckbreth, Combust. Sci. Tech. **66** 163-179 (1989).

9. "Infrared multiphoton dissociation of RDX in a molecular beam", X. Zhao, E. Hintsa, and Y.T. Lee, J. Chem. Phys. **88(2)**, 801-810 (1988)

10. "Spectral Studies of Solid Propellant Combustion IV. Absorption and Burn Rate Results for M43, XM39, and M10 Propellants", M. Warfield Teague, Gurbax Singh, and John A. Vanderhoff, Technical Report ARL-TR-180 August 1993.

11. "Multichannel Infrared Absorption Spectroscopy Applied to Solid Propellant Flames", S.H. Modiano and J.A. Vanderhoff, Proceedings of the 30th JANNAF Combustion Subcommittee Meeting, Hyatt Regency Hotel and NPGS Monterey, CA 15-19 November 1993, Vol. II, p.227.

12. "Improvements in Infrared Absorption of Solid Propellant Flames", S.H. Modiano and J.A. Vanderhoff, Proceedings of the 31st JANNAF Combustion Subcommittee Meeting, Vol. II, page 325-332 (1994), CPIA publication 620.

13. "A Comparison of the Physical and Chemical Processes Governing the CO_2 Laser-Induced Pyrolysis and Deflagration of XM39 and M43", B.L. Fetherolf, T.A. Litzinger, Y-C.Lu and K.K. Kuo, Proceedings of the 30th JANNAF Combustion Subcommittee Meeting, Vol. II, page 183-193 (1993), CPIA publication 606.

14. "A Study of Gas-Phase Processes During the Deflagration of RDX Composite Propellants Using a Triple Quadrupole Mass Spectrometer", T.A. Litzinger, Y.J. Lee, and C.J. Tang, Proceedings of the 31st JANNAF Combustion Subcommittee Meeting, Vol. II, page 307-316 (1994), CPIA publication 620.

15. Many various E-Mail private communications.

16. "Scanning Calorimetric Determination of Vapor-Phase Kinetics Data", R.N. Rogers and G.W. Daub, Analy. Chem. **45**, pp.596-600 (1973).

17. Summary workshop report of the Kinetic and Related Aspects of Propellant Combustion Chemistry Panel, held at the 31st JANNAF Combustion Subcommittee Meeting, presented at the 32nd JANNAF Combustion Subcommittee Meeting, Huntsville, AL, Oct. 1995.

18. "Advanced Oxidizers Diffusion Flame Structure", Tim Parr and Donna Hanson-Parr, Proceedings of the 31st JANNAF Combustion Subcommittee Meeting Vol. II, p. 333-344 (1994) CPIA publication 620.

19. "Advanced Diagnostic Techniques for NonSteady Burning of Solid Propellants", T.P. Parr and D.M. Hanson-Parr in Progress in Astronautics and Aeronautics, edited by L. De Luca, E.W. Price, and M. Summerfield, Volume 143, pp. 261-324, (1992).

20. "Nitramine Flame Structure as a Function of Pressure", T.P. Parr and D.M. Hanson-Parr, Proceedings of the 26th JANNAF Combustion Committee Meeting, CPIA pub. 529, Vol. I, p 27, (1989).

21. "Temperature and Species Profiles in Propellant Ignition and Combustion", T.P. Parr and D.M. Hanson-Parr, Proceedings of the 24th JANNAF Combustion Committee Meeting Vol. I, p 367 (1987) CPIA publication 476.

22. "The Application of Imaging Laser Induced Fluorescence to the Measurement of HMX and Aluminized Propellant Ignition and Deflagration Flame Structure", T.P. Parr and D.M. Hanson-Parr, Proceedings of the 23rd JANNAF Combustion Committee Meeting Vol. I, p 249 (1986) CPIA publication 457.

23. "Rapidly Sequenced Pairs of Two-Dimensional Images of OH Laser Induced Fluorescence in a Flame", M.J. Dyer and D.R. Crosley, Optics Letters **9**, 217 (1984).

24. "An Eigenvalue Approach for Computing the Burning Rate of RDX Propellants", K. Prasad and M.Smooke, presented at the 32nd JANNAF Combustion Subcommittee Meeting, Huntsville, AL, Oct. 1995.

25. "Laboratory Ignition and Combustion Data and the Determination of Global Kinetics and Energetic Parameters of RDX", A.I. Atwood, P.O. Curran and, and C.F. Price, presented at the 32nd JANNAF Combustion Subcommittee Meeting, Huntsville, AL, Oct. 1995.

26. "Further Improvements to FTIR Absorption Spectroscopy of Propellant Flames for Profiling of Species and Temperature", C.F. Mallery and S.T. Tynell, presented at the 32nd JANNAF Combustion Subcommittee Meeting, Huntsville, AL, Oct. 1995.

27. "Combustion of Ammonium Perchlorate - Polymer Sandwiches", E.W. Price et. al., Comb. Flame **63**, 381 (1986).

28. "Dependence of Burning Rate of AP-Polymer Sandwiches on Thickness of Binder Laminae", E.W. Price and J.K. Sambamurthi, Proceedings of the 20th JANNAF Combustion Subcommittee Meeting, (1983) CPIA Publication 383.

29 "Effect of Types of Binder and Burning Rate Catalysts on Edge Burning AP-binder-AP Sandwiches", E.W. Price, C. Markou, and R.K. Sigman, Proceedings of the 26th JANNAF Combustion Subcommittee Meeting, Vol. II, p 93 (1989), CPIA Publication 529.

30. "Role of the Leading Edge of the Diffusion Flame in Combustion of Solid Propellants", E.W. Price et. al., Proceedings of the 27th JANNAF Combustion Subcommittee Meeting, Vol. III, p. 31 (1990), CPIA publication 557.

31. "Propellant Diffusion Flame Structure", Tim Parr and Donna Hanson-Parr, Proceedings of the 28th JANNAF Combustion Subcommittee Meeting, Volume III, pp. 359-368, October 28 - November 1, 1991, CPIA pub. 573.

32. "Combustion Calculations for Composite Solid Propellants", M.W. Beckstead, Proceedings of the 13th JANNAF Combustion Meeting, Vol. II, p 299 (1976), CPIA publication 281.

33. "Nitramine Composite Solid Propellant Modeling", F.S. Blomshield, NWC TP-6992 (1989).

34. "The Deflagration of Solid Propellant Oxidizers", Edward T. McHale and Guenther von Elbe, Comb. Sci. Tech. **2**, pp. 227-237 (1970).

NITRIC OXIDE INTERACTIONS WITH C_2 HYDROCARBON SPECIES

B. A. WILLIAMS[†], L. PASTERNACK
Chemistry Division, Naval Research Laboratory, Washington, DC 20375-5342
†National Research Council Postdoctoral Fellow 1992-1995

ABSTRACT

We have investigated the combustion chemistry of nitric oxide doped into premixed 10 Torr flames of $CH_4/O_2/N_2$, $C_2H_6/O_2/N_2$, $C_2H_4/O_2/N_2$, and $C_2H_2/O_2/N_2$ which had similar peak temperatures. Profiles of OH, CH, $^3C_2$, CN, NCO, NH, and NO were recorded by laser-induced fluorescence and compared among the different fuels, indicating differences in hydrocarbon flame structure and NO reactivity in these environments. Concentrations of nitrogen intermediates are comparable for the methane, ethane, and ethylene flames, but are approximately a factor of three higher with acetylene, indicating much greater NO reactivity in this flame.

INTRODUCTION

Reactions of nitric oxide with hydrocarbon flame species are important for predicting and reducing NO_x emissions as well as in modeling the combustion of energetic materials in which nitrogen oxides are primary oxidants. There has been considerable work in recent years measuring the kinetics of NO reactions with CH_i radicals, and efforts to model the behavior of NO and NO_2 in methane flames have been largely successful [1,2]. For flames of higher hydrocarbons the situation is much different. Reactions of NO with radical species containing two or more carbon atoms remain largely uninvestigated, and consequently flame systems such as C_2H_2/NO_2 are poorly understood by comparison [2,3]. Reactions of C_2 species with NO have practical importance for propellants containing a polymeric hydrocarbon binder and a nitrogen-based energetic material. Pyrolysis of the binder will produce unsaturated C_2 and C_3 hydrocarbon fragments which in this environment are oxidized largely by NO. Since binders can substantially affect the burn rate of propellants, understanding this chemistry is critical to accurately predicting the propellant's behavior.

Here we report studies of premixed, stoichiometric, 10 Torr flames of methane, ethane, ethylene, and acetylene seeded with NO. Relative concentrations of the species OH, CH, $^3C_2$, CN, NCO, NH, and NO have been recorded by laser-induced fluorescence (LIF) for the four flames. The relative amounts of nitrogen intermediates in these flames indicate the reactivity of NO with the pool of hydrocarbon radicals produced by each of these fuels.

EXPERIMENT

The general details of the experimental equipment and techniques have been described previously [1]. The flames were operated in a stainless steel McKenna burner with flow rates listed in Table I. The oxygen flow rate was the same for all flames; the fuel flows were chosen to maintain an equivalence ratio of unity. Since an undiluted C_2H_2/O_2 flame is much hotter than a CH_4/O_2 flame,

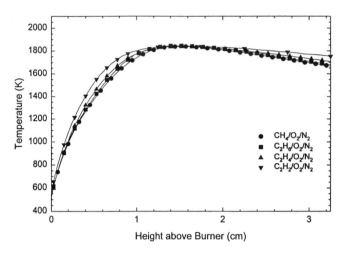

Fig. 1 - Temperature profiles of the four flames studied.

flame, a meaningful comparison between the different fuels requires minimizing the temperature difference. Flow rates of nitrogen diluent were chosen to maintain identical adiabatic flame temperatures at 10 Torr. The NO flow rates were the same for all flames; for recording profiles of OH, CH, and $^3C_2$, NO was not added except to investigate the sensitivity of these species to added NO.

Temperature profiles are shown in Fig. 1. Temperature profiles were recorded by an Al_2O_3 coated Pt/Rh thermocouple. Temperature determinations by LIF of OH in the methane flame were used to infer a radiation correction (about 120 K in the hottest region), which was applied to the thermocouple measurements in all the flames. The peak temperatures in all the flames are essentially identical; the profiles are similar, although the flame fronts of the acetylene flame and to a lesser extent the ethylene flame lie closer to the burner than those of methane and ethane. The profiles of OH are consistent with the temperature measurements; the position at which the OH mole fraction reaches one-half its maximum value occurs earlier than in the acetylene and ethylene flames.

Table I - Flame Compositions

Fuel Gas:	$\underline{CH_4}$	$\underline{C_2H_6}$	$\underline{C_2H_4}$	$\underline{C_2H_2}$
Fuel Flow Rate (sccm)	400	228	267	320
Oxygen Flow Rate	800	800	800	800
Nitrogen Flow Rate	528	628	1008	1680
Nitric Oxide Flow Rate	18	18	18	18

RESULTS

Species Profiles

In the figures shown below, LIF signals have been converted into relative mole fractions by correcting for changes in density and thermal populations of the probed spectroscopic transitions [1]. These corrections are based on fits to the temperature profiles of Fig. 1. Profiles were recorded for each of the different flames in succession, so that while the absolute concentrations of the radical species are not determined, the relative amounts for each of the flames may be compared. Since each of the fuels produces a different pool of hydrocarbon radicals during combustion, information about the NO reaction mechanisms can be inferred from such comparisons.

CH

Although CH is a minor species in terms of hydrocarbon flame chemistry, it is important in both the formation and removal of NO [4], due to the reactions:

$$CH + NO \rightarrow HCN + O \qquad (1)$$
$$CH + N_2 \rightarrow HCN + N \qquad (2)$$

In methane flames, most recent kinetic models predict that CH and CH_2 play the principal roles in NO consumption [2,4]. It is not known which are the principal NO removal pathways in acetylene flames, but since the CH concentration is greater than in the methane flame, one would expect Reaction (1) to be important in this environment as well.

Profiles of CH are shown in Fig. 2. The peak mole fraction of CH is greater by a factor of two to three in the acetylene flame than in the flames of the other fuels. The CH peak is also closer to the burner, in accord with the OH profiles. The CH profiles in the methane and ethane flames are very similar, and that of the ethylene flame more closely resembles the saturated fuels than it does acetylene.

$^3C_2$

Emission from C_2 is a pronounced feature of many types of flames, but the chemical processes leading to its formation are not well understood [5], and most kinetic models of flame combustion do not consider any reactions of C_2. C_2 has two low-lying states; the triplet state which was probed in the current study is slightly higher in energy than the singlet ground state [6]. The energy difference is small enough that both states will be thermally populated at combustion temperatures.

The amount of C_2 varies widely between the different flames (Fig. 3). The acetylene flame contains more than ten times as much $^3C_2$ as do the flames of the other fuels. The ethane and ethylene flames contain (somewhat surprisingly) almost equal peak C_2 concentrations. Methane produces much less C_2 than any of the other fuels. The significance of C_2 in NO consumption is unclear; the two species are known to have a very rapid reaction rate [7] but calculating the amount of NO consumed by this reaction requires knowing the concentrations of C_2. The C_2 + NO reaction may be significant in the acetylene flame, but seems unlikely to be important for the other fuels.

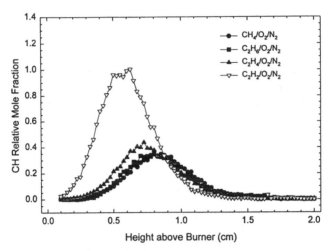

Fig. 2 - Profiles of CH mole fraction for the four flames studied.

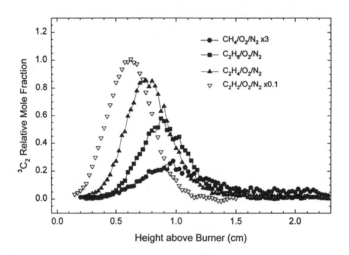

Fig. 3 - Profiles of $^3C_2$ mole fraction for the four flames studied

CN and NCO

These two species are closely linked in the kinetic pathways of nitrogen [4] . Nitric oxide consumed in the reaction zone usually forms HCN via Reaction (1) and analogous reactions with other hydrocarbon radicals. CN and NCO are the major products of HCN removal. CN profiles are shown in Fig. 4; the NCO profiles are very similar as a consequence of the close linkage between these two species. The peak concentrations of CN are fairly similar for all the flames except for acetylene, where it is about four times higher. The ratio of CN concentrations is higher than the ratio for CH, which indicates that other species besides CH contribute to the greater reactivity of NO in the acetylene flame.

NH

In most kinetic schemes, NH is thought to be formed from NCO by the reaction
$$H + NCO \rightarrow HN + CO \qquad (3)$$
suggesting that the relative NH concentrations in the different flames should closely parallel those of NCO and CN. Most kinetic models also predict this behavior. The NH profiles in Fig. 5 show that this is not the case for the acetylene flame. The peak NH concentration in the acetylene flame is about 50% higher than for ethylene, as compared to the factor of four found for CN and NCO. Evidently, there are reactions not correctly included in current models which act to reduce the NH concentration in the acetylene flame.

CONCLUSIONS

The acetylene flame seems to be unlike that of any of the other fuels studied. One might expect that as the degree of unsaturation of the fuel increases, a predictable trend should be observed. Ethylene might thus be expected to behave as an intermediate case between ethane and acetylene. While NO does react slightly more in the ethylene flame than in ethane, the difference is insignificant compared to the difference between ethylene and acetylene. Acetylene might thus be atypical of the behavior of higher hydrocarbons in general. A binder material such as polybutadiene could conceivably depolymerize into either ethylenic or acetylenic fragments; knowledge of the decomposition process is required to correctly predict the gas phase kinetics.

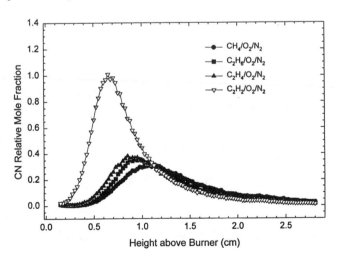

Fig. 4 - Profiles of CN mole fraction for the four flames studied

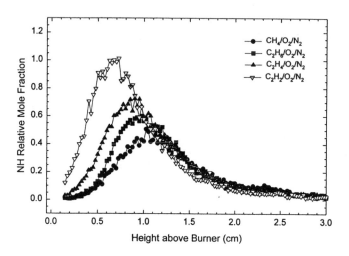

Fig. 5 - Profiles of NH mole fraction for the four flames studied

NO is much more reactive in the acetylene flame than with any of the other fuels. Partly this may be a consequence of the higher CH concentration, but the data indicates that other reaction partners also make a contribution. Reaction of NO with C_2 may be significant in this flame system. The observation that the relative concentration of NH does not follow that of CN or NCO in the acetylene flame is unexpected. The reaction pathways normally assumed for nitrogen species do not predict this to be the case, and new classes of reactions may have to be considered in the acetylene flame.

REFERENCES

1. B. A. Williams and J. W. Fleming, *Combustion and Flame*, **98**, 93 (1994).

2. R. P. Lindstedt, F. C. Lockwood, and M. A. Selim, Thermofluids Report TF/95/3, Mechanical Engineering Department, Imperial College.

3. B. A. Williams and J. W. Fleming, 31st JANNAF Combustion Subcommittee Meeting (CPIA, Columbia, MD, 1995), Vol. II, pp. 377-386.

4. J. A. Miller and C. T. Bowman, *Prog. Energy Combust. Sci.* **15**, 287 (1989).

5. A. G. Gaydon, *The Spectroscopy of Flames* (Chapman and Hall, London, 1957).

6. K. P. Huber and G. Herzberg, *Energy Levels of Diatomic Molecules* (Van Nostran Reinhold, New York, 1979).

7. H. Reisler, M. S. Mengir, and C. Wittig, *J. Chem. Phys.* **73**, 2280 (1980).

PRODUCTS AND KINETICS OF FLASH PYROLYSIS OF PEG:
A MINIMUM SMOKE BINDER

H. ARISAWA and T. B. BRILL
Department of Chemistry and Biochemistry, University of Delaware, Newark, DE 19716

ABSTRACT

Flash pyrolysis of polyethyleneglycol by T-Jump/FTIR spectroscopy to temperatures of the surface during combustion reveals that volatile products arise from approximately equal amounts of C-O and C-C homolysis. Nine volatile products are discussed. The average number of repeating units in the volatile oligomers is 2.5. A shift in product distribution occurs at 420-480 °C resulting from a change in the polymer structure. Below 420°C, di- and mono-ether oligomers and diethyleneglycol dominate. Above 480 °C, the mono-ethers and ethyleneglycol dominate. The Arrhenius constants for decomposition reflect this difference: Ea=8.8 kcal mol^{-1}, ln (A, s^{-1}) =2.0 at 370-420 °C and Ea=19 kcal mol^{-1}, ln (A, s^{-1})=10 at 480-550 °C.

INTRODUCTION

The most important ingredient in determining the combustion characteristics of a solid propellant is the energetic oxidizer or monopropellant. The binder, which contributes both mechanical strength and fuel in the propellant, is less well studied, and is, arguably, of lesser importance in the combustion process. However, an accurate model of combustion or explosion of a solid formulation must contain details about the pyrolysis chemistry of the binder. In particular, the species that are formed upon fast pyrolysis and rate of pyrolysis under combustion conditions must be known.

A binder/plasticizer in minimum smoke propellant formulations is polyethyleneglycol (PEG): $HOCH_2CH_2 - (OCH_2CH_2)_n - OCH_2CH_2OH$. Previous studies of pyrolysis of PEG primarily employ GC/MS or MS to identify volatile products of pyrolysis at a heating rate of degrees per minute. The most prevalent conclusion is that random O-C homolysis dominates C-C bond homolysis [1-4]. Random C-C and C-O homolysis both occur [5-8], although C-C homolysis is thought to initiate the process [5], and β-H transfer is thought to be a factor following O-C homolysis [1]. The rate of weight loss from PEG and related polyalkylethers at a heating rate of a few degrees per minute yields activation energies (Ea) of 26-49 kcal mol^{-1} [1, 4, 5, 9-10] with the majority of values at 45-49 kcal mol^{-1}.

This article briefly describes extensive work on the kinetics and species from a film of PEG which was heated at a rate exceeding 600 °C sec^{-1} to temperatures in the 370-550 °C range under 2 atm Ar. T-Jump/FTIR spectroscopy was employed [11]. Consequently, this study differs from previous studies in that a much faster heating rate, higher temperature range, and higher pressure were used. These conditions are much closer to those of combustion. A more detailed exposition will be presented elsewhere [12].

EXPERIMENTAL

A film of PEG consisting of about 1.5 mg of monomodal PEG (Mn=2000) was cast by melting PEG, transferring it with a pipette, and then remelting it on the filament to spread as a film.

Mat. Res. Soc. Symp. Proc. Vol. 418 © 1996 Materials Research Society

Owing to the extensive discussion required to describe the experimental details and data analysis properly, only a flow chart (Figure 1) will be presented to outline the procedures. T-Jump/FTIR spectroscopy has been described and modeled before [11, 13]. The partial least squares (PLS) method [14, 15], problem of collinearity of spectra [16, 17], and quadratic programming[17] were developed by others and were adapted here to the characterization of flash pyrolysis of PEG. The product of Figure 1 is the concentrations of the major products (Table 1) as a function of time and temperature.

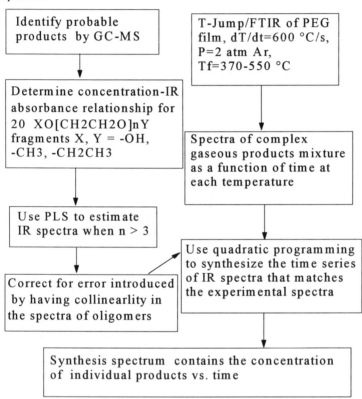

Figure 1. A summary of the experimental and data analysis procedures for species and kinetics of PEG pyrolysis.

Table 1. Quantified products liberated by PEG (Figure 1 procedure)

Monoether = $CH_3O[CH_2CH_2O]_n H + CH_3CH_2O[CH_2CH_2O]_n H$

Diether = $CH_3O[CH_2CH_2O]_n CH_3 + CH_3O[CH_2CH_2O]_n CH_2CH_3 +$
$CH_3CH_2O[CH_2CH_2O]_n CH_2CH_3$

Ethyleneglycol, Diethyleneglycol, 1,3-Dioxolane, 2-Methoxy-1,3-dioxolane, Methyl vinyl ether, Ethyl vinyl ether, 2-Butenal

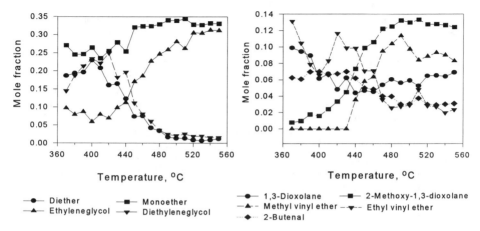

Figure 2. Temperature dependence of mole fractions of major products

Figure 3. Temperature dependence of mole fractions of minor products

RESULTS AND DISCUSSION

Nature of the products

Figures 2 and 3 summarize the mole fractions of the major and minor products, respectively, from flash pyrolysis of PEG at filament set temperatures(T_f) in the 370-550 °C range. Only the major products are discussed further here because they constitute nearly 70 % of the products. A significant quantity of glassy residue remains on the filament in the lower temperature range. Above 500 °C, no residue remains. In Figure 2, the mono- and diether-terminated oligomers along with diethyleneglycol dominate in the 370-420 °C range. Dramatic changes occur in the balance of all products in Figures 2 and 3 in the 420-480 °C range. Among the major products, ethyleneglycol becomes favored over diethyleneglycol, and the monoethers become favored over the diethers at higher temperatures. The dominance of the monoethers and dihydroxyl-terminated products is consistent with notions that C-O homolysis is favored over C-C homolysis[1-5]. Statistically, however, the dihydroxyl-terminated products, the monoethers, and diether should appear in a 10:19:7 ratio if C-O and C-C homolysis occur equally frequently. In Figure 2, the mole fractions averaged over the 370-550 °C range of the dihydroxyl products (ethyleneglycol and diethyleneglycol), the monoether products, and the diether products are 0.15±0.05 : 0.30±0.05 : 0.12±0.10. These ratios are approximately equal to the theoretical ratios when the differences caused by the temperature effect are used to be define the uncertainty. Thus, the C-O and C-C homolysis processes appear to occur to roughly equal extents under flash heating conditions. The apparent preference for hydroxyl-terminated products is at least partly statistical in origin as opposed to indicating a preference for C-O homolysis.

The shift in the product distribution at 420-480 °C may be related to decomposition of the parent PEG along with the formation of the glassy residue at temperature below range, while decomposition of both PEG and the residue occurs above this range. Both the apparence and hardness of the residue suggest that it may have higher molecular weight than the parent PEG sample. Although the reasons are not obvious for why the concentrations of the pyrolysis products depend on this change in structure of PEG, the evidence reveals that they are. Moreover, the kinetics discussed in following section reveal that a discontinuity exists in the 420-480 °C range. The preference for ethyleneglycol over diethyleneglycol at higher temperature can be attributed to the preference to form monomer units over dimer units. This pattern is reminiscent, for example, of that of hydroxyl-terminated polybutadiene[12].

The control voltage difference trace in Figure 4 has no negative inflections during the decomposition of PEG. Therefore, the overall decomposition process of the condensed phase is never exothermic. Instead, the positive inflection observed indicates that the process is actually endothermic. The concentration profile of the monoether products, which are major products (Figure 2), reveals that they form early in the process. The average degree of polymerization, DPn, (n in Table 1) increases with time, but saturates at 2.5 suggesting that steady state is reached in the pyrolysis characteristics of the condensed phase. This value of DPn was independent of temperature in the 450-550 °C range.

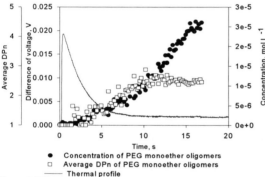

Figure 4. Difference of control voltage trace and rates of growth of monoether oligomer products and the average DPn at 450 °C

- ● Concentration of PEG monoether oligomers
- □ Average DPn of PEG monoether oligomers
- —— Thermal profile

Kinetics

All previous kinetic studies of PEG [1,4,9-10] were conducted at heating rates of degrees per minute and at lower temperature than those used here. Figure 5 is a panel of five Arrhenius plots for most of the products from flash pyrolysis of PEG. Table 2 lists the Arrhenius parameters derived from the slopes (k) of concentration vs. time data of the type shown in Figure 4. The zero-order rate equation (1) was used.

$$d\left[C(t)/C_\infty\right]/dt = k \qquad (1)$$

The activation energies are lower than the 26-49 kcal mol^{-1} previously obtained at low heating rates. Under the more aggressive conditions used in this work, the rate of evolution of most products, as evidenced by the low Ea and ln A values, is consistent with control by the evaporation/desorption rates as opposed to the rate of the condensed phase reactions. 2-Methoxy-1,3-dioxolane is different in that in the lower temperature range, the Arrhenius parameters are relatively high. This finding can be explained by the fact that, unlike the other compounds, the formation of this product involves many reactions (C-O homolysis, intramolecular cyclization, and several radical recombinations). These multiple steps together are probably responsible for the high values of the Arrhenius constants. On the other hand, above 470 °C, the rate of appearance of 2-methoxy-1,3-dioxolane in the gas phase resembles those of the other compounds, suggesting desorption control.

It is necessary to sum all of the gaseous concentrations to produce global Arrhenius constants for the overall pyrolysis rate of PEG at high heating rates in this temperature range (Figure 6).

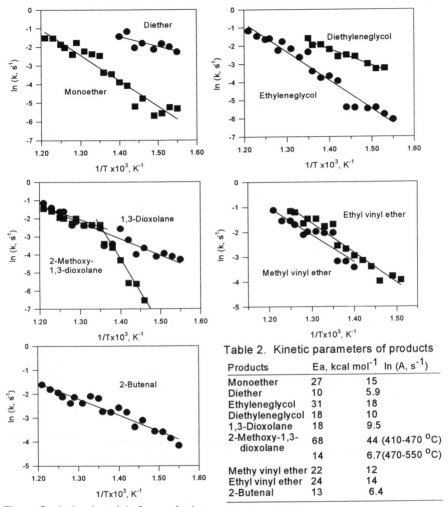

Table 2. Kinetic parameters of products

Products	Ea, kcal mol^{-1}	ln (A, s^{-1})
Monoether	27	15
Diether	10	5.9
Ethyleneglycol	31	18
Diethyleneglycol	18	10
1,3-Dioxolane	18	9.5
2-Methoxy-1,3-dioxolane	68	44 (410-470 ^{o}C)
	14	6.7(470-550 ^{o}C)
Methy vinyl ether	22	12
Ethyl vinyl ether	24	14
2-Butenal	13	6.4

Figure 5. Arrhenius plots for products

At 370-420 °C, the rather low values suggest that the rate of appearance of products is dominated by the evaporation/desorption rate of the relatively abundant products in this range (monoether, diether oligomers, diethyleneglycol, and 1,3-dioxolane). An ill-defined discontinuity then is evident in the 420-480 °C range. This temperature range coincides with a switch in the dominance of the lower temperature products in Figures 2 and 3 to those prevalent at higher temperature. Therefore, at 480-530 °C, the low but different values of Arrhenius constants suggest that the desorption/evaporation rate of a different set of products (monoether oligomers, ethyleneglycol, and 2-methoxy-1,3-dioxolane) dominates. The apparent change in the polymer structure is responsible for this change.

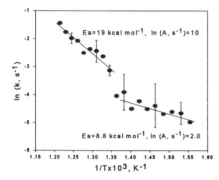

Figure. 6 Arrhenius plot for global kinetics

This study identifies the gaseous species and their rates of formation from flash pyrolysis of a film of PEG. The temperatures reached resemble those of the surface during combustion. Desorption of the various sets of products appears to control the rate of gaseous product evolution. This information helps to advance combustion modeling of propellants containing PEG by providing the identity and rates of formation of individual products.

ACKNOWLEDGMENTS

We are grateful to the Air Force Office of Scientific Research, Aerospace Sciences, for support of this work on F49620-94-1-0053, and to the Technical Research and Development Institute, Japan Defense Agency.

REFERENCES

1. S. L. Madorsky and S. Straus, J. Polym. Sci. **36**, p. 183(1959).
2. R. P. Lattimer, H. Münster and H. Budzikiewicz, Int. J. Mass. Spectrom. Ion. Processes, **90**, p. 119(1989).
3. M. M. Fares, J. Hacaloglu and S. Suzer, Eur. Polym. J., **30**, p. 845(1994).
4. M. Ishikawa, The Bulletin of Aichi Univ. of Education, **19**(Natural Science), p. 19(1970).
5. A. B. Blyumenfel'd and B. M. Kovarskaya, Vysokomol. soyed., **A12**, p. 633(1970).
6. N. Grassie and G. A. Perdomo Mendoza, Polym. Deg. Stab. **9**, p. 155(1984).
7. G. G. Cameron, M. D. Ingram, M. Younus Qureshi and H. M. Gearing, Eur. Polym. J., **25**, p. 779(1989).
8. K. J. Voorhees, S. F. Baugh and D. N. Stevenson, J. Anal. Appl. Pyrolysis, **30**, p. 47(1994).
9. R. Audebert and C. Aubineau, Eur. Polym. J., **6**, p. 965(1970).
10. E. Calahorra, M. Cortazar and G. M. Guzmán, J. Polym. Sci. Polym. Lett. Ed., **23**, p. 257(1985).
11. T. B. Brill, P. J. Brush, K. J. James, J. E. Shepherd and K. J. Pfeiffer, Appl. Spectrosc., **46**, p.900(1992).
12. H. Arisawa and T. B. Brill, Combust. Flame, to be published.
13. J. E. Shepherd and T. B. Brill, 10th Symposium on Detonation, Office of Naval Research, Arlington VA, p.849(1993)
14. P. Geladi and B. R. Kowalski, Anal. Chim. Acta, **185**, p.1(1986).
15 B. M. Wise, PLS Toolbox for use with MATLAB®, 1415 Wright Avenue, Richland, WA 99352.
16 K. R. Beebe and B. R. Kowalski, Anal. Chem., **59**, p.1007A(1987).
17. A. Grace, Optimization toolbox for use with MATLAB®, The MathWorks, Inc., 1994.

KINETICS AND SPECIES OF FLASH PYROLYSIS OF
CELLULOSE ACETATE BUTYRATE: THE BINDER OF LOVA

P. E. GONGWER, H. ARISAWA, and T. B. BRILL
Department of Chemistry and Biochemistry, University of Delaware, Newark, DE 19716

ABSTRACT

The principal binder of many LOVA propellants is cellulose acetate butyrate (CAB). By the use of T-Jump/FTIR spectroscopy, CAB was flash-pyrolyzed to set temperatures in the 465-600°C range, while rapid-scan IR spectra were used to identify the main decomposition products and to measure the rate of formation of each product as a function of temperature. Eleven specific products, which include oligomers of CAB, acids, aldehydes, ketenes, esters, CO_2 and CO, were quantified by chemometric procedures. The ketenes are the most novel products. The Arrhenius parameters reveal that below $510 \pm 20°C$, the rate of product evolution is controlled mainly by condensed phase reactions. Above $510 \pm 20°C$, the rate of product evolution is controlled by desorption/evaporation of the volatile products.

INTRODUCTION

Cellulose acetate butyrate (CAB) is a thermoplastic binder/fuel which comprises about 12% of LOVA propellant formulations [1]. LOVA is the acronym for "low vulnerability ammunition" in which enhanced safety is achieved by reducing the tendency for undesired ignition. The choice of CAB as an ingredient is partly based on considerations of the mechanical and processing properties of the propellant. However, CAB is an active fuel during the combustion event. As such, its decomposition kinetics and the identities of the volatile species under high heating rate and high temperature conditions are essential elements in modeling of the combustion of CAB.

Although much research has been reported on the kinetics of slow [2-10] and fast [11] decomposition of pure cellulose, similar studies of non-energetic derivatives of cellulose have been confined primarily to cellulose triacetate (CA). The volatile species [12, 13], nature of residue [14], mechanism [13, 15], and kinetics [13] of slow decomposition of CA have been reported. No similar studies of CAB are available. Because of the need for a kinetic and species characterization of rapid decomposition of CAB in combustion modeling, an investigation of flash pyrolysis of films of CAB was undertaken by using T-Jump/FTIR spectroscopy [16]. Chemometric procedures were used to deconvolute the IR spectra which were recorded on the composite product mixture. The identities and rates of formation of the individual, dominant, volatile products were determined in near real-time by FTIR spectroscopy.

EXPERIMENTAL

T-Jump/FTIR spectroscopy, which has been described and modeled before [16, 17], was used to flash-pyrolyze CAB at a controlled heating rate to controlled final temperatures, while recording near real-time IR spectra of the vaporized products 3mm above the surface. To accomplish this experiment, an approximately 0.05 mg sample of CAB was placed uniformly on the center of the platinum filament of the T-Jump pyroprobe. The probe was then inserted into a gas-tight spectroscopy cell. The CAB was flash-pyrolyzed under a pressure of 1 atm Ar, at a heating rate of about 600°C/s, to final temperatures in the 465 - 600°C range.

Probable identities of the pyrolysis products were determined by using GC-MS and from previous studies [11-13]. To resolve the composition of the pyrolysis gas mixture, the concentration-IR absorbance relationships (Lambert-Beer Law) of the individual products first must be determined. This relationship for the products that are stable liquids at 25°C was determined by placing an excess amount of liquid in the reaction cell and measuring the vapor-phase IR spectrum at the liquid-vapor equilibrium condition. The details of this technique are described elsewhere [18]. The Lambert-Beer relationship for formaldehyde was determined by flash pyrolyzing paraformaldehyde and taking the ratio of the absorbance spectrum to that published [19]. Other products that are stable gases at 25°C were calibrated by using the spectra of various pressures in the reaction cell. The IR spectra of the transient products containing the ketene functional group were obtained by using a pyrolysis set-up similar to that of Fisher, et al. [20]. Acetic anhydride was pyrolyzed on the platinum filament at 510°C for about 1 hour. The resulting spectrum was converted to concentration by taking the ratio of the $>C=C=O$ peak at ~2150 cm^{-1} and the CO_2 peak at 2349 cm^{-1}, whose intensity was calibrated above. The assumption is that H_2C_2O (ketene) and CO_2 have similar absorbance intensities because they are isoelectric molecules. The spectrum of ethyl ketene was obtained and calibrated similarly, except that the starting material used was butyric anhydride, the pyrolysis temperature was 450°C, and the experiment was conducted over a two minute period because ethyl ketene is unstable. CAB oligomers were calibrated by depositing a known amount of CAB in an acetone solution onto BaF_2 and allowing the acetone to evaporate. The IR beam diameter was then fixed by an apeture and the amount of sample in the path of the beam was calculated. The resulting IR absorbance spectrum was converted to concentration by dividing the amount of sample in the IR beam by the volume of the reaction cell.

After correcting the baseline for slope changes resulting from light scattering, the absorptivity-concentration relationships of the products were used in conjunction with non-negative least squares regression to resolve the concentrations of each product in the pyrolysis gas spectrum. Since spectra were recorded at 0.1 sec intervals, the time and temperature dependences of the products were extracted. Equation 1 gives the matrix equation

$$m = Ec\text{-}r(t) \qquad (1)$$

used to solve for these concentrations, where c is the concentration of an individual component, E is the calibration matrix for each product, $r(t)$ is the IR spectrum of the pyrolysis products at time t, and m is the residual matrix which is minimized in the optimization procedure. Non-negative least squares is a condition restriction that can be used because $c \geq 0$. Details of this method applied in the manner used here are presented in detail elsewhere [18].

RESULTS AND DISCUSSION

Products of Pyrolysis

As in the structure shown here, the sample of CAB contains randomly substituted butyryl (1.66), acetyl (0.96) and hydroxyl (0.38) groups at the 3 R sites. The potential major pyrolysis products of CAB were deduced by comparison of previous GC/MS data for CA [11-13]. Many of the products found in these previous slow heating studies and time-delay analyses were

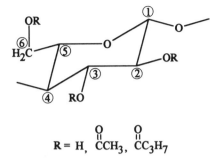

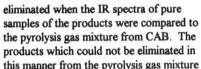

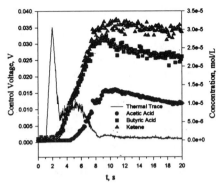

eliminated when the IR spectra of pure samples of the products were compared to the pyrolysis gas mixture from CAB. The products which could not be eliminated in this manner from the pyrolysis gas mixture

Figure 1. Rate of formation of selected pyrolysis products along with the corresponding thermal trace of the condensed phase.

were calibrated as described in the experimental section and used in equation 1. Water was not found to be a quantifiable product. The spectrum of H_2O did appear sporadically, but this often occurs as a response to noise in the non-negative least squares regression routine. Figure 1 shows the rate of formation of several of the volatile products as a function of time when CAB was flash-heated to 525°C. The control voltage trace of the platinum filament exhibits no negative inflections which indicates that the overall decomposition process of CAB is endothermic.

The mole fractions of the final concentrations of the major and minor volatile products from CAB are shown in Figures 2 and 3. The spectrum of CAB was included in the minimization of **m** in equation 1 because very weak CAB-like absorbances appeared in the vapor phase. It is probable that these absorbances are oligomers of CAB which have sufficient volatility to vaporize under the heating conditions used.

CO and CO_2 could form from the CO and OCO linkages of the anhydroglucose backbone [15], and by the decarboxylation or decarbonylation of the ester and aldehyde molecules formed. The origin of the volatile esters in which C(5)-C(6) homolysis occurs, and carboxylic acids in

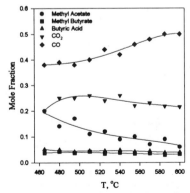

Figure 2. Mole fractions of major products

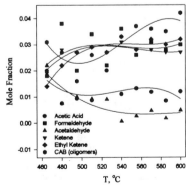

Figure 3. Mole fractions of minor products

235

which C(2,3)-O and C(6)-O homolysis occurs is readily associated with recognizable pendant groups of CAB. H-transfer is also required to stabilize these products. Butyric acid has about twice the concentration of acetic acid in accordance with the approximate relative concentration of these groups in the parent CAB sample.

The most interesting products are ketene and ethyl ketene [$CH_3CH_2C(H)=C=O$]. Figure 4 shows the resolution of ketene, ethyl ketene, and CO from the 2050 - 2200 cm^{-1} region of the spectrum. Also shown is the residual **m** (equation 1) after the solution. These two ketene products probably arise from O-C homolysis of the O-C(O)R pendant groups, which liberates -C(O)R. This fragment can rearrange to ketene when R = CH_3- and ethyl ketene when R = C_3H_7-. $CH_3C(O)H$ could result from O-C(O)CH_3 homolysis, as does ketene, but rather than losing H to form $H_2C=C=O$, the -C(O)CH_3 fragment could abstract H to form $CH_3C(O)H$. H_2CO plausibly arises from the non-esterified -OH groups of CAB.

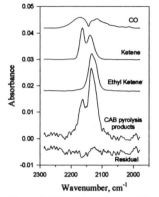

Figure 4. The deconvolution of the pyrolysis gases of CAB into the ketene components and CO

A residue remains on the filament, which is brown at 525°C, but brown-black with a sooty appearance at 540°C. The formation of H- and O-rich volatile products in Figures 2 and 3 will leave a carbon-rich residue.

Kinetics

Figures 5, 6 and 7 are panels of Arrhenius plots deduced for the major quantified volatile products. Figure 8 shows the Arrhenius plot constructed from the sum of the concentrations of all volatile products. The rate constants, k, used for these plots were derived from the zeroth-order behavior in Figure 1 by using equation 2. C_∞ was the maximum concentration of

$$d[C(t)/C_\infty]/dt = k \qquad (2)$$

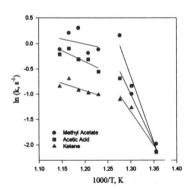

Figure 5. Arrhenius plots for the rates of evolution of products derived from the acetyl group

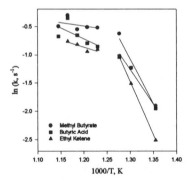

Figure 6. Arrhenius plots for the rates of evolution of products derived from the butyryl group

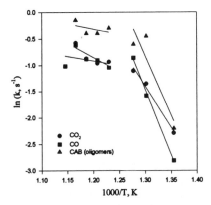

Figure 7. Arrhenius plots for the rates of evolution of CO_2, CO, and CAB (oligomers)

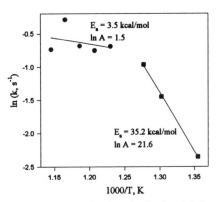

Figure 8. Arrhenius plot for the global rates of evolution of all products

Table 1. Arrhenius Constants for Flash Pyrolysis of CAB[a]

Compound	465 - 510°C		510 - 600°C	
	E_a	lnA	E_a	lnA
Methyl Acetate	51.9	33.3	4.0	2.4
Methyl Butyrate	31.3	19.4	2.4	1.0
Acetic Acid	39.0	24.5	8.5	4.8
Butyric Acid	24.2	14.6	7.7	3.9
Formaldehyde	5.5	2.4	5.5	2.4
Acetaldehyde	14.3	8.3	14.3	8.3
Carbon Dioxide	30.9	18.8	10.7	5.6
Carbon Monoxide	49.2	30.7	3.4	1.1
Ketene	26.6	16.0	5.8	2.6
Ethyl ketene	37.4	23.0	5.0	2.2
CAB(oligomers)	44.7	28.4	4.0	2.1

[a] E_a in kcal/mol; A in s^{-1}.

each product at the end of the pyrolysis run. Table 1 contains the resulting Arrhenius constants. The most notable feature is the existence of a slope break in the $510 \pm 20°C$ range. Below this temperature range the Arrhenius constants are characteristic of control by condensed phase reaction processes. Above this temperature range, the much smaller values obtained are similar to those expected of control by the rate of desorption/evaporation of the volatile decomposition products. The same behavior is found with hydroxyl-terminated polybutadiene (HTPB) [18] in a similar temperature range, but not for polyethylene glycol (PEG) [21]. HTPB, like CAB, decomposes by the occurrence of many competitive reactions, whereas PEG exhibits mostly radical chain cleavage to release oligomers of PEG.

CONCLUSIONS

Depending on the surface temperature of CAB-containing LOVA propellants, the rate of pyrolysis of CAB will be dominated either by the condensed phase reaction rates or the desorption/evaporation rates. The mole fractions, however, are not strongly dependent on temperature in the 465 - 600°C range.

ACKNOWLEDGMENTS

We are grateful for financial support from the Pennsylvania State University on ARO-URI grant DAAL03-92-G-0118. The CAB sample was provided by Dr. Rose Pesce-Rodriguez (Army Research Lab).

REFERENCES

1.   R. A. Pesce-Rodriguez, C. S. Miser, K. L. McNesby, R. A. Fifer, S. Kessel, and B. D. Strauss, Appl. Spectrosc., **46**, p.1143 (1992).
2.   A. E. Lipska and W. J. Parker, J. Appl. Polymer Sci., **10**, p.1439 (1966).
3.   A. E. Lipska and F. A. Wodley, J. Appl. Polymer Sci., **13**, p.851 (1969).
4.   F. Shafizadeh and A. G. Bradbury, J. Appl. Polymer Sci., **23**, p.1431 (1979).
5.   A. G. Bradbury, Y. Sakai, and F. Shafizadeh, J. Appl. Polymer Sci., **23**, p.3271 (1979).
6.   A. Broido and M. Weinstein, Comb. Sci. and Tech., **1**, p.279 (1970).
7.   P. C. Lewellen, W. A. Peters, and J. B. Howard, 16th Symposium (Int.) on Combustion, p.1471, The Combustion Institute, Pittsburgh, 1977.
8.   F. E. Rogers and T. J. Ohlemiller, Comb. Sci. and Tech., **24**, p.129 (1980).
9.   G. Varhegyi, M. J. Antal, Jr., T. Szekely, and P. Szabo, Energy and Fuels, **3**, p.329 (1989).
10. I. Milosavljevic and E. M. Suuberg, Preprints of Papers - American Chemical Society, Division of Fuel Chemistry, **39(3)**, p.860 (1994).
11. Z. Ozturk and J. F. Merklin, Biomass and Bioenergy, **5(6)**, p.437 (1993).
12. A. Scotney, Europ. Polym. J., **8**, p.163 (1972)
13. W. P. Brown and C. F. H. Tipper, J. Appl. Polym. Sci., **22**, p.1459 (1978).
14. A. Scotney, Europ. Polym. J., **8**, p.175 (1972).
15. A. Scotney, Europ. Polym. J., **8**, p.185 (1972).
16. T. B. Brill, P. J. Brush, K. J. James, J. E. Shepherd and K. J. Pfeiffer, Appl. Spectrosc., **46**, p.900 (1992).
17. J. E. Shepherd and T. B. Brill, 10th International Symposium on Detonation, Office of Naval Research, Arlington VA, p.849 (1993).
18. H. Arisawa and T. B. Brill, Combust. Flame, in press.
19. T. Nakanaga, S. Kondo and S. Saeki, J. Chem. Phys., **76(8)**, 15 Apr. 1982, p.3860.
20. G. J. Fisher, A. F. MacLean and A. W. Schnizer, J. Org. Chem., **18**, p.1055 (1953).
21. H. Arisawa and T. B. Brill, Decomposition Combustion and Detonation Chemistry of Energetic Materials, Materials Research Society, Pittsburgh, PA, 1995, preceding paper.

NEW ENERGETIC EPOXY BINDERS

S.R. JAIN, S. AMANULLA
AE Department, Indian Institute of Science, Bangalore 560 012, INDIA
sampat @ aero.iisc.ernet.in

ABSTRACT

A new class of epoxy resins having N-N bonds in the backbone has been synthesized with a view to explore their properties as energetic binders. The N-epoxidation of bis-dicarbonylhydrazones of adipic, azelaic and sebacic dihydrazides results in the formation of viscous resins having epoxide end groups. The resins have been characterized by the elemental and end group analyses, IR and NMR spectra. Relevant properties for their use as binders in solid propellants, such as thermal stability, heat of combustion, burn rate and performance parameters of AP-based propellant systems, have been evaluated. A significant increase in the burn rate of AP-based propellants noticed, is perhaps related to the exothermicity of the binder decomposition and the reactivity of N-N bonds with perchloric acid formed during the combustion of AP.

INTRODUCTION

Energetic compounds having N-N bonds have long been used in propellant formulations. The extreme reactivity of hydrazine and its methyl derivatives, with liquid oxidizers has been utilized in developing self-igniting (hypergolic) biliquid propellant engines. Several solid derivatives of hydrazine ignite instantaneously on coming in contact with liquid oxidizers and have been considered for hybrid propellant systems [1,2]. It is anticipated that the ignition and combustion characteristics of solid propellants could also be altered by using N-N bonded pre-polymers as binders. However, resinous pre-polymers having N-N bonds in their backbones and suitable end groups for curing after loading with solid oxidizers, appear to have never been synthesized [3]. A series of N-N bonded resins based on carbono- and thiocarbonohydrazones having epoxy end groups were, therefore, prepared in our laboratory, specifically to use them as binders for powder propellant compositions [4]. It was indeed observed that fuel composites prepared with these epoxies as binders exhibit superior ignition characteristics and, in fact, self-ignite with oxidizers like HNO_3 in the hybrid propellant mode [5]. When used in solid propellants, the burn rate of ammonium perchlorate (AP) based systems was found to be significantly enhanced as compared to those processed with the conventional polybutadiene binders [6]. These studies clearly point to the potential use of N-N bonded epoxy binders in the field of solid propellants. However, their use as binders resulted in somewhat hard structure of the propellant grain. Herein we report the synthesis of a yet new series of N-N bonded epoxides which are based on bis-carbonylhydrazones with varying number of spacer (CH_2) groups to provide the desired flexibility to the backbone. An account of the resin synthesis and characterization, and their use as binders in solid propellant systems is presented in this paper.

EXPERIMENT

Synthesis

The synthesis of bis-carbonylhydrazone epoxides was carried out using the methyl esters of aliphatic dicarboxylic acids as starting materials. The methyl ester on reacting with hydrazine hydrate yields the corresponding dihydrazide. The dihydrazide is subsequently condensed with an aldehyde/ketone to form the bis-hydrazone. The bis-carbonylhydrazones, thus prepared, are epoxidized by refluxing in excess epichlorohydrin and treating with aqueous sodium hydroxide (44%), to yield the corresponding epoxy resins. These processes may be described by the following scheme of reactions:

$$H_3CO\overset{O}{\underset{}{\overset{\|}{C}}}\text{-}(CH_2)_x\text{-}\overset{O}{\underset{}{\overset{\|}{C}}}\text{-}OCH_3 + H_2NNH_2$$

$$\Big\downarrow \text{ -CH}_3\text{OH}$$

$$H_2NNH\overset{O}{\underset{}{\overset{\|}{C}}}\text{-}(CH_2)_x\text{-}\overset{O}{\underset{}{\overset{\|}{C}}}\text{-}NHNH_2$$

$$\underset{R\text{-}\overset{R'}{\underset{}{\overset{|}{C}}}=O}{\Big\downarrow} \text{ -H}_2\text{O}$$

$$R\text{-}\overset{R'}{\underset{}{\overset{|}{C}}}=N\text{-}NH\text{-}\overset{O}{\underset{}{\overset{\|}{C}}}\text{-}(CH_2)_x\text{-}\overset{O}{\underset{}{\overset{\|}{C}}}\text{-}NH\text{-}N=\overset{R'}{\underset{}{\overset{|}{C}}}\text{-}R$$

$$\underset{Cl\text{-}CH_2\text{-}\overset{O}{\overset{}{CH}}\text{-}CH_2/NaOH}{\Big\downarrow} \text{ -NaCl/H}_2\text{O}$$

$$\overset{O}{\underset{}{\overset{}{CH_2}}}\text{-}CH\text{-}CH_2\text{-}[-N\text{-}\overset{O}{\underset{}{\overset{\|}{C}}}\text{-}(CH_2)_x\text{-}\overset{O}{\underset{}{\overset{\|}{C}}}\text{-}N\text{-}CH_2\text{-}\overset{OH}{\underset{}{\overset{|}{CH}}}\text{-}CH_2\text{-}]_n\text{-}N\text{-}\overset{O}{\underset{}{\overset{\|}{C}}}\text{-}(CH_2)_x\text{-}\overset{O}{\underset{}{\overset{\|}{C}}}\text{-}N\text{-}CH_2\text{-}CH\text{-}\overset{O}{\underset{}{\overset{}{CH_2}}}$$

Where, X=4, R=CH₃, R'=CH₂CH₃, Diepoxide of butanone(adipic)hydrazone (DEBuAH)
X=8, R=CH₃, R'=CH₂CH₃, Diepoxide of butanone(sebacic)hydrazone (DEBuSH)
X=7, R=H, R'=C₅H₄O, Diepoxide of furfural(azelaic)hydrazone (DEFAzH)
X=8, R=H, R'=C₅H₄O, Diepoxide of furfural(sebacic)hydrazone (DEFSH)

Propellant Processing

Ammonium perchlorate based propellant samples were processed with the following composition: AP, 80%; N-N bonded epoxy binder, 16%; curing agent (diaminodiphenylmethane) and plasticizer (dimethyl sebacate), 4%; by weight. Using press-molding technique individual void free strands of size 80x7x5 mm could be prepared by curing the mix at 80°C for 24 hrs. Strands for the control propellant with carboxyl terminated polybutadiene (CTPB) binder, and also those having different oxidizers, such as potassium perchlorate (KP), were processed under identical experimental conditions. For burn rate measurements, the strands were coated with a slurry of titanium dioxide in Araldite to inhibit the side burning.

Characterization

The resins were characterized by the elemental analysis, epoxy content and IR and NMR spectra. A simultaneous DTA-TG unit (Shimadzu DT-40) was used to determine the thermal characteristics. The heats of combustion data were obtained by using an adiabatic bomb calorimeter under oxygen pressure (30 KSC). Linear burn rates of the propellant strands were determined in nitrogen atmosphere at various pressures using a strand burner assembly. Theoretical performance parameters of the propellant systems were evaluated using a NASA SP-273 program.

RESULTS

It is seen that the N-H protons of bis-carbonylhydrazones react readily with epichlorohydrin resulting in the formation of N-epoxides. Apparently, the mechanism of synthesis is similar to that occurring in the epoxidation of amines [7] to give glycidylamines. Epichlorohydrin reacts with the N-H protons forming chlorohydrinamine in the first step; the dehydrochlorination by sodium hydroxide treatment results in the formation of epoxy ring in the second step. The elemental analysis of the resins given in Table I corresponds closely to the expected carbon and hydrogen contents. The observed epoxy equivalent values show some degree of homopolymerization.

Table I: Elemental Analysis, Epoxy Equivalent and Heat of Combustion Data

Resin	Color	% Carbon		% Hydrogen		Epoxy Equivalent (WPE)	Heat of Combustion (Kcal/gm)
		Calcd.	Obsd.	Calcd.	Obsd.		
DEBuAH	Brown	60.90	59.10	8.63	8.50	298	6.64
DEBuSH	Brown	64.00	62.58	9.33	9.20	328	7.13
DEFAzH	Brown	61.98	60.92	6.61	6.58	358	6.32
DEFSH	Brown	62.65	61.12	6.82	6.74	390	6.45

The epoxidation of the N-H protons is evident from both the ^{1}H NMR and IR spectral data (Table II). The ^{1}H NMR spectra of bis-carbonylhydrazones show -NH proton

Table II: Infrared and ^{1}H NMR Spectral Data

Resin	Absorption Frequency, ν (cm^{-1})/Chemical Shift δ (ppm)
DEBuAH	IR: 3300 (OH), 2900 (CH$_2$ str.), 1710 (C=O), 1640 (C=N), 830, 900 (epoxy), 740 (-[CH$_2$]$_4$-); ^{1}H NMR: 1.07, 1.90, 2.25 (CH, butanone), 1.70 (-[CH$_2$]$_4$-), 2.50-3.50 (CH, epoxy).
DEBuSH	IR: 3300(OH), 2910 (CH$_2$ str.), 1720 (C=O), 1640 (C=N), 840, 900 (epoxy), 740 (-[CH$_2$]$_8$-); ^{1}H NMR: 1.08, 1.90, 2.30 (CH, butanone), 1.65 (-[CH$_2$]$_8$-), 2.50-3.45 (CH, epoxy).
DEFAzH	IR: 3380 (OH), 2910 (CH$_2$ str.), 1720 (C=O), 1640 (C=N), 830, 920 (epoxy), 740 (-[CH$_2$]$_7$-); ^{1}H NMR: 1.38 (-[CH$_2$]$_7$-), 2.80-3.80 (CH, epoxy), 6.65-7.70 (CH, furfural), 8.07 (CH, ald.).
DEFSH	IR: 3400 (OH), 2900 (CH$_2$ str.), 1710 (C=O), 1640 (C=N), 840, 910 (epoxy), 740 (-[CH$_2$]$_8$-); ^{1}H NMR: 1.32 (-[CH$_2$]$_8$-), 2.70-3.75 (CH, epoxy), 6.75-7.72 (CH, furfural), 8.03 (CH, ald.).

resonances around 10.8 to 11.6 ppm [8]. These resonances are found to be absent in the corresponding resins. Instead, the spectra show resonances in the region 2.5 to 3.8 ppm, which could be attributed to the epoxide protons. Other resonances are observed virtually unaltered in the resins, e.g. the methylene and furfural proton resonances both in the hydrazone and its resin occur around 1.07 to 2.25 and 6.65 to 7.72 ppm respectively.

The IR spectra of the resins show almost all the major expected absorptions. The split band in the region 2910-2900 cm^{-1} could be attributed to the presence of epoxy groups, arising due to the stretching of CH and CH$_2$ groups contained in the epoxide ring. Other characteristic absorptions of the epoxide group appear in the 1100 cm^{-1} region as a result of ring breathing or symmetric stretching vibration and at 895 cm^{-1} and 830 cm^{-1} due to asymmetric stretching vibrations [9] in these resins. The carbonyl (C=O) absorption around 1710 cm^{-1}, and C=N absorption at 1640 cm^{-1} appear as expected. A broad absorption peak around 3400 cm^{-1} indicates the presence of OH groups, possibly due to oligomer formation. The bending vibrations of -(CH$_2$- groups at 740 cm^{-1} were present in the spectra of the resins and the parent hydrazones. As expected, the -NH stretching vibrations which appear in the parent hydrazones around 3240 cm^{-1} were found to be absent in the corresponding resin spectra.

The simultaneous DTA-TG, carried out in nitrogen atmosphere at a heating rate of 10°C/min, show that all these resins decompose exothermically, as expected (Fig 1). The

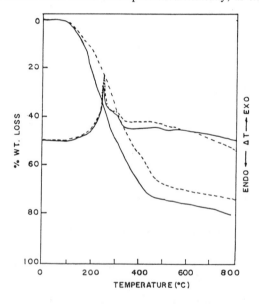

Fig. 1: Simultaneous DTA-TG of (–) DEBuSH and (- -) DEFSH Resins.

weight loss of <5% which occurs below 120°C could be due to the evaporation of the residual epichlorohydrin (bp 115°C) although no endotherm is seen in the DTA thermogram. Apparently, the resins start decomposing above 120°C resulting in continuous weight loss. The decomposition exotherm peaks at 220°C. A shoulder is seen at the end of the exotherm (280°C);

242

the TG shows about 50% weight loss at this stage. The next weight loss transition is observed at around 430°C which corresponds to a minor exothermic hump in the DTA. At the end of the broad exotherm (720°C) the weight loss observed is about 75%.

A useful property of these resins is the convenient viscosity at room temperature which permits processing of compositions having high solid loadings. Compositions having 80% solids could be processed easily using these resins as binders and cured with amine curatives at slightly elevated temperatures. As shown in Fig 2, the burn rate of AP-based

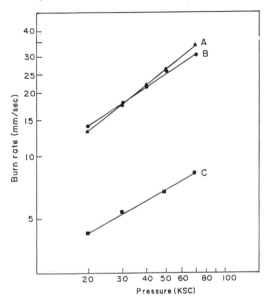

Fig. 2: Strand Burn Rates of AP-Propellants Processed with (A) DEFSH (B) DEBuSH and (C) CTPB Binders.

propellants increases with chamber pressure following the Vieille's law with pressure exponent equal to 0.70, when these resins are used as binders. Virtually no change in burn rate is observed with the change in the resin substituent. However, there is significant increase in the burn rate (3 to 4 fold) as compared to that of the propellant strands processed with the conventional polybutadiene binder (CTPB). This finding has specific importance considering the fact that the enhancement of burn rate of AP based propellants is usually achieved by cumbersome methods. While the metal oxide catalysts add unwanted dead weight, the use of ultra fine AP powder (the other method of enhancing the burn rate) causes production and processing difficulties. Enhancement of burn rate by simply changing the binder is obviously an advantageous method. The burn rate studies further show that while the effect of these binders on AP-based systems is remarkable, no noticeable change is seen on KP-based systems. For instance, the burn rate of KP-based propellant at 50 KSC (12.4 mm/sec) is almost equal to that observed when CTPB binder is used, instead. This observation suggests the involvement of $HClO_4$ in the combustion process. As cited earlier, the N-N bonded compounds react with extreme rapidity with acids, often resulting in ignition [1,2]. The reaction of perchloric acid, the primary product of AP decomposition, with the N-N bonded binder thus perhaps results in augmenting the burn rate; no acidic species is expected to be produced by the decomposition of KP.

The rocket performance parameters (Table III) of propellants based on AP/N-N bonded binders show that the theoretical specific impulse and the chamber temperature are comparable to those obtained with the AP/CTPB system. It is to be noticed that a lesser amount of the

Table III: Theoretical Rocket Performance Parameters of Propellants Based on AP/N-N Bonded Epoxy Binder Systems (P_c=70 KSC., P_e=1 KSC).

Performance Parameter	AP/DEBuAH 86.3/13.7	AP/DEBuSH 87.1/12.9	AP/DEFAzH 85.3/14.7	AP/DEFSH 85.6/14.4	AP/CTPB 86.5/13.5*
T_c(K)	2891	2900	2941	2941	2946
M_c	27.19	27.09	27.63	27.59	25.25
C^*(m/sec)	1475	1481	1478	1479	1529
I_{sp}(sec)	243.8	244.7	244.8	245.1	247.8

⋆ - maximum practical solid loading

oxidizer required permits the processing of stoichiometric amount of the solid oxidizer loading with these binder systems which is not possible with CTPB. The use of N-N bonded binders thus, not only results in a significant enhancement in the burn rate but also in clean burning (stoichiometrically balanced combustion) of the propellant system.

CONCLUSIONS

Bis-dicarbonylhydrazones, on epoxidation yield epoxy terminated N-N bonded resins, which have convenient viscosities and could be used as binders for powder compositions for hybrid propellant systems. The use of these resins as binders in AP-based solid propellants results in significant enhancement of the burn rate.

REFERENCES

1. S. Jain, P. Krishna and V. Paiverneker, J. Spacecraft Rockets 16, p.69 (1979).
2. G. Rajendran and S. Jain, Fuel 63, p.709 (1984).
3. S. Jain, K. Sridhara and P. Thangamathesvaran, Prog. Polym. Sci. 18, p.997 (1993).
4. P. Thangamathesvaran and S. Jain, J. Polym. Sci. Part A: 29, p.261 (1991).
5. P. Thangamathesvaran and S. Jain, J. Aero. Soc. India 45, p.194 (1994).
6. S. Jain and S. Amanulla, (Unpublished Results).
7. W. Potter, Epoxy Resins, Springer-Verlag, New York, 1970 p.24.
8. G. Rajendran and S.Jain, Indian J. Chem. 24B, p.680 (1985).
9. H. Lee and K. Neville, Handbook of Epoxy Resins, McGraw-Hill, London, 1967 p.4-45.

COMBUSTION CHEMISTRY OF ENERGETIC MATERIALS STUDIED BY PROBING MASS SPECTROMETRY

O.P.KOROBEINICHEV, L.V.KUIBIDA, A.A.PALETSKY, A.G.SHMAKOV
Institute of Chemical Kinetics and Combustion, Siberian Branch Russian Academy of Sciences, 630090 Novosibirsk, Russia, korobein@kinetics.nsk.su

ABSTRACT

The methods of probing mass spectrometry (PMS) for diagnostic of flames and for the study of kinetics and mechanism of the thermal decomposition products of energetic materials (EM) are described. Several types of instruments based on microprobe and molecular beam mass spectrometric sampling have been developed. Time of flight mass spectrometer has been used. Apparatuses for high (10 atm) and low (<1 atm) pressure have been developed for the study of combustion and decomposition of EM by PMS "in situ". Several examples are presented to demonstrate application of PMS method for the study of EM flame structure, thermal decomposition and dynamic of ignition. Experimental data on decomposition of double base propellants ammonium dinitramide, ammonium perchlorate are presented.

INTRODUCTION

The main source of our knowledge on the combustion chemistry of energetic materials (EM) is the results of flame structure studies: spatial distributions of temperature and species concentration in flames. The main methods applied to the investigation of chemical structure of EM flames are the following: 1) probing mass spectrometry; 2) spectroscopic methods-absorption and emission, Laser Induced Fluorescence (LIF), Spontaneous Raman Scattering (SRS), Coherent Anti-Stokes Raman Spectroscopy (CARS). Until recently there were a few works on EM flame structure. However, the improvement of experimental technique, the development of works on flame structure modeling and the rise of interest to EM combustion chemistry, increased the number of works in this field [1-7].

At present one of the most effective and universal experimental technique for studying EM flame structure is the method of mass spectrometric probing of EM flames (MSPEMF) which was improved in [4]. It allows the detection of all stable species present in the flame as well as the structure of EM flames with the resolution sufficient to study EM combustion. The PMS method consists in the following: a burning strand of EM moves with a speed exceeding the burning rate toward a probe so, that a probe is continuously sampling gaseous species from all the zones including those adjacent to the burning surface. The sample is transported to an ion source of a time-of-flight (TOF) or quadrupole (QMS) mass spectrometer. Mass spectra of samples are recorded with simultaneous filming of the probe and burning surface. The data allow identification of stable components, determination of their concentrations and spatial distributions, i.e. the to study of flame microstructure. Method of probing mass spectrometry is successfully applied also for the study of kinetics and mechanism of EM thermal decomposition using flow reactor, which contains the heated specimen of EM and is linked to the inlet system of TOFMS. It allows to receive information about the products of EM decomposition as well as the rate of evolving each product as a function of time. The knowledge of the kinetics and mechanism EM thermal decomposition at high temperature provides a basis for the elaboration of EM combustion model. The purpose of this work is to describe mass spectrometric probing technique, to demonstrate some examples of its applications for study EM flame structure, kinetic and mechanism of EM thermal decomposition and ignition dynamics.

EXPERIMENT

Two types of apparatuses have been developed to study flame structure. The sample is transported to an ion source as a molecular flow using a microprobe with the inlet orifice of 10-20 μm in the first type of setup and as a molecular beam using a sonic probe with the inlet orifice of 20-200 μm in the second type. The former setup has a high spatial resolution and just slightly disturbs the flame allowing study of the flame with a narrow combustion zone up to 0.1 mm. However, in this case radicals recombine and quenching can become a problem. The latter setup with molecular beam mass spectrometric (MBMS) sampling allows detection of radicals, but more strongly disturbs the flame, and therefore, has a reduced spatial resolution.

Fig.1 shows the MBMS system [1-3], which has been used to examine flame structures for nitramines (RDX and HMX), double base propellants, zone of flameless ADN combustion. It includes: an apparatus for probing a flame containing a molecular beam sampling system, a time-of-flight mass spectrometer (TOFMS) type MSKh-4 as a detector, a combustion chamber, a scanning system, a data-acquisition system and an experiment controller based on CAMAC equipment and a computer. A sample produces a molecular beam which passes to an ion source.

The flame is sampled with a probe (3), a 25 mm high cone with a 50° external angle, a 40° internal angle, and a 0.1 mm diameter orifice at the apex. Gas expansion in the cone results in a supersonic jet directed into a skimmer chamber (5), evacuated by an oil diffusion pump BN-3 (500 l/s). A stainless steel skimmer (8), a 50-mm-high cone with a 40° external angle, a 30° internal angle, and a 2 mm diameter orifice at the apex is behind the sampling cone. The skimmer is designed so that only a supersonic jet core enters the collimator chamber (6), evacuated by an oil diffusion pump NO5 (500 l/s). Pressure in the skimmer chamber is $2 \cdot 10^{-3}$ Torr. The molecular beam from the skimmer passes through a 4 mm diameter orifice of collimator (9) to the detector chamber with the TOFMS ion source. Background gas pressure in the collimator chamber is 10^{-5} Torr. An electromagnetic chopper (11) capable of cutting off the molecular beam and a beam modulator consisting in a slotted disk (10) rotated by DG-2TA engine with an adjustable frequency from 1 to 100 Hz are found in the collimator chamber. Ion source (4) of MSKh-4 TOFMS is evacuated by a heteroion pump NORD-250 (250 l/s), and the drift tube (7) of MSKh-4 - by a turbomolecular pump TMN-200. Pressure in the ion source is less than 10^{-6} Torr. The distance between the orifices varies: the skimmer-collimator one is 310 mm; the collimator-detector one is 50 mm; and the skimmer-sampling cone one is varied from 5 to 25 mm by changing the flange to which the probe is attached. The ignition spiral (12) is automatically removed from the combustion zone after ignition. To scan EM flame a control system and a stepper motor (13) are required. The burning strand (14) is moved with a motor (13). Thermocouple (15) serves to measure temperature profile in flame. A strand is moved at a speed less than 10 mm/s and is driven by a stepper motor, step - 1.25 μm. The data acquisition and control system consists of an Elektronika-60 microcomputer, a CAMAC apparatus, a double-beam oscilloscope, a scanner control, and a printer. To study the flame structure at high pressure (10 atm) by MBMS quartz probe with the inner angle of 40° with the orifice of 20 μm and wall thickness near the probe tip of 25 μm has been used. The study of $H_2/O_2/Ar$ (0.1/0.05/0.85) flame structure stabilized on flat burner at 10 atm has been carried out to test this system. The velocity of fresh mixture at the exit from the burner was 11.2 cm/s. The results of measurement of profiles

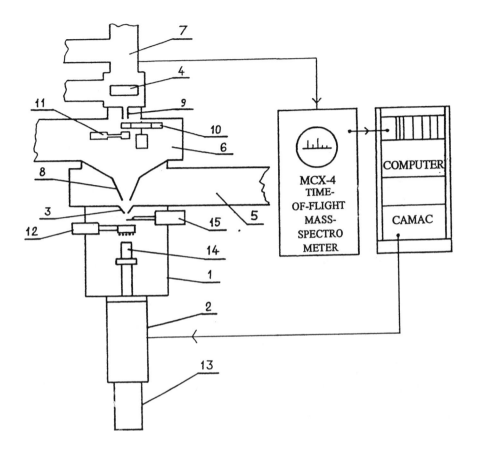

Fig. 1. MBMS system for studying the flame structure of solid propellants with TOFMS.
1) combustion chamber; 2) scanning system; 3) probe; 4) ion source; 5) skimmer chamber;
6) collimator; 10) slotted disk; 11) electromagnetic chopper; 12) ignition spiral;
13) stepper motor; 14) burning strand; 15) thermocouple.

species concentrations in above flame shown on Fig.2 demonstrate the possibility of MBMS application for the study of flame structure at high pressure. The width of flame zone (approximately 0.7 mm) is much more the spatial resolution of probe (approximately 0.15 mm). The data obtained and the accumulated experience allow us to hope for a successful application of this MBMS setup to the study of flame structure of some EM with flame zone width of part of mm. at high pressure.

For the investigation of double-base propellant flames at high pressure about 10-20 atm the setup of the first type with microprobe sampling was used. The combustion products were sampled using a metallic probe (probe N2), and then transported through a tube with diameter 3 mm and length 1 m to the probe N1 of setup shown on Fig.1. Several types of probe N2, differing

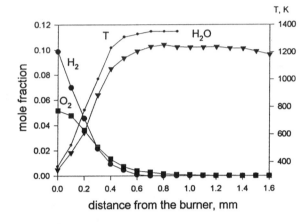

Fig. 2. Profiles of stable species and temperature in
10%H₂/5%O₂/85%Ar flame at 10 atm.

in material (stainless steel, aluminum with a protective film of Al_2O_3), orifice size 0.5-0.12 mm and interior opening 10-40° have been used. The time to transport the sample from the probe N2 to the probe N1 varied (depending on the length of transport tube and inlet orifice of probe N2) from 1.5 to 0.15 s. In some cases the strand was placed at a distance of 5 mm. from the probe so that the combustion product flow was directed along the axis of the probe N1. In the other cases the strand was placed at the distance of 5 mm from the probe N2 so that the strand and probe axis were perpendicular to each other. This made it possible to decrease the probability of blocking the probe orifice with soot combustion products.

Temperature profiles have been measured in some cases by an immovable thermocouple, in others cases - thermocouple moved toward burning surface of a strand with a scanning device at the rate, exceeding the burning rate. Pt-PtRh(10%) and W-WRe thermocouples with the diameter of wire 20-30 μm were used. The construction of Pt-PtRh(10%) moving thermocouple is shown on Fig.3 The time constant for Pt-PtRh thermocouple of diameter 20 μm, length 3 mm has been found 0.004 s.

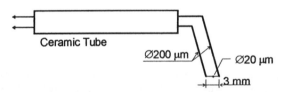

Fig. 3. Construction of thermocouple unit.

Quantitative aspects of the results of MS sampling investigations of the EM flame structure are essentially defined by the accuracy of determination of sampling technique errors. In the case of flames with narrow zones of combustion (0.1 mm) the zone width is comparable with the external diameter of a probe tip. This case was studied in [10]. When appropriate correction is made the error in determining the concentration profiles by the probe method is small enough.

The setup for mass spectrometric investigations of the kinetics of thermal decomposition of energetic materials under the conditions approximately similar to those present in the condensed phase in the vicinity of the burning surface and the method itself are detailed in [1,9]. The installation based on molecular beam mass spectrometric setup, explaining the principle of the method, is represented on Fig.4. The EM specimen to be studied (several mg) is clipped between

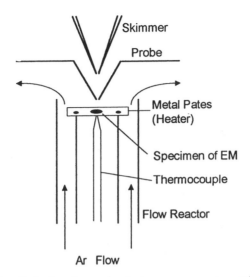

Fig. 4. Installation based on molecular beam mass spectrometric setup for
study of kinetics and mechanism of the EM thermal decomposition.

electric current heated metal plates and installed in a flow reactor nearby the probe tip orifice. A molecular beam, entering ion source of the time-of-flight or quadruple mass spectrometer, is formed from the withdrawn sample. Gas carrier Ar was allowed to pass through the reactor at the rate high enough to provide the intensities of mass peaks of EM decomposition products in samples mass spectra to be proportional to the rates of decomposition resulting in these products. Another cells for heating can be used as well, but in the above case the heating rate is maximal and equal to 20 - 200 degrees /s. Use of special equipment for a high rate of EM specimens heating allows investigation of EM decomposition at high temperatures close to those on EM burning surface. Methods of EM decomposition investigations using Rapid-Scan FTIR spectroscopy, SMATCH/FTIR, and T-Jump/FTIR spectroscopy developed by T. Brill [10] provides also an important information on the kinetics and mechanism of EM thermal decomposition under the conditions, close to the combustion ones. The above methods and mass spectrometric ones compliment each other. Their combination allows a more comprehensive idea of the mechanism of EM decomposition.

RESULTS

Double base propellants flames.
Fig.5 shows the dependencies of mass peak intensities (with different m/e) in mass spectra of the samples taken from the double-base propellant "N" flame (at 1 atm Ar and initial temperature of 120°C) on the distance to the burning surface L (determined from the corresponding time-dependence). The propellant "N" composition is the following: nitrocellulose - 57%, nitroglycerine - 28%, nitrotoluene - 11%, plasticizer - 4%. The strand was placed in the combustion chamber N1 at the distance of 5 mm from the probe. The burning surface was removed from the inlet orifice with the rate equal to the burning rate of the strand. The moments

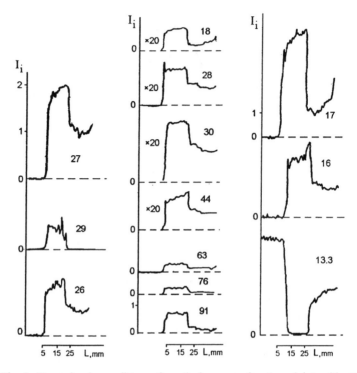

Fig. 5. Dependencies on distance from the burner surface L peak intensities I_i (in arbitrary units) in mass spectra of samples taken from double base propellant N flame under 1 atm. Ar.

of abrupt decrease and increase of the peak with m/e=13.3 (Ar^{+3}) correspond to the ignition and extinction of the strand. The propellant combustion products contain: $CO+N_2$ (m/e=28), NO (30), CO_2 (44), H_2O (18), CH_2O (29), HCN (27), CH_4 (16), dinitrotoluene (63, 91), nitroglycerine (76). Fig.6 presents the time-dependence of the mass peak intensities with m/e=40 (Ar), 27 (HCN), 28 ($CO+N_2$), 44 (CO_2) in the sample mass spectra taken from the flame of propellant "N" strand burning at 20 atm. in the combustion chamber N2. The sample temperature under stationary conditions was 2300K. Sharp splashes of the peak intensity with m/e=27 (HCN) were recorded at the strand ignition and extinction. In addition to the dependencies in Fig.6 similar data were obtained for the mass peaks with m/e=12 (CO, CO_2), 14 (N_2). The peak intensities in mass spectra were measured and the inlet system was calibrated against the combustion products of N_2, CO, CO_2 mixed up with argon. The contributions made by CO and N_2 to peak with m/e=28 were determined from the fragment ions peaks with 12 and 14 in the CO and N_2 mass spectra and in that of sample taken from the flame. Using experimental data on mass spectra of the samples taken from the "N" propellant flame at 20 atm and the calibration results the relative mole fraction of N_2, CO, CO_2 have been obtained.

Those of H_2 and H_2O were determined in terms of material balance equation. The experimental results are listed in Table 1 (α_i - mole fraction).

Table I. The mole fractions of double-base propellant "N" combustion products at 20 atm.

α_i	CO	CO_2	N_2	H_2	H_2O
experiment	0.448	0.0914	0.115	0.192	0.154
calculation	0.472	0.0674	0.115	0.168	0.178

Table I also gives the calculation results for the equilibrium combustion product composition at T=2300K. There is some discrepancy in these data probably due to either the presence of carbon black in the combustion products (neglected in the material balance equation) or the absence of equilibrium.

Similar studies have been performed at p=20 atm for the double base propellant N2, of a higher heat of explosion than "N" propellant and the following composition: nitrocellulose - 58%, nitroglycerin - 40%, centralite - 2%. A fair agreement is observed between the mole fractions of the combustion products of propellant N2 - the experimental (the product combustion temperature was 2700K) and the calculated equilibrium at 2700K (Table II).

Table II. The mole fractions of double-base propellant N2 combustion products at 20 atm.

α_i	CO	CO_2	N_2	H_2	H_2O
experiment	0.348	0.164	0.135	0.0745	0.279
calculation	0.35	0.161	0.135	0.073	0.28

ADN flameless deflagration.

Ammonium dinitramide (ADN) is a new energetic material which can be used as an oxidizer in solid rocket propellants with high environmental safety [13]. ADN decomposition chemistry was studied by the method T-Jump_FTIR at 260°C under 1 atm. Ar [14]. The decomposition process becomes strongly exothermic as the first gas products are detected. The first detected products are mostly HNO_3, NH_3 and N_2O in roughly similar amounts. Minor quantities of NO_2, ammonium nitrate (AN) and H_2O are also present in the initial spectrum. Gas product ratio has also been obtained on a laser-pyrolyzed strand of ADN with the use of a quartz microprobe mass spectrometer MPMS [15] and are comparable to those obtained by IR spectroscopy. In [15] the ADN material was received as a crystalline powder and was pressed into cylindrical pellets 0.64 cm in diameter, 0.4-0.9 cm long with a consolidation pressure of about 500 atm. Pellet densities were approximately of 1.5 g/cm^3. Regression rates of ADN pellets under 1 atm increased from 0.5 cm/s to 0.7 cm/s as the incident heat flux was increased from 50 W/cm^2 to 300 W/cm^2. Temperature

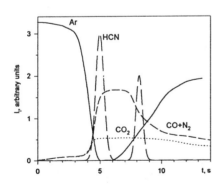

Fig. 6. Time dependence behavior of mass peak intensities of samples taken from double base propellant N flame under 20 atm. Ar at ignition and combustion.

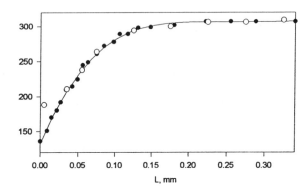

Fig. 7. Temperature profiles near the AND burning surface at P=1 atm. for
two different experiments (● - exp. N1, ○ - exp. N2)

measurements using fine-wire thermocouples indicated a surface temperature of 570-580K.
MPMS technique did not allow HNO_3 detection in products of laser pyrolysis of ADN.

Self-sustaining deflagration of ADN under 1 atm have been studied in this work by MBMS.
ADN used in this work have been synthesized at Zelinsky Institute of Organic Chemistry Russian
Academy of Sciences by the method described in [16]. The purity of ADN is 97%. The main
impurity is ammonium nitrate (AN, about 3%). Melting starts at 90-94°C. Decomposition starts at
130-135°C. Specific density of ADN crystal is 1.82 g/cm³. ADN strands 21 mm in length, 10 mm
in diameter, of specific density of 1.79 g/cm³ pressed using the pressure of 7000 atm. They burn
steadily and flamelessly under 1 atm. at the rate of 3.3-3.4 mm/s. Temperature profiles of burning
ADN have been measured using a thermocouple shown in Fig.3, moved toward ADN burning
surface with a scanning device at the rate, exceeding ADN burning rate. The temperature profiles
are shown in Fig.7.

The position of burning surface is unknown. A narrow zone 0.1 mm in width with a final
temperature of 315°C was found to exist nearby ADN burning surface (most probably in K-phase
reaction layer). The temperature is increasing slowly outward from the burning surface, ranging
up to 410°C at the distance of 2 mm.

Data on mass spectrometric investigation of ADN burning zone structure nearby the burning
surface have been obtained using quartz probe of 100-microns orifice. Two types of experiments
have been conducted: a) A strand mounted at the distance of 1 mm from the probe is immovable.
Burning surface is moving outward from the probe at ADN burning rate. b) Burning strand is
moving toward the probe at the rate exceeding ADN burning rate. In the experiment of the first
type the following masses have been detected in mass spectra of samples taken from combustion
products: 46 (NO_2, HNO_3), 30 (NO, N_2O, NO_2, HNO_3), 44 (N_2O), 17 (NH_3, H_2O), 28 (N_2, N_2O),
18 (H_2O), 63 (HNO_3). Profiles of mass peaks intensities were measured. They were variously
falling outward from the burning surface (from 1 to 10 mm), which is attributable to: a) HNO_3
and NH_3 interaction to yield ammonium nitrate, b) clogging probe orifice with ammonium nitrate
particles.

We supposed the detection of peak of mass 124, attributable to molecular ion ADN, and
those of 106 and 107 attributable to molecular and fragmentary ions of dinitramide.

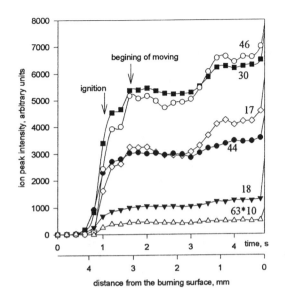

Fig. 8. The behavior of mass spectral peaks intensities during the probing of burning ADN.

Simultaneously the pressure in ion source as a function of time has been measured. While the burning surface was approaching the probe, the mass peaks 28, 44, 18, 30, 46, 63, 17 slowly increased and at the some moment the latter and the pressure increased abruptly more than 10 times. We considered it to be the moment of the probe contact with the burning surface. Before this moment peaks of mass 106, 107, 124 have not been detected. A good repeatable measurements for the ratio of all peak intensities near the burning surface have not been received. Fig.8 shows profiles of mass peak intensities for one of the experiments. The largest discrepancy (about 20-30%) was observed for peaks of mass 46, 17, 63. The mass spectra reveal the following approximate product ratios of $10NO_2:5NH_3:3N_2O:2NO:N_2:HNO_3$.

Their comparison with the data of [14,15] on thermal decomposition and laser-pyrolysis of ADN (received from another source and possibly having distinctive characteristics) showed some distinction from them. It is further proposed to increase the accuracy of measurements of ADN deflagration products composition basing on the more precise measurement of sensitivity coefficients of all species.

Thermal decomposition of ammonium perchlorate.

Experiments on mass spectrometric investigation of ammonium perchlorate (AP) decomposition (powder with the particles of 5 μm and tablets of 25 μm in thickness) under high rate heating (50 degrees/s) were conducted on setup with a molecular beam sampling system, providing analysis components undetectable using MPMS (first of all vapors of $NH_3 + HClO_4$) to be found in decomposition products. The procedure capability of detecting $HClO_4$ vapors and measuring their concentration opens up the way of detection and investigation of AP dissociative sublimation, taking place concurrently with the process of decomposition. The experimental results are represented in Fig.9. Both products of AP thermal decomposition (such as HCl, Cl_2, ClOH, ClO_2) and the product of AP dissociative sublimation ($HClO_4$) were found. At the heating rate of 50 degrees/s the rate of the first decomposition stage is maximal at 430°C, and of the second decomposition stage - at 560°C. The concurrent process (AP sublimation) occurs in one stage. The sublimation (peak intensity of mass 83, which is fragmentary ion of perchlorice acid) is maximal approximately at the same temperature as the rate of the second stage, but its maximum is at the lower temperature than the maximum of the first stage, where sublimation can be neglected.

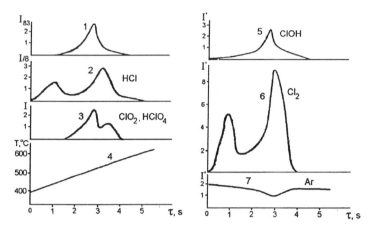

Fig. 9. Dependencies on time of mass peaks intensities (in arbitrary units) m/e=83 (I-HClO₄), 36 (2-HCl). 67 (3-HClO₄, ClO₂), 52 (5-ClOH), 70 (6-Cl₂), 20 (7-Ar) and temperature (4) on the decomposition of AP tablets of 25 μm in width at 1 atm. and at the heating rate of ~50 deg/s in argon flow on the installation with molecular beam sampling.

CONCLUSIONS

The method of flames investigation by probing mass spectrometry is an important instrument for the study of chemistry of energetic materials combustion and thermal decomposition. It is a good addition to the spectroscopic methods of flames investigation (EMC, LIF, CARS, etc.) and such methods of the investigation of EM thermal decomposition under the conditions close to those of combustion as SMATCH/FTIR, T-Jump/FTIR. These types of experiments will lead to a better basic understanding of solid propellants combustion, which in turn facilitates propellants formulation and the prediction of propellants ballistic behavior.

ACKNOWLEDGMENT

This work was supported by the US Army MICOM and NASA under Contract NDAAHO195CR141 in the part connected with ADN study and by ISF under Grants NRC100, RC1300 in the part connected with the development of MBMS setup for the study of flame structure at high pressure. Authors express gratitude to Dr.W.D.Stephens and Dr.J.Carver for support and discussion.

REFERENCES

1. O.P.Korobeinichev, Combustion, Explosion and Shock Waves, **23**, 565 (1988).
2. O.P.Korobeinichev, Pure & Appl. Chem., **65** (2), 269 (1993).
3. O.P.Korobeinichev, L.V.Kuibida, in: Prog. Astron. and Aeron., New York, **88**, (1983), p.197-207.

4. O.P.Korobeinichev, A.G.Tereshchenko, Dokl. Akad. Nauk USSR, **231** (5), 1159 (1976) (in Russian).

5. T.A.Litzinger, Y.J.Lee, C.J.Tang, in: Application of Free-Jet, Molecular beam, Mass Spectrometric Sampling, Proceedings Workshop, Estes Park, Colorado, October 11-14, 1994, The National Renewable Energy Laboratory, 1994, p.128-135.

6. D.Hanson-Parr, T.Par, to appear in The Proceedings of 25th International Symposium on Combustion (the Combustion Institute) (1994).

7. T.Edwards, paper presented at International workshop on transient combustion and stability of solid propellant, Milan, Italy, November 12-14 (1990).

8. O.P.Korobeinichev, A.G.Tereshchenko, I.D.Emel'ynov, A.L.Rudnitskii, S.Yu.Fedorov, L.V.Kuibida, V.V.Lotov, Fiz. Gorenia Vzryva, **21** (5), 22 (1985).

9. O.P.Korobeinichev, G.I.Anisiforov, A.G.Tereshchenko, AIAA Journal **13** (5), 628 (1975).

10. T.Brill, Prog. Energy Combust. Sci. **18**, 91 (1992).

11. Z.Pak. AIAA Paper. 93-1755, June 1993.

12. T.B.Brill, P.J.Brush and D.G.Patil, Combustion and Flame, **92**, 178 (1993).

13. B.L.Fetherolf and T.A.Litzinger in Proceeding of the 29th JANNAF Combustion Meeting (CPIA Publ., 1992), p.329-338.

14. A.O.Luk'yanov, V.P.Gorelik and V.A.Tartakovsky, Izv. Akad. Nauk., Ser Khim. **1**, 94 (1994).

ATOMIC FORCE MICROSCOPY OF HOT SPOT REACTION SITES IN IMPACTED RDX AND LASER HEATED AP

J. SHARMA*, C.S. COFFEY**, A.L. RAMASWAMY***, R.W. ARMSTRONG***
* Naval Surface Warfare Center, Carderock Division, Silver Spring, MD
** Naval Surface Warfare Center, Indian Head Division, Silver Spring, MD
*** University of Maryland, College Park, MD

ABSTRACT

An atomic force microscope (AFM) has been used to reveal residual sub-micron sized decomposition sites in drop weight impacted RDX and laser irradiated AP crystals. In impacted RDX, the small and early reaction sites observed are hemispherical craters, ranging in size from 20-300 nm. The smallest reaction site encompassed about 10,000 molecules with an expected energy evolution of 2×10^{-14} J. On a somewhat larger scale hillocks of 200-800 nm were observed, their shape giving evidence of internal reaction and hot spot melting. Dislocation densities as high as 5×10^{12} per cm^2 were observed in sub-ignited RDX. High resolution AFM images of the RDX lattice structure indicate molecular rotation as well as displacements at dislocation sites. In AP, after nanosecond pulsed laser irradiation, reaction sites were trumpet shaped with a smallest size of approximately 50 nm. Most sites contained a crystallographically oriented central square lid formed above the surrounding crystal surface, probably relating to the orthorhombic to cubic phase transition documented in micron scale cracking patterns observed at the laser heated sites.

INTRODUCTION

No matter how energy is supplied to an energetic material, action starts in microscopically small sites, some of which can precariously develop into hot spots and then cause violent action. The idea of such hot spots in explosives was originally proposed by Bowden and Yoffe [1] and they suggested a size of 0.1-10 microns for them. Previously, the observations of the reaction sites and hot spots was limited to micron size due to the limitations of a scanning electron microscope. The advent of atomic force microscopy(AFM) has opened up the possibility of investigating them down to the molecular scale and consequently to their earliest stages. Understanding the creation of reaction sites and evolution of hot spots is important as it can lead to improvements in the safety aspects of explosives handling and storage.

It is believed that crystal defects provide sites for hot spots so they control the sensitivity of an explosive. However many varieties of defects, small and large exist and as to which defects play important role under what conditions has not been determined. For example in all crystals grown from solution, there are solvent inclusions of microns size. Furthermore there are always tiny crystallites of secondary growth on the surface of crystals. We do not know if any of these are critical. Dislocations and twinning are often blamed for high sensitivity but so far we have no measure of their role. In order to make the explosives less sensitive and less prone to accidents, it is necessary to determine which kind of crystal defects are critical and need to be reduced.

Earlier AFM results on RDX were presented in Reference [2], where it was shown that impacted RDX exhibits as high as 5×10^{12} dislocations per cm^2. In the present paper results on molecular displacements, reaction sites and hot spots of RDX will be discussed. In AP, the

257

phase transition from orthorhombic to cubic form has been found to be associated with the formation of decomposition sites, and it will be discussed.

EXPERIMENT

In order to observe smallest size reaction sites and hot spots, impacted samples of RDX have been investigated by using an AFM. Sugar-like crystals of RDX, 100-300 microns in size, were sandwiched between sheets of heat sensitive film or between cleaved sheets of mica. A 10 KG weight was dropped in the BIC machine of C.S. Coffey [3], from a chosen height and the samples were picked up at different distances from the center of the impact pattern. Both "go" and "no go" samples were examined. The samples were investigated in AFM within a few days of impacting. For the studies in ammonium perchlorate (AP), a single crystal of AP was irradiated on (001) crystal plane with a 8 ns pulse from a Nd/Yag laser (1064 nm in wavelength) with a 0.25 J of energy focussed over an area of 2 mm diameter.

A Digital Instruments Nanoscope II, Scanning Probe Microscope was used in the present work. The measurements were carried out in air at room temperature, using a 100 or 200 micron cantilever with a 3 micron oxygen sharpened silicon nitride tip. As mentioned in the earlier paper [2], the cantilever was used in repulsive mode and rather a high force of 10^{-8} - 10^{-7} N was used to obtain the images, in height mode.

RESULTS AND DISCUSSION

In order to see what is the effect of mild impact on the molecular arrangements of RDX, some samples of "no go" impact on heat sensitive film, from a drop height of 20 cm only, were imaged and compared with the surface of a control crystal. Small area images with large magnification were used. Figure 1(a) shows the surface molecular pattern of a control crystal, on a (001) or (210) plane. Each hill represents a molecule and the arrays are quite regular. Figures 1(b, c and d) show the effects of the mild impact in the form of molecular disarrays. Lateral as well as vertical displacements of approximately 20 % from expected positions are measured. In Figure 1(d) the vertical scale was so adjusted that some molecules are contained in the height and some are not. This was done to bring out the variation in heights of the molecules. The chopped heads of some of the molecules prove that they are displaced up and down in the vertical direction also. This leads to the conclusion that the periodicity is randomly reduced by the lateral and vertical displacements of molecules. In doing so the molecules seem to climb on each other and alter their distances in a haphazard way. A closer look at Figure 1 (b) and (c) shows changes in the apparent widths presented by some of the molecules. It appears that they have changed their orientations by rotating into new equilibria. RDX is not a spherically symmetrical molecule and that is the reason why this re-orientation can be seen. This significant phenomenon involving re-orientation of the molecules in a deformed crystal is being reported for the first time. It means that after the disarray, when the intermolecular distances have changed, some of the molecules accommodate to new configuration by re-orienting themselves. In the modelling of dislocations and disruptions, such an event has not been considered so far. In the future this should be considered and this re-orientation will apply not only to RDX but to all molecular solids. One clear message from these images is that RDX behaves like a molecular solid, which it is. The molecule as an entity can move around. Even at a modest impact the surface periodicity is considerably destroyed.

Reaction sites in RDX were observed in the sample from a "go" experiment, in which the weight was dropped from a height of 50 cm. Some of the effected RDX, which had survived

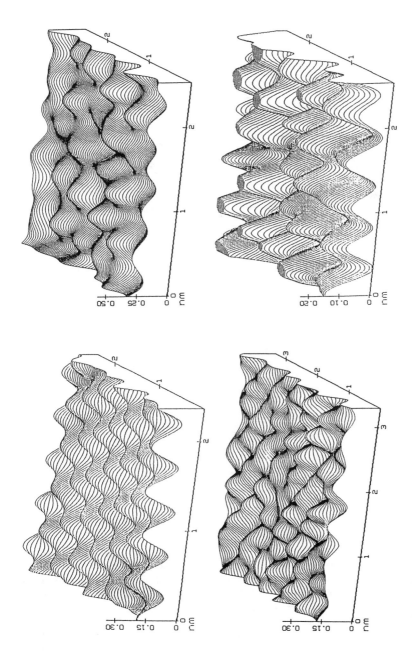

Figure 1. The AFM images (line plots after Fourier filtering) of RDX surfaces, showing (a) perfect array of molecules on a control crystal expected to be (001) or (2T0) plane, (b,c and d) disarray on mildly impacted RDX particles. In (d) the z-scale has been adjusted to bring out z-directional displacements, indicated by chopped heads.

and remained stuck to the mica sheet proved interesting. The sample showed evidence of violent action even though it had been selected from the central part of the impact pattern, where usually action does not start. The reaction sites were in the form of craters, some shown in Figure 2

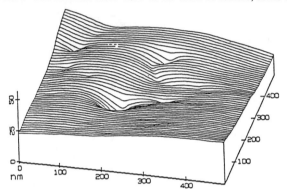

Figure 2. Crater formation observed on an RDX particle picked from the central part of the impact pattern of a 50 cm "go" test. The craters ranged from 20-100 nm in diameter approximately.

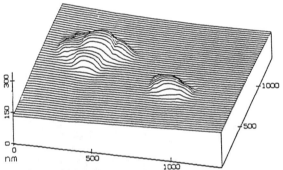

Figure 3. Magnified images of hillock-like reaction centers, showing custard appearance, probably caused by non-uniform internal reaction and ballooning. They range from 200-1000 nm in size.

and they were found to range between 20 and 100 nm in diameter (approximately). The larger ones showed hemispherical shape whereas the small craters were somewhat trumpet shaped. It appears that at these sites the RDX reacted and disappeared. The craters showed rounded edges. The 20 nm reaction site represents the smallest reaction site to-date reported for RDX. A 20 nm diameter hemispherical reaction site encompasses about 10,000 molecules, with an estimated release of only 2×10^{-14} J of energy. Thus AFM work provides us with a number representing the smallest reaction site and the least amount of energy required to start a hot spot. This size agrees well with the calculations of Armstrong et al [4] who arrived at the number of 3,500 molecules per atomic plane for the smallest critical reaction site in RDX formed in 40 ns as characteristic time. From Boddington's [5] relationship, the estimated temperature for this hot spot would be about 900 K. The 20 nm size would also signify the minimum size which needs to be activated for the development of a hot spot. This implies that smaller volume of material would rather lose energy to the lattice than grow into a reaction center. Probably it can be argued that crystal defects smaller than this size may also not contribute to sensitivity. This finding supports the arguments of Mallay et al [6] given to explain why a localized thermal spike of 10^4 K from a fission fragment over a smaller radius than 20 nm, does not ignite an explosive.

Larger reaction sites and of different shape than those mentioned in the above paragraph were exhibited by the RDX sample selected from the middle of the radius of the impact pattern. These reaction sites were in the shape of hillocks ranging over 200-1000 nm in diameter, as shown in Figure 3. A closer look of the uneven shape would indicate that these hillocks may have arisen from melting, internal reaction and some ballooning. The hillocks show custard appearance, indicating reaction of different intensity in different parts of the hillock. The base of these hillocks is about ten times larger in diameter than the craters discussed previously. Consequently the volume would be about a thousand times larger, which could happen from solid to gas conversion during ballooning. Attempts were made to grind off the heads of these hillocks by the scanning tip, in order to determine if they were hollow, but they did not succeed. However some large hillocks of approximately 500 nm diameter did show craters. This would support the idea that gas generation could be taking place. It is not possible to pin down from these results whether the hot spot is produced from adiabatic compression of trapped gases, friction or from deformation of the lattice and consequent production of dislocations and deformations. It might be mentioned that in more severely affected areas of the same sample large volcano like hillocks were observed 1000 - 2000 nm in diameter with evidence of liquid being sputtered out. (Figure 1 of Reference 2.) It is not surprising that melting is observed, since the melting point of RDX is lower than but very close to the decomposition temperature. Often it is believed that the ignition of RDX is preceded by local melting, at least in proton beam irradiated RDX and HMX, bubbly froth-like melts were clearly observed. [7].

In contrast to RDX the results on laser heated AP, were very different. A connection has been made between AFM observations and scanning electron microscope results obtained on laser damaged AP material [8]. Figure 4, (in which white means up and black means down,) shows a large area affected by the laser. A crack, on the (010) plane which follows $\pm[100]$ orthorhombic direction of the crystal, is seen with surrounding area full of reaction sites. The crack was also visible in an optical microscope, which is not surprising since it is approximatley one micron wide, quite close to the wavelength of visible light. At some places (top right of Figure 4) the white specks form almost a lace-like area of dense reaction. White beads, as reaction sites, can be seen all over. At some places the beads seem to reside in dark craters. The magnified images in Figures 5, 6, and 7 bring out the fact that the white beads are actually square lid-like structures and some of them are sitting on top of craters from which they are formed and which are approximately 100-200 nm in diameter. The square lids are about 100 nm on each side and 20 - 50 nm high. The square lids were found to be all oriented in the same direction. It appears that during the heating of AP by the laser, energy got localized at tiny spots and produced the trumpet shaped craters from decomposition. The craters are the surface counterparts of the gas bubble generated by the decomposition. The square blocks provide clear evidence of the cubic (rocksalt) morphology. Since AP has a reversible phase transition from orthorhombic to cubic form at 240°C, the formation of the square lids implies that such a temperature has been reached. Boggs and Kraeutle [9], have reported the growth of square and rectangular holes on the surface of thermally decomposed AP. Since they used an SEM to observe them, they could

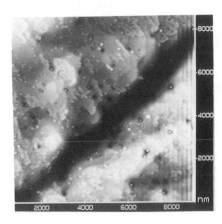

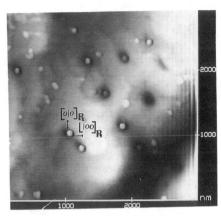

Figure 4. An AFM image of laser damaged AP crystal face, showing a crack and also a beady deposit of the reaction product. Its pile-up at some places gives lacy appearance.

Figure 5. Magnified image shows that the white beads are rather square blocks or lids, evidently of AP transformed into cubic phase. Some sitting on craters.

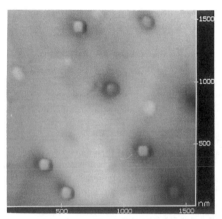

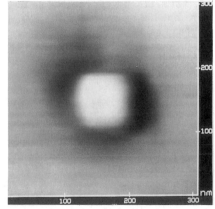

Figure 6. Further magnification reveals that the square blocks are similarly oriented.

Figure 7. Magnified image of a square lid, showing the peripheral area depressed.

see the square and rectangular holes only after they grew to larger size. The AFM images provide finer details. Figure 8 shows that the area close to the square lid is depressed, as if sunk-in and that the lid sticks out above the surrounding area, as if it is buoyed up. It is conceivable that the area near the cube gets hot enough to melt or at least becomes soft so that it forms a depression. Alternatively the depressed area may simply be caused by the very local loss of material in the decomposition process. Figure 9 shows a cross-section of a typical crater, which is more or less bowl shaped. This shape may arise from dissociative sublimation of AP, as described by Boggs and Kraeutle [9]. Probably the reaction starts at a depth of about 20 nm below the surface of the crystal, corresponding to the lower tip of the crater.

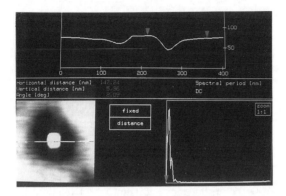

Figure 8. A section of the square lid and the surrounding depressed area of AP is shown. The crater is typically about 200 nm in diameter and the square lid is about 100 nm on each side, with a height of 50 nm

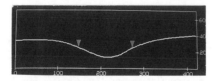

Figure 9. The section of a left-over reaction crater of AP, from which the square lid had moved out, shows a shallow bowl-shaped bottom. Moving out of the square lid seems frequent.

A closer look into the one micron wide crystal cracks, (Figure 4), shows that there are a good number of cubes inside the cracks also. Furthermore the edges of the cracks shows square gliches. This leads us to believe that the phase transition is taking place in the cracks also, and the cubic geometry causes them to have zig-zag steps.

It has been found that in areas of more violent action the lids are no more in the shape of square blocks. They became tapered probably from dissociative sublimation, but were still lodged in craters.

CONCLUSIONS:

The present work shows that the smallest size for reaction sites in RDX is 20 nm. If this size indeed represents the smallest possible reaction site, then it is an interesting finding. The absence of smaller sites may imply that when energy is deposited over smaller than this critical area, it gets dissipated in the lattice without producing reaction sites and hot spots. If at least about 10,000 molecules can be involved then a reaction site can be formed, which may develop into a hot spot. This also implies that a minimal amount of 0.5×10^{-14} J of activation energy is required for any action to be started. Again, if 20 nm is the smallest size for growth, then crystal defects of smaller size may not play any role in the sensitivity behavior of RDX. In that case, one should look for the role of larger defects or the collection of smaller defects such as dislocations into microscopic shear bands.

In the case of laser irradiated AP, at the level of energy deposited in the present experiment, it appears that decomposition and the phase transformation from orthorhombic to cubic form are major events. Cubes are produced all over. At many sites the material used to form the cubes leaves behind trumpet or bowl shaped craters. In most of the craters one can still observe the square blocks produced by them and they are all oriented in the same direction. This means that the original orthorhombic crystal still retains control on the line up of the square lids and thus appears to be intimately involved in the decomposition process. One interpretation made from the shape of the area near the square lids is, that the local spots has at least undergone softening, so that it has picked up the imprint of the square it has supported.

REFERENCES

1. F.P. Bowden and Y.D. Yoffe, Initiation and Growth of Explosion in Liquids and Solids, Cambridge University Press, London, 1952. pp. 64-65.

2. J. Sharma and C.S. Coffey, APS-Shock Compression of Condensed Matter - 1995 Seattle, Wa. (In publication)

3. C.S. Coffey and S.J. Jacobs, J. App. Phys. **52**, p. 6991 (1981).

4. R.W. Armstrong, C.S. Coffey, V.F. DeVost and W.L. Elban, J. App. Phys. **68** (3), p. 679 (1990).

5. T. Boddington, Ninth Symposium (International) on Combustion, Academic Press, New York, 963, p. 287.

6. J. F. Mallay, H.J. Prask and J. Cerny, Nature, Vol. 203, No. 4944, pp. 473-476 (1964).

7. J. Sharma and B.C. Beard, Effects of Particle Beams on Explosives, Technical Report, NAVSWC TR 91-682,(1991), Naval Surface Warfare Center, Silver Spring,MD.

8. A.L. Ramaswamy, H. Shin, R.W. Armstrong , C.H. Lee and J. Sharma, J. Mat. Soc. (submitted for publication).

9. T.L. Boggs and K.J. Kraeutle, Combustion Science and Technology, **1**, pp. 75-93 (1969).

Part IV

Initiation and Detonation

PROBING DETONATION PHYSICS AND CHEMISTRY USING MOLECULAR DYNAMICS AND QUANTUM CHEMISTRY TECHNIQUES

M. D. Cook, J. Fellows, and P. J. Haskins
Defence Research Agency, Fort Halstead, Sevenoaks, Kent TN14 7BP, UK,
mdcook@taz.dra.hmg.gb

ABSTRACT

Modern quantum chemistry and molecular dynamics computer codes are powerful tools with which to study the physics and chemistry of energetic materials at the molecular level. Quantum chemistry calculations, on one or two energetic molecules, can give valuable information about the initial steps in their decomposition. Molecular dynamics calculations, even with empirical potentials, can yield important information about the physical processes involved in the initiation and growth of reaction of energetic materials. The combination of Molecular dynamics and quantum chemistry techniques offers the potential to probe energetic material reaction chemistry in real systems, in some detail, in the near future. Such an approach is vital if we are to be able to create new realistic macroscopic models within hydrocodes that can describe the initiation and growth of reaction in explosives. This paper gives an overview of the approach being adopted at DRA Fort Halstead to understanding energetic materials at the molecular level. In particular, the use of quantum chemistry and Molecular dynamics to help construct new macroscopic models will be discussed.

INTRODUCTION

Designing realistic macroscopic models to describe the ignition and growth of reaction behaviour of energetic materials is a major goal. This can only be achieved if there is a good understanding of the underlying microscopic processes that determine the macroscopic behaviour. At the microscopic level, many authors have published decomposition schemes for a variety of energetic materials, but very few have been able to offer conclusive experimental evidence to support their predictions. Likewise, there is a wealth of literature that describes experiments aimed at attempting to understand underlying mechanisms for the behaviour of energetic materials, but that have only led to engineering fits to the macroscopic behaviour. More recently, there has been a growth in real-time spectroscopy experiments on shocked energetic materials[1,2,3]. However, this is a difficult area of research, and the technology required to resolve each individual reaction step has yet to be developed. Unfortunately, the time-scales between the microscopic and the macroscopic are very different and it is not easy to bridge the gap. A detailed understanding at the molecular level will provide the foundation for a deeper understanding at the macroscopic level. However, it is clear that an understanding of the relationship between the microscopic, the intermediate mesoscopic level, and the macroscopic, is required in order to obtain a good predictive capability. At present, it is not entirely clear what form a mesoscopic model might take, but it must be able to account for hotspot phenomena in a realistic manner.

When we talk about understanding energetic materials, what we really mean is develop realistic models that can reproduce and predict explosive response to a variety of stimuli. To-date there has been much success at developing finite element models to describe the overall

Mat. Res. Soc. Symp. Proc. Vol. 418 © 1996 Materials Research Society

macroscopic behaviour, particularly for systems that are already detonating. These models have given good engineering capability. In order to understand all responses of energetic materials, the physics and the chemistry, and their inter-play, needs to be addressed in far greater detail than has been the case to-date. The role of chemistry in particular has been grossly underestimated. It appears that many workers in the field of energetic materials often over-look the fact that, although it is possible to describe the behaviour of a detonating ideal explosive by physics alone, it is chemistry that is responsible for the phenomenon. This becomes very apparent when considering the processes of initiation and growth of reaction in these materials. This has led, over the last few years, to an increased effort to attempt to understand the initiation and growth of reaction in energetic materials at the molecular level. Modern quantum chemistry techniques coupled with increasingly powerful super computers are opening up new avenues of research. Only a few years ago, it was very time consuming to model even simple bond scission processes in small molecules. It is now possible to model the chemistry of small clusters of quite large molecules. The quantum chemistry codes have become much more robust and can yield a wealth of structural and spectroscopic data which is invaluable to the experimentalist. Furthermore, we can use these powerful tools to explore possible reactions, particularly in the early stages of the decomposition of individual energetic materials. This is very useful since there has been much debate over the type of reactions that might be responsible for the initiation and growth of reaction process. For example, it is still not certain whether or not the initial reactions are unimolecular or bimolecular, whether they produce radicals or ions. It is generally accepted however, that once the chemistry is underway, free radical reactions, particularly through chain branching reactions, form the bulk of the chemistry; and it is termination reactions of free radicals that produce the products and release much of the heat to drive the detonation process.

A further exciting development, in the last few years, is the formulation of Molecular Dynamics (MD) computer codes that incorporate reactive potentials[4,5]. Although these codes are limited to-date to model systems, notably diatomics, they are already giving a wealth of information on the relation of the microscopic molecular behaviour with the mesoscopic and macroscopic regimes. This sort of approach should allow macroscopic models of ignition and growth of energetic materials, to be developed in a much more fundamental and systematic manner.

In this paper, we address the use of ab-initio quantum chemistry techniques to study the primary reactions in model energetic materials. Recent results of molecular dynamics calculations employing empirical many-body potentials to study reactive energetic model systems are also outlined. The way in which these molecular level techniques can aid the development of macroscopic ignition and growth models, that can be implemented in hydrocodes, is discussed.

QUANTUM CHEMISTRY CALCULATIONS ON ENERGETIC MATERIALS

The majority of energetic materials containing C, H, N, and O atoms can be broadly divided into three classes by the type of functional group responsible for conferring energetic properties. These principle groups are nitro, nitramine and nitrate-ester. Compounds containing nitro groups are generally the most chemically insensitive, while those containing the nitrate ester moiety, the most sensitive. In the work that we present here, we report results for nitromethane, the simplest of the nitro class of explosives. Previous work[6] has also addressed methyl nitramine and methyl nitrate, the simplest members of the nitramine and nitrate ester classes.

We have previously reported the results for unimolecular bond scission processes in nitromethane, methyl nitramine and methyl nitrate[6]. The weakest bond in all three compounds is the X-NO$_2$ bond. For nitromethane the C-N bond strength is calculated to be 65.6 kcal/mol, whereas, the N-NO$_2$ in methyl nitramine is 47 kcal/mol. The O-NO$_2$ group in methyl nitrate has a bond strength of only 38 kcal/mol. These bond strengths are consistent with the observed trend in sensitivity. The fact that there is a correlation between the X-NO$_2$ bonds in these classes of compounds and their sensitivity suggests that a unimolecular bond scission process might be the rate controlling reaction in these compounds. This is quite believable in a 'cook-off' type environment where, the energetic material is heated relatively slowly from the outside, and the molecules obtain translational energy through the heat conduction process. In this case the translation energy can be easily transferred into vibrational modes, which if sufficiently large, can cause unimolecular bond scission. On-the-other-hand, in a shock process the energetic material is rapidly compressed to a high density in the region of the shock, and although the material is also heated rapidly, there is a high probability that a bi- or even a tri-molecular process may be dominant. In order to shed some light on this dilemma, we have performed potential energy calculations on the interaction of two nitromethane molecules.

All the quantum chemistry calculations have been carried out using the GAUSSIAN 92 suite of computer codes[7]. The calculations have either been run on a CRAY YMP 4-32, a Silicon Graphics Power Challenge or on a Silicon Graphics Indy Workstation. Most of the calculations reported in this paper employed a 6-31g** basis set. Many of the calculations were carried out using Density Functional Theory with the Becke-3LYP parameterisation. A few calculations were performed at post SCF levels using many-body perturbation theory at the fourth order (MP4) including single, double, triple and quadruple excitations.

POTENTIAL ENERGY SURFACE CALCULATIONS ON NITROMETHANE

A number of starting geometry's have been considered including a linear arrangement (fig 1.), and side-on arrangements (figs 2 and 3.). Constraints were imposed in the Z-matrix through judicious use of dummy atoms for all arrangements. Constraints were necessary in these calculations in order to simulate the effects of pressure. For the linear arrangement, where the nitromethane molecules were arranged head-to-tail, the C-N--C-N back-bone was initially placed in line. Preliminary calculations were carried out with no constraints on the system. A full geometry optimisation was first carried out with the two molecules separated at a distance of ca. 3.5 A^o (between the carbon atom of one molecule and the nitrogen atom of the other). This calculation provided a baseline energy and starting geometry for subsequent calculations. The effect of compression (as from a passing shock wave) was simulated by incrementally reducing the distance between the two molecules, using the N$_1$-C$_2$ (where the designation 1 refers to one molecule and 2 refers to the other) distance as the reaction co-ordinate, and optimising the remaining geometry at each point. These preliminary calculations showed that if no constraints were imposed, the system orientated itself to reduce the overall energy, resulting in the two molecules twisting out of alignment.

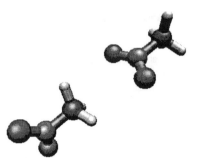

Fig 1. Two nitromethane molecules in a
linear arrangement.

Fig 2. Two nitromethane molecules in a side-on arrangement
whereby the CN--CN backbone is held a rectangular
geometry

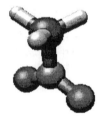

Fig 3. Two nitromethane molecules arranged with
the CN--CN backbone in a cross arrangement.

Calculations were then carried out on the linear system where increasingly severe constraints were applied to simulate the effect of neighbouring nitromethane molecules which, under shock compression, might be expected to offer resistance to lateral motion. Initially, the nitromethane molecules were arranged head-to-tail, with the C-N--C-N backbone held rigidly in line, by means of dummy atoms. As before the N_1-C_2 distance was incrementally reduced and the remaining geometry optimised. A plot of energy against distance for this set of calculations is shown in fig 4. Initially, as the distance between the molecules was decreased, a rapid rise in the energy of the system was observed. It was noticed, that the methyl group facing the nitro-group of the nitromethane molecule in front gradually turned planar at carbon, as the distance between the molecules was reduced. Frequency calculations at the equilibrium distance (ca. 3.5A°), and down to 2.1A° showed that the Raman frequency for the C-H stretch of this methyl group was shifted to higher frequencies as the distance between the molecules was reduced. The shift in Raman frequency of the methyl group under shock compression can be easily understood by considering simple hybridisation theory. The bonds making up the C-H and C-N around carbon are SP^3 hybridised. As the nitromethane molecules are moved closer together under compression, the H-C-H angle increases and tends towards trigonal symmetry. As it does so, the bonding around carbon tends towards SP^2 for the C-H bonds, and P for the C-N bond. Since the percentage of 'S' character in the C-H bonds has increased, the bond shortens and stiffens.

Energy / au

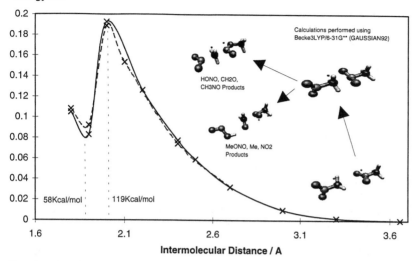

Fig 4. Potential energy surface for the bi-molecular reaction of nitromethane in an end-on configuration. The dotted line represents the potential surface to form MeNO, CH_2O and HONO; the solid line is for the nitro-nitrite isomerisation reaction.

This effect has been observed in shock experiments[2]. At a separation of 2.0A°, a peak in the potential energy surface was obtained, further reduction of the distance between the molecules resulted in a reaction which led to MeNO, CH_2O and HONO as products. The overall endothermicity for this process was found to be 58 kcal/mol with an activation energy of 119 kcal/mol. More recent calculations have shown that the potential surface in this constrained system is quite complex, and it is possible to over-shoot the bi-molecular reaction described above and jump to a similar energy state in which an isomerisation reaction is observed resulting in one nitromethane molecule being rearranged to form methyl nitrite. The other molecule is observed to undergo bond scission resulting in methyl and nitro radicals.

Calculations were also carried out where the nitromethane molecules were positioned side-by-side with methyl groups adjacent to nitro groups. The C-N--C-N backbone was arranged such that the atoms formed the corners of a rectangle. This arrangement was held ridged throughout the calculation, only the shape of the rectangle being allowed to change as the distance between the two neighbours was reduced. Once again the distance was incrementally reduced and the remaining geometry reoptimised at each point. No significant reaction was observed in this system and the energy continued to climb rapidly as the distance between the molecules was reduced. Eventually, when the distance between the two molecules had been reduced to 1.6A° an exchange reaction was observed, in which the methyl and nitro groups swapped their bonding partners. This reaction only resulted in two nitromethane molecules orthogonal to their original position. It was concluded that this was probably not a significant reaction as it appeared to offer only a very high energy pathway, and there was no chemical change in the system.

A further side-on arrangement was tried whereby, the two nitromethane molecules presented themselves cross-on to one another (see fig 3.), the centre of alignment occurring at the centre of the C-N bond of both molecules. In this set of calculations, the molecules were held rigidly orthogonal to one another as the distance between them was incrementally reduced. Interestingly, the products of this interaction were two methyl and two nitro radicals, the same products as the uni-molecular decomposition route. Thus, it is possible that a bi-molecular process, which is favoured at high pressures, may to some degree mimic the unimolecular mechanism. The activation energy for this process was found to be 130 kcal/mol.

MOLECULAR DYNAMICS

Molecular orbital calculations can only be carried out on at most a few molecules, and then can only give the thermodynamics of possible decomposition reactions in energetic materials. Although it is possible to derive an approximation to the kinetics by means of statistical mechanics, the applicability of such techniques to studying condensed phased reactions is questionable. The only satisfactory solution is to obtain kinetic data from molecular dynamics simulations of large ensembles of molecules. In these codes, systems consisting of thousands of molecules can be studied, and bulk material properties calculated. However, at present it is only possible to use empirical potentials to describe the interactions between atoms and molecules, usually for model systems. In order to calculate kinetic data on real explosive molecules, either empirical potentials or, more preferably, ab-initio potentials which describe accurately the interaction of these molecules will have to be used.

Molecular dynamics calculations on model systems using empirical potentials can, and have been shown to be extremely useful in understanding the general chemical-physical processes occurring at a molecular level when energetic systems undergo reaction. It has been demonstrated, that the essential behaviour of a model explosive can be simulated using an empirical potential which has an implicit 3-body term to mimic chemistry. Molecular dynamics calculations on model explosive systems show all the pertinent features that would be expected of real systems. For example, if in the simulation, reaction is initiated by means of a 'flyer plate' of molecules, then a critical velocity is required to obtain growth to detonation. The initial shock wave that passes into unreacted material on impact, is followed by a zone of highly compressed material, which in turn is followed by the reaction zone, where most of the chemistry takes place. The reaction zone is succeeded by the expanding gaseous products. A steady velocity of detonation is observed, and if the reaction is over-driven, then the system still settles to the characteristic velocity of the model. These features are generally observed for real explosive systems.

We have previously published details of our own molecular dynamics code DYNAMITE[8] and its subsequent use to study detonation phenomena. In this paper, we only give a brief outline of the work reported elsewhere[5]. We have studied the initiation, growth of reaction and steady-state characteristics of a model diatomic system. The influence of exothermicity, activation barriers, and initial shock conditions in both 2D and 3D versions of the code have all been addressed. A further set of calculations have also been performed on a modified version of the code to examine thermal initiation (cook-off).

Initiation in homogeneous explosives is experimentally observed to a show a weak dependence on pulse duration except for very short high-energy pulses, but to require a minimum particle velocity. For homogeneous systems, it has been widely shown that initiation can be described by use of thermal explosion theory. We have demonstrated previously, that our model is consistent with this behaviour, with a strong dependence on pulse duration for short pulses, and weak dependence for longer pulses. Reasonable fits to the MD results were obtained with a simple thermal explosion model based on a single-step exothermic Arrhenius reaction. Overall, both the 2D and 3D calculations gave very similar results. As would be expected for a homogeneous system the Walker-Wasley critical energy (P^2t) criterion does not provide a good fit to the initiation threshold curved obtained in the simulations. It is interesting to note however, that it does give an approximate fit to the shorter duration pulse results.

Experiments carried out to study thermal initiation showed reasonable thermal reaction kinetics. The simulations were carried out on a square/cubic box of molecules at various initial temperatures. This model was used to study the reaction processes occurring, the times to reaction, and to obtain the specific heat (C_v). A wide variety of reactions have been observed including unimolecular, bimolecular, concerted trimolecular, and back reactions. A particular feature worth noting is that classic "induction time" behaviour is observed at the lower temperatures.

IGNITION AND GROWTH MODELS

The development of suitable routines to model the ignition and growth of reaction within hydrocodes, is of fundamental importance, particularly to hazard prediction. Whilst there are a number of ignition and growth routines which are widely used, these methods do not provide any understanding and can only be used as engineering tools.

We have begun a programme of work aimed at constructing an ignition and growth of reaction model for explosives based on well-founded physical and chemical principles. The model is basically homogeneous with heterogeneity covered in a separate 'hotspot' routine. The reason for this is that any model must be able to describe all explosive behaviour, including homogeneous; besides which, all explosives whether heterogeneous or not, exhibit some homogeneous characteristics. Hotspots merely serve to make the basic homogeneous model more sensitive, in certain circumstances than otherwise would be the case. For example, when very strong shocks are involved, apparently heterogeneous materials can exhibit homogeneous behaviour[9].

The underlying homogeneous model comprises a number of coupled Arrhenius kinetic steps which are, at present, calibrated to experimental data for each type of energetic molecule. The model typically has an initial endothermic Arrhenius step followed by one or more exothermic steps. It is envisaged that general principles and direct calibration of the kinetic scheme will, eventually, be derived from MD studies. Equations-of-state are required for both the unreacted explosive and the product gases. A particular feature of the model is that it employs temperature dependent kinetics, whereas others are pressure dependent; it is switched on throughout the calculation, and has no arbitrary ignition criteria.

The current hotspot model is still under development, but is based on adiabatic bubble collapse. Heat from the collapsing bubble is assumed to be transferred to the surrounding energetic material, and if sufficient heat is produced this can lead to an accelerated burn of the energetic material away from the pore. If sufficient numbers of pores are present this can lead to growth of reaction, and ultimately detonation in the energetic material. While the model is still in development, it is interesting to note that no arbitrary parameters have been employed so far although, there is still much work to be done. In the future, it is our intention to develop a number of hotspot models based on other mechanisms such as shear and friction.

CONCLUSIONS

In this paper we have outlined a number of possible mechanisms to account for the initial steps in the decomposition of nitromethane at the molecular level. In particular we have demonstrated the possibility of bi-molecular processes, one leading to bond scission of the two reacting nitromethane molecules, and another to the products MeNO, CH_2O and HONO. Such processes may well be important in shock initiation of nitromethane and indeed, nitro compounds in general. We have also outlined the results from our molecular dynamics calculations, and have discussed the increasing importance of this technique to understanding the initiation and growth to detonation, as well as the detonation process itself, at the molecular level. Information from our molecular dynamics studies are also helping us to formulate new ignition and growth models based on sound physical and chemical principles. Although it is still in its infancy, the new ignition and growth model that we are developing at Fort Halstead shows much promise, and if progress continues at the present rate, should provide the foundation for a predictive capability for the reaction of energetic materials to various stimuli, in the near future.

REFERENCES

1. G. I. Pangilinan, "Time -resolved Molecular Changes in Shocked Sensitized Nitromethane Undergoing Chemical Reactions", in the Proceedings of the APS Topical Conference "Shock Compression of Condensed Matter" held in Seattle, Washington, July 1995.

2. J. M. Winey, Y. M. Gupta, "Time-Resolved Raman Spectroscopy to Examine the Shock-Induced Decomposition of Nitromethane", in the Proceedings of the APS Topical Conference "Shock Compression of Condensed Matter" held in Seattle, Washington, July 1995.

3. T. P. Russell, T. M. Allen and Y. M. Gupta,"High-Pressure Time Resolved Optical spectroscopy of the Deflagration Chemistry of Energetic Materials in a High Pressure Anvil Cell", in the Proceedings of the APS Topical Conference "Shock Compression of Condensed Matter" held in Seattle, Washington, July 1995.

4. C. T. White, "Atomistic Simulations as a Nanoscale Probe of Shock-Induced Chemistry", in the Proceedings of the APS Topical Conference "Shock Compression of Condensed Matter" held in Seattle, Washington, July 1995.

5. P. J. Haskins and M. D. Cook, "Molecular Dynamics Studies of Thermal and Shock Initiation in Energetic Materials", in the Proceedings of the APS Topical Conference "Shock Compression of Condensed Matter" held in Seattle, Washington, July 1995.

6. M. D. Cook & P. J. Haskins,"Decomposition Mechanisms and Chemical Sensitisation in Nitro, Nitramine and Nitrate Explosives"., Proceedings of the Nineth Symposium (International) on Detonation, Portland, Oregon, 28 Aug - 1 Sept 1989, Vol II, 1027.

7. GAUSSIAN 92, M. J. Frisch, M. Head-Gordon, G.W. Trucks, J. B. Foresman, H. B. Schlegel, K. Raghavachari, M. A. Robb, J. S. Binkley, C. Gonzalez, D. J. Defrees, D. J. Fox, R. A. Whiteside, R. Seeger, C. F. Melius, J. Baker, R. L. Martin, L. R. Kahn, J. J. P. Stewart, S. Topiol, and J. A. Pople, Gaussian Inc., Pittsburgh, PA, 1992.

8. P. J. Haskins M. D. Cook, "Molecular Dynamics Studies of Shock Initiation in a Model Energetic Material", Proceedings of the joint International Association for Research and Advancement of High Pressure Science and Technology and American Physical Society Topical Group on Shock Compression of Condensed Matter Conference, held at Colorado Springs, Colorado, 28 June - 2 July 1993, p1341-1344.

9. P. K. Tang, "A Study of the Role of Homogeneous Process in Heterogeneous Explosives", Proceedings of the Tenth International Detonation Symposium, Boston, Massachusetts, 12-16 July 1993.

EFFECTS OF NANOSCALE VOIDS ON THE SENSITIVITY
OF MODEL ENERGETIC MATERIALS

C. T. WHITE[A], J. J. C. BARRETT[A], J. W. MINTMIRE[A], M. L. ELERT[B],
and D. H. ROBERTSON[C]
[A] Naval Research Laboratory, Washington, DC 20375
[B] U. S. Naval Academy, Annapolis, MD 21402
[C] Indiana University–Purdue University at Indianapolis, Indianapolis, IN 46202

ABSTRACT

Because of its importance in designing safer, more reliable explosives the shock to detonation transition in condensed phase energetic materials has long been a subject of experimental and theoretical study. This transition is thought to involve local *hot-spots* which represent regions in the material which couple efficiently to the shock wave leading to a locally higher temperature and ultimately initiation. However, how at the atomic scale energy is transferred from the shock front into these local *"hot spots"* remains a key question to be answered in studies of the predetonation process. In this paper we report results of molecular dynamics simulations that suggest that even nanometer scale defects can play an important role in the shock to detonation transition.

INTRODUCTION

In a classic paper, Campbell *et al.* [1] reported experimental results which indicated that strong shock waves propagating in energetic materials can interact with density inhomogeneities such as voids leading to a local decomposition of the explosive and subsequent exothermic chemical reactions. These exothermic chemical reactions can in turn strengthen the shock wave to the point that detonation ultimately begins. We have reported results of molecular dynamics (MD) simulations which showed that even nanometer scale voids interacting with strong shock waves can lead to significant local heating of the material. Although these earlier simulations suggested possible atomic-scale modes of coupling shock wave energy to internal degrees of freedom through material defects they employed potentials that did not allow for any chemical reactions. [2] Hence the possibility of shock induced chemistry caused by the defects could not be studied. Herein we report simulations of nanometer scale voids interacting with a strong shock wave. These simulations employ potentials that do allow for shock-induced chemistry. [3-6] We find that the presence of nanometer size defects can have a profound effect on the initiation characteristics of our model materials.

In the next section we briefly describe our model and approach. In Section III we present our results. Finally, Section IV provides some conclusions.

MODEL

To meet the need for potentials in molecular dynamics simulations of shock induced chemistry which produce reasonable reaction energetics while remaining computationally simple enough for large-scale molecular dynamics simulations, we have developed a reactive empirical bond order (REBO) approach which uses a potential form similar to the form introduced by Tersoff in his study of silicon [7] and Robertson et al. [8] in their study of shock induced phase transitions. In this approach, the interaction between a pair of atoms is given in the form of a simple pair potential, but the attractive term in the potential, which represents chemical bonding, is given according to the number of near neighbors already present. Thus an isolated pair of atoms will experience the full attractive potential, while atoms already engaged in bonding to other near neighbors will be more weakly attracted to each other. Depending on the choice of parameters, an REBO potential can be made to favor any desired valence, from diatomic molecules to metallic systems.

Starting from this basic formalism we have developed over the last several years several REBO potentials especially tailored to the study of shock-induced chemistry and related phenomena. [3-7] These many body potentials are capable of simultaneously following the dynamics of thousands of atoms in a rapidly changing environment, while including the possibility of exothermic chemical reactions proceeding along chemically reasonable reaction paths from cold solid-state reactants to hot gas-phase molecular products. They also incorporate the strong intramolecular forces that bind atoms into molecules and the weak intermolecular forces that bind molecules into solids. These potentials have been shown to yield chemically sustained shock waves that have properties that are consistent with continuum theory. [4-6] Herein we report results for strong shock waves interacting with nanometer size defects in a REBO based model of ozone. The parameters characterizing this model have been reported elsewhere. [6]

RESULTS AND DISCUSSION

To begin to explore the possible role of nanometer scale defects on the shock to detonation transition within our models we have carried out simulations of the effects of nanoscale voids on the initiation characteristics of our recently developed model of ozone in two dimensions. To perform these simulations, a crystal of the ozone reactant was set up in the xy plane. Next, nanometer diameter voids were induced randomly in the material so to reduce its density by 10 %. Then strong shock waves were produced in the defected material by striking the molecular solid with a high velocity flyer plate made of crystalline ozone. Finally, the particle dynamics were followed using a predictor-corrector method. [9] Periodic boundary conditions were applied in the y direction to simulate the behavior of an infinitely wide system.

Our principal result in these ongoing studies is summarized in Figure 1. In this figure we depict the wavefront position versus time for two different simulations both begun with a four-layer plate moving at 3.7 km/sec. The lower of the two curves in Figure 1 corresponds to a simulation begun assuming a defect free crystal. Although some initial chemistry was begun in this simulation, it quickly ceases and the shock wave begins to slow as shown by the decreasing slope of this curve.

The behavior of the void containing material, however, is entirely different. In this system a detonation begins leading to an initially increasing but soon constant shock front velocity. These results demonstrate the importance of nanoscale defects in the initiation characteristics of this model.

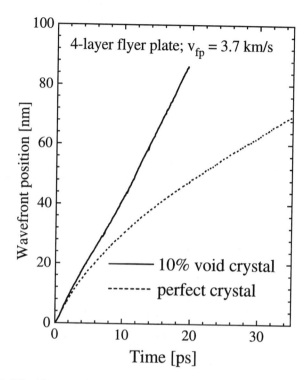

Figure 1. Wavefront position versus time for an ozone crystal with (upper curve) and without (lower curve) defects.

IV. CONCLUSIONS

The results reported in this paper demonstrate that nanometer size void defects can have severe effects on the shock to detonation threshold of our model ozone system. However, we have yet to demonstrate that the reduction in threshold that we see is not simply a density effect. We strongly suspect, however, that this is not the case based concurrent simulations of strong shock interacting with nanometer scale cracks. In these simulations we find that the initiation characteristics of the model are dependent on the crack width suggesting that there is also a similar, geometry related mechanism, at work in this system as well.

ACKNOWLEDGMENTS

JJCB was supported by an NRC residence Associateship at the Naval Research Laboratory. MLE acknowledges support from the NRL-USNA Cooperative Program for Scientific Interchange. This work was supported by the Office of Naval Research (ONR) through the Naval Research Laboratory and through the ONR contract # N00014-96-WX-20330.

REFERENCES

[1] A. W. Campbell, W. C. Davis, J. B. Ramsay, and J. R. Travis, *Phys. Fluids* **4**, 511 (1961).

[2] J. W. Mintmire, D. H. Robertson, and C. T. White, *Phys. Rev. B* **49**, 14859 (1994).

[3] D. W. Brenner, M. L. Elert, and C. T. White, *Shock Compression of Condensed Matter - 1989*, S. C. Schmidt, J. N. Johnson, and L. W. Davison Elsevier Science Publishers p. 263 1989.

[4] C. T. White, D. H. Robertson, M. L. Elert, and D. W. Brenner, in *Microscopic Simulations of Complex Hydrodynamic Phenomena*, M. Mareschal, and B. L. Holian, Eds., New York: Plenum Press, 1992, p. 111.

[5] D. W. Brenner, D. H. Robertson, M. L. Elert, and C. T. White, *Phys. Rev. Lett.* **70**, 2174 (1993).

[6] J. J. C. Barrett, D. H. Robertson, D. W. Brenner, and C. T. White, this conference.

[7] J. Tersoff, *Phys. Rev. Lett.* **56**, 632 (1986); *Phys. Rev. B* **37**, 6991 (1988).

[8] D. H. Robertson, D. W. Brenner, and C. T. White, *Phys. Rev. Lett.* **67**, 3132 (1991).

[9] C. W. Gear, *Numerical Initial Value Problems in Ordinary Differential Equations*, Englewood Cliffs: Prentice-Hall, 1971, p. 148.

HOT SPOTS IN A MOLECULAR SOLID UNDER RAPID COMPRESSION: ENERGY SHARING AMONG THE *T-R-V* DEGREES OF FREEDOM.

D. H. TSAI
10400 Lloyd Road, Potomac, MD 20854

ABSTRACT

Sharing of the kinetic energy among the translational-rotational-vibrational (*T-R-V*) degrees of freedom in a molecular crystal, following excitation of the *T* degrees of freedom, has been studied by means of molecular dynamics. The model was similar to that employed in [2]. The *T* degrees of freedom were energized by adding a velocity distribution to the (100) molecular lattice planes so as to excite certain normal modes in the [100] direction. The kinetic energy components in *T,R,V* and the rates of energy sharing between *T-V* and *T-R* were determined. The results were then compared with those from rapid compression of the same model, with or without defects.

INTRODUCTION

In earlier work [1-2], the heating of hot spots in a rapidly compressed crystal was found to be substantially enhanced by structural relaxation in the crystal, especially around defects. This relaxation process caused a part of the strain energy due to mechanical compression, especially around the defects, to be locally converted to kinetic energy and heat. Moreover, because the relaxation process could propagate to other parts of the crystal, the hot spot could be considerably larger than the size of the initial defects.

In addition to these mechanisms of hot spot heating, the sharing of the kinetic energy among the *T-R-V* degrees of freedom in the molecules is equally of interest, because it is the *V* degrees of freedom that are ultimately responsible for the initiation of chemical reactions. This report summarizes a model study of the energy sharing problem by means of molecular dynamics (MD). This problem has been discussed by several authors from the quantum mechanical viewpoint of multiphonon "up-pumping" through the "doorway" modes [3-5].

MODEL

The molecular model was similar to that of [2]. It was an fcc lattice with 4096 atoms (in $32 \times 16 \times 16$ lattice planes) in Cartesian coordinates, with the X,Y,Z axes in the [100], [010] and [001] directions, respectively, and with periodic boundaries in all three directions. The atoms, all of one kind, were grouped into 512 molecules. Each molecule was made up of eight atoms in two adjacent atomic X-Y planes (Fig. 1). The molecules in adjacent *pairs* of X-Y planes also formed an fcc crystal. Morse potentials were used to represent the interactions between atoms i and j:

$$V(R_{ij}) = \epsilon A\{\exp[-2\alpha(R-R_1)]-2\exp[-\alpha(R-R_1)]\}.$$

Here, $\epsilon=R_1=1.0$, $\alpha=ln2/0.1$, A is the well depth: $A=A_m$ if i and j belonged to the same molecule, and $A=A_n$ if they belonged to different molecules. Three sets of $A_m:A_n$ were used: (A) 4:0.5, (B) 2:0.5, and (C) 1:1. The stiffness ($\partial^2 V/\partial R^2$) of the intra- and intermolecular potentials, relative to one another, was then 8:1, 4:1, and 1:1. The box-like molecules of Fig. 1 were not in a configuration that was at the global minimum of the potential energy, but the molecular crystal constructed with these potentials proved to be stable in the range of temperatures and volumes investigated here. With potentials (B) and "reasonable" values for the size and mass of the atoms (nitrogen), the unit of time τ was 2.6×10^{-13} s, the unit of energy ϵ was 0.6×10^{-13} ergs, the highest intermolecular frequency ω was 0.92×10^{13} hertz, yielding a Debye temperature Θ_D of 4.4×10^2 K. Other details of the model and the method of computation were the same as those described in [2].

281

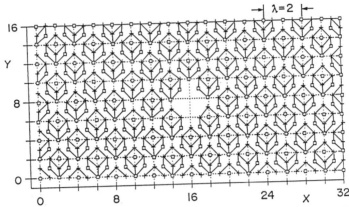

FIG. 1 Lattice configuration showing box-like molecules in two middle X-Y planes with a vacancy in the middle. Atoms in plane 7 are shown as (□), in plane 8 as (+).

Partition of the kinetic energy E_k into T-R-V degrees of freedom — From a given set of initial and boundary conditions, the trajectories and the energies of all atoms, as independent particles and as functions of time, were first obtained from MD calculations. The data were then processed according to the molecular grouping of the atoms. With eight atoms in each molecule, E_k for each molecule may be partitioned into three T, three R, and eighteen V degrees of freedom. The following notations are used:

E_{kx} = kinetic energy associated with U_x of an atom, averaged over a molecule.

T_x = average translational kinetic energy of a molecule associated with the *rigid-body* motion of its center of mass (CM) in the X direction, per atom.

R_x = average rotational kinetic energy of a molecule associated with the *rigid-body* rotation about the X axis through the CM of the molecule, per atom.

V_x = average vibrational kinetic energy of a molecule associated with velocities in the X direction, per atom.

Similarly for the energies with subscripts y and z. Subscript t denotes the sum of the x,y,z components. Thus $E_{kt}=E_{kx}+E_{ky}+E_{kz}=T_t+R_t+V_t$. E_{kx} and T_x (and for subscripts y and z) were obtained directly from the MD results at each time step. Next, the total angular momenta of each molecule, in rigid-body rotation, about the three axes passing through the CM of the molecule were obtained, and R_x was set equal to the x component of the kinetic energy associated with the x component of the angular momentum. Similarly for R_y, R_z. Then, $V_x=E_{kx}-T_x-R_x$, and similarly for V_y, V_z. But here the choice of the rotational axes was arbitrary [6], hence the partition of R_x and V_x, etc., also was arbitrary. This means that equipartition of R and V into the x,y,z components at thermal equilibrium was only approximate, but equipartition of E_k and T always obtained. It was found that when the system was heated from an equilibrium state, by scaling the velocities by a small increment each time step, the equipartition of E_k into T, R, and V was not disturbed.

Excitation of T degrees of freedom — After thermal equilibration to an initial temperature E_{ki}, at zero and at 10% volume compression in the [100] direction, a sinusoidal mass velocity distribution was added to all (100) molecular lattice planes. The shortest wave length, $\lambda=2$ (Fig. 1), of this disturbance corresponded to adding U_x and $-U_x$, respectively, to alternate molecular planes. The energy added was $\Delta T_x=(mU_x^2)/2$ per atom. The subsequent oscillations of these planes, at the highest intermolecular frequency, simulated the normal mode oscillations which a planar shock front in the X direction would excite. Relaxation of

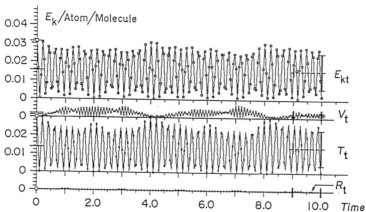

FIG. 2 Energy sharing among T-R-V in the lattice after excitation of T by $\Delta T_x = 0.0312$. Perfect lattice, interaction (B), zero compression, $X_s = 0.7004$, $Y_s = 0.7008$, $Z_s = 0.7023$, $E_{ki} = 0.133 \times 10^{-8}$, and $\lambda = 2$. The zero axes of T_t, V_t, and E_{kt} have been shifted upward.

ΔT_x would then give a direct measure of the energy sharing among the T-R and T-V degrees of freedom.

Simulation of shock compression — As in [2], rapid volume compression (by 10%) was applied by shrinking the X coordinates of the atoms by a scale factor X_s at a uniform rate over a time interval $\tau = 1$. This was approximately equal to the transit time of a sound wave through a typical dimension of a defect. Increasing the rate by a factor of two did not appreciably alter the observed relaxation of E_k. When the system was compressed at these rates from an equilibrium state, and if the compression did not give rise to structural relaxation, the equipartition distribution of E_k and T was not disturbed. This method of compression saved a great deal of computation. It was considered to be a reasonable simulation of shock compression, because the various relaxation processes occurred mostly after the end of compression, and were slow compared with the compression process.

RESULTS

Effect of excitation wave length λ — The relaxation of ΔT_x was found to depend on both λ and E_{ki}. Fig. 2 shows the energy sharing process following the addition of $\Delta T_x = 0.0312$ per atom, at $\lambda = 2$, with interaction (B) and $E_{ki} \approx 0$. There was very little energy sharing between T-R and T-V. These results followed directly from the non-interacting nature of the normal modes, and from the mismatch of the inter- and intramolecular frequencies, when the lattice was at a low temperature, and the interactions were nearly harmonic.

Actually, some exchange did occur between T and V in Fig. 2, in both directions, even from the beginning. After $\tau = 16$ (not shown), the energy sharing accelerated because of the accumulated effects of anharmonicity and/or umklapp scattering [7]. At $\tau \approx 18$, isolated peaks in E_k and T appeared, indicating that the anharmonicity was causing solitons to develop. At $\tau = 30$, the maximum time of this computation, equipartition of the energies was still far from being achieved. Similar results were obtained for interactions (A) and (C).

At longer λ, solitons developed more quickly. At $\lambda = 16$, for example, well-formed solitons were observed after the first oscillation of the lattice. Again there was very little thermalization within the time of the computation.

Effect of E_{ki} — As E_{ki} was raised, the energy sharing became more rapid. Table I lists the results obtained over a range of E_{ki}. The equilibration of T_x, T_y, T_z in one case, B5, with $\Delta T_x = 0.5$, $\lambda = 2$, $E_{ki} = 0.757$, and 10% compression in the X direction, is depicted in Fig. 3. The relaxation time τ_r was taken to be the time to re-establish equilibrium, i.e., when the fluctuations in T_t first returned to the expected fluctuations for the system at equilibrium. The equilibration of T-R-V for this case, not shown here because of space limitation, was entirely consistent with these results.

Table I shows that the decrease in τ_r with increasing E_{ki} occurred in all three sets of interactions. But at about the same E_{ki}, τ_r was not much affected by the difference in the relative stiffness of the potentials. Now, with interaction (C), the inter- and intramolecular potentials were identical, hence the more rapid energy sharing of ΔT_x at higher E_{ki} could only be due to the higher anharmonic and/or umklapp scattering of the molecular vibrations [7],

Table I. Relaxation time τ_r of T_x at different E_{ki} and ΔT_x. $\lambda = 2$.

Zero Compression					10% compression in X direction				
Run	A_m:A_n	ΔT_x	E_{ki}	Relaxation time, τ_r	Run	A_m:A_n	ΔT_x	E_{ki}	Relaxation time, τ_r
A1	4:0.5	0.5	0.770	2.8	A4	4:0.5	0.5	0.750	2.8
A2	4:0.5	0.5	1.026	1.6	A5	4:0.5	0.5	1.002	1.4
A3	4:0.5	0.5	1.533	1.2	A6	4:0.5	0.5	1.547	1.2
					B4	2:0.5	0.5	0.152	6.0
B1	2:0.5	0.5	0.701	2.0	B5	2:0.5	0.5	0.757	2.4
B2	2:0.5	0.5	1.025	1.6	B6	2:0.5	0.5	1.016	2.0
					B7	2:0.5	0.78	1.016	1.8
B3	2:0.5	0.5	1.528	1.2	B8	2:0.5	0.5	1.539	1.0
C1	1:1	0.5	0.081	8.0					
C2	1:1	0.5	0.789	2.8	C6	1:1	0.5	0.790	3.2
C3	1:1	0.245	1.030	1.8					
C4	1:1	0.5	1.030	2.0	C7	1:1	0.5	1.030	2.4
C5	1:1	0.5	1.583	1.2	C8	1:1	0.5	1.528	1.6

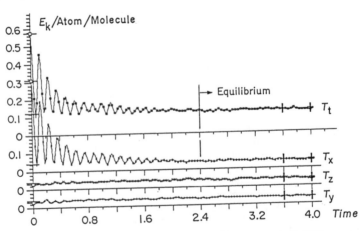

FIG. 3 Equilibration of T degrees of freedom in case B5 of Table I. 10% compression in the X direction, $X_s = 0.6304$, $Y_s = 0.7008$, $Z_s = 0.7023$, $E_{ki} = 0.757$, $\Delta T_x = 0.5$.

impurity and boundary scattering being absent. By analogy, the thermalization process for (A) and (B) also may be attributed to anharmonicity and umklapp scattering. These results indicate that in the range of intramolecular potentials examined here, the softer inter-molecular phonons did not have much of a role in the energy exchange among the molecules. In one case that was comparable with C2, but with an intramolecular potential 25 times stiffer, τ_r remained about the same as that of C2. Still stiffer potentials would make the calculation difficult, because very small time steps would be required to maintain computational accuracy. For this reason, very stiff potentials were not investigated here.

Role of a Vacancy Defect under Rapid Compression — Fig. 4 shows the distribution of the average E_{kt} in the crystal with a single vacancy defect when the crystal was compressed 10% in the X direction in one unit of time. The equipartition of T-R-V was not much disturbed until the onset of substantial structural relaxation around $\tau=3$. The structural relaxation caused the average T_t to increase. As structural relaxation continued, a series of shear bands were generated. Equipartition of T-R-V was essentially achieved at $\tau=15$, which was also the end of structural relaxation. With $E_{ki}=0.217$ for this run, Table I shows that the relaxation

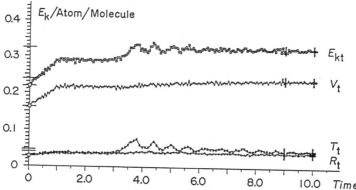

FIG. 4 Energy sharing among T-R-V in the lattice with a single vacancy, when X_s was compressed from 0.7004 to 0.6304 in one unit of time. Interaction (B), $E_{ki}=0.217$.

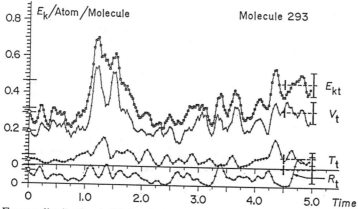

FIG. 5 Energy distribution in Molecule 293 of Fig. 4, showing the heating of V from the expansion of the molecule into the vacancy.

time for ΔT_x should to be less than 6.0. Thus the structural relaxation process was slower than the energy sharing process for this case.

A closer examination of the results (Fig. 4) revealed that structural relaxation actually had started around $\tau=1$, when some molecules next to the vacancy expanded rapidly into the vacancy. The expansion directly converted a part of the strain energy into the V degrees of freedom in these molecules. An example is shown in Fig. 5 for Molecule 293, in which the peak V was more than twice the average V in Fig. 4, even at $\tau\approx10$.

SUMMARY AND CONCLUSIONS

The sharing of energy in the T-R-V degrees of freedom in a crystal (Fig. 1) following the addition of ΔT_x has been studied by means of MD modeling over a range of temperatures (E_{ki}), wave lengths (λ) of ΔT_x, and 1D compression of the crystal. The mechanism of energy sharing was found to be by anharmonic and/or umklapp scattering of molecular vibrations. At low E_{ki}, energy sharing was slow (Fig. 2), and equipartition of T-R-V required more than hundreds of molecular vibrations. During this time, soliton-like molecular vibrations developed. At higher E_{ki}, the relaxation time of ΔT_x decreased with increasing E_{ki}, and equipartition of T-R-V was achieved in 50 to 10 molecular vibrations (Fig. 3, Table I). The stiffness of the molecules, in the range examined here, had only a small effect on the relaxation time.

When the model contained a vacancy defect and was rapidly compressed, relaxation of the lattice structure occurred, accompanied by shear-band formation. This process energized the T degrees of freedom, which equilibrated with R and V at rates faster than the structural relaxation process (Fig. 4). But the expansion of some of the compressed molecules into the vacancy directly and rapidly energized the V degrees of freedom (Fig. 5).

The hot spot heating from structural relaxation observed here involved the conversion of potential energy to kinetic energy, under essentially "static" conditions. In shock compression, however, there is the additional (and prior) step of conversion of the kinetic energy associated with the mass velocity U_p to the potential energy of compression, and the increased T from interplanar oscillations, under "dynamic" conditions. The energy sharing among T-R-V under these "dynamic" conditions remains to be investigated.

ACKNOWLEDGMENTS — I thank R.W. Armstrong, J.F. Belak, Z.Y. Chen, J.P. Cui, and S.F. Trevino for helpful discussions in the course of this work. Travel support was provided by the Office of Naval Research through Contract No. N00014-90-J-1388.

REFERENCES

1. D.H. Tsai, J. Chem. Phys. **95**, 4797 (1991); in *Structure and Properties of Energetic Materials*, edited by D.H. Liebenberg, R.W. Armstrong, and J.J. Gilman, Mater. Res. Soc. Proc. Vol. 296, (Materials Research Society, Pittsburgh, PA, 1993) pp. 113–120.
2. D.H. Tsai and R.W. Armstrong, J. Phys. Chem. **98**, 10997 (1994); "Molecular Dynamics Modeling of Hot Spots in Monatomic and Molecular Crystals under Rapid Compression in Different Crystal Directions," presented at the International Conference on Shock Waves in Condensed Matter, St. Petersburg, Russia, July 18–22, 1994, to be published in the Russian J. Phys. Chem.
3. D.D. Dlott and M.D. Fayer, J. Chem. Phys. **92**, 3796 (1990).
4. A. Tokmakoff, D.D. Dlott, and M.D. Fayer, J. Phys. Chem. **97**, 1901 (1994).
5. L.E. Fried and A.J. Ruggiero, J. Phys. Chem. **98**, 9786 (1994).
6. E.B. Wilson, Jr., J.C. Decius, and P.C. Cross, *Molecular Vibrations* (Dover, New York, 1955) Chaps. 2,11.
7. N.W. Ashcroft and N.D. Mermin, *Solid State Physics* (Holt, Rinehart and Winston, New York, 1976) Chap. 25.

An *Ab Initio* Investigation of Crystalline PETN

A. Barry Kunz

Department of Electrical Engineering, Michigan Technological University, Houghton, MI 49931 U.S.A., albert@mtu.edu

ABSTRACT

Solid energetic substances have long played an important technological role as explosives and also as fuels. Much of the research on the solid phases has been concentrated on ground state properties, and also on the chemistry of the molecules comprising the solid. In addition, significant understanding of the detonation properties has been obtained by semi-empirical continuum mechanical modeling. Traditional solid state studies of this important class of materials have mostly been ignored. This may be due to the apparent success of the semi-empirical models in describing the detonation properties, as well as the practical difficulties in performing band theoretical studies. Recent interest in more fundamental questions relating to the basic properties of these systems as materials, coupled with a desire to probe fundamental questions relating to the initiation and sustaining of the chemical reactions leading to combustion/detonation is generating significant interest in the basic solid state properties of such energetic systems. In particular, recent analysis of detonation by J. J. Gilman emphasizes the need to include excitation of the electronic system in obtaining an understanding. In this manuscript, the basic solid state properties of PETN are considered.

INTRODUCTION

There has been a lengthy history of interest in the properties of energetic systems undergoing combustion, or detonation. The present study limits considerations to solid phase systems, and to solid pentaerythritol tetranitrate (PETN) in particular. Most of the understanding achieved for solid energetic explosives has been obtained using semi-empirical continuum models. In addition much of the non-empirical modeling that has been performed has been focused on properties of the molecule in question, generally in a ground state configuration. This lack of fundamental quantum mechanical studies of the solid phase is easily understood, when one recognizes that mankind has been successfully constructing explosive, and pyrotechnic devices for many centuries before the technological methodology needed to understand such systems would be available. In addition, there is a second, and equally serious problem, the limitations placed upon such studies by available computational machinery, coupled with the complexity of most useful energetic solids. A PETN unit is $C_5H_8N_4O_{12}$, and in the solid there are two such units in a single cell. The crystal structure is a tetragonal one, and the space group is $P\overline{4}2_1c$ [1]. As an illustration of the complexity, consider the following data pertaining to the computations . The unit cell contains a total of 58 atoms, and this unit cell possesses a total of 324 electrons, of which only 84 are core electrons. A split valence basis set, as used here consists of 410 atomic orbitals. Despite such difficulties, the study of the solid state properties of energetic solids is desirable, as these would allow density , geometry, and binding energy determinations, as well as a determination of the effects of crystal structure on the energy levels of the molecules from which the solid is composed.

It is a long held belief that initiation (of combustion or detonation) in energetic solids is associated with small regions of the solid termed "hot spots" [2,3]. The evidence for the existence of these regions is appealing, although circumstantial. However, an atomistic description of what constitutes a "hot spot" is not generally agreed upon. In the laboratory, initiation may be accomplished by impacting a pellet of the energetic solid with a hard, massive flyer plate. It is generally seen that the mechanical energy needed to initiate reaction is significantly less than the thermal energy needed for initiation; hence the hypothesis of "hot spots" [4,5].

This is in reasonable accord with the theoretical predictions of S. F. Coffey [6]. Coffey argues that hot spots in crystals are associated with regions of high shear stress. Normally such regions occur at the surface of the sample, but may be internally generated as well. Such regions are normally accompanied by a high density of dislocations. The Coffey model is far less specific in providing the precise physical effect of such regions on the molecules constituting the solid. In addition to the incompleteness of the Coffey model, this theory does not provide a convincing model for the presence of the charges found in the fractoemission from energetics [7]. Additionally, some studies performed by Kunz and Beck [8] provide strong evidence that the presence of charges inside an energetic solid can provide significant diminution of the strength of molecular bonds, and in some instances even cause their dissociation. Here again the source of the charges is left unidentified.

J. J. Gilman has recently subjected the question of shock sensitivity and initiation in energetic solids to scrutiny, and observes a number of significant factors indicating that excited electronic states play a role in such phenomena. [9]. Gilman observes: 1) Traditionally, the shock velocity associated with detonation is about 7 km/s (0.7×10^{14} Angstroms/s); 2) The shock fronts are sharp (5 Angstroms); 3) The rise times are short (70 femtoseconds); 4) The lattice planes at the front are highly compressed (50%); 5) The fronts are optically opaque; 6) There are very high peak pressures at the shock front (Mbar range); 7) The front is far from thermodynamic equilibrium. This set of observations is used to downplay the hot spot mechanisms, especially those that require the transport of solid defects as being incompatible with the observed shock velocities. This speed is also not compatible with ground state thermal chemical models involving phonons/vibrons on similar grounds. Gilman argues that shock initiation is related to the metalization of the solid and that sensitivity correlates with the formation of delocalized electrons. The studies here evaluate several properties of PETN which are related to the Gilman model and this is fully discussed in the conclusion.

In this manuscript, an attempt is made to initiate a process for the detailed understanding of the solid state properties of PETN. This is in hope of eventually developing a fundamental understanding, by which conduction electrons might be generated, as well as a model for the resulting solid state chemistry. The case of PETN is particularly attractive in that detailed studies due to Dick [10] indicate that the shock sensitivity to detonation exhibits a strong directional dependence. In fact several directions for impact do not progress to detonation even though the shock wave fully crosses the crystal.

METHODOLOGY

The methods of computation used here are based upon the normal non-relativistic Hartree-Fock (HF) approximation. In such a case, if one makes the normal Born-Oppenheimer approximation, the total system Hamiltonian is:

$$H = -\frac{\hbar^2}{2m}\sum_{i=1}^{n}\nabla^2 - \sum_{i=1}^{n}\sum_{I=1}^{N}\frac{e^2 Z_I}{\left|\vec{r}_i - \vec{R}_I\right|} + \frac{1}{2}\sum_{i=1}^{n}\sum_{j=1}^{n}\cdot\frac{e^2}{\left|\vec{r}_i - \vec{r}_j\right|} + V_{NN}$$

(1)

Ideally, one would like to solve the time independent Schroedinger equation for this Hamiltonian,

$$H\Psi_i = E_i\Psi_i$$

(2)

However, for a system containing n electrons and N nuclei, such an exact solution remains impossible. In this system the ith electron has a position given by $\vec{r}_i$, including spin degrees of freedom as needed. $\vec{R}_I$ designates the position of the Ith nucleus, with atomic number Z_I. V_{NN} is the nuclear repulsion energy. The electron has charge e, and mass m. The atomic system of units is used, for which $e=m=\hbar=1$.

The HF model consists of approximating Ψ_i by an antisymmetrized product of one electron orbitals, $\psi_j(\vec{r}_j)$:

$$\Psi_i(\vec{r}_1,----,\vec{r}_n) \approx \tilde{A}\prod_{j=1}^{n}\psi_j(\vec{r}_j)$$

(3)

If the one electron orbitals are chosen variationally, the one electron orbitals are defined by the HF equation,

$$F(\rho)\psi_i(\vec{r}_i) = \varepsilon_i\psi_i(\vec{r}_i),$$

$$\rho(\vec{r}_1,\vec{r}_2) = \sum_{i\le n}\psi_i^+(\vec{r}_1)\psi_i(\vec{r}_2)$$

$$F(\rho) = -\frac{\hbar^2}{2m}\nabla^2 - \sum_{I=1}^{N}\frac{e^2 Z_I}{\left|\vec{r} - \vec{R}_I\right|} + e^2\int\frac{\rho(\vec{r}_2,\vec{r}_2)}{\left|\vec{r} - \vec{r}_2\right|}d\vec{r}_2$$

$$-e^2\frac{\rho(\vec{r},\vec{r}_2)}{\left|\vec{r} - \vec{r}_2\right|}\hat{P}(\vec{r}_2,\vec{r})$$

(4)

$\hat{P}$ is the operator that interchanges the designated coordinates. This is the system of equations which is termed the canonical HF equations, and is the system solved by matrix techniques for the infinite, pure, perfect, periodic lattice problem (the band structure problem) by the code **CRYSTAL92** [11].

COMPUTATIONAL RESULTS

The HF method used here ignores correlation effects, and an initial test of the method for PETN was seen as advisable. To this end, the crystalline density of PETN is computed. The "experimental" geometry of the PETN molecule in the crystalline environment is used here. The solid state parameters of the crystalline environment are varied to achieve a minimum energy for the PETN solid. In this case, distortions from a trigonal geometry were permitted, although such distortions were not found to produce any energy lowering. The actual experimental determination of the atom positions within the molecule is obtained from X-ray diffraction measurement and the locations of the Hydrogen atoms are not accurately found. The Hydrogen positions used here are for the ideal C-H$_2$ system. The relaxation of the various crystalline axes, in this case the a and c axes indicate that the energy minimum is achieved for a slight compression of the lattice. The total energy as a function of cell volume is given in Figure 1A for the hydrostatic compression. The computed unit cell volume is found to be about 2.5% greater than experiment. This agreement with experiment is acceptable. The basis set used here is a Gaussian 6-21g set. This study assumes that the molecular distortion modes are much stiffer than are the solid state modes. This is probably valid for small lattice distortions. None-the-less, as a check on this assumption, the molecular bond lengths were allowed to vary in proportion to the variation in solid state parameters. This study produces a significantly higher energy for each distorted configuration than did the study with the rigid molecule. Therefore, based on energetic considerations, the rigid molecule is considered the more acceptable of the two models and is used in the later studies.

This study is then enhanced to look at the P-V data for PETN as a function of several variables. In particular the data of Dick for detonation demonstrates a strong directional dependence for sensitivity [10]. It is attempted to see if any of the solid state properties mirror the observations of Dick. Initial studies are performed for compressions along the a and c axes. Dick finds detonation occurring for a 12.4 GPa shock along the c axis but not along the a axis. The Volume-Energy results were computed for these two cases initially, and using the appropriate low temperature formula, P=-dU/dV, the P-V curve was computed for both cases. These results are seen in Figure 1B. Clearly, it requires a much higher pressure to produce a given volume change for the a axis than for the c axis.

Clearly the above study differentiates strongly between the a and c axes as far a sensitivity to volume change as a function of volume compression. However, the relationship of this to detonation is not established. This is partially demonstrated by a study of the uppermost bonding orbitals, those at the top of the PETN valence band, which lies at -0.405 A.U., (HOMO), and the lowest anti-bonding states, at the bottom of the conduction band (LUMO). These are studied as a function of pressure for a and c-axis compression as well. The basic electronic energy structure is first determined for PETN and shown as a density of states vrs. energy plot in Figure 1C. Then the sensitivity of the HOMO to uniaxial distortion is shown in Figure 1D. The sensitivity of the LUMO is not shown, as it is much less sensitive to lattice distortion than is the HOMO. The LUMO decreases with pressure, whereas the HOMO increases in energy with pressure indicating a closing of the band gap. One may attempt to relate this to detonation sensitivity using the hypothesis of J. J. Gilman. In this hypothesis, the assumption is made that detonation is due to the energetic substance undergoing an electronic phase transition from insulator to metal. This requires a band closing, and absent the augmentation of the present calculations to include correlation (a process that is very expensive to accomplish, and is being undertaken), the actual metalization pressure is not easily predicted. The important trend is possible to identify in the present case, as the gap closes for less pressure for c-axis compression than for a-axis compression.

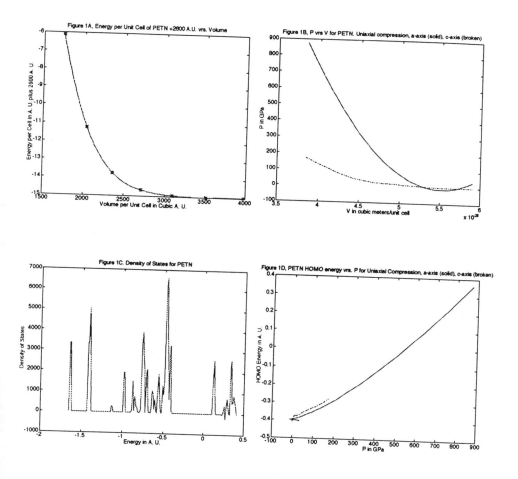

Figure 1. Various aspects of the solid state properties of PETN are shown in this Figure. In Figure 1A, the dependence of crystalline energy on volume is seen for a hydrostatic compression. In Figure 1B the Pressure-Volume dependence of PETN is seen for uniaxial compression along both the crystalline a-axis and c-axis. In Figure 1C, the electronic density of states is given. In Figure 1D the dependence of the LUMO level on pressure is seen for compressions of the a-axis and the c-axis.

CONCLUSIONS

The study presented here establishes the ability to perform complicated lattice optimization studies on solids such as PETN at the HF level. The ability to compute the pressure-volume dependence of several modes of uniaxial distortion, along with hydrostatic compressions is also demonstrated. Additionally, the ability to examine changes in crystalline energy levels with such distortions is also found.

Most importantly, these results are in accord with the experimental directional dependence of the sensitivity of PETN to detonation. This result is seen both in the greater sensitivity to volume distortion for the c axis, as well as the sensitivity to distortion of the electronic levels for the c axis, as demonstrated by the rate of gap closure. This is by comparison to the a axis. This result is consistent with the metalization hypothesis due to Gilman, and lends this hypothesis some support. This support is not unambiguous however, in that there are a number of other routes to solid state chemical reactions which may also be consistent with the current calculation, and not specifically examined here. Much further study will be required to establish any mechanism as definitive. It is not beyond the realm of possibility that there are several mechanisms ultimately responsible for detonation initiation that are in effect, and are also consistent with the present modeling. Further studies should include the role of defects on the gap closure and related issues with distortion for instance.

ACKNOWLEDGMENTS

The author wishes to express his appreciation to Dr. Richard S. Miller for support of this research, and to Dr. Herman Ammon, of the University of Maryland, and Dr. Alan Pinkerton, of the University of Toledo, for their advice and encouragement. Research supported in part by the U. S. Navy, Office of Naval Research under grant N00014-91-J-1953.

REFERENCES:

1. J. J. Dick, J. Appl. Phys., **53**, 6161, (1982).
2. R. D. Bardo , *Shock Waves in Condensed Matter*, Y. M. Gupta, ed. 843, (Plenum Press, New York, 1985).
3. W. L. Elban , R. G. Rosemeir , K. C. You , and R. W. Armstrong , *Chemical Propulsion Information Agency*, **404**, 81, (1984).
4. G. E. Duval , *Shock Waves in Condensed Matter*, Y. Gupta, ed., 1, (Plenum Press, New York, 1985).
5. J. K. Dienes , Chemical Propulsion Information Agency, **404**, 19, (1984).
6. S. F. Coffey, *Phys Rev*, **B24**, 6984, (1981).
7. T. Dickenson , *Chemical Propulsion Information Agency*, **404**, 153, (1984).
8. A. B. Kunz , and D. R. Beck , *Phys. Rev.*, **B36**, 7580, (1987).
9. J. J. Gilman, Chemical Propulsion Information Agency, **589**, 379, (1992); Phil. Mag. **B67**, 207, (1993).
10. J. J. Dick, App. Phys. Lett., **44**, 859, (1984).
11. R. Dovesi , V. R. Saunders , and C. Roetti, *CRYSTAL92 User* Documentation, University of Torino, and SERC Daresbury Laboratory, (1992).

MOLECULAR DYNAMICS CALCULATIONS ON THE PROPERTIES OF THE REACTION ZONE IN LIQUID EXPLOSIVE

L.SOULARD

Commissariat à l'Energie Atomique - Vaujours Moronvilliers, BP 7, 77181, Courtry, France

ABSTRACT

We present the results of molecular dynamics calculations concerning the structure and the thermodynamic state of the reaction zone in the liquid explosive NO under shock sollicitation.We show that the compression of explosive is due to inertial shock followed by endothermic decomposition. Then exothermic reactions occur and cause a slow decreasing of pressure until the end of the reaction zone. Analysis of particle velocities distribution and interatomic forces shows that only a partial equilibrium is reached in the reaction zone. As the relaxation time of particles velocities is very small compared to the characteristic time of chemical reactions, thermal equilibrium is always reached and the various molecular species are isothermal. It is not the case of interparticle forces and pressure in the usual thermodynamic way cannot be defined : the reaction zone is not isobaric. Our conclusion is that an equation of state cannot be defined in the reaction zone.

INTRODUCTION

Every macroscopic model which describes the properties of an explosive under shock is in fact an average of many microscopic processes. For the model concerning the shock to detonation transition, the main process occurs in the reaction zone. Unfortunately, the analysis of the molecular characteristics of the reaction zone is very difficult in both experimental and theoretical approaches. As a matter of fact, usual energetic molecules are generally complicated from the chemical point of view. Morover, the processes in the reaction zone are very fast and occur in a small region which moves very rapidly. Thus, many aspects of reaction zone in macroscopic models of detonation are described with a purely empirical formulation.

An interresting way to describe the main properties of the reaction zone is by molecular dynamics calculations because both time and spatial scales of this method are very well suited to look at reaction zone. Nevertheless, reactive systems which are used in molecular dynamics calculations are generally simple to avoid long numerical calculations [1,2,3,4,5] and the physical meaning of the results can be examined. To reconcile physical aspects and practical problems, we have chosen a very simple but real explosive, the nitrogen monoxide NO.

The study of NO is relatively simple in the isolated molecule case, but the problem is much more complicated in a hot and dense liquid phase as behind a shock front. Thus, the NO modelization we use in this study is simple and numerical results must be considered only qualitatively.

The first part of this article is on the modelization of NO which is used in a molecular dynamics code. Thermodynamic and chemical state of the reaction zone versus shock intensity is analysed in the second part.

MODEL FOR NO

Potential functions

Molecular dynamics is the resolution of the classical equation of motion for many interactive particles (atoms or molecules). Thus, it is necessary to define before any calculation the corresponding interparticle forces, that is the interparticle potential functions.

Our system is a schematic view of NO. By hypothesis, the mechanism of decomposition is :

$$NO \rightleftarrows \lambda_1 O_2 + \lambda_2 N_2 + \lambda_3 N + \lambda_4 O$$

Equilibrium bond distances and energies for NO, O_2 and N_2 in the fundamental electronic state are deduced by quantum chemical calculations performed with GAUSSIAN 92™ code. Results (Table 1) are in good agreement with experimental data. [4].

Table 1. Equilibrium bond distances and energies for NO, O_2 and N_2.

	Calculated equilibrium lengths (m)	Calculated bond energies (J)
NO	1.15×10^{-10}	1.058×10^{-18}
N_2	1.10×10^{-10}	1.584×10^{-18}
O_2	1.21×10^{-10}	0.834×10^{-18}

These results are then used in a 3D molecular dynamics code via a purely empirical function $\varepsilon_1(r)$ (equation 1), where r is the interatomic distance :

$$\left. \begin{array}{l} \varepsilon_1(r) = \sum_{k=0}^{3} a_k \left(\dfrac{r}{r_0} \right)^{k-n} \quad , \quad r \leq r_0 \\ \varepsilon_1(r) = 0 \quad , \quad r \geq r_0 \end{array} \right\} \quad (1)$$

r_0 is a constant, n and a_k are calculated with the equilibrium length and bond energy (Table 2). Morover, we impose that $\varepsilon_1(r)$ and its two first derivatives are zero for $r = r_0$.

Table 2. Coefficients of potential functions $\varepsilon_1(r)$

coefficient	NO	O_2	N_2
r_c (m)	2.8×10^{-10}	2.8×10^{-10}	2.8×10^{-10}
a_0 (J)	0.926×10^{-16}	1.136×10^{-16}	0.915×10^{-16}
a_1 (J)	-1.855×10^{-16}	-2.220×10^{-16}	-1.879×10^{-16}
a_2 (J)	1.188×10^{-16}	1.402×10^{-16}	1.220×10^{-16}
a_3 (J)	-0.257×10^{-16}	0.301×10^{-16}	-0.266×10^{-16}

Intermolecular interactions are described with an empirical function $\varepsilon_2(r)$ (equation 2) identical for all configurations. :

$$\varepsilon_2(r) = b_1 \left[1 - \frac{b_2}{r} \right]^{b_3} \quad , \quad r \leq b_2$$
$$\varepsilon_2(r) = 0 \quad , \quad r \geq b_2 \tag{2}$$

with $b_1 = 10^{-16}$ J, $b_2 = 2.5 \times 10^{-10}$ m and $b_3 = 6$.

Initial conditions of the molecular dynamics calculation

The nitrogen monoxide is liquid at 169K with a density $\rho_0 = 1269$ kg/m^3. The initial positions of particles correspond to a regular orthorhombic structure (a = b = 4.206×10^{-10} m, c = 3.025×10^{-10} m). The overall system is obtained by translation of the unit cell along the three cartesian axes x, y, z. Finally, the initial geometry is a parallelepiped oriented along the x axis. Initial particle velocities are deduced from a Maxwell-Boltzmann distribution corresponding to 100K. All momenta are zero.

The limiting conditions of parallelepiped are periodic for the faces which are parallel to the x axis and mirror for the other faces. Because the $\varepsilon_2(r)$ function is always repulsive, the initial regular structure is unstable and relaxation of the system generates a disordened phase as in a liquid. We note a small cooling of the system because the initial potential energy is minimum at initial time. After the relaxation phase, a shock wave is loaded by putting the left mirror in motion with various velocity u : 1000m/s, 2000m/s, 3000m/s, 4000m/s, 5000m/s and 6000m/s. The calculation is stopped when the shock wave reaches the right mirror.

RESULTS.

Reactivity versus shock velocity

The shock waves corresponding to the three lower u velocities do not cause the decomposition reaction. A few reactions of decomposition occur when $u=4000$m/s, but only the two last shock waves ($u=5000$m/s and $u=6000$m/s) induce a decomposition (Figure 1). A same conclusion was obtained by one dimensional calculation [4].

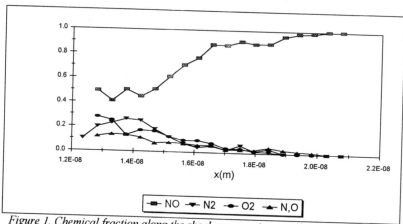

Figure 1. Chemical fraction along the shock propagation axis for u=6000m/s and
$t = 4.956 \times 10^{-12}$ s

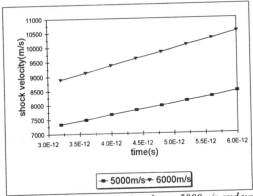

Figure 2. Shock velocity versus time for u=5000m/s and u=6000m/s.

The shock velocity is constant when the system is inert according to the general properties of an inert sustained shock. On the other hand, we observe an acceleration of the reactive shock wave because the exothermic decomposition of NO converts potential energy into kinetic energy (Figure 2). Because the phenomenon is qualitatively the same in the two reactive cases, we consider now only the u=6000m/s shock wave.

<u>Definitions</u>

Calculations of thermodynamic parameters are made at a given time by cutting along the xx' axis the system in slices with a thickness large enough to permit statistical physics calculations, but not too large to conserve a good spatial resolution along the axis. It is important to note that calculations and analysis of thermodynamics properties of the reaction zone are only semi-quantitative because the number of particles in each slice is relatively small. Morover, time average is not possible because the system is not ergodic.

The temperature T of a slice including enough particles for a good statistical representation is easly calculated by usual statistical physics if the time to reach velocity equilibrium is short with respect to the characteristic time of evolution of the overall system. To decide if the system is or is not at velocity equilibrium (that is thermal equilibrium), a simple method is to compare the numerical velocities distribution and the theoretical distribution calculated with equilibrium hypothesis. In our case, a good agreement is observed behind the shock front (the region corresponding just to the front is a non-equilibrium zone by definition).

In the same way, pressure cannot be define if interparticle forces are not in an equilibrium state. Nevertheless, shock relations permit defining a parameter η which is independent of the equilibrium state of interparticle forces

. If the system is isotropic and initially at rest, we have the following relation (equation 3)

$$\rho_{v_0} Dv_x = \frac{1}{3V} \sum_{i=1}^{N_v} \sum_{j=1}^{N} f_{ij} \bullet r_{ij} = \eta \qquad (3)$$

where v_x is the average particle velocity in the slice, N_V is the number of particles included in the slice, f_{ij} and r_{ij} interaction force and distance between particles i and j, respectively.

For an inert and sustained shock, the system behind the front is at equilibrium and relation (3) is the usual shock relation between shock velocity, average particle velocity, density at rest and pressure. Note that relation (3) is also the virial theorem, where the perfect gas term does not appear because for each slice the limit conditions are periodic. Finally, the density ρ and the specific energy e can always be calculated independently of the equilibrium state.

Thermodynamic structure of reactive shock

The most classical representation of the structure of the reaction zone is the ZND hypothesis. In this model, the initial shock (the so called ZND peak) corresponds to the point of inert or reactive shock polar for the shock velocity D. The ZND peak is followed by reaction zone where irreversible chemical decompositions occur. The final state corresponds to the end of the reaction zone, or, in other words, to the chemical equilibrium point. Comparison (Figure 3) of curves $\eta(x)$ and η_{shock} deduced from u, D and Hugoniot relations, we observe that, for each time, η_{shock} is always lower than maximum values of $\eta(x)$.

Superposition of $\eta(x)$ and chemical composition of the system shows a very thin zone $(\approx 10^{-9}\, m)$ before the maximum value of $\eta(x)$ in which endothermic chemical reactions (e.g. $NO \rightarrow N + O$) are the most important from the energetic point of view. Then, because the potential function which describes interactions between free atoms is purely repulsive, the sum of forces and then η, must be increased at the begining of decomposition. Note that the corresponding time is very small : $10^{-13}\, s$. After this time, molecular oxygen and molecular nitrogen formation involve a decrease of interaction forces, and then a decrease of η until chemical equilibrium (Figure 4).

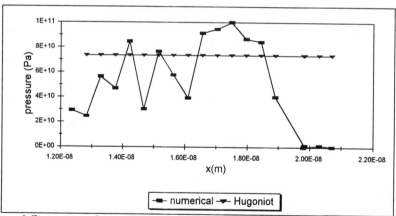

Figure 3. Comparison of η profile deduced from numerical calculations and η_{shock} ; deduced from Hugoniot relation for u=6000m/s.

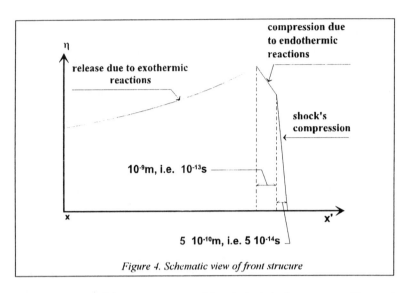

Figure 4. Schematic view of front strucure

The evolution of $v_x(x)$ component of particle velocity is in the same way. The maximum value corresponds to the shock front and is greater than piston's velocity u, because the shock is a reactive shock. On the other hand, the mirror condition behind the shock implies that limit value of $v_x(x)$ is u. Then, $v_x(x)$ decreases along the reaction zone and because total energy must be conservative, specific energy increases.

Partial thermodynamic parameters

Macroscopic models for chemical decomposition under shock require generally an equation of state for the reaction zone. Unfortunately, experimental studies give only shock polar of inert explosives and equation of state (isentropic release) of detonation products. Thus, the equation of state of the reaction zone is deduced from experimental values and an hypothetic mixture law. For that, it must be supposed that two thermodynamic parameters are equal. Further mixture laws exist in the litterature : (pressure, temperature), (pressure, density), etc.

The aim of this part of the paper is to propose an analysis of characteristics of the reaction zone in term of partial thermodynamics parameters to give some informations on the mixture law for an homogeneous explosive.

From statistical definition of ρ, e, η and T, it is easy to define partial parameters $\rho_\alpha, e_\alpha, \eta_\alpha$ and T_α for each molecular or atomic species α so that the total thermodynamics parameters are linear functions of partial parameters. The components of a chemical equilibrium system are all at the same pressure and same temperature. On the other hand, the reaction zone is by definition in non chemical and thermodynamic equilibrium so that parameters such as temperature or pressure cannot be defined as usual. Nevertheless, we have shown that the relaxation time of particle velocities is very small with regard to the overall characteristic time evolution of the system. Thus, it is possible, within a good approximation to suppose thermal equilibrium at any time. In this case, partial temperatures for each chemical species must be the same for a given time. Numerical profiles of partial temperatures confirm this analysis (figure 5) and the reaction zone can be supposed isothermic.

The non chemical equilibrium inside the reaction zone is due to the lack interparticle equilibrium. In this condition; η_α are not equal in the general case and the reaction zone is not an isobaric system. At the end of the reaction zone, the system reachs equilibrium and the mixture must be lead to an isobaric system. Numerical results confirm this analysis (figure 6).

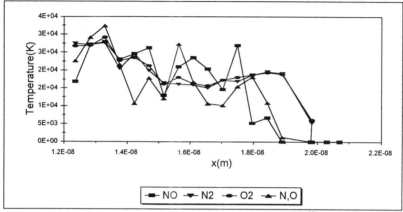

Figure 5. Partial temperatures at $t = 4.956 \times 10^{-12}$ s

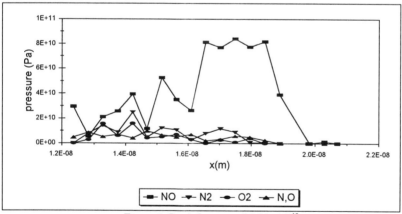

Figure 6. Partial η at $t = 4.956 \times 10^{-12}$ s.

Then, in our case, it is impossible to define an equation of state in the usual way. Nevertheless, the use of Arrhénius law for kinetic formulation is correct because thermal equilibrium is always reached.

CONCLUSIONS

This study gives some informations on the stucture of the reaction zone with a realistic chemical decomposition. The two parts structure of the shock front results from initially endothermic chemical process (NO break). This step must be obtained in more complexe explosives with a probably largest time scale. Behind the shock front, evolution of thermodynamic parameters results from exothermic reactions (O_2 and N_2 formation).

The thermodynamics of reaction zone is complex. An equation of state can 't be use in usual way because the time scale of chemical processes is the time scale of the overall process (non-isobaric system). Nevertheless, the relaxation of particle velocities is very fast and reaches thermal equilibrium at each point of the reaction zone. Then the mixture can be supposed isothermal.

REFERENCES

1.   S.G. Lambrakos, M.Peyrard, E.S. Oran and J.P. Boris, Phys. Rev. B, **39**(2), 993 (1989).
2.   D.H. Robertson, D.W. Brenner and T.C. White, Phys. Rev. Lett., **67** (22), 3132 (1991).
3.   J.J. Erpenbeck, Phys. Rev. A, **46**(10), 6406 (1992).
4.   M.E.Elert, D.M. Deaven, P.W. Brenner and T.C. White,Phys. Rev. B, **39**(2), 1453 (1989).
5.   L.Soulard, in : *Approches microscopiques et macroscopiques des détonations*, St. Malo (1994).

SIMULATIONS OF OZONE DETONATION USING A REACTIVE EMPIRICAL BOND ORDER (REBO) POTENTIAL FOR THE OXYGEN SYSTEM

J. J. C. Barrett, D. H. Robertson,[†] D. W. Brenner, and C. T. White
Chemistry Division, Naval Research Laboratory, Washington, DC 20375–5320
[†] Department of Chemistry, University of Indiana - Purdue University at Indianapolis, Indianapolis, IN 46202

ABSTRACT

The short length and time scales associated with chemical detonations make these processes excellent candidates for study by MD simulation. Potentials used in these simulations must have sufficient flexibility to describe gas-phase properties of isolated reactant and product molecules, high density material generated under shock compression, and allow smooth adjustment of bonding forces during chemical reaction. The REBO formalism has been shown to provide these characteristics and allow the treatment of a sufficient number atoms for sufficiently long times to demonstrate a chemically-sustained shock wave (CSSW). In this paper we present a REBO potential describing the oxygen system for use in MD simulation of detonation in an ozone molecular solid. The potential reproduces spectroscopic properties of isolated gas-phase O_2 and O_3. It also describes an ozone molecular solid with density and speed of sound within physical norms. We observe detonation characteristics that depend on crystallographic orientation in simulations using a three dimensional ozone molecular crystal.

INTRODUCTION

In recent years this group has been exploring shock wave processes and shock-induced chemistry using molecular dynamics (MD) simulation.[1-4] To accurately describe the complex chemistry and physics in these large systems it was necessary to develop potential energy functions capable of calculating the forces between the thousands of atoms required to model these processes within a reasonable time frame. In addition the potential must be sophisticated enough to incorporate the essential physics of the problem. That is give reasonable bonding properties for the molecules involved and a realistic description of the chemistry which occurs during a detonation. In this regard we have developed a series of REBO potentials for diatomic molecular solids.[1-3] We have shown that MD simulations using these REBO potentials quantitatively describe such complex hydrodynamic phenomena as shock wave splitting due to polymorphic phase transitions in some of these diatomic molecular models.[2,5]

We have used the REBO formalism to describe a model diatomic energetic material.[2,3] This AB molecular solid has been shown to detonate under appropriate conditions and have detonations properties intrinsic to the model assumed and are consistent with experimental results.[3] We have also adapted the REBO form to describe the complex chemistry which occurs in a wide variety of hydrocarbons (alkanes, alkenes, alkynes, and radicals).[6] This potential has been used to simulate reactions which occur during hypervelocity impacts of C_{60} with diamond surfaces,[7] the initial stages of C_{60} formation,[8] and nanoindentation, friction, and adhesion experiments.[9-15]

Recently we have been applying the REBO formalism to describe shock induced chemistry in systems more complex than the simple diatomic models discussed earlier. In this work we model detonation chemistry in an ozone molecular solid.[16] The REBO potential

Mat. Res. Soc. Symp. Proc. Vol. 418 © 1996 Materials Research Society

used in these studies gives an accurate description of the isolated O_2 and O_3 molecules, it also gives a stable crystal structure for a three dimensional (3D) O_3 molecular solid. In this paper we present a REBO potential for the oxygen system and results of our simulations of shock induced chemistry using a 3D ozone molecular crystal.

METHODS

In developing this potential our goals are to provided an accurate description of the individual molecules as well as the crystalline solids. We also require a potential that gives a realistic description of the bonding forces between atom pairs during reaction while being simple enough to treat a large enough number of molecules for sufficiently long times to model shock induced processes. We accomplish these goals by using the REBO form employed in our earlier studies of diatomic systems. The general form for the total potential energy, V_{tot}, is given by

$$V_{tot} = \frac{1}{2}(V_{bond} + V_{vdW}),$$
(1)

where V_{bond} is the covalent and V_{vdW} is the van der Waals contribution to the potential energy.[17] The van der Waals potential (vdW) is given by

$$V_{vdW} = \sum_{i,j\neq i} V(R_{ij}),$$
(2)

with R_{ij} the distance between atoms i and j. The function $V(R_{ij})$ by is given by either

$$V(R_{ij}) = \varepsilon \left\{ \left(\frac{\sigma}{R_{ij}}\right)^{12} - \left(\frac{\sigma}{R_{ij}}\right)^6 \right\}$$
(3a)

or

$$V(R_{ij}) = \sum_{n=0}^{3} P_n \cdot R_{ij}{}^n,$$
(3b)

depending on the distance R_{ij}. The Lennard-Jones (L-J) parameters, of eq. (3a), are taken from earlier work on solid O_2,[18,19] while eq. (3b) allows for a smooth transition from the vdW to the covalent potential and truncation of the long-range attractive forces at computationally tractable distances.

The covalent energy function is written as

$$V_{bond} = \sum_{i,j\neq i} \left[V_r(R_{ij}) - \beta_{ij} \cdot V_a(R_{ij})\right] F(R_{ij}),$$
(4)

where V_r and V_a represent the repulsive and attractive parts of V_{bond}, respectively. Equation (4) is patterned after a form originally introduced by Tersoff to describe static properties of silicon.[20,21] The term β_{ij} in eq. (4) represents a many-body coupling which modifies the bonding strength between atom pair ij depending on the local bonding environment of each atom. The many-body coupling can be tailored to describe highly coordinated metals,[22] tetrahedrally bonded semiconductors[23,24] or low coordination molecular solids with a few strong bonds.[1-3] Herein, the parameters and functions entering eq. (4) are chosen to allow a smooth evolution of the bonding characteristics of oxygen in going from molecular O_3 to O_2 during chemical reaction with further perturbation to bonding character as the system goes to a high density material under shock compression.

The bond order function β_{ij} is taken as

$$\beta_{ij} = \left[1 + \sum_{k \neq i,j} F_i(R_{ik}) \cdot G_i(\theta_{kij})\right]^{-\frac{1}{2}}$$
$$+ \left[1 + \sum_{l \neq i,j} F_j(R_{jl}) \cdot G_j(\theta_{ijl})\right]^{-\frac{1}{2}} \tag{5}$$
$$+ 2\Re(F_i, F_j),$$

where θ_{ijk} is the angle formed by atoms i, j, k. The cutoff function, F, truncates the long range attractive tail of the covalent bonding function, while G provides bond angle functionality. The $\Re$ term is a bicubic spline as a function of F_i and F_j which preferentially stabilizes cyclic structures of O_3 and O_4. The value of $\Re$ is smoothly interpolated from a value of 0.0 for isolated O_2 and O_3 to -0.6 for highly coordinated oxygen.

Table I. Spectroscopic properties of O_2 and O_3 calculated with this REBO potential compared to experimental and quantum calculations.

Molecule	REBO	Experimental[a]	
O_2	5.2140	5.2026	
O_3	6.2744	6.3432	
Constant	**REBO**	**Experimental**	**Theory**
$R_e(O_2)$	1.2070	1.208[a]	1.2073[b]
$R_e(O_3)$	1.2803	1.271[c]	1.2717[b]
θ_{O-O-O}	116.73	116.8[c]	116.78[b]
Vibrational Mode	**REBO[d]**	**Experimental[e]**	**Theory[b]**
O_2	1499.5	1580.4	
O_3 - symm. stretch	1247.9	1103	1098
O_3 - asym. stretch	1229.3	1042	1043
O_3 - bend	802.3	701	707

a. reference 29; *b.* reference 28; *c.* reference 26; *d.* from harmonic force constants; *e.* reference 24

The function F, which enters into eqs. (4) and (5), is given by

$$F(R_{ij}) = \sum_{l=0}^{3} P_l \cdot R_{ij}{}^l, \tag{6}$$

where coefficients, P_l, depend on the distance R_{ij}. The function G is a polynomial expansion in $cos(\theta)$, given by

$$G(\theta_{ijk}) = \sum_{m=0}^{3} P_m \cdot cos(\theta_{ijk})^m, \tag{7}$$

where coefficients, P_m, depend on the angle, θ_{ijk}. The attractive and repulsive parts of the covalent potential are given by

$$V_a(R_{ij}) = S \frac{D_e}{2(S-1)} exp\left\{-\alpha\sqrt{\frac{2}{S}}(R_{ij} - R_e)\right\} \tag{8}$$

and

$$V_r(R_{ij}) = \frac{D_e}{(S-1)} exp\left\{-\alpha\sqrt{2S}\ (R_{ij} - R_e)\right\}. \qquad (9)$$

The parameters D_e, R_e and α are the usual Morse parameters, while S introduces additional flexibility into the Morse-type function.[20,21]

In Table I we show a comparison of spectroscopic properties for O_2 and O_3 calculated using this REBO potential with experimental[25-28] and quantum[29] results. Total binding energies, bonds lengths, and the O_3 bond angle agree to within 1%. The vibrational frequencies (to harmonic approximation) agree to within 15% of the experimental values. The stable 3D ozone crystal structure that we obtain using this potential is shown in figure 1 and the crystallographic parameters are given in the caption. The density for this crystal is 1.559 g/cc which is in reasonable agreement with the density of 1.614 g/cc for liquid ozone at 78 K.[30.]

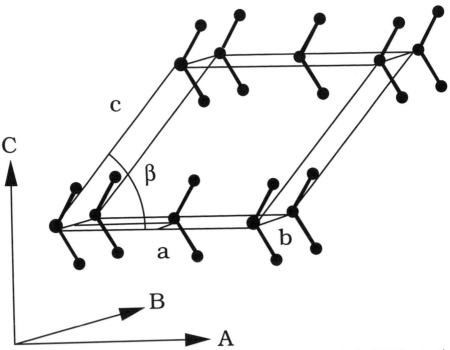

Figure 1. Three dimensional ozone crystal as determined with the REBO potential described herein. a = 6.1845Å; b = 3.5706 Å; c = 6.1993 Å; β = 48.312°

The MD simulations are carried out using a standard[31] third-order predictor-corrector algorithm with variable step size to solve Hamilton's equations of motion for a system of N-oxygen atoms. Periodic boundary conditions are applied to the coordinates perpendicular to the direction of shock-front propagation to simulate an crystal of infinite extent. For a detonating system with 17 ps simulation time N is approximately 10,000.

RESULTS

The initial setup for a simulation initiated with a flyer plate along the C-axis is shown in figure 2. The flyer plate is composed of crystalline O_3 and is given an initial velocity, v_{fp}, directed toward the crystal. It is found that for an eight layer flyer plate the minimum v_{fp} required to initiate a detonation along this axis is 4.0 km/s. The overall exothermic reaction driving this detonation is given by

$$2O_3 \rightarrow 3O_2 \quad , \quad \Delta H = -298.0\frac{\text{kJ}}{\text{mol}}, \tag{10}$$

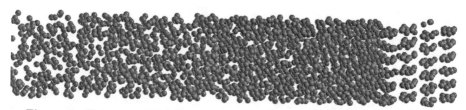

Figure 2. Upper panel: Initial setup for 3D ozone shock simulation with flyer plate. Lower panel: 3D ozone system under detonation conditions, simulation time, 17 ps.

In the lower panel of figure 2 we show the system initiated in this manner after 17 ps simulations time. The high density region at the detonation front is clearly seen in this figure. The figure also shows that as we move further behind the front the particle density begins to decrease. This is quantified, along with several other detonation properties in the figure 3.

Under steady-flow conditions, in the reference frame of a stationary shock front, the Rankine-Hugoniot relation for the conservation of mass requires that the quantity

$$\bar{\rho} \equiv \rho u, \tag{11}$$

where ρ is the density and u the local particle velocity, be conserved across a planar shock front.[32] In figure 3a we plot $\bar{\rho}$ relative to shock front position for simulation time of 17 ps. The degree to which $\bar{\rho}$ equals its value in the unshocked material shows how close the simulation is approaching steady-flow with an unchanging profile in that region. Figure 3a shows that the CSSW in this system stabilizes at the front reaching near steady-flow conditions up to 25 nm behind the front in as little as 17 ps. Figure 3 also shows that the density, ρ, (figure 3b) and local particle velocity along the crystal axis, v_x, (figure 3c) rise sharply at the shock front, plateau, then fall as O_3 reacts and detonation products expand.

305

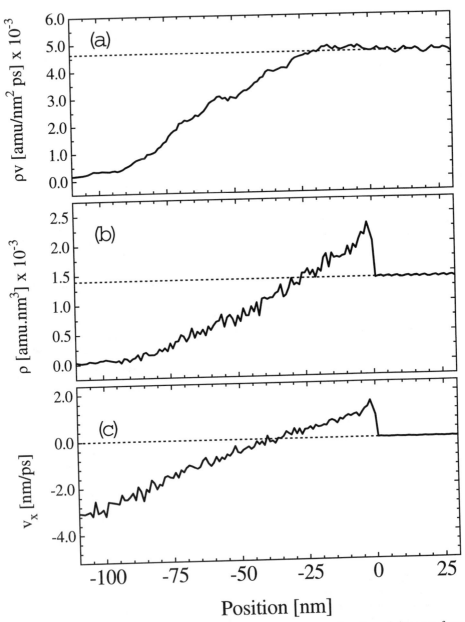

Figure 3. Detonation properties for ozone detonation as a function of distance from shock front. (a) $\bar{\rho}$, (b) denisty, ρ, (c) velocity component in the direction of shock front propagation, v_x.

Attempting to initiate a detonation along the B-axis reveals an orientational effect in the detonations properties of this system. Using the same conditions that initiated a detonation along the C-axis fails to initiate a detonation in the system along the B-axis. This is illustrated by plotting the wave front as a function of time for both cases in figure 4. In the case for flyer plate impact along the C-axis, after a short induction period the wave front propagates down the crystal at a constant velocity, (detonation velocity of 3.5 km/s) indicating a sustained detonation. Whereas the plot for the impact along the B-axis shows a downward curvature, indicating that initiation has failed. This result may be interpreted as an indication that the mechanism which couples the shock wave energy to internal modes of the O_3 molecule are more efficient along the C-axis than the B-axis resulting in a lower initiation threshold for the C-axis.

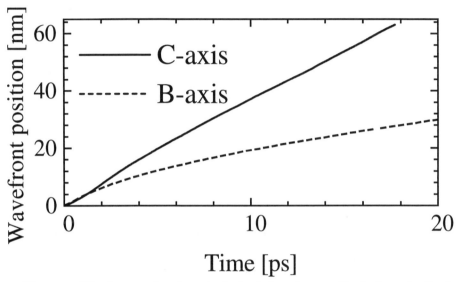

Figure 4. Wavefront vs time from shock simulation of ozone. Shock along the C-axis initiates a detonation in the system, while shock along the B-axis under the same conditions fails to initiate a detonation.

CONCLUSIONS

In this paper we present a potential energy surface for the oxygen system within the reactive empirical bond order (REBO) formalism. We have shown that this potential has sufficient flexibility to reproduce spectroscopic and structural characteristics for individual, gas-phase O_2 and O_3 molecules. In addition the REBO potential allows for condensation of the system into a molecular solid giving a 3D ozone molecular crystal with a density in agreement with literature values. Simulations with the 3D ozone solid show a threshold to detonation behavior which is seen in experimental studies of energetic materials. The detonation velocity in the 3D system along the C-axis is determined to be 3.5 km/s. In addition we show a crystallographic orientational dependence to the initiation threshold.

ACKNOWLEDGEMENTS

This work was supported in part by ONR through NRL and ONR contract #N00014-96-WX-20330. We also thank ONR and NRL for grants of computer resources. JJCB acknowledges a NRL-NRC Postdoctoral Research Associateship.

REFERNCES

1. D. W. Brenner, C. T. White, M. L. Elert, and F. E. Walker, *Int. J. Quant. Chem.: Quant. Chem. Symp.*, **23**, 333 (1989)
2. C. T. White, D. H. Robertson, M. L. Elert, and D. W. Brenner, *Microscopic Simulations of Hydrodynamic Phenomena*, eds. M. Mareschal and B. L. Holian, Plenum Press, New York, p. 111 (1992)
3. D. W. Brenner, D. H. Robertson, M. L. Elert, and C. T. White, *Phys. Rev. Lett.*, **70**, 2174 (1993)
4. M. L. Elert, D. M. Deaven, D. W. Brenner, and C. T. White, *Phys. Rev. B*, **39**, 1453 (1989)
5. D. H. Robertson, D. W. Brenner, and C. T. White, *Phys. Rev. Lett.*, **67**, 3132 (1991)
6. D. W. Brenner, *Phys. Rev. B*, **42**, 9458 (1990)
7. R. C. Mowrey, D. W. Brenner, J. W. Mintmire, and C. T. White, *J. Phys. Chem.*, **95**, 7138 (1991)
8. D. H. Robertson, D. W. Brenner, and C. T. White, *J. Phys. Chem.*, **96**, 6133 (1992)
9. J. A. Harrison, C. T. White, R. J. Colton, and D. W. Brenner, *Surf. Sci.*, **271**, 57 (1992)
10. *ibid.*, *Phys. Rev. B*, **46**, 9700 (1992)
11. *ibid.*, *MRS Bulletin*, **18**, 50 (1993)
12. *ibid.*, *J. Phys. Chem.*, **97**, 6573 (1993)
13. *ibid.*, *Wear*, **168**, 127 (1993)
14. S. B. Sinnott, R. J. Colton, C. T. White, and D. W. Brenner, *Surface Science*, **316**, 1055 (1994)
15. J. A. Harrison and D. W. Brenner, *J. Am. Chem. Soc.*, **116**, 10399 (1994)
16. J. J. C. Barrett, D. H. Robertson, D. W. Brenner, and C. T. White, *Phys. Rev. B*, submitted.
17. See authors for details on the parameterization of the potential.
18. K. Kobashi, M. L. Kline, and V. Chandrasekharan, *J. Chem. Phys.*, **71**, 843 (1979)
19. J. C. Lauffer and G. E. Leroi, *J. Chem. Phys.*, **55**, 993 (1971)
20. J. Tersoff, *Phys. Rev. B*, **37**, 6991 (1988)
21. J. Tersoff, *Phys. Rev. Lett.*, **56**, 632 (1986)
22. G. C. Abell, *Phys. Rev. B*, **31**, 6184 (1985)
23. J. Tersoff, *Phys. Rev. B*, **39**, 5566 (1989)
24. T. Ito, K. E. Khor, and S. Das Sharma, *Phys. Rev. B*, **41**, 3893 (1990)
25. T. Tanaka and Y. Morino, *J. Mol. Spectrosc.*, **33**, 538 (1970)
26. R. H. Hughes, *J. Chem. Phys.*, **24**, 131 (1956)
27. A. Barbé, C. Secroun, and P. Jouve, *J. Mol. Spectrosc.*, **49**, 171 (1974)
28. *CRC Handbook of Chemistry and Physics*, 64th ed., p. F-173, CRC Press, Boca Raton, FL, (1983)
29. J. N. Murrell, K. S. Sorbie, and A. J. C. Varandas, *Mol. Phys.*, **32**, 1359 (1976)
30. *CRC Handbook of Chemistry and Physics*, 64th ed., p. B-118, CRC Press, Boca Raton, FL, (1983)
31. C. W. Gear, *Numerical Initial Value Problems in Ordinary Differential Equations* Prentice Hall, Englewood Cliffs, NJ, (1971).
32. H. Eyring, R. E. Powell, G. H. Duffey, and R. B. Parlin, *Chem. Revs.*, **45**, 69 (1949)

MOLECULAR DYNAMICS STUDY OF THE EFFECT OF VARYING EXOTHERMICITY ON THE PROPERTIES OF CONDENSED-PHASE DETONATION

M. L. ELERT *, D. H. ROBERTSON **, and C. T. WHITE ***
*Chemistry Dept., U. S. Naval Academy, Annapolis, MD 21402, elert@nadn.navy.mil
**Dept. of Chemistry, Indiana U. - Purdue U., Indianapolis, IN 46202
***Chemistry Division, Code 6179, Naval Research Laboratory, Washington DC 20375

ABSTRACT

To investigate the role of exothermicity on the properties of a chemically sustained shock waves, a series of two-dimensional molecular dynamics simulations was carried out in which the exothermicity was systematically varied. The simulations were based on a model diatomic system which has been previously shown to produce reasonable values for shock wave properties. A decrease of 33% in the amount of energy released in the reaction produced a significant decrease in detonation front velocity and an increase in the impact energy necessary to initiate a sustained shock wave. Redistribution of energy between the reaction products at constant total exothermicity had a much smaller effect on the properties of the detonation front.

INTRODUCTION

Many factors influence the properties of a chemically sustained shock wave. It is well known, for example, that increased density in condensed-phase energetic materials is correlated with increased detonation velocity. It is often difficult, however, to separate the effects of density, exothermicity, reaction mechanism, and other factors when comparing a series of energetic materials, since each will differ from the others in many respects. With computer simulation, it is possible to design a series of model systems in which only a single aspect of the model is varied at a time. We report here a series of two-dimensional molecular dynamics simulations in which the amount of energy released in a model system, and the distribution of energy among the products, is varied. Effects on the initiation threshold and on the speed, peak density and temperature of the shock wave are reported.

MODEL

In this study we employ variants of a model diatomic system which has been previously shown [1] to produce stable chemically-sustained shock waves with properties consistent with those of typical condensed-phase energetic materials. The model employs a reactive empirical bond-order (REBO) potential of the type used by Tersoff in his study of silicon [2] and by Robertson et al. [3] in their study of shock-induced phase transitions. The REBO formalism efficiently incorporates many-body effects, allowing for realistic simulation of bond-forming and bond-breaking processes. In the model diatomic system used previously, two generic atom types A and B are allowed to form stable diatomic species. The reactant diatomic molecule AB is given a bond energy of 2.0 eV and the product molecules A_2 and B_2 each have a bond energy of 5.0 eV, so that the overall reaction $2AB$ $\rightarrow$ $A_2 + B_2$ releases 6.0 eV. This exothermic reaction provides the energy necessary to sustain the shock front. Details of the potential are given elsewhere.[1]

309

To perform the simulations, a crystal of the AB reactant was set up in the xy plane, and a "flyer plate" of the same material was given a large initial velocity in the positive x direction and allowed to impact the crystal from the left. In all of the simulations reported here, the flyer plate had a thickness of three molecular layers. Periodic boundary conditions were applied in the y direction, so that the model simulates the behavior of an infinitely wide system. Particle dynamics were followed using a third-order predictor-corrector method.

RESULTS

Reduced Exothermicity

In the original model of ref [1], the generic atom types A and B were chemically equivalent, in the sense that A_2 and B_2 had exactly the same bond strength. To adjust the exothermicity of the reaction, the bond energy of B_2 was changed in this study from 5.0 to 3.0 eV. Other parameters of the potential function were modified slightly in order to minimize the effect of the change in bond energy on other features of the interaction, particularly the barrier heights for three-body reactions. In the original model, the barrier height for both of the collinear three-body reactions involving AB as a reactant was .071 eV. The revised potential produced a barrier height of .083 eV for the reaction $AB + B \rightarrow A + B_2$, and .012 eV for $A + AB \rightarrow A_2 + B$. Although the latter represents a large percentage change, this barrier is already so low that, at the temperatures encountered in the reaction zone, the change in reaction dynamics upon lowering the barrier still further is negligible. The only significant change in the revised model, therefore, is that the energy change ΔE for the three-body reaction $AB + B \rightarrow A + B_2$ changes from -3.0 eV to -1.0 eV.

The decrease in energy release had significant effects on the properties of the detonating system. In the original model, with the flyer plate thickness employed here, the minimum flyer plate impact velocity which would produce a sustained shock front was 5.2 km/sec. In the revised model, the threshold increased to 6.1 km/sec. As shown in figure 1, a lower impact velocity produced some initial chemical reaction and a propagating shock front, but the energy release was too low to sustain the reaction. Under these circumstances chemical reactions just behind the front gradually ceased, and the shock front slowed and dissipated. Above the initiation threshold, other properties of the shock front were also affected by the change in exothermicity. The shock front velocity decreased from 9.5 km/sec to 8.3 km/sec, the temperature in the reaction zone just behind the front decreased from about 9000 K to 6500 K, and, as shown in figure 2, the density in the reaction zone dropped by about 50 amu/nm^2.

An Endothermic Reaction

In the model just discussed, both of the three-body reactions involving AB were still exothermic. To investigate the possible role of an endothermic reaction in modifying the reaction dynamics, a third model was developed in which the bond energies of A_2 and B_2 were 7.0 and 1.0 eV, respectively. The bond energy of AB remained fixed at 2.0 eV, so in this model the three-body reaction $AB + B \rightarrow A + B_2$ was endothermic by 1.0 eV. The energy release in the overall reaction $2AB \rightarrow A_2 + B_2$, however, was still 4.0 eV as in the previous model. This ensured that any change in shock front properties could be attributed to changes in the distribution of energy among product molecules rather than a difference in the total energy released. Once again, care was taken to minimize changes in three-body barrier heights. There was of course a large,

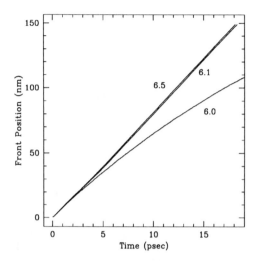

Figure 1. Shock front position vs. time for the revised model with an exothermicity of 4.0 eV. The two upper curves show a constant shock front velocity, while the lower curve represents a case in which the detonation fails to propagate. Each curve is labeled with the flyer plate velocity used to initiate the shock wave, in km/sec.

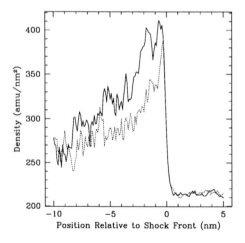

Figure 2. Density vs. distance from the shock front for the original model (solid line) and for the revised model with an exothermicity of 4.0 eV (dashed line). Both curves were calculated from atomic positions about 10 psec after the start of the simulation.

qualitative change in the nature of the AB + B → A + B$_2$ reaction; from an exothermic reaction with a barrier height of less than 0.1 eV, it became an endothermic reaction with an energy barrier of more than 1.0 eV.

This rather drastic change in the nature of the reaction dynamics had relatively little effect on the properties of the detonation. Compared to the previous reduced-exothermicity model, the initiation threshold decreased from 6.1 to 5.8 km/sec, and there was a corresponding rise in the detonation velocity from 8.3 to 8.6 km/sec. Density and temperature profiles were largely unchanged. Perhaps as a result of the extremely simple kinetics of this model system, total energy released was the only significant factor in determining the properties of the chemically sustained shock wave.

CONCLUSIONS

The preliminary results presented here demonstrate that molecular dynamics simulations can be used to probe the effect of changes in specific properties of a reactant energetic material on the features of the detonation front. In particular, the amount of energy released in a shock-induced reaction can be varied without affecting other properties of the system, such as density and reaction mechanism. As expected, a decrease in exothermicity resulted in a decreased shock velocity, lower peak density and temperature, and an increase in impact threshold. It would be of interest to lower the energy release in this model system further to find the point at which no flyer plate speed will be sufficient to initiate sustained detonation. This would afford the opportunity to study the initiation of detonation in a regime where the formation of a chemically sustained shock wave was barely supported; under these circumstances, the effects of impurities, voids, and other perturbations to the reactant crystal might be magnified.

ACKNOWLEDGMENTS

This work was supported in part by the Office of Naval Research (ONR) through the Naval Research Laboratory and through ONR contract number N00014-96-WX-20330. MLE acknowledges support from the NRL/USNA Cooperative Program for Scientific Interchange.

REFERENCES

1. D. W. Brenner, D. H. Robertson, M. L. Elert, and C. T. White, Phys. Rev. Lett. **70**, 2174 (1993).

2. J. Tersoff, Phys. Rev. Lett. **56**, 632 (1986); Phys. Rev. B **37**, 6991 (1988).

3. D. H. Robertson, D. W. Brenner, and C. T. White, Phys. Rev. Lett. **67**, 3132 (1991).

REACTION RATE MODELING IN THE DEFLAGRATION TO DETONATION TRANSITION OF GRANULAR ENERGETIC MATERIALS

S. F. SON*, B. W ASAY, J. B. BDZIL, and E. M. KOBER
Los Alamos National Laboratory, Los Alamos, NM 87545, *son@lanl.gov

ABSTRACT

The problem of accidental initiation of detonation in granular material has been the initial focus of the Los Alamos explosives safety program. Preexisting models of deflagration-to-detonation transition (DDT) in granular explosives, especially the Baer & Nunziato (BN) model, have been examined. The main focus of this paper is the reaction rate model. Comparison with experiments are made using the BN rate model. Many features are replicated by the simulations. However, some qualitative features, such as inert plug formation in DDT tube-test experiments and other trends, are not produced in the simulations. By modifying the reaction rate model we show inert plug formation that more closely replicates the qualitative features of experimental observations. Additional improvements to the rate modeling are suggested.

INTRODUCTION

Weak shock and thermal insults, rather than strong shocks, are more likely accident scenarios in which energetic materials undergo detonation. The possible paths that lead to detonation involve a variety of processes ranging from relatively slow, subsonic processes such as fracture, heat conduction, compaction and deflagration to fast, supersonic processes such as shock waves and detonation. Consequently, DDT events encompass a wide variety of temporal and spatial scales not found in shock initiation.

Various models have been used in the study of shock initiation. In contrast to DDT events, shock initiation of detonation is nearly a single process phenomenon. As a result, models were developed and calibrated for a narrow range of conditions. Generally, these models were built around a variety of equation of state assumptions with discrepancies in the description absorbed into the reaction rate model. Multiphase and energy localization effects were not considered explicitly. Consequently, these models have difficulty predicting complex initiation involving ramped waves, multiple shocks, or weak impacts. This leads one to adopt a different modeling philosophy than the one commonly used to formulate shock-initiation models to describe DDT.

Accidental initiation of detonation in granular materials is the initial focus of this program. Granular materials can be well-characterized and are assumed to simulate damaged explosives or propellants. A considerable amount of work considered the DDT of granular materials during the 1980s [1-7]. In the area of theory and modeling, the work of Baer and Nunziato has received the most thorough development and represents the current status of DDT modeling [5, 6]. The objective of this paper is to evaluate this model, explore some improvements, and suggest future work in this area. Many observed qualitative features are replicated by the simulations. However, some qualitative features, such as inert plug formation in DDT tube-test experiments and other trends, are clearly not produced in the simulations. Using modified rate models we show inert plug formation that more closely replicates the qualitative features of experimental observations.

MODEL

High pressure permeability measurements of inert materials similar to granular explosives were made recently [8]. The extremely low measured permeabilities motivated the development of a new one-velocity (large-drag limit), two-phase continuum mixture model [9-11]. This model is both computationally more efficient and easier to implement in a multi-dimensional, multi-material hydrocodes and was used to evaluate current rate models against experimental data.

Large Drag Model

The basic hydrodynamic equations of our two-phase, large-drag model consists of separate phase mass and energy conservation equations, a single mixture momentum equation, and an equation for the volume fraction [5, 13]. The theory is based on the continuum theory of mixtures. The equations are

solid mass

$$\frac{D}{Dt}(\phi_s\rho_s)+(\phi_s\rho_s)\vec{\nabla}\cdot\vec{u}=C,$$ (1)

gas mass

$$\frac{D}{Dt}(\phi_g\rho_g)+(\phi_g\rho_g)\vec{\nabla}\cdot\vec{u}=-C,$$ (2)

momentum

$$\frac{D\vec{u}}{Dt}+\frac{1}{\rho}\vec{\nabla}p=0,$$ (3)

solid energy

$$(\phi_s e_s)\frac{De_s}{Dt}+(\phi_s e_s)\vec{\nabla}\cdot\vec{u}=\mathcal{E},$$ (4)

gas energy

$$(\phi_g\rho_g)\frac{De_g}{Dt}+(\phi_g\rho_g)\vec{\nabla}\cdot\vec{u}=-\mathcal{E}+(e_g-e_s)C,$$ (5)

volume fraction

$$\frac{D\phi_s}{Dt}=\mathcal{F}+C/\rho_s.$$ (6)

The saturation condition is $\phi_s+\phi_g=1$. Subscripts s and g refer to the solid and gas phase, respectively. The Lagrangian derivative is $D/Dt\equiv\partial/\partial t+\vec{u}\cdot\vec{\nabla}$. The state variables are density (ρ_i), specific internal energy (e_i), pressure (p_i), temperature (T_i), and volume fraction (ϕ_i) where i is s or g. The mixture variables are the particle velocity ($\vec{u}=(\phi_s\rho_s\vec{u}_s+\phi_g\rho_g\vec{u}_g)$), the density ($\rho=\phi_s\rho_s+\phi_g\rho_g$) and the pressure ($p=\phi_s p_s+\phi_g p_g$).

We have analyzed the structure of two-phase, two-velocity shocks in the limit of large drag [11]. Large O(1) differences in the velocity are principally found in these zones that comprises a two-phase shock. Elsewhere in the flow, velocity differences are small and effectively approximated by a combined pressure-density term which follows from the asymptotic limit of large drag. Although thin, this drag-relaxtion zone does not integrate to a simple jump condition; a structure problem needs to be solved. Based on these results, we have developed the following regularization for shocks [10]

$$p_g\Rightarrow p_g+\left(\frac{v_r\rho_g}{\rho+\phi_g\rho_g(v_r-1)}\right)Q,$$ (7)

where the regularization constant, v_r, sets the partition of energy between the solid and gas. Q is the viscous pressure for the mixture, taken as the artificial viscosity used by the hydrocode. We find that the reactive simulations are insensitive to the value of v_r assumed, especially when the gas density is low.

The Hayes equation of state (EOS) is assumed for the solid (*e.g.*, [5])

$$T_s(e_s, \rho_s) = T_{s0} + \frac{e_s}{C_{vs}} + \frac{1}{C_{vs}}\left((t_3 - p_{s0}/\rho_{s0})(1 - v_s) + G(v_s, t_4)\right), \tag{8}$$

$$p_s(T_s, \rho_s) = p_{s0} + (T_s - T_{s0})C_{vs}g_v + \rho_{s0}\frac{dG}{dv_s}, \tag{9}$$

where $G(v_s, t_4) \equiv -t_4\left[v_s^{1-n} - (n-1)(1-v_s) - 1\right]$ and $v_s \equiv \rho_{s0}/\rho_s$. The Jones-Wilkins-Lee (JWL) EOS is assumed for the gas (e.g., [5])

$$T_g(e_g, \rho_g) = T_{g0} + \frac{1}{C_{vg}}(e_g + \Delta H) - \frac{1}{C_{vg}\phi_{s0}\rho_{s0}}\left(\frac{A}{R_1} + \frac{B}{R_2}\right), \tag{10}$$

$$p_g(T_g, \rho_g) = T_g\omega C_{vg}\rho_g + \mathcal{A} + \mathcal{B}, \tag{11}$$

where $\mathcal{A}(\rho_g) \equiv A\exp(-R_1\phi_{s0}\rho_{s0}/\rho_g)$ and $\mathcal{B}(\rho_g) \equiv B\exp(-R_2\phi_{s0}\rho_{s0}/\rho_g)$.

<u>Baer & Nunziato Model for the Energy and Compaction Interaction Terms</u>

In this section the models for the interaction terms used by BN [6] for granular HMX are presented. The source terms appearing in Eqs. 1-6 are the rate of interphase energy exchange ($\mathcal{E}$), compaction ($\mathcal{F}$), and volumetric mass exchange ($\mathcal{C}$). In BNs calculations the compaction work term in the energy equation was neglected [12], so

$$\mathcal{E} = \mathcal{H}(T_s - T_g), \tag{12}$$

where $\mathcal{H}$ is the volumetric heat transfer coefficient. Here the heat transfer (Eq. 12) is driven by the difference in bulk temperatures. In the published form (see Eq. 45 in ref. [5]) there is an additional term to partition energy as a result of compaction work. For the compaction work form suggested by Baer & Nunziato much of the compaction work goes into the gas [9, 10, 13]. When the gas has little mass and low heat capacity (typical case) anomalously high temperatures in the gas occur. However, simply neglecting compaction work is not justified. Other modeling choices are available. The volumetric heat transfer coefficient, $\mathcal{H}$, is assumed to have the form, $\mathcal{H} = (3k_H\phi_g)/a_p^2$ where a_p is an effective particle radius, taken to be $a_p = a_{p0}\min(1,(\phi_s/\phi_{s0})^{1/3})$. The compaction law used by BN [12] is

$$\mathcal{F} = \begin{cases} \dfrac{\phi_s\phi_g}{\mu_c}(p_s - \beta - p_g), & p_s - \beta \geq 0 \\[2ex] \dfrac{\phi_s\phi_g}{\mu_c}(-p_g), & p_s - \beta < 0 \end{cases} \tag{13}$$

where μ_c is the compaction viscosity. The intragranular stress or configuration pressure, β, used for HMX is

$$\beta(\phi_s) = \begin{cases} (\phi_s - \phi_{s0})\tau \ln(\phi_g)/\phi_g, & \phi_s - \phi_{s0} \geq 0 \\ 0, & \phi_s - \phi_{s0} < 0 \end{cases} \tag{14}$$

The volumetric mass exchange term, C, will be discussed in the following section.

BAER & NUNZIATO REACTION RATE MODEL

In the framework of the continuum theory of mixtures the smallest volume resolved contains many particles. Since the modeling structure is based on this continuum scale, the reaction rate source term, C, must be introduced as a subscale model. In other words detailed chemistry and full resolution of the temperature variations on the subscale can not be directly incorporated in this structure. Simplified subscale modeling must be introduced. One such model has been proposed by Baer & Nunziato [6]. This model is adopted here as a first step and is evaluated against some experimental results.

The volumetric rate of mass exchange, C, is a model for reaction, modified with switches that attempt to impart the flavor of a compaction supported ignition mechanism via an induction variable, I, and a grain interface temperature function, ζ. The induction time equation is

$$\frac{DI}{Dt} = k_I \left[(p - p_0)/D_p \right]^2 (1 - I), \tag{15}$$

where $0 \leq I \leq 1$. This equation is simply a timer that triggers a change in the amount of gas combustion energy released, ΔH, given by $\Delta H = H_{pyr}, I < 1/2$; $\Delta H = H_{det}, I \geq 1/2$. This is meant to model the transition from very little heat release in the gas to vigorous heat release.

The evolution for the surface temperature function is described by,

$$\frac{D\zeta}{Dt} = \mathcal{H}(T_g - T_{int})/(C_{vs}\phi_s\rho_s), \tag{16}$$

where $\zeta(T_{int}, T_g, T_s)$ is related to the interface temperature T_{int} by

$$T_{int} = \frac{5(\zeta + T_s) + BiT_g}{(5 + Bi)(1 + Bi)} + \frac{Bi\left(T_s - Bi\zeta + \sqrt{Z^2 - |T_g - T_s|^2}\right)}{(1 + Bi)}, \tag{17}$$

where $Z \equiv \max(Bi\zeta, 0) + |T_g - T_s|$. BN use $Bi \equiv 10k_g/k_s$ in calculations [12]. This equation was derived assuming energy exchange by only convective heat transfer between the phases. Heating by compaction and compression was neglected. The interface temperature, T_{int}, is used only to trigger the principal reaction in C,

$$C = -H(T_{int} - T_{ign})\, 3(\phi_s\rho_s) k_p \left(\frac{p_g}{D_p}\right)^{bn} \frac{(1-f)^{2/3}}{a_p} - H(\phi_s - \phi_{s0}) k_h \left[\frac{(p - p_0)}{D_p}\right]^2 \rho_s (1 - \phi_{s0}) f \tag{18}$$

where $f \equiv \max(\phi_s - \phi_{s0}, 0)/(1 - \phi_{s0})$, k_p and k_h are rate constants, and $H(x) = 0, x < 0$; $H(x) = 1, x \geq 0$. The first term is the main reaction term. It is motivated by the steady burning of energetic materials that exhibit regression rates proportional to the pressure

316

raised to a power (Saint Robert's burning-rate law). The second term is a "hot-spot like" term that is triggered by any amount of compaction.

These "kinetics" exhibit two paths to vigorous reaction. The first is that the surface temperature reaches the ignition temperature first, triggering the vigorous reaction term which raises the pressure causing the induction timer to quickly advance past the criteria ($I > 0.5$). The second (most common mode in tube test experiments) is that the induction timer, I, advances past the critical value ($I > 0.5$) and the gas energy state changes. Often this change in energy state is sufficient to raise the surface temperature above the critical temperature because of the increase in gas temperature and again vigorous reaction is achieved. However, in some low speed piston situations the change in energy state is not sufficient to drive the surface temperature above the critical value and only the weak part of the reaction occurs (see ref. [6]).

Table 1. Default Parameters ($\phi_{s0} = 0.73$, HMX)

Symbol	Value	Symbol	Value	Symbol	Value
ΔH_{det}	7.796×10^6 J / kg	D_p	100 MPa	n	9.8
ΔH_{pyr}	2.30×10^6 J / kg	C_{vs}	1500 J / kg K	R_1	4.2
k_H	0.094 W / m K	C_{vg}	2380 J / kg K	R_2	1.0
k_s	0.423 W / m K	k_h	100 s^{-1}	ω	0.25
t_3	4.97×10^5 J / kg	k_I	2×10^4 s^{-1}	a_{p0}	$75 \, \mu m$
t_3	8.24×10^4 J / kg	a_t	13.5 GPa	v_r	1.0
ρ_{s0}	1900 kg / m^3	τ	12 MPa	P_{g0}, P_{s0}	10^5 Pa
g_v	2100 kg / m^3	μ_c	100 kg / m s	T_{g0}, T_{s0}	300 K
A	270 GPa	B	−496 MPa	T_{ign}	525 K
k_p	$\begin{cases} 0.1227 \text{ m / s, } p_g < 69 \text{ MPa} \\ 0.1139 \text{ m / s, } p_g \geq 69 \text{ MPa} \end{cases}$	b_n	$\begin{cases} 1.0, \, p_g < 69 \text{ MPa} \\ 0.8, \, p_g \geq 69 \text{ MPa} \end{cases}$		

The parameter values used by BN to model coarse grain granular HMX (class D) at $\phi_{s0} = 0.73$ are given in Table 1. Using these parameters, all qualitative features were replicated and fair quantitative comparison with BN's published 1D calculations [6] were obtained. The ignition locus differed by a maximum of $4 \, \mu s$ through a $60 \, \mu s$ simulation. Quantitative agreement was found between MESA and 1D MacCormack's method based codes integrating either the large-drag limit equations or the full BN equations (2-velocity).

TUBE TESTS

The DDT behavior of various granular explosives has been studied using x-radiograph and various pin techniques [1, 2]. In these tube tests the ignition of the bed is accomplished by driving a piston into the bed. A schematic of such an experiment is shown in Fig. 1. In some of the experiments thin lead foils were placed in the granular bed and x-radiograph snapshots were taken to measure the density profile at various times. Typical radiographs are shown in Fig. 2. The figure shows radiographs taken at four times. Careful measurement of the foil locations determine the density profile of the granular HMX at these times. The static (a) is an exposure showing the position of the equally spaced lead foils before the HMX in the burn chamber is lit. The first dynamic radiograph (b) shows the piston moving into the tube and the first indications of the initial compaction wave. The second dynamic radiograph (c) was taken after the transition to detonation has occurred (the detonation wave is between foils 8 and 9). At this instant in time, foils 3-6 are closely spaced and the tube wall is less expanded indicating a high density region with little reaction (a "plug"). The final radiograph shown (d) shows similar characteristics, except the detonation front is between foils 11 and 12. The foils appear to remain flat at all times.

We have simulated 1D tube-test experiments using both the two-velocity with large-drag models to further justify the use of the large-drag model [10]. Two-velocity simulations were found to compare favorably with large-drag model simulations when realistic drag coefficients are used.

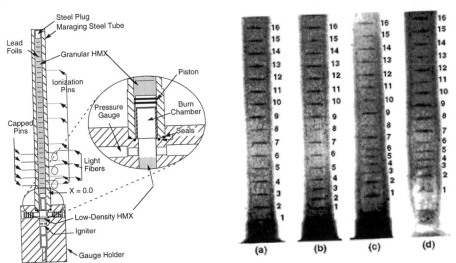

Fig. 1   Schematic of tube test experiment. This figure is taken from ref. [2].

Fig. 2   Flash radiographs (30 *ns* exposure) of shot B9153. This figure is taken from ref. [2].

2D Steel Tube Simulations

The parameters listed in Table 1 were calibrated using a 1D model to an experiment of Sandusky & Bernecker [1]. 2D simulations of this tube test experiment with Lexan were performed and large 2D effects were apparent [14]. To match the general x-t features of the Sandusky & Bernecker experiment using 2D simulations, the model needs to be re-calibrated.

An important requirement of a DDT model is that it reproduce observed qualitative features and trends. For example, experiments have exhibited plugs, as discussed above. A calculation of the tube-test experiment performed by McAfee *et al.* [2] that exhibited plug formation is shown in Fig. 3. Parameters were chosen to match Shot E5586, as closely as possible. The piston is initially at rest and is driven by burning low-density HMX in the burn chamber (see Fig. 1). The tube is high-strength steel (Vascomax with an elastic-plastic strength model). The granular HMX has an initial solid volume fraction of 0.64. For this simulation the following are assumed: $C_{vg} = 2422$ J / kg K, $A = 228$ GPa, and $B = 458$ MPa. The remaining parameters assumed are listed in Table 1.

The lead wave does not exhibit much curvature because the high strength steel provides strong confinement. Only small wall expansion near the lead wave is observed. This flat compaction wave structure is consistent with radiographs that show little curvature in the lead foils (see Fig. 2, for example). The amount of the deformation in the tube walls is not incompatible with experimental observations, except plugs are not formed in the simulation. The lack of plug formation is a qualitative deficiency of the rate model assumed. This will be demonstrated later.

Figure 4 shows a comparison between 1D and 2D simulations for the tube-test experiment performed by McAfee *et al.* [2]. The profiles from the 2D simulations are taken at the centerline. At early times there is very little difference between 1D and 2D profiles. In fact, the profiles shown at 30 μs for the 1D and 2D simulations are indistinguishable in the figure. This is because the high strength steel provides strong confinement, especially at lower pressures. At longer times, the qualitative features are the same but quantitative differences are observed.

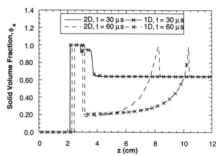

Fig. 4 Calculated solid volume fraction profiles for 1D and 2D simulations.

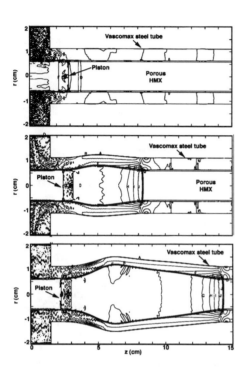

Fig. 3 Calculated pressure contours at 30 (a), 60 (b), and 90 μs (c) from start of simulation for McAfee *et al.* tube test experiment [2] ($\Delta r = \Delta z = 200\ \mu m$). The maximum pressure contour "J" is 0.881, 1.87, and 1.82 GPa for (a), (b), and (c) respectively. $\Delta p = 145$, 283, 282 MPa for (a), (b), and (c) respectively.

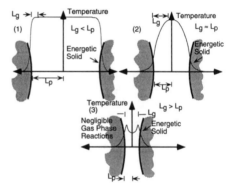

Fig. 5 Schematic of the interaction between the characteristic pore size (L_p) and gas-phase preheat & reaction zone (L_g).

PLUG FORMATION AND OTHER QUALITATIVE FEATURES OF DDT

Even with the detailed experimental evidence available (*e.g.* [2]), the exact nature and role of the observed plug remains unclear. In fact, there are indications that plugs may not always form [15], with higher burning rate explosives being less likely to exhibit plugs than lower burning rate materials [16]. Previous DDT models, such the BN model described above, have produced high density regions; however, these plugs are quickly consumed from below by a vigorous reactive wave that evolves to full detonation.

Here we suggest a scenario for plug formation and demonstrate it with simulations. The essence of the idea is that the plug forms when there is no longer adequate intergranular pore space to yield significant gas-phase reactions (cf. Chapt. 3 of [15]). In relatively slow compaction processes these gas-phase reactions are needed to provide conductive feedback of energy to the solid to sustain vigorous pyrolysis. In very rapid compaction processes, energy dissipated to particle surfaces (i.e., friction, plastic work, etc.) is sufficient to yield vigorous pyrolysis of the solid without significant conductive feedback from the gas phase. This latter process initiates rapid reaction at the top of the plug, leading to full detonation.

Simple modifications are made to the basic framework of the BN "kinetics" model. Instead of Eq. (13), here we assume $\mathcal{F} = (\phi_s \phi_g / \mu_c)(\max(p_s, \beta) - \max(p_g, \beta))$ [9]. The subscripts "s" and "g" refer to solid and gas, respectively. The configuration pressure, β, is modeled using $\beta = \min(-\max(\phi_s - \phi_{s0}, 0)\tau \ln(\phi_g) / \phi_g, \beta_{\max})$ [9]. Vigorous reaction significantly affects interphase heat transfer. This has been ignored previously. To crudely account for this, $k_H \equiv H(I - 0.5)k_1 + (1 - H(I - 0.5))k_2$ where I is the induction time variable ($0 \le I \le 1$), H is the Heaviside step function, and k_1 & k_2 are constants. The source term assumed for the energy equation is $\mathcal{E} = -p_g \mathcal{F} - (T_{int} - T_g)\mathcal{H}$. The first term corresponds to depositing all compaction work with the solid [9, 10, 13], and the second is Newton's law for heating the solid where T_{int} is the interface temperature and is used instead of the bulk solid temperature in the BN model.

Heating of the Solid

We need an estimate for T_{int} to be used in the heat transfer potential and as an ignition criterion. The solid energy equation can be written as,

$$\frac{De_s}{Dt} = \frac{p_s}{\rho_s^2}\frac{D\rho_s}{Dt} + \frac{(p_s - p_g)\mathcal{F}}{(\phi_s \rho_s)} + \frac{(T_g - T_{int})\mathcal{H}}{(\phi_s \rho_s)}. \tag{19}$$

The terms on the right hand side of Eq. (19) correspond to heating (increase in internal energy) via compression, compaction (frictional, plastic work, etc.), and heat transfer processes, respectively. This heating is not uniform; "hot spots" form which act as reaction centers. As a first approximation we assume compaction heating is deposited at the particle interface (in addition to convective heat transfer), and compressive heating is uniformly distributed. We employ integral methods (e.g., [17]) to obtain an expression for the interface temperature. The temperature profile in the particle is assumed to be of the form $(T(r) - T_c) / (T_{int} - T_c) = (r / a_{p0})^m$, where T_c is the centerline temperature and m is a constant. We obtain the bulk temperature, T_s, in terms of T_c and T_{int} by integrating the assumed temperature profile to yield $T_s = (mT_c + 3T_{int}) / (m + 3)$. Before significant reaction occurs, the boundary condition at the interface of a spherical particle is $k_s(\partial T / \partial r)_{r=a_p} = (a_p / 3\phi_s)((T_g - T_{int})\mathcal{H} + (p_s - p_g)\mathcal{F})$. Applying this boundary condition to eliminate T_c gives

$$T_{int} = \frac{(m+3)T_s + 3BiT_g}{m + 3Bi + 3} + \frac{a_{p0}a_p(p_s - p_g)\mathcal{F}}{3\phi_s k_s(m + 3Bi + 3)}, \tag{20}$$

where here $Bi \equiv ha_{p0} / 3k_s = k_H a_{p0} / 3a_p k_s$. When vigorous reaction begins, this expression is no longer appropriate because reaction was neglected in its derivation. Therefore, T_{int} is not allowed to exceed 600 K.

Effect of Pore Size on Ignition

Ignition and deflagration of energetic materials are greatly influenced by the gas-phase zone (preheat and reaction zone), often called the stand-off distance, L_g. The pore size, L_p, in a damaged explosive is estimated to vary from about 0.2 to 15 μm for porosities of 0.01 to 0.4, assuming an effective particle size of 35 μm. The stand-off distance of HMX is estimated to vary from about 0.5 to 50 μm for pressures of 100 to 0.1 MPa, respectively (*e.g.* ref. [18]). Therefore, conditions *can* exist where gas-phase ignition would be inhibited by the pore size ($L_p < L_g$); as well as, the case where $L_p > L_g$ (see Fig. 5) (*cf.* Chapt. 3 of ref. [15]). In Fig. 5, three conditions are depicted: (1) high pressure and low ϕ_s, (2) intermediate pressure and intermediate ϕ_s and (3) low pressure and high ϕ_s.

To investigate the effects of pore size on ignition and burning we consider the effect of an additional ignition constraint for gas-phase driven combustion to occur, $L_p / L_g > const$. L_g in the simplest approximation scales with $k_g / (\rho_s r_b C_{vg})$ (*e.g.* [18]) where k_g, ρ_s, and C_{vg} are the conductivity of the gas, density of the solid and heat capacity of the gas, respectively. In steady gas-phase driven combustion $r_b \sim k_p (p_g / D_p)^{b_n}$ where b_n is a constant and D_p is a scaling constant, so $L_g \sim k_g / (\rho_s k_p (p_g / D_p)^{b_n} C_{vg})$. A measure of the pore size is $L_p \sim \sqrt{\kappa}$ where κ is the permeability. Assuming the correlation for permeability adopted by Baer & Nunziato [5], we find $L_p \sim a_p \phi_g^{2.25}$. Thus, $L_p / L_g \sim a_p \phi_g^{2.25} (\rho_s k_p (p_g / D_p)^{b_n} C_{vg}) / k_g$. Lumping nearly constant parameters into a specified threshold parameter, R_{k0}, we require for gas-phase ignition that $R_k \equiv a_p \phi_g^{2.25} p_g^{b_n} > R_{k0}$. This gas-phase zone ignition constraint would be applicable to conditions where ignition leads to gas-phase driven combustion. Gas-phase driven combustion is a condition where conductive heating from the gas-phase zone plays a critical role in sustaining the pyrolysis of the solid. A plug forms when the pore size is so small that gas-phase reactions are quenched as a result of heat loss to the pore walls. As a consequence, ignition fails and gas is not generated to resist the compaction process, which yields a high density plug.

Modified Rate Model

The gas-phase driven mode can not be the *only* mode of ignition and burning in explosives since undamaged heterogeneous plastic-bonded explosives with very little void can be shock initiated. Initiation of single crystal HMX requires extremely high pressure shocks. Therefore, localized heterogeneous heating is important. Here we modify the BN reaction rate, albeit crudely, to account for localized compaction heating leading to reaction with the addition of the third term in the following rate expression.

$$
\begin{aligned}
C = &-3\mathrm{H}\left(T_{int} - T_{ign}\right) \mathrm{H}\left(R_k - R_{k0}\right) k_p\left(\phi_s \rho_s\right)\left(\frac{p_g}{D_p}\right)^{b_n} \frac{(1-f)^{2/3}}{a_p} \\
&-\mathrm{H}\left(\phi_s - \phi_{s0}\right) k_h \left[\frac{(p - p_0)}{D_p}\right]^2 \rho_s\left(1 - \phi_{s0}\right)f - \mathrm{H}\left(W_c - W_{c0}\right) k_{cw}\phi_s
\end{aligned}
\tag{21}
$$

where here $f \equiv \left(\phi_s - \phi_{s0}\right) / \left(1 - \phi_{s0}\right)$ is the rate of compaction work (second term on right hand side of Eq. (2)), and W_{c0} is a threshold rate of compaction work. The parameters k_p, k_h, and k_{cw} are rate constants. The first two terms are adopted from [6] with slight modification. A key modification is the extra gas-phase ignition requirement, $\mathrm{H}\left(R_k - R_{k0}\right)$, that has been added to the first term. Results qualitatively similar to those obtained by [6] are obtained when $R_k > R_{k0}$ ($L_p \gg L_g$) everywhere. If $R_k < R_{k0}$ ($L_p \ll L_g$) everywhere the main reaction term (first term in Eq. 3) does not contribute. The induction time equation is also adopted (see Eq. 16) with slight modification (the evolution of the induction time variable, I, is not allowed when $R_k < R_{k0}$). The numerical calculations, shown in the following section, utilize MacCormack's method with time splitting applied to the source terms.

<u>Simulations of Plug Formation</u>

In the simulations presented we assume the following parameters unless otherwise specified (see Table 1 for other parameters): $k_{cw} = 2.5 \times 10^8$ kg / m^3 s, $W_{c0} = 60$ GJ / kg, $R_{k0} = 0.2$ m(Pa)^{bn}, $m = 20$, $B = 496$ MPa (corrected sign), $\beta_{max} = 500$ MPa, $k_I = 1.2 \times 10^5$ s^{-1}, $k_1 = 0.094$ W / m K, $k_2 = 9.4$ W / m K, and $u_p = 100$ m / s (u_p is the piston speed). We first consider the case where $W_{c0} \to \infty$; that is, compaction initiated reaction is suppressed. Under these conditions a plug will form, but detonation will not initiate on top of the plug. At early times the rate of compaction work is seen corresponding to the initial compaction wave (see Fig. 6). After an ignition delay, vigorous reaction begins near the piston. This produces a secondary compaction wave, as indicated by the rate of compaction work profile at $t = 30$ μs. The initial overtakes the secondary compaction wave and then a combined compaction wave forms. The maximum rate of compaction work increases throughout this simulation.

The inset in Fig. 6 shows the x-t plot for this simulation. Two leading nearly linear wave loci are evident (neither attributable to detonation). In experiments, x-t traces like this have been interpreted as an initial compaction wave (labeled "1") followed by a weak compaction driven reactive wave, or "compaction-driven" regime (labeled "3"). To investigate whether "3" is a self-propagating wave driven by the weak reactions another simulation was considered where the reaction was inhibited in the high density region. Shutting off reaction completely in the high density region ($k_h = 0$ where $R_k < R_{k0}$) yields the same x-t characteristics. Therefore in these simulations the second slope, "3", is simply a stronger compaction wave driven by burning near the piston rather than weak reaction in the plug region and not a "compaction-driven" self-propagating wave.

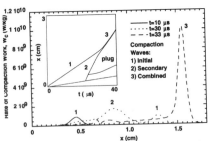

Fig. 6 Evolution of the rate of compaction work when "compaction driven" reaction is not allowed. Inset is the x-t plot for this simulation.

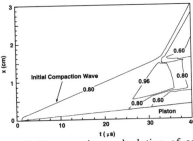

Fig. 7 Distance-time calculation of solid volume fraction, ϕ_s, of a DDT event in a 1D tube test.

Figure 7 illustrates the case when compaction initiated reaction is allowed. The "inert" plug, supported by burning near the piston, forms an accelerating effective piston. In this accelerating system, the compaction wave at the head of the inert plug strengthens with time, as discussed above, and initiation occurs where $W_c > W_{c0}$ (in this case, near the point where the secondary compaction wave overtakes the first). These calculations are qualitatively similar to experimental observation (cf. [2]). If the burning rate of the explosive is increased (corresponding to small L_g) by a factor of 10 (all else the same) detonation does occur; but, an "inert" plug is *not* formed since $R_k > R_{k0}$ *everywhere*. This is in agreement with observations [16]. Further, if burning rates are very slow (corresponding to large L_g) detonation is not readily achieved. This is the case for insensitive explosives [19].

FUTURE RESEARCH DIRECTIONS

There are several areas that merit further attention. DDT in granular materials is a hideously complex process; however, models of DDT would be of value in safety calculations if only qualitative trends and features were predicted faithfully. Models calibrated to give very good quantitatively comparison to a set of experiments over a narrow parameter space are of little worth in safety problems where a wide range of situations may arise. Suggested areas for attention are:

1) More fundamental simplified reaction rate modeling should be considered. Reaction rate models in continuum-based models must continue to use simplified models because reaction occurs on the subscale and can not be simulated directly in multi-dimensional engineering calculations. Simple ignition criterion such as a critical interface temperature and burning laws based on "p^n" laws that are only valid for steady burning conditions are not likely to be adequate. Under conditions of rapid pressurization, the unsteady burning rate can potentially deviate significantly from the steady burning rate law. Further, these burning laws are obtained under conditions with gas phase heating to the solid playing a significant role. Under conditions where the gas phase can not play a significant role, the ignition and burning needs to be modeled differently. Analytical techniques such as high activation energy asymptotics and integral methods may be useful tools in developing more fundamental reaction-rate submodels.

2) Current hot spot modeling lacks physical basis and qualities. For example, the weak "hot spot like" reaction proposed by BN proceeds as long as the solid volume fraction is above the original value. There is no possibility for the reaction to quench. A hot-spot model should take into account the competing effects of heating or cooling of the particles by compaction, deconsolidation, compression, expansion, chemical energy release, convection heat transfer, and thermal conduction into the particle.

3) There is a need to better understand and model the heating of energetic materials by compaction. The relative roles of friction, plastic work and cracking need to be better understood and how this heating can be better modeled on the continuum scale.

4) It is uncertain how good an approximation the convective heat transfer coefficient is; especially when reaction occurs. It is a poor approximation to assume that heat transfer correlations for inert heating continue to be valid when reaction begins.

5) Unusual modes of propagation need to be better understood. For example, under some conditions relative flow may play a role and convective burning may be important [8]. The role and nature of compaction-driven reaction waves also merits further study.

6) In current modeling an effective spherical particle is assumed and the diameter is assumed to be related to solid volume fraction by a simple algebraic relationship. An important question to address is how to better model the particle size distribution as the material is crushed and burned.

7) Generally, combustion experiments of energetic materials are performed in environments where the thickness of the gas phase zone is much smaller than surrounding confinement. In a bed of damaged explosive the pore size may be smaller than the gas phase which would affect the

burning significantly. The ignition and burning inside a small void (pore) in energetic solids needs detailed study.

SUMMARY

The published BN "kinitics" scheme was used as an initial reaction model. We find it does not reproduce some trends and qualitative features observed in DDT experiments. The reaction rate modeling in granular energetic materials is difficult and merits detailed study. This work has demonstrated a physically motivated scenario for plug formation. Observed characteristics and trends are reproduced, including burning rate trends. Instead of a self-propagating reaction wave, these results indicate that the "compaction driven burning" regime of a DDT event may simply be a strengthened compaction wave augmented by burning near the piston. The reaction modeling remains deficient in numerous ways and is a current area of attention.

ACKNOWLEDGMENT

The authors would like to acknowledge support from Los Alamos National Laboratory which is supported by the U. S. Department of Energy under contract W-7405-ENG-36.

REFERENCES

1. Sandusky, H.W. and R.R. Bernecker, Proc. 8th (Int.) Det. Symp. , p. pp. 881-891, (1985).
2. McAfee, J.M., et al., Proc. 9th (Int.) Det. Symp. , p. 256-266, (1989).
3. Gokhale, S.S. and H. Krier, Prog. Combust. Sci 8, p. 1-39, (1982).
4. Butler, P.B. and H. Krier, Combustion and Flame 63, p. 31-48, (1986).
5. Baer, M.R. and J.W. Nunziato, International Journal of Multiphase Flow 12(6), p. 861-889, (1986).
6. Baer, M.R. and J.W. Nunziato, Proc. 9th (Int.) Det. Symp. , p. 293-305, (1989).
7. Powers, J.M., D.S. Stewart, and H. Krier, Combustion and Flame 80, p. 264-279, (1990).
8. Asay, B.W., S.F. Son, and J.B. Bdzil, submitted to Int. J. Multiphase Flow , (1995).
9. Menikoff, R., et al., in preparation , (1995).
10. Son, S.F., et al., in preparation , (1995).
11. Kapila, A.K., et al., in preparation , (1995).
12. Baer, M.R., private communication , (1992).
13. Bdzil, J.B. and S.F. Son, LANL LA-12794, (1995).
14. Son, S.F., J.B. Bdzil, and E.M. Kober, presented at JANNAF Propulsion Systems Hazards Subcommittee Meeting , (1995).
15. Belyaev, A.F., et al., Transition From Deflagration to Detonation in Condensed Phases, (Israel Program for Scientific Translations, Jerusalem, 1975).
16. Kondrikov, B.N., personal communication , (1994).
17. Goodman, T.R., Application of Integral Methods to Transient Nonlinear Heat Transfer, Advances in Heat Transfer, eds. J. T. F. Irvine and J.P. Hartnett. Vol. 1. (Academic Press, New York, 1975),p. 51-122.
18. Brewster, M.Q., et al., presented at the 31st Joint Propulsion Conference, AIAA 95-2859 , (1995).
19. Asay, B.W. and J.M. McAfee, 10th Symp. (Int.) on Det. , (1993).

A SIMPLIFIED METHOD FOR DETERMINING REACTIVE RATE PARAMETERS FOR REACTION IGNITION AND GROWTH IN EXPLOSIVES

PHILIP J. MILLER
Engineering Sciences Branch, Research and Technology Division
Naval Air Warfare Center, Weapons Division,
China Lake, CA 93555-6001

ABSTRACT

A simplified method for determining the reactive rate parameters for the ignition and growth model is presented. This simplified ignition and growth (SIG) method consists of only two adjustable parameters, the ignition (I) and growth (G) rate constants. The parameters are determined by iterating these variables in DYNA2D hydrocode simulations of the failure diameter and the gap test sensitivity until the experimental values are reproduced. Examples of four widely different explosives were evaluated using the SIG model. The observed embedded gauge stress-time profiles for these explosives are compared to those calculated by the SIG equation and the results are described.

INTRODUCTION

The ignition and growth concept of shock initiation and detonation wave propagation in heterogeneous solid explosives has been described by Lee and Tarver [1]. It has been applied through the use of the two-dimensional, finite element, Lagrangian hydrodynamic code, DYNA2D [2], to a variety of PBX-type explosives and propellants [3,4]. The two-dimensional experimental data included failure radius of cylinderical rate sticks, corner turning, flyer plate initiation with short and long pulse durations, and the production of diverging or failing detonation waves by smaller radius flyer plates.

A reactive flow hydrodyamic computer code model normally consists of: an unreacted explosive equation of state, a reactive product equation of state, a reaction rate law that governs the chemical conversion of exposive molecules to reaction product molecules, and a set of mixture equations to describe the states attained as the reactions proceed. The unreacted equation of state is normalized to shock Hugoniot data. The reaction product equation of state is derived from expansion data, such as that obtained in a cylinder test. The reaction rates are usually inferred from embedded gauge and/or laser interferometric measurements of pressure and/or particle velocity histories. The ignition and growth reactive flow model has been used to analyze a great deal of one- and two-dimensional shock initiation and self sustaining detonation data. In this reactive flow formulation, the unreactive and product equation of state are both JWL (Gruneisen) forms:

$$P = A\exp(-R_1V) + B\exp(-R_2V) + \omega C_v T/V , \tag{1}$$

where P is pressure; V is relative volume; T is temperature; and A, B, R_1, R_2, ω (the Gruneisen coefficient); and C_v (the average heat capacity) are constants. The ignition and growth reaction rate law is of the form:

$$\partial F/\partial t = I(1-F)^b(\rho/\rho_0-1-a)^x + G_1(1-F)^cF^dP^y + G_2(1-F)^eF^gP^z , \tag{2}$$

325

where F is the fraction reacted; t is time; ρ_0 is the initial density; ρ is the current density; p is the pressure in Mbars; and I, G_1, G_2, b, x, a, c, d, y, e, g, and z are constants. The first term in Equation (2) controls the initial rate of reaction ignited during shock compression and is limited to fraction reacted F ≤ Figmax. The second term in Equation (2) is used to simulate the relatively slow growth of hot spot reactions during low pressure shock initiation calculations, and the third term is used to rapidly complete the shock to detonation transition in those calculations.

In the development stage of new explosives and propellants it is not always feasible or practical to obtain the necessary data to develop this model to its fullest, particularily for the development of insensitive munitions. Unfortunately, it is this characterization that often determines whether or not one proceeds with the development of a particular explosive. In this paper a much simpler ignition and growth (SIG) model, that is based upon the failure diameter and gap test sensitvity rather than the more sophisticated embedded gage or particle velocity measurements, is described. The rate equation developed here is similar to that of Tarver and Hallquist, reference [3], but it differs in the method of determining the parameters. The SIG rate equation, based primarily on measured gap test critical pressures and at least, a best guess of the failure diameter, will be demonstrated to be nearly indistingishable in its ability to reproduce embedded gauge data with that of the more complicated three term ignition and growth model.

REACTION RATE EQUATION

The rate equation used in this paper has the form [3],

$$\partial F/\partial t = I(1-F)^{2/3}(\rho/\rho_0 - 1)^4 + G(1-F)^{2/3}F^{2/9}P^Z , \tag{3}$$

where F is the mass fraction of the explosive that has reacted, ρ/ρ_0 is the density of the shocked explosived divided by the initial density, P is the pressure in megabars, and I, G, and Z are constants. The exponents for the density, (1-F), and F terms are fixed. Their dependencies are based strictly on geometric and experimental considerations (reference [1]). Initially, the pressure dependency was set at P^2 (in order to correspond to a P^2t relationship) however it was found that for explosives with failure diameters less than 1 cm that a P^3 dependency resulted in computer simulations of the gap test and failure diameter that were much less difficult to obtain than if P^2 were used. However, it must be pointed out that both P^2 and P^3 worked equally well in reproducing the experimental data. In addition, no restrictions were placed on the extent of reaction by the ignition term.

To determine the reaction rate constants for ignition (I) and growth (G), DYNA2D hydrocode simulations of the reported failure diameter and gap test results for a specific explosive were repeatedly performed by varing the values of I and G until the experimental results were satisfactorily reproduced. It was found that, while I and G are not independent, the growth rate constant G affected the failure diameter most strongly, and that the initiation rate constant I mostly affected the gap test sensitvity. These parameters, along with the given detonation velocity, are the determining factors for being able to reproduce these experimental two-dimensional results. Validation of this method for determining the reactive rate parameters for a given explosive can be easily carried out by comparing the simulated stress-time profiles to the experimental curves obtained from embedded gauge experiments, if they have been obtained.

EVALUATION OF THE SIG CHEMICAL RATE EQUATION

Explosives Investigated

Four explosives were investigated using this simplified ignition and growth (SIG) model. They were PBX 9404 (94% HMX, 3% nitrocellulose and 3% tris-chloroethylphosphate), PBXN-110 (88% HMX and 12% HTPB binder), PBXN-111 (20% RDX, 43% AP, 25% Al and 12% HTPB) and PBXW-126 (20% RDX, 22% NTO, 20% AP, 26% Al and 12% HTPB). The unreactive equation of state Hugoniots and the JWL equation of state parameters for their reaction products for the four explosives are listed in Table I. Included in the table are the experimental failure diameters and gap test sensitivities. Also, in Table I. are the evaluated chemical reaction rate parameters I and G determined in this study using the SIG equation (3) and the input equation of state data.

Embedded Gauge Stress-Time Profiles

Figure 1. shows the experimental pressure-time histories of the four explosives investigated here. It also shows the calculated profiles using equation (3) and the ignition (I) and growth (G) rate parameters derived from the failure diameters and gap tests of the explosives. The overall gross features of the profiles appear to compare quite well. However, several facts need to be noted. The calculated profiles are more sensitive to the gap test results than to the failure diameter. Therefore, the explosives used in the embedded gauge measurements should be nearly identical as possible, if not from the same batch, as that used to obtain the gap test sensitivities. As mentioned previously, the two explosives PBXN-110 and PBX 9404, which have small failure diameters were calculated using a P^3 dependency. The simulations could have been carried out using a P^2 dependency obtaining nearly the same results. However, in this case the value of I becomes a very small number in order to reproduce the data. The result is that the value of I approaches that of the flucuations of the noise level in the calculations. Also, the calculated profiles become too sensitive to small changes in I. Thus, it was easier to use the P^3 dependency to reproduce the experimental data.

A similar investigation by Murphy and co-workers [11] using simulations of wedge tests and failure diameter tests has been reported. They, however, place an arbitrary maximum on F_{igmax} the fraction of sample consumed by the ignition term. They, also, use an estimated minimum for the compression thresold in the ignition term. The use of either of these conditions causes their model parameters for I and G to differ from the results here. In addition, the determination of the exponent for the pressure dependence was based on previous investigations using the complete three term I & G model. Here, also the choice is rather arbritary.

CONCLUSIONS

These results are encouraging in that a fairly easy and an inexpensive method for developing the ignition and growth reactive flow model is demonstrated. Unfortunately, the results are disappointing since the rate equation is pressure driven and it was demonstrated that the use of the embedded gauge profiles, failure diameter, and gap test sensitivity do not determine the pressure dependency unambiguously. The exponent of P in the growth term to a great extent determines the fraction of material consumed in the ignition processes, or the number density of ignition sites formed. These are expected to vary widely between small and large failure diameter explosive compostions. The ignition rate and the growth rate are both pressure

Table I. JWL Equation of State and Reaction Rate Parameters

Explosive	PBX 9404[3,5]	PBXN-110[5-7]	PBXN-111[5,8]	PBXW-126[9]
Unreactive Hugoniot				
ρ_0 g/cc	1.842	1.67	1.78	1.80
C_0 cm/µsec	0.2492	0.199(0.240)[+]	0.2117(0.280)	0.220
S_1 mm/µsec	2.093	3.05(1.27)	2.758(1.38)	2.00
Γ	0.8867	0.8	0.9	0.912
c_v Mbar/K	2.781×10^{-5}	2.40×10^{-5}	2.39×10^{-5}	2.5×10^{-5}
Y_0 Mbar	0.002	0.002	0.002	0.005
G Mbar	0.0454	0.0354	0.0354	0.0354
Reaction Product JWL Parameters				
D_{cj} cm/µsec	0.880	0.8311	0.5775	0.6468
P_{cj} Mbar	0.398	0.292	0.130	0.170
E_0 Mbar-cc/cc	0.102	0.087	0.054	0.080
c_v Mbar/K	1.0×10^{-5}	1.0×10^{-5}	1.0×10^{-5}	1.0×10^{-5}
A Mbar	8.524	4.69924	6.4493[++]	6.735567
B Mbar	0.1802	0.00106	0.0039774	0.011804
R_1	4.6	3.86	5.4	4.97
R_2	1.3	1.0	1.03	1.00
ω	0.38	0.40	0.16	0.25
Reaction Rate Parameters				
z	3	3	2	2
I µsec^{-1}	250.	33.	3.	6.5
G µsec^{-1}	2000.	600.	200	70
Failure Dia.	0.6 mm	6.0 mm	37 mm	37 mm[++]
Gap Test Sens[*]	14 kbar	30 kbar	47 kbar	58 kbar[++]

[+] Data in parentheses from wedge test data, references [7] and [10].
[++] Unpublished work, NSWC, White Oak Laboratory.
[*] Pressure in PMMA gap material.

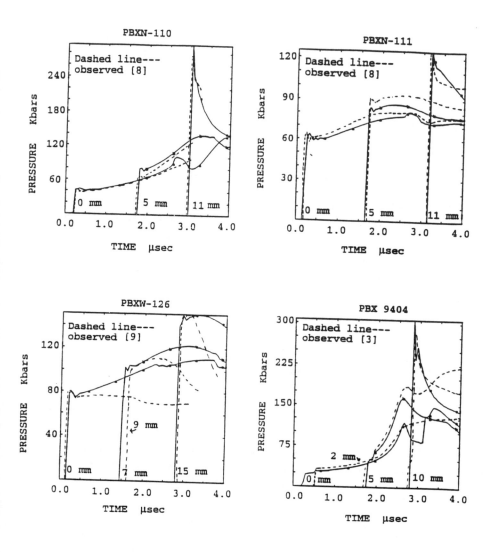

Figure 1. Calculated stress-time profiles. The details of the
embedded gauge experiments are given in the cited
references. The gauge positions are marked in mm.

sensitive and to fully develop a rate equation that more closely represents the shock to detonation processes, it is necessary that an experiment or experiments be found that reflect these pressure dependencies more accurately. Perhaps the pressure effects on reaction front radius of curvature and/or the reaction zone thickness can be useful in further characterizing the chemical reaction rate kinetics of detonation, this needs to be further investigated.

ACKNOWLEDGMENTS

The Author acknowleges the ONR 6.2 Undersea Weaponry Program and the NAVSEA Insensitive Munitions Advanced Development (IMAD) program for support of this effort.

REFERENCES

1. E. L. Lee and C. M. Tarver, Phys. Fluids 23, 2362(1980).

2. J. O. Hallquist, User's Manual for DYNA2D-an Explict Two-Dimensional Hydrodynamic Finite Element Code with Interactive Zoining," Lawrence Livermore National Laboratory Report UCID-18756, July 1980.

3. C. M. Tarver and J. O. Hallquist, Seventh Symposium (International) on Detonation, Naval Surface Weapons Center NSWC MP-82-334, Annapolis, MD, 1981, p.488.

4. C. M. Tarver and L. G. Green, Ninth Symposium (International) on Detonation, Office of the Chief of Naval Research, OCNR 113291-7, Portland, OR, 1989, p.701.

5. T. N. Hall and J. R. Holden, "Navy Explosives HandBook", NSWC MP 88-116, 1998.

6. G. T. Sutherland, in Shock Compression of Condensed Matter 1989, eds. S. C. Schmidt, J. N. Johnson, and L. W. Davison (Elsevier Science Publishers B. V., 1990) p.721.

7. T. Schilling and S. Martin, "Shock Initiation of PBXW-113 Using the Wedge Technique," NWCTP 6961, Naval Air Warfare Center, China Lake, CA, 1988.

8. G. T. Sutherland, J. W. Forbes, E. R. Lemar, and P. J. Miller, " Shock Response of Several U. S. Navy Plastic Bonded Explosives", Bull. Am. Phys. Soc. Vol.40, No. 6, p. 1399, July 1995. Presented at the 1995 APS Topical Conference on Shock Compression of Condensed Matter, Seattle,WA.

9. M. J. Murphy, R. L. Simpson, P. A. Urtiew, P. C. Souers, F. Garcia, and R. G. Garza, "The Initiation and Detonation Behaviour of PBXW-126, a RDX/AP/Al/NTO Composite Explosive", Bull. Am. Phys. Soc. Vol.40, No. 6, p. 1399, July 1995. Presented at the 1995 APS Topical Conference on Shock Compression of Condensed Matter, Seattle,WA.

10. J. C. Dallman, Unpublished, "Wedge Tests of PBXW-115 Explosive", LANL,Feb.17 ,1987.

11. M. J. Murphy, E. L. Lee, A. M. Weston, and A. E. Williams, "Modeling Shock Initiation In Composition B," Tenth International Detonation Symposium, Boston, July 1993, Office of Naval Research, ONR 33395-12, p.963, 1995.

INITIATION OF CRYSTALLINE EXPLOSIVES BY SHOCK OR IMPACT

C. S. COFFEY
Indian Head Division, Naval Surface Warfare Center, Silver Spring, MD. 20903-5640.

ABSTRACT

The central issue determining initiation response of explosive crystals to shock or impact is the generation of the energy necessary to cause molecular dissociation. This controls the initial rate of release of chemical energy. Here, the energy dissipated in the crystals as they undergo plastic deformation due to shock or impact is taken to be the source of the energy required to achieve initiation. This determines initiation sensitivity to shock or impact. Several problems are discussed including initiation of detonation and the von Neumann pressure spike structure of a detonation wave.

INTRODUCTION

The plastic deformation that occurs in crystalline solids during shock or impact involves the creation and motion of dislocations. Associated with this plastic deformation is the energy dissipation due to the lattice perturbations produced by these moving dislocations. These processes are particularly important to the initiation of chemical reaction in crystalline explosives for this dissipation is a means by which energy can be introduced into the molecules of the solid and reaction initiated during plastic flow. In earlier work we have described the energy dissipation rate and localization due to moving dislocations as well as the associated plastic deformation rate due to the applied shear stress introduced by a shock or impact.[1-4] Here this work is extended and applied to the initiation of reaction in crystalline explosives.

The conventional theories of detonation of Chapman and Jouguet, and of Zeldovich, von Neumann and Doering are only approximations and do not address the means by which energy is generated and transferred to the molecules by the driving shock wave and how detonation is initiated.[5] Similarly, the processes responsible for the initiation of crystalline explosives during the more modest stimuli of impact have never been adequately set forth. The same uncertainties apply to the initiation of the chemical reactions that have been observed in a myriad of intermetallic and crystal compacts that have been subjected to rapid plastic deformation by shock or impact.

PLASTIC DEFORMATION

The motion of dislocations in a crystalline solid subjected to an applied shear stress involves switching the electron bonds of the atoms or molecules immediately ahead of the dislocation core and attaching them to the atoms or molecules of the core. This is a "seamless" process, for the lattice must return to its undisturbed state after the passage of the dislocation. The switching of the electron bonds is a quantum mechanical process and can be treated in terms of tunneling.[6] The dislocation velocity can be written as $v = Tv_0$ where T is the probability that tunneling occurs and v_0 is the shear wave speed. In the asymptotic limits $T \to 0$ as $\tau \to 0$ and $T \to 1$ for $\tau \gg \tau_0$ where τ is the applied shear stress and τ_0 is a characteristic yield stress of the material. This gives the correct asymptotic behavior for the dislocation velocity.

Dislocations are generated in sources such as grain boundaries and impurities in response to the applied shear stress. A simple generic source has been proposed in which the source has an average size, l_0, beyond which the newly created dislocation pair must move before the next dislocation pair can be created. This allows the average rate of creating dislocations by a single source to be expressed as[3]

$$\frac{dn}{dt} = 2\,\frac{v}{l_0}\,p_c(\tau) \tag{1}$$

The factor of 2 accounts for the formation of a dislocation pair of opposite sign with each creation event and $p_c(\tau)$ is the probability of creating a new dislocation pair at a shear stress τ. For large shear stress, $\tau \gg \tau_0$, $p_c \approx 1$, similarly when $\tau \ll \tau_0$, $p_c \to 0$.

The plastic strain rate, can be expressed in terms of the Orowon relation $d\gamma_p/dt = Nbv$ where b is the Burgers length and N is the number of moving dislocations obtained by integrating equation (1) with respect to time and multiplying the result by the number of active dislocation sources N_s per unit area. This yields the following expression for the plastic strain rate[6]

$$\frac{d\gamma_p}{dt} \approx 2\,T(\tau)^2\,\frac{v_0^2 b}{l_0}\,p_c(\tau)\,N_s\,t \tag{2}$$

where t is the time measured from the onset of the shear stress.

The plastic deformation rate given in equation (2) admits two time regimes of interest determined by the time required for a newly created dislocation to move away from its source, at a velocity v, and encounter an obstacle which forces it to stop. Let L_0 be the average distance between the dislocation source and an obstacle such as a grain boundary, then at early times after the application of a shock or impact when $t < L_0/v$ the number of moving dislocations increases with time and the plastic strain rate is given by equation (2). However, at later times when $t \geq L_0/v$, the moving dislocations will begin to encounter obstacles and their motion halted. A near steady state is reached where the number of newly created dislocations is offset by the number of recently stopped dislocations and the number of moving dislocations approaches a constant. For this case the plastic deformation is determined by substituting the time $t = L_0/v$ into equation (2)

Both the dislocation velocity and the plastic strain rate are determined in large part by the behavior of the tunneling probability. In this a major role is played by the effective mass of the tunneling particle. For most materials the effective mass of the bonding electron will be quite large, $m \gg$ electron mass, since the electron is usually securely attached to its parent atom or molecule. Thus, for most materials, tunneling and plastic flow will be quite small at low shear stress. However, for a relatively large shear stress, approaching the yield strength of the material, the energy of the bonding electrons will increase to near the energy of the top of the core-lattice potential well. For this case the effective mass of the tunneling particle is not of great importance and $T \to 1$. This asymptotic behavior is universal for all crystalline solids that plastically deform by creating and moving dislocations.

ENERGY DISSIPATION DURING PLASTIC FLOW

A dislocation moving through the lattice with a velocity v will encounter and perturb the lattice potential at a rate of $v_0 T/d$ times per second, where d is the lattice spacing. These perturbations generate a spectrum of phonons centered in a band about $\omega = 2\pi v_0 T/d$. Typically, for most explosive crystals $v_0 \approx 2 \times 10^3$ m/s and $d \approx 5 \times 10^{-10}$m, so that the center frequency is somewhat greater than $\omega = 10^{13} T$ rad/s. The band extends on either side of center frequency to an extent that is dependent on the shape of the lattice potential. These lattice perturbations will cause energy dissipation to occur at a rate[1-4]

$$\frac{dE}{dt} = \frac{4\pi\Gamma G^2 N}{d}T + N\sum \hbar\omega_{j,j-1}\sum_{f,u}\left|\sum_l \frac{\langle f|H'|l\rangle\langle l|H'|u\rangle}{E_l - E_u - \hbar\sum \omega_{j,j-1}}\right|^2 - K\frac{dT}{dx}. \quad (3)$$

where Γ is a constant, G is the shear modulus and H' couples the dislocation lattice perturbations to the internal molecular vibrational modes.

For molecular crystals the denominator of the second term in (3) contains the difference between the energy levels of the internal vibration modes of the molecules in the crystal and the phonon energies generated by the moving dislocations. The internal molecular vibrational modes of most explosive molecules have frequencies of the order of $\omega_{j,j-1} \approx 10^{13}$ rad/s so that these internal vibrational modes can be resonantly excited by the optical phonons in the spectrum generated by the fast moving dislocations driven by shear due to shock. This results in very rapid molecular dissociation and energy release associated with detonation. The remaining terms in equation (3) describe energy dissipation due to impact, $T \ll 1$, and will not be considered here.

PLASTIC DEFORMATION AND INITIATION

It is convenient to map the energy dissipation rate onto the plastic deformation rate as shown in Figure (1) which while general is specific to HMX. The oscillations in the transition region are due to the quantum mechanical nature of tunneling.[6] This allows the rate of initiation of chemical reaction to be related to the rate of plastic flow. The plastic deformation rate has been divided into four regions according to the magnitude of the tunneling coefficient. Regions I and II apply to the conditions where T and $d\gamma_p/dt$ are small so that no reaction occurs in region I while only impact driven reactions occur in region II. Region IV applies to the high shear stress regime where $T \to 1$ and the energy dissipation rate is dominated by the very rapid multi-phonon excitation of the internal modes associated with detonation. Region III applies to the transition between the relatively mild reactions of impact and the very rapid reactions of detonation. This is where explosive sensitivity and hazards are determined. A similar transition region exists between Regions I and II at the onset of impact initiation and while important will not be discussed here.

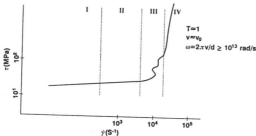

I   IMPACT REGION, NO INITIATION.
II   IMPACT INITIATION REGION.
III   TRANSITION REGION, SHOCK SENSITIVITY DETERMINED.
IV   SHOCK INITIATION, DETONATION REGION, T≈1.

Figure (1). Superposition of plastic deformation rate and initiation behavior.

The shear stress at which rapid plastic flow first occurs in most explosive crystals is quite low, less than 100 MPa and this is the onset of asymptotic Region IV where T → 1 and the fast reactions of detonation first occur. Because of the asymptotic behavior of the tunneling probability, detonation occurs for all higher shear stress levels. Regions II and III are where the initiation of reaction due to impact occurs and where the transition between impact initiation and the initiation of detonation takes place. The hazards associated with explosives are readily seen here because typically only a small increase in the shear stress is required to cause the transition from a mild impact driven reaction of Region II to a very rapid and violent reaction of Region IV.

MACROSCOPIC SCALE CALCULATIONS

As noted at the outset, the conventional theories of detonation fail to take into account the energy needed to initiate molecular dissociation and reaction. Here, the concern will only be with the asymptotic region T → 1 which allows the applied shear stress to determine the macroscopic condition of detonation. The amplitude of the shear stress can be written as

$$\tau = (\vec{dr} \cdot \vec{v}) P \tag{4}$$

In order for a steady state detonation to develop, initiation must proceed at a constant rate on the surface of the initiation portion of the detonation wave. To achieve this constant rate of molecular excitation and dissociation requires that the shear stress be a constant over the entire surface of the initiation front so that $d\tau = 0$ or

$$d\tau = \vec{dr} \cdot \vec{\nabla}\tau = 0 \tag{5}$$

Combining equations (4) and (5) gives the condition for a steady state detonation wave as the Laplace equation

$$\nabla^2 P = 0 \tag{6}$$

For a cylindrical charge this has the solution

$$P(r, z) = P_0 J_0(\alpha r) e^{\alpha z} \tag{7}$$

where $J_0(\alpha r)$ is the zero order Bessel function. The quantity P_0 is the pressure amplitude of the steady detonation wave on the axis of the charge. The magnitude of the shear stress is determined from equations (5) and (7)

$$\tau = \alpha P_0 e^{\alpha z} (J_1(\alpha r)^2 + J_0(\alpha r)^2)^{1/2} dr \tag{8}$$

For an aggregate of randomly oriented explosives crystals it is only necessary that the magnitude of the shear stress be a constant on the initiation surface at the front of the steady state detonation wave. Thus, the magnitude of the shear stress on the axis of the charge, r =

0, z = 0, and at any other point (r,z) on the constant shear/initiation surface are equal

$$\tau(r=0, z=0) = \tau(r, z) \tag{9}$$

which reduces to

$$e^{-\alpha z} = (J_1(\alpha r)^2 + J_0(\alpha r)^2)^{1/2} \tag{10}$$

Equation (10) determines the initiation front.

Since the amplitude of the shear stress is a constant on the steady state initiation surface, the energy dissipation rate must also be a constant on this surface. For the current case of large amplitude shear stress, $T \rightarrow 1$ on the initiation surface so that the energy dissipation rate is dominated by resonant multi-phonon excitation. This results in very rapid molecular dissociation and release of chemical energy which is manifested by a very rapid increase in pressure that occurs uniformly over the entire initiation front. The peak pressure occurs in the time required to achieve substantial multi-phonon driven dissociation, ≈ 100 pico-seconds, and may account for the leading edge of the shock wave in the ZND Theory.[4,5] Because the solid explosive is being consumed, the rate of energy release and the pressure behind the initiation front must quickly decrease toward lower levels more typical of the steady state. This process, shown in Figure (2a), accounts for the von Neumann pressure spike that characterizes the initiation portion of a detonation. Since it closely follows the initiation front the von Neumann spike is also described by Equation (10)

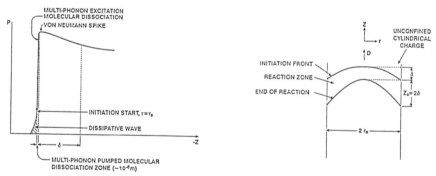

Figures (2a,b). Initiation front profiles for detonation in a cylindrical charge.

It is possible to relate the initiation front with the pressure at the end of the reaction zone. Let δ be the length of the reaction zone as measured along the axis of the cylindrical charge. Then equations (7) and (10) are satisfied simultaneously when the initiation front at the edge of the charge lags that at the center by an amount δ and the pressure at the end of the reaction zone at the edge of the charge lags that at the center by an amount 2δ. This is shown schematically in Figure (2b). These arguments assume that equilibrium is established at the end of the reaction zone and neglect the free surface boundary conditions at the edge of the charge.

The critical shear needed to initiate detonation can be estimated from equation (8) by setting r = 0 and z = 0, so that $\tau_c \approx \alpha P_0 \ell$ where $\ell \approx dr$ and is equal to the average particle size across which the shear is developed. P_0 is the Chapman Jouguet pressure. Recently, data from a number of experiments has been successfully fitted to the pressure profile of equations (7) and (10).[7] When the reaction length, δ, is obtained from the measured pressure profile, equation (10) determines the quantity α which is a measure of the shear stress due to the pressure gradient. Table (I) lists the quantities α, δ and τ_c for several explosives.

Table (I). Critical shear stress, τ_c, needed for steady detonation.

Material	δ(mm)	α(mm^{-1})	P_0(GPa)	τ_c(MPa)
PBXN-110	1.3	.012	29.1	29
PBXN-111	2.7	.0324	12.3	39
Comp B	.93	.0113	26.4	30

The quantities α and δ have been obtained from experimental data by E. R. Lemar with the assumption that the measured δ was the reaction zone length. The particle size was $\ell = 10^{-4}$m.

Detonation failure occurs when the rate of energy dissipation in the initiation front fall below the levels needed to sustain the detonation. Failure is determined when equation (10) cannot be satisfied for the condition of a reaction zone length $z = \delta$ at the edge of the charge, $r = r_0$.

CONCLUSIONS

The initiation of chemical reactions in explosive crystals during shock or impact is postulated to occur due to the energy dissipation and localization that takes place during plastic deformation. This accounts for the profile of the detonation wave including the von Neumann spike and provides a basis for treating initiation due to impact and the transition from impact initiation to detonation.

ACKNOWLEDGEMENTS

The author wishes to acknowledge the past support of the Office of Naval Research and the NSWC IR Program and wishes to thank Dr. D. H. Liebenberg, for his help and encouragement.

REFERENCES

1. Coffey, C. S., Phys. Rev. B **24**, 6984 (1981).

2. Coffey, C. S., Phys. Rev. B **32**, 5335 (1984).

3. Coffey, C. S., J. Appl. Phys. **70** (8), 4248 (1991).

4. Coffey, C. S., in Structure and Properties of Energetic Materials, MRS Symposium **296**, 63 (1992).

5. Fickett, W. and Davis, W. C., In Detonation, (University of California Press, 1979).

6. Coffey, C. S., Phys. Rev. B. **49**, 208 (1994).

7. Lemar, E. R. and Forbes J. W., in Shock Waves in Condensed Matter, 1995, to be published.

ULTRAFAST DYNAMICS OF SHOCK WAVES AND SHOCKED ENERGETIC MATERIALS

DAVID E. HARE *, I-Y. SANDY LEE **, JEFFREY R. HILL*, JENS FRANKEN *, ,
HONOH SUZUKI ***, BRUCE J. BAER ****, ERIC L. CHRONISTER ****, DANA D.
DLOTT*

*School of Chemical Sciences, University of Illinois, Urbana, IL 61801. Correspondence to Dana D. Dlott, Box 37 Noyes Lab, 505 S. Mathews Ave., Urbana, IL 61801. d-dlott @ UIUC.EDU
**Jet Propulsion Laboratory, California Institute of Technology, Pasadena, CA 91109
***Department of Chemistry, Kyushu University, Fukuoka 812, Japan
****Department of Chemistry, University of California, Riverside, CA 92521

ABSTRACT

Experimental measurements of material effects induced by the passage of sharp shock fronts require techniques which provide high temporal resolution and high spatial resolution. Since typical shock velocities are a few microns per nanosecond, sub-nanosecond probing requires sub-micron spatial resolution. In our experiments, the required temporal resolution is furnished using picosecond laser generated shock waves and picosecond spectroscopy. The spatial resolution is furnished by engineering nanometer scale structures into our shock target arrays. In one technique, absorption transients in the spectrum of a thin layer of molecules, termed an optical nanogauge, are investigated. Shock-induced molecular energy transfer processes are observed in condensed matter for the first time. In a second technique, sub-micron particles of an energetic material are shocked and probed using ps coherent Raman spectroscopy. This probing technique permits the instantaneous measurement of the temperature, pressure and composition of an energetic material under dynamic shock loading.

INTRODUCTION

In this paper, we describe the results of two different kinds of experiments for investigating the dynamics of shocked molecular materials on the sub nanosecond time scale. These experiments are designed to probe some of the most fundamental questions about the behavior of materials immediately after the passage of a fast risetime shock front. In addition, there is a practical motivation behind our work--a fundamental understanding of shock compression dynamics is thought necessary to understand the impact ignition sensitivity of energetic materials, and to engineer safer materials using principles of rational design.

In the past, many shock wave experiments have been performed using picosecond laser pulses [1]. Until now, none of these methods were capable of measuring the picosecond time scale dynamical behavior of molecules induced by the passage of a shock front. The key issue has been the lack of high spatial resolution in prior experiments. In condensed phases, typical shock velocities are a few $\mu m/ns$. Most optical probing methods interrogate a volume whose smallest dimension is typically many μm. In that case, even the shortest probe pulses would probe an average over a region of molecules shocked during a time window of perhaps many ns. On the other hand, if some method could be found to localize the probed volume to a region of nm scale, it would become possible to investigate molecules which were shocked within a time window on the order of ps. In our experiments, sub-μm probing is accomplished by building very small structures in our shock samples. Our experiments are the first to provide the necessary spatial,

337

spectral and temporal resolution to probe the detailed behavior of molecular materials, including high explosives, immediately behind a solid-state shock front.

Consider Fig. 1, a solid-state shock wave moving at a constant velocity U_s in a constant-velocity reference frame. A small volume element of dimension δd, located a distance d behind the front, contains material which was shocked during the time interval $\{t, t + \delta t\}$, where t = d/U_s. Therefore, resolving the short-time response of material immediately following the passage of a shock front requires very high temporal and spatial resolution.

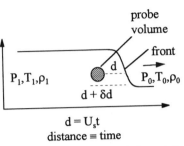

$$d = U_s t$$
$$\text{distance} \equiv \text{time}$$

Figure 1. High time resolution of the effects of a shock front passing through material requires a probing technique with high space resolution, e.g. 50 ps time resolution implies ~250 nm spatial resolution.

It is useful to discuss the dynamics of the shock front in terms of the equilibrium properties of a material with a steady-state shock moving through it. Ahead of the front, the material is at initial temperature, pressure and density (T_0, P_0, ρ_0). Well behind the front, the material is in equilibrium at (T_1, P_1, ρ_1). For many materials, shock impact measurements have been used to determine the relationship between P and ρ, which is termed the Hugoniot relation [2]. Knowing the Hugoniot for a particular material, a measurement of shock velocity U_s unambiguously determines P and ρ. Extensive Hugoniot data is tabulated in the LASL handbook [3]. Understanding the temperature increase across the front, $\Delta T = T_2 - T_1$, presents a much more difficult problem, which is one of the primary subjects of our research. When a material is shocked, it undergoes an *irreversible adiabatic compression* [2]. When the shocked material is allowed to unload, the rarefaction shock process is a *reversible adiabatic expansion* [2]. For a particular shock, the equilibrium temperature increase can be computed using the relation [2,4],

$$T_1 \approx T_0 \exp(\Gamma \Delta V) + \Delta T_{irr}, \tag{1}$$

where Γ is the bulk Grüneisen parameter, and ΔV is the relative volume change, $\Delta V = 1-(\rho_0/\rho_1)$. The first term on the right hand side of Eq. (1) is the temperature increase which would result from a reversible adiabatic compression. The second term is the excess temperature increase due to the irreversible nature of shock compression. This term depends on the details of shock compression (e.g. ΔT_{irr} is greater for a single-stage compression than for a multi-stage compression [2]). Theoretical calculations of ΔT using Eq. (1) are usually quite inaccurate. ΔT is exponentially dependent on Γ, and Γ is not known to high accuracy for any but the simplest materials. Γ may have a V and T dependence which is not considered in Eq. (1). Finally, ΔT_{irr} depends on the details of the shock path. Below we will show how our ps laser experiments can be used to measure T_1 and ΔT_{irr}. When our samples are shocked, they jump from T_0 to temperature T_1, given by Eq. (1). These temperatures can be measured *in situ* using techniques of time-resolved vibrational spectroscopy. When our samples are subsequently unloaded, the cooling during unloading is precisely equal to the first term on the right hand side of Eq. (1), so the final temperature after the loading-unloading cycle, also measured by vibrational spectroscopy, is $T_0 + \Delta T_{irr}$.

It is possible to calculate Γ from first principles. In molecular solids, it is a very complicated problem in anharmonic lattice dynamics, which is only as good as available intermolecular potentials permit [5], highlighting the importance of measuring Γ directly. Such calculations begin by computing the volume-dependence of the phonon and vibrational mode frequencies. For the ith mode, there is a mode Grüneisen parameter γ_i,

$$\gamma_i = \partial \ln \omega_i / \partial \ln V . \qquad (2)$$

In molecular materials, $\gamma_i \sim 1\text{-}5$ for phonons, which are collective excitations involving translations and librations of entire molecules, and $\gamma_i \sim 10^{-2}$ for intramolecular vibrational excitations, which are distortions such as stretching or bending of molecules [6]. The bulk Grüneisen parameter Γ is given by [6],

$$\Gamma = \sum_i \gamma_i C_{v,i} / \sum_i C_{v,i} , \qquad (3)$$

where $C_{v,i}$ is the heat capacity of the ith mode. Equation (3) shows the bulk Grüneisen parameter is a weighted average of mode Grüneisen parameters, weighted by the mode heat capacities.

Although a shock eventually creates a bulk temperature jump to final equilibrium temperature T_1, immediately behind the front the energy deposited by the front is not in a thermal (statistical) distribution. When Eq. (1) is expanded according to Eq. (3), it can be seen that the passage of the shock front will deposit more energy in states with larger values of γ_i. In other words, the shock initially excites phonons to a greater extent than vibrations. Certain lower energy vibrations, especially large amplitude bending modes may also be efficiently pumped by shock, but higher frequency vibrations are not. This point has been discussed by several authors in the last few years [7-10].

The nonequilibrium energy distribution immediately behind the shock front gives rise to a phenomenon termed "overheating" [10,11]. To understand overheating, it is useful to introduce the term *quasitemperature*. At equilibrium, the occupation number of a state with energy hv is given by the Planck distribution, $n_v = [\exp(hv/k_BT)\text{-}1]^{-1}$ where k_B is Boltzmann's constant. To characterize a nonequilibrium energy distribution, one needs a set of occupation numbers $\{n_v\}$. A nonequilibrium distribution could be characterized alternatively by a set of mode quasitemperatures $\{\theta_v\}$, defined by the relation [10], $n_v = [\exp(hv/k\theta_v)\text{-}1]^{-1}$. Overheating means that after the passage of the shock front, the quasitemperatures of some of the initially excited states greatly exceed T_1 for a brief period of time. Behind a shock, overheating in certain states may be quite intense. For example, in theoretical calculations of a 4.7 GPa shock in naphthalene, which produced a net temperature jump from 300K to 650K, the phonons were briefly overheated to a quasitemperature of 2500 deg, which persisted for a few tens of ps [10]. Experimental techniques need to be developed to directly measure the time-dependent behavior of various mode quasitemperature $\theta_v(t)$, immediately behind a shock front [10].

Well behind the shock front, overheating ceases to exist as shock-induced nonequilibrium energy distributions relax to equilibrium temperatures. In the case of a hot impurity domain, overheating diminishes by thermal conduction from the impurity domain to the bulk. In the case of hot phonons or a specific hot vibration, overheating diminishes due to vibrational energy transfer. The origin of mechanical energy transfer is anharmonic coupling [5], which provides a mechanism to randomize energy among a material's phonons and vibrational states.

Returning to the question of energetic material sensitivities mentioned at the beginning of this section, shock heating can have a substantial effect on the likelihood of energetic material ignition. Chemical reactions are extremely sensitive to temperature. All other things being equal, a given shock will produce a larger equilibrium temperature increase in a material with a larger Γ or a larger compressibility. Thus materials which are highly incompressible, or which have highly harmonic phonons are heated less by shock. Understanding the relationships in molecular materials between the compressibility or the Grüneisen parameter, and the structure of the molecule or the unit cell might someday help designers engineer safer insensitive explosives.

The role of overheating in initiation of solids is intimately tied to the formation of hot spots. A pile of tinder will not ignite if the temperature is uniformly raised a few degrees, but if all that energy is concentrated into a single spark, a fire can result. Any mechanism which might concentrate the energy of a shock into a hot spot could drastically affect sensitivity. There are a variety of scenarios where overheating can cause hot spots to be formed. In systems containing impurities, an impurity domain with a larger value of Γ will be initially heated to a temperature greater than T_1. Even if the impurity and bulk have equal Γ, an impurity domain which is more compressible than the bulk will be temporarily overheated by the passage of a shock front. Differences in the microscopic rate constants for various mechanical energy transfer processes can lead to the temporary concentration of energy in an impurity domain, or in a particular vibrational level of the bulk molecules [10]. For example, the sea of delocalized overheated phonons produced behind the shock front can transfer energy efficiently to the vibrations of a defect-perturbed domain, where the phonon-to-vibration transfer rate is only very slightly larger than in the bulk, temporarily producing a hot spot [10].

Until now, no experimental techniques existed to investigate overheating behind a solid state shock front. Given what is presently known about ambient temperature energy transfer processes in materials consisting of large molecules, overheating effects are expected to vanish in perhaps 100 ps (~500 nm length scale) [10]. Even a problem as seemingly straightforward as determining the bulk temperature rise in an energetic material following a shock has proven problematic due to the lack of good fast diagnostics for measuring temperature inside a shocked material. Another difficulty which arises is the difficulty in separating the contributions from shock-induced heating and from heating caused by chemical reactions in the energetic material. In the next two sections, we will describe the first experiments which allow the direct investigation of these interesting phenomena.

EXPERIMENT

Two somewhat different experimental setups were used. Both experiments have several elements in common. Picosecond pulses from Nd:YAG lasers are used to generate shock waves in precise microfabricated layered shock target array assemblies. A single array consists of thousands or millions of identical, individual shock target elements. The arrays are scanned through intersections of several laser pulses so that a fresh target element is presented at every new laser shot, permitting efficient signal averaging at a high repetition rate (up to 1,000 shocks per second). Targets are designed so that a volume being probed has some sub-micron dimension, to provide the necessary spatial resolution. Shock effects are probed using optical absorption or coherent anti-Stokes Raman (CARS) spectroscopy. Experiments performed using a static high pressure cell (diamond anvil) at the University of California, Riverside, were used to determine the static response of the absorption or Raman spectrum to equilibrium high pressure and temperature.

OPTICAL NANOGAUGE EXPERIMENTS

Our optical nanogauge experiments [12] are diagrammed in Fig. 2. A multilayer target is fabricated by sputtering and spin coating. A ps pulse, whose diameter is ~250 μm, produces a hot plasma in the 4 μm thick Al layer. Rapid expansion of the plasma drives a shock wave through the transparent polymer target. For the first few ns, this shock remains planar and it propagates at a constant velocity through the target (c.f. fig. 6). A ps white light pulse enters the polymer and passes through the nanogauge twice before entering a microscope and an optical spectrograph, where the instantaneous absorption spectrum of the nanogauge is recorded. The nanogauge is an ~300 nm thick layer of PMMA containing R640 dye. Because the nanogauge is the only component of the target with an intense absorbance near 560 nm, this arrangement localizes the probed region to an ~300 nm thick volume. As the shock front passes through the nanogauge, the dye spectrum is perturbed, which can be interpreted to reveal the dynamics of the dye layer induced by the passage of the shock front.

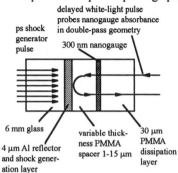

The spectral response of the nanogauge material to static temperature and pressure is illustrated in Fig. 3. These are complicated molecular systems, and it is difficult to cleanly separate P and T effects. Although it is a bit of an oversimplification, our results indicate *the spectral peak red shifts with increasing pressure*. The absorption does not shift noticeably with temperature, but due to hot-band transitions, *the spectrum develops an increased red edge absorption with increasing temperature*.

Figure 2. Schematic diagram of nanogauge experiment, from ref. 12. The gauge consists of R640 dye in PMMA, which has an intense absorbance near 560 nm.

Some typical spectra of shocked nanogauges, spaced 8 μm from the metal shock generation layer, are shown in Fig. 4. In these panels, we use time t = 0 to denote the time when the shock front is midway through the nanogauge (*vide infra*). (The shock front reaches the nanogauge about 2.5 ns after the shock-generation pulse is absorbed by the sample, as seen in Fig. 6.) At t = -70 ps (Fig. 4a), the spectrum (*dashed curve*) is slightly changed from the spectrum (*solid curve*) where no shock pulse was used. These changes are not due to shock-induced effects on the nanogauge. They are caused by shock-induced changes in the reflectivity and refractive index of the metal layer and polymer spacer layer [12]. This -70 ps dashed curve, which represents the nanogauge spectrum just before the shock front hits, is reproduced in each panel as a reference spectrum. At t = 30 ps, (Fig. 4c), the front has just passed through the gauge. Notice the red shifted absorption spectrum and the substantially increased red edge absorption intensity. Between 30 ps and 2 ns, the spectral redshift does not change, but the red edge absorption intensity decreases. Some of these time-dependent effects

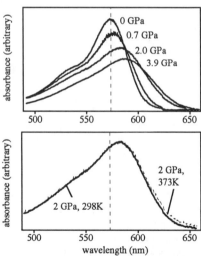

Figure 3. (*top*) Pressure dependence of nanogauge spectra in a diamond anvil cell. Increasing pressure induces a red shift. (*bottom*) Temperature dependence at P = 2 GPa. Increasing temperature 75 degrees induces a small increase in the red absorption. Adapted from ref. 12.

341

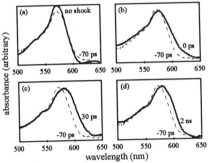

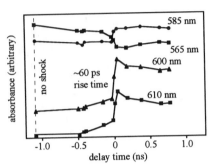

Figure 4. Spectra of a nanogauge during the passage of a shock front. Time t = 0 indicates the front is midway through the 300 nm thick gauge layer. Adapted from ref. 12.

Figure 5. Time dependence of dye absorption at different wavelengths. An ~60 ps risetime is observed. Adapted from ref. 12.

can be seen more easily in Fig. 5, a plot of the time-dependent absorption intensity at different wavelengths. The 600 nm data in Fig. 5 show a fast ~60 ps rise. The midpoint of this rise is taken to indicate the front midway through the nanogauge and is thus used to denote t = 0. The absorption decrease at 565 nm and simultaneous increase at 585 nm, respectively on the blue and red edges of the dye absorption, indicate a fast redshift due to the pressure jump at the front. The risetime of the absorbance changes appears to be limited by the transit time of the shock through the 300 nm thick layer, so it is possible thinner layers could give even better time resolution.

We fabricated a series of target arrays with different thickness spacer layers [12]. The shock front arrival time at the nanogauge versus thickness gives the shock velocity, as shown in Fig. 6. According to reference data [3], this velocity in PMMA corresponds to a shock pressure of P = 2.1 GPa. In Fig. 7, we compare the ps time scale spectral shift due to shock compression to the static shift at P = 2.0 GPa in the diamond anvil. The shifts are identical, showing the nanogauge absorption shift can be used to measure the shock front pressure with ps time resolution.

Notice in Fig. 5 how the 600 nm data rises to a peak and then declines slightly on the ~100 ps time scale. This ultrafast rise and fall is related to the rise and fall of the red absorption edge seen in Figs. 4(c) and 4(d). Occurring on the red absorption edge, this effect indicates an ~100 ps time scale shock-induced overheating process in the dye molecules. A 2 GPa shock is expected to produce a bulk temperature jump in PMMA of ~100 deg [12]. Figure 3 shows the extent of red edge broadening which results from a 75 deg temperature increase. This broadening is far less than observed near t=0 under shock conditions (Fig. 4c), indicating the peak quasitemperature of the dye molecules is *temporarily much greater than 100 deg above ambient*. However at present our signal-to-noise ratio is not sufficient to say much more about the details of the energy transfer

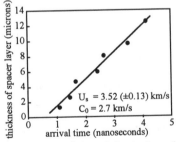

Figure 6. Velocity of shock through target arrays is constant in time, indicating a planar shock wave whose velocity U_s is about 20% greater than C_0, the acoustic velocity in PMMA. This velocity indicates a shock pressure of P = 2.1 GPa. From ref. 12.

process which causes dye molecule overheating. We cannot yet tell, for example, whether the PMMA is undergoing some complicated energy transfer process which affects the dye spectrum, whether the entire dye molecule is being overheated relative to the PMMA, or whether some specific vibrational levels of the dye, which contribute greatly to hot-band absorptions, are being efficiently pumped by the shock front.

Our embedded optical nanogauges can measure the location of a solid-state shock front with a time resolution of ≈60 ps, a considerable improvement over all prior picosecond laser or embedded gauge measurements [12]. Even faster measurements might be possible with thinner gauges. Accurate measurements of shock wave velocity were used to deduce a shock pressure of P = 2.1 GPa.

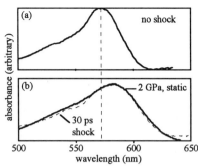

Figure 7. (a) Absorption spectrum of dye in nanogauge at ambient pressure. (b) Comparison of absorption spectrum under static P = 2.0 GPa conditions, to spectrum obtained at 30 ps delay time.

The picosecond time scale peak redshift of the nanogauge dye matches the value obtained in 2 GPa static experiments, showing the nanogauge absorption shift is an accurate way of measuring very fast shock-induced pressure increases. Measurements of fast transients on the red absorption edge of the dye after the passage of the shock front indicate we are observing ultrafast molecular dynamics behind the shock front, which are most likely due to some kind of transient overheating followed by fast (~100 ps) cooling of the dye molecules. In future work, we intend to improve the signal-to-noise of these measurements by using a multichannel detection system, which will help us investigate the details of the overheating process. We also intend to fabricate samples with several nanogauges, to investigate how propagating through thin layers of material can affect the rise time of the shock front.

PICOSECOND CARS EXPERIMENTS ON ENERGETIC MATERIALS

Concurrent with the development of the nanogauge technique has been the development in our lab of a somewhat different method of shock generation, and a very powerful method of probing shock dynamics using ps coherent anti-Stokes Raman spectroscopy (ps CARS). This method is illustrated in Fig. 8. The target array consists of a glass substrate with a few μm-thick layer of PMMA in which small (< 1 μm) crystalline grains of an energetic material are embedded. In our initial experiments [4], the material was TATB (triamino trinitro benzene), an insensitive high explosive. The PMMA is doped with a near-IR absorbing dye [14], which is pumped by a ps duration optical pulse from a YAG laser, of wavelength $\lambda = 1$ μm. The pump pulse produces a large temperature jump (typically hundreds of

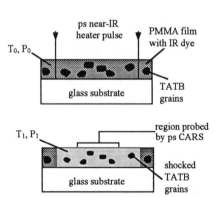

Figure 8. A 150 ps pulse is used to heat a PMMA layer doped with near-IR absorbing dye. Pressure build-up in the PMMA shock compresses grains of TATB for a few ns. The vibrational spectra of TATB and PMMA are probed using time-delayed ps CARS pulses.

343

degrees [13,14]) in the dyed PMMA, but not in the TATB, which does not absorb at that wavelength. The temperature jump process is isochoric, since it occurs faster than the PMMA layer can undergo thermal expansion [14]. Pressures up to 5 GPa have been generated in our apparatus, which endure until the PMMA releases the pressure via expansion on the nanosecond time scale [14].

When the pressure in the PMMA surrounding the TATB crystallites is suddenly increased, the grains undergo shock compression. The timescale of shock compression is determined by a dimension of the sample. It is interesting how this works out in the system diagrammed in Fig. 8. The ps pulse generates an instantaneous pressure jump everywhere in the sample except in the TATB. The rise time of the shock compression of TATB is thus determined by the size of the embedded grains. Accounting for shock propagation and ringing up in the particle, a <1 μm diameter particle in PMMA undergoes compression in about 100 ps [4]. The size of the PMMA sample does not explicitly determine the shock risetime in the grains, but it does determine the duration of the shock loading, since the time required for shock unloading is roughly the acoustic transit time across the thin dimension of the PMMA layer [14]. It is important to note that thermal conduction from hot PMMA to cold TATB is slower than the shock loading and unloading. A typical time constant for conductive heating of the TATB grains is ~50 ns [4]. Thus any effects we see on the ~1 ns time scale are due to shock compression, as opposed to laser heating [4].

The apparatus used in these experiments is shown in Fig. 9. It consists of a high repetition rate Nd:YAG laser which produces a ps duration optical shock pulse in the near-IR, and a pair of dye lasers which produce a pair of synchronized CARS pulses [16]. One dye pulse has a narrow optical bandwidth (~1 cm^{-1}) and the second has a broad bandwidth (~150 cm^{-1}). Using an optical array detector, on even a single shot an entire CARS spectrum is generated using this broadband, multiplex technique. In a typical experiment, the lasers pulse at a few hundred shots per second, and a spectrum is generated by averaging for a few tens of seconds. The large area shock target array (~250 cm^2) is scanned using a motorized XY positioner [14].

We have been able to incorporate many different kinds of crystalline energetic materials into PMMA thin films, including RDX, TATB, NTO, and PETN [4]. The CARS spectra of two of these PMMA films, containing either TATB or NTO, are shown in Fig. 10.

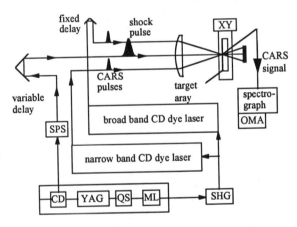

Figure 9. Apparatus used for ps CARS studies of shocked materials. A near-IR pulse from the YAG laser generates the shock. A synchronized pair of pulses from two dye lasers generates a ps CARS spectrum, which is detected using a spectrograph and optical array detector. Key: CD = cavity dumper, QS = Q-switch, ML = mode locker, SHG = frequency doubler crystal, OMA = optical multichannel detector.

The CARS technique provides high resolution vibrational spectra. The frequency shifts of sharp Raman transitions can be used to infer the temperature and pressure of a material. The magnitudes and directions (i.e. redshift, blueshift) of vibrational shifts depend on a subtle balancing of attractive and repulsive intermolecular forces, which cannot yet be accurately predicted from first principles at the present time, but which can be measured quite accurately. In prior work, we showed how the PMMA Raman transition at ~810 cm^{-1} (see Fig. 10) could be used to accurately determine T and P in PMMA during laser temperature and pressure jumps [14].

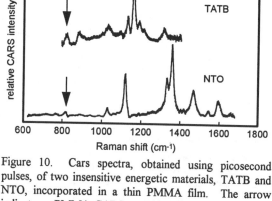

Figure 10. Cars spectra, obtained using picosecond pulses, of two insensitive energetic materials, TATB and NTO, incorporated in a thin PMMA film. The arrow indicates a PMMA CARS transition which is used for T and P calibration.

Some time-resolved spectra of TATB in PMMA are shown in Fig. 11. Static high pressure experiments [16] show the larger peak hardly shifts with pressure at all, and the smaller peak blueshifts with increasing pressure. Temperature dependent measurements were made using a 150 ns duration near-IR pulse which is too long to produce high pressure [4]. With increasing T, both peaks broaden substantially, the larger peak blueshifts and the smaller peak redshifts. Under shock conditions, produced by a ps duration near-IR pulse, both Raman peaks blueshift, but they do not broaden much. Using simultaneous measurements of both peak

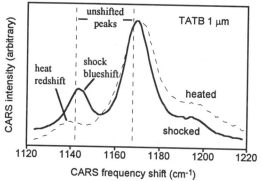

Figure 11. CARS spectra of TATB heated by a long duration (~150 ns) optical pulse, or shocked by a ps duration pulse. The dashed lines indicate the position of TATB spectral peaks at ambient T and P. From ref. 4.

frequencies and T and P calibration data, it is possible to instantaneously determine T and P in TATB with ps CARS. The data shown here were taken with an intensified diode array which does not give very good spectral resolution (an image is blurred among ~7 channels), so the values of T and P given here are rough estimates. A new CCD detector is being installed now which will greatly improve spectral resolution.

An example of this method is shown in Fig. 12. The "no shock" spectrum (representing ambient conditions), and the -370 ps (negative time denotes ps CARS probe prior to the shock pulse) spectra, are identical. At 300 ps, shock loading is near maximum. At this time, the crystals have been heated by the irreversible process of shock compression (see Eq. (1)). Both peaks are blueshifted at this time. Determining T_1 and P_1 in this material is simplified because the larger peak is insensitive to P [16]. Its blueshift gives an estimate of $T_1 \sim 150°C$. From the shift of the smaller peak in Fig. 12, which is responsive to both T and P, an estimate of $P_1 \sim 1.2$ GPa is obtained. By 1.4 ns, the PMMA has unloaded. The unloading process is a reversible adiabatic expansion. The temperature change during unloading is given by the first term in Eq. (1). After unloading, the redshift of the smaller peak in Fig. 12 shows the TATB is approximately 50 deg above ambient temperature. At this time, TATB has been subjected to a cycle of irreversible shock loading and reversible unloading, and the residual temperature increase in the TATB is simply the ΔT_{irr} term in Eq. (1). These preliminary measurements show how it is possible to determine ΔT_{irr} and Γ in Eq. (1), using ps CARS temperature measurements of two peaks at two times during the shock cycle [4]. Because the time-scales of these measurements are faster than any chemical reactions occur (if significant chemical reactions had occurred, the TATB peaks would diminish and new

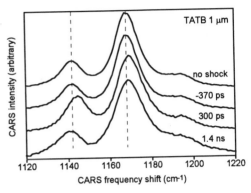

Figure 12. Ps CARS spectra of 1 μm diameter grains of TATB during shock compression. Time 300 ps is near the maximum shock loading point. Time 1.4 ns is after most of the shock in the PMMA has unloaded. At this time, the redshift of the smaller peak is due to heat remaining in the particle after a cycle of irreversible shock and reversible unloading. Dashed vertical lines indicate unshifted TATB peak locations. From ref. 4.

peaks would appear), these fundamentally important properties of energetic materials and of the irreversible shock process can be directly determined.

CONCLUSIONS

Two new techniques capable of measuring the behavior of shocked molecular materials on the ≤ 100 ps time scale have been described. With the experience gained so far, an understanding of the advantages and disadvantages of both techniques is beginning to emerge. The nanogauge technique is advantageous in that it provides information about the propagation of shock waves, the risetime of the shock front, and energy transfer dynamics of molecules immediately behind the front. The shock process is particularly simple and well-characterized, in that it involves a planar, single-stage, constant velocity compression. Disadvantages of the nanogauge method are the difficulties of precisely fabricating the multilayer target array with layers which are very flat and uniform, and the need to use a very thin gauge layer. In the very thin gauge, practical considerations presently limit us to the study of the optical absorption of strongly colored dye molecules. Energetic materials are mostly crystalline, which are hard to incorporate into multi-layer thin film assemblies and are ordinarily colorless materials.

Disadvantages of the embedded crystal method are the lack of information about shock propagation or dynamics immediately behind the front. The shock process is not very well characterized since the particles are subjected to a nonspherical compression and there are some complicated reverberations (ringing up) due to shock impedance mismatchs between PMMA and crystals. But there are many advantages. It is very easy to fabricate the arrays, which have just a single layer whose thickness need not be controlled very accurately. The time response is limited by the particle size, not by the size of the PMMA film, so it is possible to attain high time resolution even when the optical pathlength of material being studied is large. This feature permits vibrational spectroscopy, which is typically less sensitive than optical absorption, to be used. Vibrational spectroscopy is highly advantageous, since it permits instantaneous probing of T, P, and sample composition.

We have made the very first observations of shock-wave induced molecular dynamics occurring on the sub 100 ps time scale in molecular materials using the optical nanogauge technique. A better knowledge of ultrafast energy transfer processes behind the front is expected to lead to a better understanding of shock-initiation processes in energetic materials. Using embedded particles, we have demonstrated sub 100 ps time scale measurements of T and P in a shocked energetic material, which should ultimately allow us to accurately measure the heat generated in an irreversible shock compression process, on a time scale fast relative to that required for chemistry. The ps CARS technique is clearly able to address the important problems of understanding the initial steps in energetic material chemistry during shock initiation.

ACKNOWLEDGMENT

The work at Illinois was supported by Air Force Office of Scientific Research contract F49620-94-1-0108, Army Research Office contract DAAH 04-93-G-0016, and National Science Foundation grant DMR 94-04806. Work at the University of California was supported by National Science Foundation grant CHE-94-00542. One of us (H. S.) thanks the Taro Yamashita Foundation for their financial support in 1994-5.

REFERENCES

1. See, e.g., A. L. Huston, B. L. Justus and A. J. Campillo, Chem. Phys. Lett. 118, 267 (1985); B. L. Justus, A. L. Huston and A. J. Campillo, ibid, 128, 274 (1986); X. Z. Lu, R. Rao, B. Willman, S. Lee, A. G. Doukas, and R. R. Alfano, Phys. Rev. B, 35, 7515 (1987); K. P. Leung, S. S. Yao, A. G. Doukas, R. R. Alfano, and P. Harris, Phys. Rev. B31, 942 (1985).

2. Y. B. Zel'dovich and Y. P. Raiser, *Physics of Shock Waves and High-Temperature Hydrodynamic Phenomena*, (Academic Press, New York, 1966), Vols. 1 and 2.

3. *LASL Shock Hugoniot Data*, S. P. Marsh, Ed. (University of California, Berkeley, 1980).

4. D. E. Hare, J. Franken and D. D. Dlott, Chem. Phys. Lett. 244, 224 (1995).

5. S. Califano, V. Schettino, and N. Neto, Lattice Dynamics of Molecular Crystals, (Springer, Berlin, 1981).

6. A. J. Pertsin and A. I. Kitaigorodsky, The Atom-atom Potential Method: Applications to Organic Molecular Solids, (Springer-Verlag, Berlin, 1987).

7. D. J. Pastine, D. J. Edwards, H. D. Jones, C. T. Richmond and K. Kim, in High-Pressure Science and Technology, v. 2, edited by K. D. Timmerhaus and M. S. Barber (Plenum, New York, 1979), p. 364.

8. C. S. Coffey, and E. T. Toton, J. Chem. Phys. 76, 949 (1982); F. J. Zerilli,; and E. T. Toton, Phys. Rev. B 29, 5891 (1984).

9. R. D. Bardo, Int. J. Quantum Chem. Symp. 20, 455 (1986).

10. D. D. Dlott and M. D. Fayer, J. Chem. Phys. 92, 3798 (1990); A. Tokmakoff, M. D. Fayer, and D. D. Dlott, J. Phys. Chem. 97, 1901 (1993).

11. S. Chen, X. Hong, J. R. Hill and D. D. Dlott, J. Phys. Chem. 99, pp. 4525-4530 (1995).

12. I-Y. S. Lee, J. R. Hill, H. Suzuki, B. J. Baer, E. L. Chronister, and D. D. Dlott, J. Chem. Phys (in press).

13. S. Chen, I-Y. S. Lee, W. Tolbert, X. Wen and D. D. Dlott, J. Phys. Chem., 96, pp. 7178-7186 (1992).

14. D. E. Hare, J. Franken and D. D. Dlott, J. Appl. Phys. 77, pp. 5950-5960 (1995).

15. D. E. Hare and D. D. Dlott, Appl. Phys. Lett., 64, pp. 715-717 (1994).

16. W. M. Trott and A. M. Renlund, J. Phys. Chem. 92, 5921 (1988); S. K. Satija, B. Swanson, J. Eckert and J. A. Goldstone, J. Phys. Chem. 95, 10103 (1991).

INVESTIGATION OF SHOCK-INDUCED CHEMICAL DECOMPOSITION OF SENSITIZED NITROMETHANE THROUGH TIME-RESOLVED RAMAN SPECTROSCOPY

G.I. Pangilinan, C.P. Constantinou, Y.A. Gruzdkov, and Y.M. Gupta
Shock Dynamics Center and Department of Physics, Washington State University, Pullman, Washington 99164 - 2814

ABSTRACT

Molecular processes associated with shock induced chemical decomposition of a mixture of nitromethane with ethylenediamine (0.1 wt%)are examined using time-resolved, Raman scattering. When shocked by stepwise loading to 14.2 GPa pressure, changes in the nitromethane vibrational modes and the spectral background characterize the onset of reaction. The CN stretch mode softens and disappears even as the NO_2 and CH_3 stretch modes, though modified, retain their identities. The shape and intensity of the spectral background also shows changes characteristic of reaction. Changes in the background, which are observed even at lower peak pressures of 11.4 GPa, are assigned to luminescence from reaction intermediates. The implications of these results to various molecular models of sensitization are discussed.

INTRODUCTION

A good understanding of molecular processes in shock-induced chemical reactions in condensed materials is important to problems related to detonation and impact sensitivity. [1-3] Experiments that directly probe these molecular processes are central to current research on energetic materials. Results from these experiments complement continuum studies, and together with advances in theory and computational methods can provide a detailed understanding of shock-induced reactions.

We have utilized time-resolved UV-vis absorption [4] and Raman spectroscopies [5] to probe the electronic and vibrational changes that occur in shocked pure nitromethane and sensitized nitromethane (nitromethane with 0.1 wt% ethylenediamine). In pure nitromethane shocked to 14.0 GPa, a pressure induced red-shift of the n- π^* electronic band edge, and hardening of the CN and NO_2 and CH_3 stretching modes were observed; however, no time-dependent spectroscopic changes arising from chemical reactions were observed within our experimental time window. In sensitized nitromethane shocked to as low as 10.0 GPa, an irreversible red shift of the band edge characteristic of chemical reaction onset was observed. To examine the molecular changes associated with the reaction, we have improved our time-resolved Raman spectroscopic measurements to monitor the nitromethane vibrations during the chemical reaction [6] . In this paper we discuss our time-resolved Raman experiments on sensitized nitromethane and infer plausible molecular processes during its decomposition.

The sensitization of nitromethane upon the addition of small amounts of amines is well known from continuum measurements. [7,8] Several molecular models have been proposed to explain the sensitization. Engelke [7] proposed that the nitromethyl aci-ion ($H_2C = NO_2^-$) concentration, enhanced at high pressures or upon amine addition, is responsible for the sensitization. Based on quantum chemical calculations, Cook and Haskins [8] proposed that CN breakage occurs due to a bimolecular interaction between the amine and the nitromethane molecule, and subsequently the reaction progresses autocatalytically. In 1992, Constantinou et.al. [9] proposed that a charge transfer complex, present at ambient conditions, weakens the CN bond which breaks at shock conditions. The plausibility of these models is examined here.

EXPERIMENT

Materials and Method

The samples used in the experiments were prepared from 99 +% nitromethane (CH_3NO_2, Aldrich) and 99 +% ethylenediamine ($NH_2CH_2CH_2NH_2$, Aldrich). We mixed 5 microliters of freshly distilled ethylenediamine (density = 0.899 g/cm^3) to 4.5 grams of nitromethane (density = 1.127 g/cm^3) to yield 0.1 wt% ethylenediamine/nitromethane mixtures. The experiments were carried out on the same day that the mixtures were prepared.

Figure 1 shows the experimental configuration used to obtain time-resolved Raman measurements on the shocked samples. Briefly, a sample of about 250 micrometer thickness is contained between two sapphire windows. Raman spectra are obtained continuously at 45 or 50 ns time intervals while the sample is in a state of uniaxial strain. Light from a pulsed laser (514 nm; ~120 millijoule energy selected by adjusting aperture A; ~3.5 microsecond duration) is transmitted through optical fibers and focused to a 1 millimeter diameter spot on the sample. The scattered radiation from a 400 micrometer diameter spot is collected through lenses and transmitted through optical fibers to a relay lens system. This lens system consists of 2 lenses and an aperture to optimize transmission from the optical fiber to the spectrometer (500 M SPEX, NJ) and a holographic edge filter (POC, CA) to attenuate elastically scattered light. The Raman scattered light is spectrally dispersed by the spectrometer, temporally dispersed by the streak camera (Imacon 790, Hadland) intensified by a microchannel plate intensifier and recorded on a back illuminated CCD detector (PI, CA) in an intensity-frequency-time format.

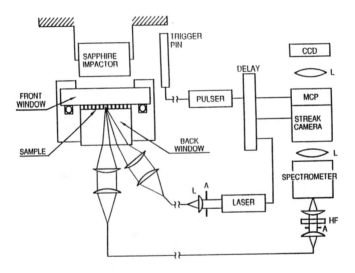

Figure 1. Experimental Configuration for the time-resolved Raman scattering measurements

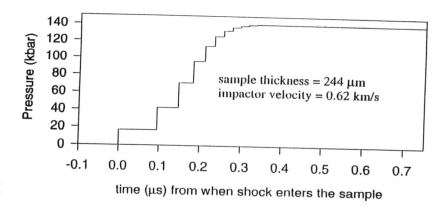

Figure 2. Typical pressure history in the stepwise loading experiments.

<u>Pressure History in the Material</u>

    The plane shock wave compressing the sample is generated by impact between a sapphire impactor and the front window. The liquid sample is subjected to step wave loading as the shock wave reverberates between the two sapphire windows to a final pressure. This pressure is maintained while Raman scattering data are collected from the center of the sample.

    The pressure history and the peak temperature in the sample were calculated using wave propagation codes and the equations of state of c-cut sapphire [10] and pure nitromethane. [11] Furthermore, with the low concentration of the amine used, the equation of state of the unreacted, sensitized nitromethane mixtures is expected to be comparable to that of pure nitromethane. The equation of state of nitromethane [11] was based on the universal liquid Hugoniot equation [12] with constants adjusted to match data from Lysne and Hardesty. [13] The specific heat was modeled based on a single Einstein oscillator, and values of $(dP/dT)_V$ extrapolated from measurements of the dielectric constant as a function of pressure. [11] A typical pressure history is shown in Figure 2. It is important to note that the peak pressure obtained in these calculations is quite accurate depending only on the shock response of sapphire; the calculated temperature however is strongly dependent on the nitromethane equation of state used.

RESULTS

    We now discuss two experiments that show evolving spectral changes that correspond to chemical reaction onset in the shocked nitromethane/ethylenediamine mixtures. In Experiment 1, we shocked sensitized nitromethane to 11.4 GPa peak pressure with a calculated temperature of 720 K. In Experiment 2, we shocked the sample with a faster impactor velocity to obtain a higher peak pressure of 14.2 GPa and calculated temperature of 769 K. [6] The pressure history for Experiment 2 is shown in Figure 2.

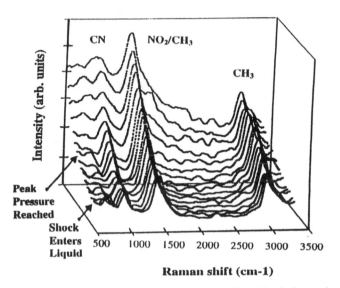

Figure 3. Three dimensional plot of the Raman spectra of sensitized nitromethane shocked to 11.4 GPa peak pressure, showing the evolution in time of the spectral background, even when pressure is constant.

Figure 3 shows successive 50 ns Raman spectra obtained from sensitized nitromethane shocked by step wave loading to 11.4 GPa (Experiment 1). At ambient pressure, three Raman bands at 917 cm^{-1}, 1400 cm^{-1}, 2968 cm^{-1} corresponding to the CN stretch, NO_2 stretch convoluted with the CH_3 bend, and the CH_3 stretch respectively are observed. During the shockup period, the energies of the Raman bands increase. Upon reaching peak pressure, and to the end of the experiment, the energies of the Raman bands are maintained. The time response of the frequency shifts therefore agrees with the pressure history in the sample. There is also an apparent broadening in the CN mode at latter times. Except for this CN broadening, the behavior of the vibrational bands are therefore similar to those in unreacted pure nitromethane shocked by stepwise loading. [5]

In contrast to the Raman spectra of unreacting material however, a broad feature from 500 - 2000 cm^{-1} grows in time after peak pressure has been reached. This time dependent change is suggestive of chemical reaction onset.

At higher shock pressures, more dramatic changes are observed in time.[6] Figure 4 shows spectra, taken at 45 ns intervals, from sensitized nitromethane shocked to 14.2 GPa (Experiment 2). The three vibrational bands are shown at ambient pressure. All of the bands are observed to harden during shockup until peak pressure is reached at 300 ns. Later in time, the CH_3 stretch mode shows softening and then disappears while the NO_2 stretch/CH_3 bend and the CH_3 stretch modes retain their identities.

The broad feature at 500 cm^{-1} - 2000 cm^{-1} also observed in Experiment 2, is evident after peak pressure is reached (t=360 ns). As in Experiment 1 the spectral background continues to grow even at constant pressure.

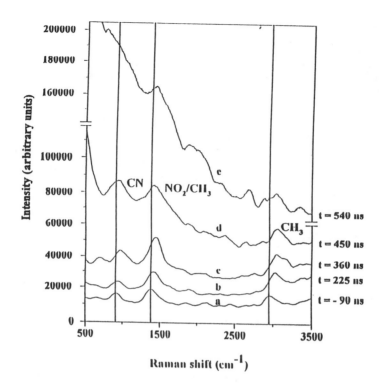

Figure 4. Representative Raman spectra of sensitized nitromethane shocked to 14.2 GPa. Vertical lines mark the ambient frequency positions of the vibrational modes. The CN vibration softens in (d) and then disappears in (e). An increase in the background similar to that in Figure 3 is also observed. Shock enters the samples at t=0, and successive spectra are vertically offset by 5000 counts for clarity. [6]

DISCUSSION AND CONCLUSIONS

The onset of chemical reaction is identified in the Raman experiments through time resolved spectral changes that do not correlate with the pressure history, in the shocked materials. The background feature around 500-2000 cm⁻¹ that grows in time, and the CN vibrational softening and its disappearance, while the pressure in the material is held constant, indicate that chemical changes are occurring in shocked sensitized nitromethane. These Raman results are consistent with previous conclusions from UV-vis absorption experiments on similarly shocked mixtures [4] where sensitized nitromethane shocked to as low as 10.0 GPa was found to undergo irreversible chemical changes.

The molecular processes that occur can be inferred from the Raman scattering results. Since the intensity of Raman scattering signals arise from an averaged contribution of all molecules that are being probed, the vibrational changes observed in Figure 4, are caused by changes in the bulk material. CN scission is therefore an important, step in the decomposition of bulk nitromethane. This bond softening observed is incompatible with the proposed enhancement of the nitromethyl aci-ion concentration. [7] The aci-ion with a CN double bond is expected to have an even higher energy CN vibration.

The background feature observed in Figures 3 and 4 probably arises from fluorescing byproducts of initial reactions involving complexes present at ambient pressures [9] or formed due to bimolecular reactions between an amine and a nitromethane molecule. [8] These byproducts and the energy released from these initial reactions then promote the reactions involving the bulk nitromethane. Because of the low concentration of amines used, vibrational spectroscopy cannot identify the nature of the complexes (charge transfer or hydrogen bonded) involved in the initial decomposition. We are currently exploring fluorescence and emission spectroscopy in an attempt to identify reaction intermediates or products.

ACKNOWLEDGMENTS

We wish to acknowledge the valuable contributions of Dr. J.M. Winey to the nitromethane project. Professor George Duvall is sincerely thanked for many helpful discussions and for his work on the nitromethane equation of state. Kurt Zimmerman and Dave Savage are thanked for their assistance with the performance of the impact experiments. This work was supported by ONR Grant N00014-93-1-0369; Dr. R.S. Miller is thanked for his strong interest in this work.

REFERENCES

1. Proceedings of the Tenth International Detonation Symposium, Office of Naval Research, Arlington, VA (1995).

2. Y.M. Gupta, J. de Physique IV, Colloque C4, 5, p. C4-345 (1995).

3. R. Cheret, Detonation of Condensed Explosives, Springer-Verlag, New York, (1993).

4. C.P. Constantinou, J.M. Winey, and Y.M. Gupta, J. Phys. Chem.98, 7767 (1994).

5. G.I. Pangilinan and Y.M. Gupta, J. Phys. Chem. 98, 4522 (1994).

6. Y.M. Gupta, G.I. Pangilinan, J.M. Winey, and C.P. Constantinou, Chem. Phys. Lett. 232, 341 (1995).

7. R. Engelke, Phys. Fluids 23, 875 (1980).
R. Engelke, W.L. Earl, and C.M. Rohlfing, Int. J. Chem. Kinet. 18, 1205 (1986).
R. Engelke, W.L. Earl, and C.M. Rohlfing, J. Chem. Phys. 84, 142, (1986).
R. Engelke, D. Schiferl, C.B. Storm, and W.L. Earl, J. Phys. Chem. 92, 6815, (1988).
R. Engelke, W.L. Earl, and C.B. Storm, Propellants, Explosives and Pyrotechnics 189, 36, (1988).

8. M.D. Cook and P.J. Haskins, Proceedings of the 9th Symposium (Int.) Detonation 1027, (1989); M.D. Cook and P.J. Haskins, Proceedings of the 10th Symposium (Int.) Detonation 870, (1995).

9. C.P. Constantinou, T. Mukundan, and M.M. Chaudri, Philos. Trans. R. Soc. **A339**, 403 (1992).

10. L.M. Barker and R.E. Hollenbach, J. Appl. Phys. **41**, 4208, (1970).

11. J.M. Winey, <u>Time-Resolved Optical Spectroscopy to Examine Shock-Induced Decomposition in Liquid Nitromethane</u>, Ph.D. Dissertation, Washington State University, (1995).

12. R.W. Woolfolk, M. Cowperthwaite, and R. Shaw, Thermochim. Acta. **5**, 409, (1973).

13. P.J. Lysne and D.R. Hardesty, J. Chem. Phys. **59**, 6512, (1973).

VIBRATIONAL ENERGY TRANSFER IN HIGH EXPLOSIVES: NITROMETHANE

XIAOYU HONG, JEFFREY R. HILL, DANA D. DLOTT
School of Chemical Sciences, University of Illinois, Urbana, IL 61801. Correspondence to Dana
D. Dlott, Box 37 Noyes Lab, 505 S. Mathews Ave., Urbana, IL 61801. d-dlott @ UIUC.EDU

ABSTRACT

Time resolved vibrational spectroscopy with picosecond tunable mid-infrared pulses is used to measure the rates and investigate the detailed mechanisms of multiphonon up-pumping and vibrational cooling in a condensed high explosive, nitromethane. Both processes occur on the ~100 ps time scale under ambient conditions. The mechanisms involve sequential climbing or descending the ladder of molecular vibrations. Efficient intermolecular vibrational energy transfer from various molecules to the symmetric stretching excitation of NO_2 is observed. The implications of these measurements for understanding shock initiation to detonation and the sensitivities of energetic materials to shock initiation are discussed briefly.

INTRODUCTION

In the initiation to detonation of condensed phase energetic materials, molecular mechanical energy transfer plays two distinct and possibly significant roles [1-3]. (a) When a shock front passes through a material, phonons are preferentially excited [1-6]. (In liquids which are not translationally invariant, the preferentially excited states are collective instantaneous normal modes [7], which for simplicity we will continue to call phonons.) Mechanical energy transfer from phonons to molecular vibrations, termed multiphonon up-pumping [1], is needed before thermochemical reactions occur [1-6], because chemical reactions require the formation of vibrationally activated complexes. (b) Most elementary steps in the chemical reactions of secondary explosives are endothermic. Virtually all the excess heat is produced in a very few highly exothermic steps [8]. These steps produce highly vibrationally excited nascent products. Energy transfer from molecular vibrations to phonons is required to continue driving the shock front through the system [3], if a sustained detonation wave is to be produced.

In equilibrium thermodynamics, the occupation number of a vibrational state with fundamental frequency v is related to the temperature T by the Planck distribution function,

$$n = [\exp(hv/k_BT)-1]^{-1} , \qquad (1)$$

where k_B is Boltzmann's constant. To describe a nonequilibrium state, a set of occupation numbers $\{n\}$ is required. It is convenient to introduce the concept of quasitemperature [1] into this discussion. The quasitemperature θ_v of a state with frequency v is defined by the relation,

$$n = [\exp(hv/k_B\theta_v)-1]^{-1} . \qquad (2)$$

In quasitemperature notation, an instantaneous nonequilibrium distribution can be specified by a set of time dependent quasitemperatures, $\{\theta_v(t)\}$.

The idea of delayed excitation behind a shock front, of vibrational states in a gas, was discussed extensively in the 1950's [9]. The extension to condensed matter, and its relevance to energetic materials was first discussed to our knowledge in 1979 by Pastine et al. [4], and substantial contributions were made by Coffey and Toton [5a], Zerilli and Toton [5b], Bardo [6], Calef and Tarver [10], Dlott and Fayer [1], and Fried and Ruggerio [11]. Until the present work, there have been no experimental measurements of vibrational energy transfer in high explosives.

Vibrational energy transfer processes can possibly affect energetic material initiation in several different ways:

(1) A finite time for phonon-to-vibration transfer may result in an induction time delay between shock front arrival and initiation [6]. The vibrationally starved zone behind the shock front is termed the up-pumping zone. The thickness of the up-pumping zone $\ell_{up} = U_s t_{up}$, where U_s is the shock velocity and t_{up} is the time constant for up-pumping [10].

(2) During the up-pumping process, an extensive delocalized sea of ultrahot phonons exists. The phonon energy can become localized and concentrated in the vibrations of defect perturbed domain sites, provided the anharmonic phonon-to-vibration coupling at the defect domain exceeds even very slightly the anharmonic coupling in the bulk material. In this case, a hot spot is formed at the defect domain. Formation of a hot spot can greatly enhance the likelihood of initiation, provided the spot is large enough, it is hot enough, and it persists long enough [1].

(3) Chemical reaction dynamics can be affected by energy transfer rates. The rate constant for any elementary step in the initiation chemistry can be expressed using an Arrhenius function, $k = A\exp(-E^*/RT)$. The preexponential factor A depends on the strength of interaction between the vibrations of the activated complex and the phonon bath. For instance, in Kramers' theory [12], there are limiting cases where A can increase or decrease with increasing t_{up}. A suggested correlation between energy transfer rates and energetic material sensitivities has been rationalized using this argument [11]. As in (2) above, a defect perturbed domain with different t_{up} could potentially form a hot spot if the chemical reaction rates in the domain were faster than in the bulk.

(4) During vibrational cooling, hot spots could be formed from a sea of vibrationally excited molecules if vibrational energy spontaneously localized on individual molecules [13]. Consider two molecules, each excited in the v = 1 vibrational state. Localizing this energy on one molecule produces one v = 2 and one v = 0 molecule. For anharmonic vibrations, the energy of the v = 2 molecule is less than the sum of the two v = 1 molecules. In a system where very many vibrationally excited molecules communicate, the excitation level on one molecule will increase until the energy decrease is balanced by the entropy decrease. In some systems, this can occur at very high excitation levels [13]. Hot spot formation by vibrational localization can occur only if the rate of vibrational transfer among excited molecules exceeds the vibrational decay rate [14].

Until recently, nothing concrete was known about the details, rates and mechanisms of multiphonon up-pumping and vibrational cooling in condensed high explosives. The experimental tools to study these processes were not sufficiently developed. We have undertaken an extensive series of investigations, using newly developed, powerful laser sources which are tunable in the mid-IR, to study energy transfer dynamics in a model high explosive system, nitromethane (NM).

EXPERIMENTAL

The experimental technique involves a ps infrared pump pulse, which creates an excess of vibrations or phonons, and a visible probe pulse [15]. The probe generates an instantaneous incoherent anti-Stokes Raman spectrum of the nitromethane. In anti-Stokes Raman scattering, the intensity of a Raman transition is proportional [15] to the occupation number n in Eq. (2). By comparing transient spectra to spectra obtained at a known (ambient) temperature, parameters such as the absolute Raman cross-section can be accounted for [2]. This technique is able to measure instantaneous absolute vibrational quasitemperatures $\theta_v(t)$ for all the Raman active vibrations of NM [2-3]. The experimental apparatus is diagrammed in Fig. 1.

A new concept had to be developed to study multiphonon up-pumping, because available infrared laser sources could not produce large concentrations of phonons. A near-IR dye termed IR-165 was dissolved in the NM. This dye could be pumped by the 1.053 μm fundamental from the YLF laser. Ultrafast (~3 ps) radiationless relaxation [16] of the dye produced a shower of phonons, which caused up-pumping of the NM. Vibrational cooling measurements [3] were made by tuning the OPA into the C-H stretching excitations of NM (~3,000 cm⁻¹). These are the highest energy molecular vibrations in NM.

MULTIPHONON UP PUMPING

Up-pumping data on ambient temperature NM are shown in Fig. 2. First, phonons are excited by the laser via the near-IR

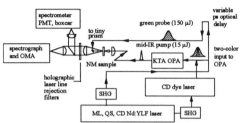

Figure 1. Apparatus for IR pump--anti-Stokes Raman probe measurements on nitromethane (NM). A mode-locked (ML), Q-switched (QS) Nd:YLF laser pumps a cavity-dumped (CD) dye laser. A frequency doubled (SHG) cavity dumped pulse (green) from the YLF laser is mixed with the dye laser in an optical parametric amplifier (KTA OPA) to produce mid-IR pulses. Some of the YLF pulse is used for anti-Stokes probing, which is detected by a multichannel detector (OMA) or photomultiplier tube (PMT). Adapted from ref. 3.

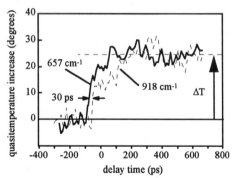

Figure 2. Multiphonon up pumping of two NM vibrations at 657 cm⁻¹ and 918 cm⁻¹. The bulk temperature jump ΔT is about 25 deg. The lower energy 657 cm⁻¹ vibration becomes excited ~30 ps before the 918 cm⁻¹ vibration. Up pumping is complete in about 100 ps. Adapted from ref. 2.

359

dye. Then molecular vibrations begin to be excited by phonons via up-pumping. Ultimately the system comes to equilibrium at a higher temperature; here about 25 degrees higher (indicated by the dashed horizontal line). We studied two Raman active vibrations. The 657 cm^{-1} vibration is a NO$_2$ rocking mode and the 918 cm^{-1} vibration is a C-N stretching mode. The lower frequency 657 cm^{-1} vibration becomes excited ~30 ps before the higher frequency 918 cm^{-1} vibration. Up pumping is complete in about 100 ps [2].

Vibrational cooling data following C-H stretching excitation at 3000 cm^{-1}, is presented in Fig. 3. The bulk temperature jump ΔT is ~ 10 degrees. Here we studied the totally symmetric nitro (NO$_2$) stretching mode at 1400 cm^{-1} and the C-N stretch at 918 cm^{-1}. The data show energy accumulates in both vibrations for a few tens of ps before a subsequent decay process. The higher energy nitro stretch is excited before the lower energy C-N stretch. The timescale of vibrational cooling is about 100 ps.

Some experiments were performed to investigate the possibilities for energy transfer and vibrational energy localization in NM. In these experiments, we used NM samples with a small amount of added alcohols such as methanol (MeOH) [14]. A ps mid-IR pulse at 3600 cm^{-1} was used to pump the methanol O-H stretch. Then NM vibrational excitation was monitored using anti-Stokes Raman spectroscopy. A reasonably efficient intermolecular energy transfer process from MeOH to the NM nitro stretch was observed (see Fig. 4). Notice the peak vibrational quasitemperature increase of the nitro stretching vibration (~80 deg) is quite a bit larger than the bulk temperature jump (~10 deg).

Using isotope substitution experiments, the detailed pathways of the intermolecular transfer process were elucidated. The dominant pathway

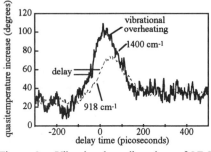

Figure 3. Vibrational cooling data of NM vibrations at 918 and 1400 cm^{-1}. The bulk temperature jump ΔT is ~ 10 degrees. The higher energy vibration becomes excited first. Both vibrations show intense, transient overheating. Cooling occurs on the 100 ps time scale. Adapted from ref. 3.

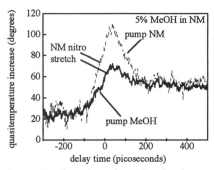

Figure 4. Time dependent vibrational quasitemperature for NM nitro stretch (1400 cm^{-1}) when NM is pumped (C-H stretch) or when methanol (MeOH) is pumped (O-H stretch). The efficiency of intermolecular energy transfer from MeOH to the nitro stretch is about 40% of the efficiency of intramolecular energy transfer from NM C-H to NM nitro stretch. Adapted from ref. 14.

is (O-H)$_{MeOH}$ → (C-H)$_{MeOH}$ → (C-H bend)$_{MeOH}$ → (NO$_2$)$_{NM}$. [14]. It was surprising to see efficient intermolecular energy transfer from C-H bending to nitro groups. In our paper [14], we suggested that other molecules with C-H bending modes, typically formed in energetic material reactions, might also be able to efficiently donate vibrational energy to nitro stretching vibrations.

DISCUSSION

These experiments provide the first snapshots of mechanical energy transfer processes in condensed high explosives containing a nitro group. Up-pumping and vibrational cooling both occur on the 100 ps time scale. (This time scale might be a somewhat faster under high pressure [2]). Energy moves up and down a vibrational ladder. That is to say, when low energy phonons are pumped, the lower energy vibrations are excited first and the higher energy vibrations are excited subsequently (Fig. 2). When high energy vibrations are pumped, the higher energy vibrations are excited first and the lower energy vibrations are excited subsequently (Fig. 3).

In earlier theoretical papers [5,6], quantum mechanical calculations were performed which suggested the up-pumping time constants in NM were on the order of 100 ns. It was suggested this quite slow up-pumping process could explain the induction time in NM, which was of similar magnitude [6]. Induction time refers to the delay time between the passage of the front and the observation of bright emission from the shocked NM. Our experiment quite clearly demonstrates that up-pumping occurs on the tens of ps time scale, which could not possibly explain the induction time. Instead the induction time is mostly likely related to the details of NM reaction kinetics--heat is not evolved until some of the very last steps in a complicated chain [8]. There is not enough room in this article to discuss in detail why these models predicted such a long up-pumping time constant. Using a great deal of oversimplification, here is an explanation:

We are concerned with processes where a large number of phonons pump a vibration. For example fourteen phonons of ~100 cm^{-1} energy might pump a 1400 cm^{-1} nitro stretching mode. Older theoretical papers [5,6] assumed the mechanism to involve high-order anharmonic multiphonon processes. In the example given, a 14-phonon process is required. The rates of such very high-order processes are ordinarily quite slow [3]. Our work shows up-pumping involves a ladder-climbing process. For example a very few phonons might first pump the 480 cm^{-1} vibration of NM. Then one or two phonons will then pump 480 cm^{-1} to 657 cm^{-1}. Then one phonon will pump 657 cm^{-1} to the next higher energy vibration, etc. This sequential (as opposed to simultaneous) up-pumping requires several individual steps, but each step is a lower-order multiphonon process. The overall time constant for several sequential lower-order steps to occur can be of order ~100 ps, very much faster than the rate of a single very high-order process [3].

Our observation of efficient intermolecular energy transfer from donor molecules to the nitro stretch of NM is interesting and suggestive of intriguing possibilities. As mentioned in the introduction, if an efficient intermolecular vibrational energy transfer channel exists, and if the vibration being pumped has a sufficiently long lifetime, vibrational energy can spontaneously localized onto a single molecule, pumping it into a very high energy vibrational state capable of undergoing a chemical reaction. Our experiments show the conditions needed for this localization exist in the ~1400 cm^{-1} nitro stretch of NM. Here we are speculating, but it is possible that the ignition of high explosives might be assisted by spontaneous localization of vibrational energy onto nitro groups. This line of reasoning might help explain why nitro groups are ubiquitous in high explosives, and why loss of NO$_2$ is often the initial step in energetic material chemistry.

Our research was supported by Air Force Office of Scientific Research contract F49620-94-1-0108, Army Research Office contract DAAH 04-93-G-0016, and National Science Foundation grant DMR 94-04806.

REFERENCES

1. D. D. Dlott and M. D. Fayer, J. Chem. Phys. 92, 3798 (1990); A. Tokmakoff, M. D. Fayer, and D. D. Dlott, J. Phys. Chem. 97, 1901 (1993).

2. S. Chen, W. A. Tolbert and D. D. Dlott, J. Phys. Chem. **98**, pp. 7759-7766 (1994).

3. S. Chen, X. Hong, J. R. Hill and D. D. Dlott, J. Phys. Chem. **99**, pp. 4525-4530 (1995).

4. D. J. Pastine, D. J. Edwards, H. D. Jones, C. T. Richmond and K. Kim, in High-Pressure Science and Technology, v. 2, edited by K. D. Timmerhaus and M. S. Barber (Plenum, New York, 1979), p. 364.

5. (a) C. S. Coffey, and E. T. Toton, J. Chem. Phys. 76, 949 (1982); (b) F. J. Zerilli,; and E. T. Toton, Phys. Rev. B 29, 5891 (1984).

6. R. D. Bardo, Int. J. Quantum Chem. Symp. 20, 455 (1986).

7. G. Seeley and T. J. Keyes, J. Chem. Phys. 91, 5581 (1989); B.-C.Xu and R. M. Stratt, J. Chem. Phys. 92, 1923 (1990).

8. C. F. Melius, J. Phys. (Paris) C4, 341 (1987).

9. Y. B. Zel'dovich and Y. P. Raiser, *Physics of Shock Waves and High-Temperature Hydrodynamic Phenomena*, (Academic Press, New York, 1966), Vols. 1 and 2.

10. C. Tarver and D. Calef, Lawrence Livermore Tech. Rept. UCRL-52000-88-1.2, Energy and Technology Review Jan-Feb 1988, 1.

11. L. E.Fried and A. J. Ruggerio, J. Phys. Chem. 98, 9786 (1994).

12. H. A. Kramers, Physica (Utrecht) 7, 284 (1940).

13. D. W. Chandler and G. E. Ewing, J. Chem. Phys. 73, 4904 (1980).

14. X. Hong, S. Chen, and D. D. Dlott, J. Phys. Chem. 99, pp. 9102-9109 (1995).

15. A. Laubereau, and W. Kaiser, Rev. Mod. Phys. 50, 607 (1978).

16. S. Chen, I-Y. S. Lee, W. Tolbert, X. Wen and D. D. Dlott, J. Phys. Chem., 96, pp. 7178-7186 (1992).

DETONATION CHEMISTRY OF GLYCIDYL AZIDE POLYMER

PING LING, JILL SAKATA, CHARLES A. WIGHT
Chemistry Department, University of Utah, Salt Lake City, UT 84112, wight@chemistry.utah.edu

ABSTRACT

The initial step of chemical reaction initiated by laser-generated shock waves has been observed in glycidyl azide polymer (GAP) in condensed phase. Shocks are generated by pulsed laser vaporization of thin aluminum films and launched into adjacent films of GAP at 77 K. Comparison of FTIR spectra obtained before and after shock passage shows that initial reaction involves elimination of molecular nitrogen from the azide functional groups of the polymer. The shock arrival time has been measured by a velocity interferometer as a function of thickness of GAP and laser fluence. The shock pressure has been calculated by using a universal liquid state Hugoniot. A simple model is proposed to calculate shock velocity and pressure as a function of laser fluence. The results are in agreement with experimental data.

INTRODUCTION

Elucidation of individual chemical steps during an explosion or detonation is important for understanding the microscopic mechanism of explosion sensitivity towards heat, spark, and impact. This research is driven in part by the need to develop insensitive explosives, propellants and pyrotechnics that are safer to manufacture, handle and store. While much of the research has focused on understanding the structural and mechanical characteristics of energetic compounds, recent progress has also been made to understand the chemical mechanisms of combustion, thermal explosion and detonation.

Recently, we developed a thin film laser pyrolysis technique for determining initial condensed phase reaction products under conditions that mimic a thermal explosion.[1] The methodology is complementary to other methods such as SMATCH/FTIR developed by Brill and co-workers,[2] in which gas phase products are detected. These methodologies have been successful in revealing initial steps in the mechanism of thermal decomposition of energetic materials like RDX [2-4] and NTO.[5] However, there are limited experiments on initial chemical steps during an impact or shock wave. In spite of some theoretical approaches that could in some extent reveal relationships between chemical reactions and propagation of detonation waves on a microscopic level, [6-13] in general it is unknown whether the chemistry of detonation is the same as that of thermal explosion. One might guess that in many cases it is, but if bimolecular reaction pathways are competitive with unimolecular reactions, then the high pressure conditions of detonation should favor the bimolecular pathways. Fundamental issues of impact sensitivity may hinge on detailed knowledge of chemical reaction mechanism associated with rapid loading of high pressure shock waves.

The basic idea of the experiment is to subject a thin film of aluminum to a strong laser pulse. Rapid absorption of laser energy causes the aluminum film to explode, generating an expanding plasma and launching a shock wave into an adjacent film of energetic material. Laser-generated shock waves have been known for a long time, and studies have shown strong shocks (up to 50 GPa) can be generated in confined films using nanosecond or picosecond laser pulses.[14-16] Shock waves of this type have been used to investigate detonation of insensitive high explosives.[17] Recently we demonstrated that laser-generated shock waves are capable of initiating chemical reaction in glycidyl azide polymer.[18] This paper will further characterize the shock strength and discuss the reaction mechanism.

EXPERIMENT

In this section, we first describe how to initiate shock waves in GAP film and detect the chemical reaction. Then we show the method of measuring shock velocity by using a velocity interferometer.

Chemical Reaction in Laser-Generated Shock Waves

The GAP was from Specialties Chemicals Division of 3M, which markets the material as an energetic binder for propellants. The material used in our study is the uncrosslinked polyol,

$$
\begin{array}{c}
N_3 \\
| \\
CH_2 \\
| \\
-(CH-CH_2-O)_n
\end{array}
$$

GAP

which is a viscous liquid at room temperature. Although the polymer is linear, it is formed from a branched initiator. The average number of OH terminating groups per molecular is 2.7, and the average molecular weight of the polymer is 700. Therefore, each polymer molecular has an average of seven azide functional groups.

Samples are prepared by placing a small drop of GAP onto a CaF_2 optical window (25mm dia., 3 mm thick). Next, a film of aluminized mylar is stretched over the surface of a second CaF_2 window, and the two halves are assembled in the manner shown schematically in Figure 1. The layered sample is mounted in an OFHC copper retainer inside a dewar vessel. This apparatus is evacuated and the sample is cooled to 77 K by pouring liquid nitrogen into the reservoir.

The cold sample is irradiated with a single pulse from a Nd:YAG laser (Continuum Surelite) at $1.064\,\mu m$. The laser pulse duration is 7 ns. The average power of the pulse is measured with a disk calorimeter (Scientech Model 38-0101) at 10 Hz pulse repetition rate. A lens is used to focus the beam to the desired fluence. The spatial profile of the beam is determined by recording the average laser power while a razor blade is scanned across the beam at the position of the sample. The data are differentiated and fit to a Gaussian function to determine the $1/e$ diameter of the beam. Laser fluence quoted here are determined by dividing the measured pulse energy by the beam area.

Each laser pulse causes a region of aluminum on mylar to explode, generating a shock wave in mylar which is transmitted to the adjacent GAP film. Following the irradiation, the sample is placed into the sample chamber of a FTIR spectrometer (Mattson RS/10000) and the IR beam directed through the hole in the aluminum layer produced by the laser. This geometry ensures that the spectrum is obtained

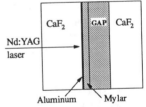

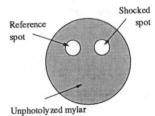

Figure 1 - Sample target assembly.

only for that portion of the GAP subjected to the shock wave. Prior to assembly of each sample, a transparent region is formed on the aluminized mylar film (again by laser ablation of the aluminum). This makes it possible to record a reference IR spectrum of GAP prior to irradiation of the sample. The spectrum of the reference spot is checked after the formation of the shocked spot in order to detect any changes in sample thickness or composition that may occur in areas remote from the irradiated spot. The thickness of the GAP film was determined to be 10±3 μm by obtaining IR spectra of GAP sample in a liquid cell with a calibrated path length.

It was found that 0.75 J/cm^2 is about the highest laser fluence that could be used in the experiment. Attempts to increase the fluence resulted in cracking of the CaF$_2$ windows. We found that propagation of shock across the interface between any two dry solid surface was unreliable. To circumvent this, all solid interfaces were filled with a thin layer of mineral oil or silicone oil.

<u>Shock Velocity Measurement</u>

A schematic diagram of the apparatus is shown in Figure 2. The 25mm dia. target array begins with 3 mm thick quartz window, which supports a 100 mm thick microscope cover slip. On one side, the cover slip has a 0.1 μm vacuum-deposited aluminum film which absorbs the laser pulse. Next comes a layer of GAP inside a spacer ring that determines the thickness of the GAP layer (15 μm to 2 mm), followed by a second quartz window that has a mirror coating on the side facing the GAP. The array is held pressure tight by clamping the entire assembly in a stainless steel cylinder held by a threaded retaining ring.

Each laser pulse is absorbed by the aluminum film on the disposable glass cover slip, generating a rapidly expanding plasma gas. The expanding plasma gas launches a shock wave into the GAP layer, setting the material into motion in the direction of laser beam. When the pressure wave reaches the opposite side, it disturbs the mirrored surface of the second quartz window. This disturbance is detected as a Doppler shift of the 632 nm He-Ne laser beam propagating in a home-built velocity interferometer. The change in fringe pattern of the interferometer is detected by a fast photomultiplier tube (Hamamatsu R928 tube wired to give a rise time of 1.5 ns), and recorded by a digital storage oscilloscope (Tektronix Model TDS 540, 1 Gsample/s at 500 MHz bandwidth).

The velocity interferometer has a delay leg of 90 to 150 cm, corresponding to a time delay of 3-5 ns. It is capable of detecting a velocity as low as about 50 m/s.[19] Usually, the particle velocity associated with a strong shock is much greater then this value. Shock fronts in condensed media typically have widths that are comparable to atomic dimensions,[7,20] and our detection system is not fast enough to record the acceleration of the mirrored surface. Also, the multilayered structure of the sample causes the velocity interferogram to have a complicated structure, and we are unable to identify the rarefaction wave in order to determine the particle velocity information. However, we are able to record the shock arrival time to within an experimental uncertainty of 2-3 ns. By changing the thickness of the spacer ring that defines the thickness of the GAP layer, the velocity of the shock is determined from the measured transit times. The "zero time" is determined by

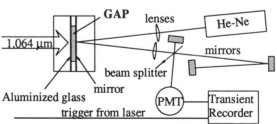

Figure 2 - Velocity interferometer setup.

omitting the GAP layer and the second quartz window from the array. In this configuration, the interferometer beam is reflected directly from the aluminum film that is vaporized by the laser pulse.

The speed of sound in GAP is measured in a similar way. The first quartz window and glass cover slip are replaced with a 1.5 mm thick aluminum desk. The shock wave induced by laser irradiation of the disk is dissipated to an ordinary sound impulse prior to reaching the GAP layer. However, the impulse is still strong enough to disturb the mirrored surface on the second window, allowing detection by the velocity interferometer. Records of transit time as a function of GAP thickness in this configuration provided an accurate method of measuring sound speed.

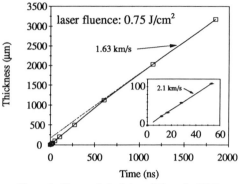

Figure 3 - Measured shock transit times for GAP.

RESULTS

<u>Shock Velocity and Pressure</u>

Experimental data show that the transit times of ordinary sound impulses through GAP layers are linearly dependent on the thickness of the layers. Linear fitting gives the sound speed in GAP to be $c_0 = 1.55 \pm 0.05$ km/s. For comparison, measurement of the sound speed in water by this method gives 1.52 ± 0.02 km/s, which is within 2% of the literature value.[21]

Transit times for laser-generated shock waves are shown in Figure 3 for data obtained at laser fluence 0.75 J/cm^2. For relatively large thicknesses (>1 mm) the shock velocity

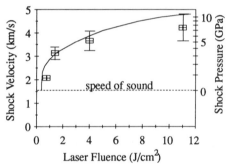

Figure 4 - Measured shock velocities and calculated shock pressures.

approaches the sound speed. However, at thicknesses less than 100 μm the velocity of shock is greater than 2 km/s, as shown in the inset of Figure 3. The linear dependence of arrival time on sample thickness shows that for sample thickness less than 100 μm, the shock velocity is constant.

Shock transit times were measured for GAP films at higher laser fluence (up to 11 J/cm^2) using spacers of 100 μm or less. The experimentally determined shock velocities are shown as a function of laser fluence in Figure 4.

The particle velocities achieved in these experiments are calculated from shock velocities by use of the universal liquid state Hugoniot equation.[22]

$$\frac{U_s}{c_0} = 1.37 - 0.37\, e^{\frac{-2U_p}{c_0}} + 1.62\, \frac{U_p}{c_0} \tag{1}$$

where c_0 is the speed of sound, and U_s and U_p are the shock and particle velocities, respectively. The shock pressure is calculated from these quantities by use of the relation

$$P_S = \rho \, U_S U_p \qquad (2)$$

where ρ is the density of GAP (1.34 g/cm^3, measured in our lab). If U_s and U_p are expressed in km/s and ρ in g/cm^3, then the resulting value of P_s is in GPa. The shock pressure calculated in this manner are shown on the right hand side of Figure 4.

The laser fluence dependence of the shock pressure is qualitatively consistent with the following simple model. Let us first express the absorbed laser fluence as the sum of two terms

$$\Phi = \phi_0 + \phi_1 \qquad (3)$$

where ϕ_0 represents the energy per unit area required to vaporize the aluminum target material, and ϕ_1 is the work associated with setting the surrounding GAP material into motion. We neglect the motion of supporting glass because the shock impedance of glass is about 6 times as that of GAP; therefore the particle velocity in glass is only about 1/6 of that in GAP. The work term is given approximately by

$$\phi_1 \approx P_S U_p \Delta t \qquad (4)$$

where Δt is the laser pulse duration (7 ns in this work). For shock velocity from zero to about 5 km/s, the Hugoniot of Eq. (1) can be rewritten in the usual linear form without much error,

$$U_S = a \, c_0 + b \, U_p \qquad (5)$$

where a=1.03, b=1.92 ±0.03. Substituting this expression into Eq.(2) we obtain

$$P_S = \rho \, (a \, c_0 + b \, U_p) \, U_p \qquad (6)$$

Solving for U_p and substituting into Eq.(4) we obtain

$$\phi_1 = \frac{1}{2} P_S \left[\left(\frac{a^2 c_0^2}{b^2} + \frac{4 P_S}{b \rho} \right)^{\frac{1}{2}} - \frac{a c_0}{b} \right] \Delta t \qquad (7)$$

For an estimate of ϕ_0, we note that the fluence required to vaporize a 0.1 µm film of Al is about 0.33 J/cm^2. The choice of this thickness Al film for the target material was based partly on the fact this layer is thicker than the optical penetration depth of Al,[16] but less than the thermal penetration depth

$$d \approx (\kappa \, \Delta t)^{\frac{1}{2}} \qquad (8)$$

during the laser pulse duration. The relationship between laser fluence and shock pressure under our experimental conditions may be written in the form

$$\Phi \ (J/cm^2) = 0.33 + \frac{1}{2} P_S \left[\left(\frac{a^2 c_0^2}{b^2} + \frac{4 P_S}{b \rho} \right)^{\frac{1}{2}} - \frac{a c_0}{b} \right] (\Delta t) \qquad (9)$$

A plot of P_s as a function of Φ calculated from this relationship is shown as the solid curve in

Figure 4. The prediction of the simple model is in reasonable agreement with the experimental results. Together, they show that shock pressures of several GPa can be generated in GAP at moderate laser fluence.

Loss of Azide Functional Group after Passage of Shock

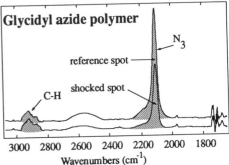

Figure 5 - IR spectra of 10 μm GAP film showing loss of azide groups following laser-induced shock.

Representative FTIR spectra of GAP before and after shock are shown in Figure 5. The most obvious difference is a 29-45% reduction in the integrated intensity of the band at 2100 cm⁻¹. This band is associated with the azide functional groups of the polymer. The band near 2900 cm⁻¹, which is associated with C-H stretching vibrations along the polymer backbone, remains essentially unchanged.

Infrared spectra of shocked samples were obtained after warming from 77 K to room temperature. This warming is accompanied by 20-30 % further reduction in the intensity of the 2100 cm⁻¹ band.

Finally, we learned that it is possible to induce sustained explosion in confined thick films of GAP by laser-generated shock waves. These experiments were conducted at room temperature. When explosion occurs, the supporting windows are fractured and large amounts of gas are released from the sample. Based on previous studies of thermal decomposition of GAP,[23-25] these gases are principally N_2, CO and C_2H_4. The entire target array is covered with burned polymer, even well outside the area of the sample irradiated by laser. The fluence threshold for explosion decreases dramatically and nonlinearly with increasing film thickness, as shown in Figure 6.

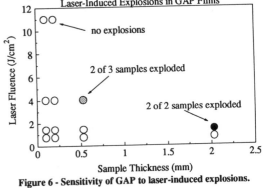

Figure 6 - Sensitivity of GAP to laser-induced explosions.

DISCUSSION

As mentioned in above, the principal observation of chemical changes after passage of shock is loss of azide functional groups as evidenced by reduction of the 2100 cm⁻¹ band.

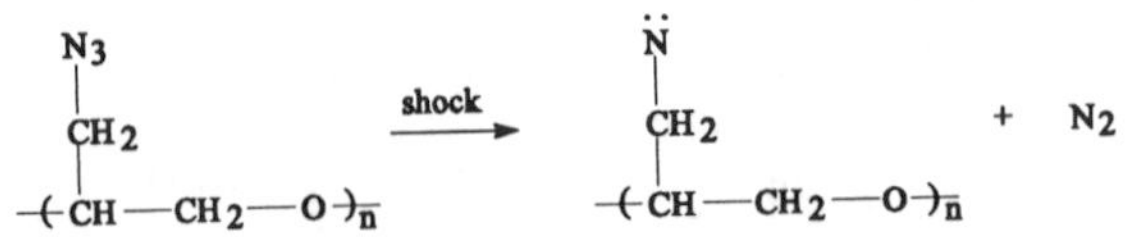

One might suppose that the nitrene radical might rearrange to an imine structure by a 1,2-

hydrogen atom shift,

Evidence for this type of rearrangement was obtained by Haas et al. in their study of thermal decomposition of GAP.[23] Although we made an extensive search for new NH stretching bands near 3400 cm^{-1}, we obtained no convincing evidence of formation of imines in our shocked samples.

The laser fluence used in the above experiment is about 0.71 J/cm^2, which corresponds to an experimental value of shock pressure 0.6-0.9 GPa.. Although this value was measured at room temperature, we may use Eq.(9) to estimate pressure at low temperature. When the samples are cooled to 77 K , usually the density and speed of sound of the sample increases, and the form of the Hugoniot may also change. According to Eq.(9), the increasing of density and speed of sound results in higher pressure generated at given laser fluence. For a worst case, let us assume that the density of GAP at 77 K increases by 10 % and speed of sound increases by 1 km/s. If the parameter b lies in the range 1 to 2, the pressure generated at 0.71 J/cm^2 will be about 20 % greater in the cold solid compared with the room temperature liquid. Therefore, the shock pressure at 77 K may be in the range around 1 GPa.

In order to look for the mechanism of losing azide functional group after passage of shock, let us first examine the possibility of reaction due to temperature increase associated with shock waves. Calculation of temperature increase due to shock compression requires quantities such as Grüneisen parameter, specific heat $C_v(T)$, as well as the Hugoniot equation. Unfortunately, these data are not available for GAP. However, a comparison with other materials might be helpful. For example, the temperature increase for TNT is about 50 K/GPa, for carbon tetrachloride about 120 K/GPa, for nitromethane about 70 K/GPa, and for water about 54 K/GPa.[26,27] The pressure of shock initating chemical reaction in our 77 K experiment is about 1 GPa, so it seems unlikely that the temperature increase will exceed 100-200 K. GAP is known to be stable to 473 K, so the driving force for eliminating azide functional group is not likely the ordinary thermal force. Shock initiated reactions require transfer of substantial amounts of mechanical energy from the shock front to the internal vibrational states of the molecules.[8-13] Dlott and co-worker have proposed that, especially in large molecular energetic materials, energy could be transferred from a shock produced phonon bath to the molecule's internal vibrations by multiphonon up-pumping. The mechanism of up-pumping is anharmonic coupling of excited phonon modes with low-frequency molecular vibrations, termed doorway modes. Once the doorway modes are excited, typically by two-phonon absorption, the molecule could absorb more phonons to reach higher energy levels, and so on [12,13] According to this model, shock energy is quickly channeled to the phonon bath with a quasitemperature θ_p before other processes can occur. If the specific heat of the phonon bath is 1/3 to 1/5 of the total specific heat,[13] then for the estimated 100 K temperature increase in GAP associated with the 1 GPa shock wave, the initial phonon bath quasitemperature will be 377 to 577 K, which corresponds to a characteristic phonon energy of 260 to 400 cm^{-1}. The polymeric nature of GAP allows for the possibility of very low-frequency doorway modes, so it is reasonable to expect that motions of the polymer backbone will allow this energy to be coupled into the higher frequency bond stretching modes, which would be required to induce loss of azide groups.

For the explosions occurring in thick samples, generally it is not possible to distinguish between a thermal explosion and detonation solely on the basis of a threshold measurement. However, the strongest evidence for detonation is the fact that sensitivity is much lower for a 2 mm thick layer as compared with a 0.5 mm layer. The entire energy of a 90 mJ laser pulse distributed over a GAP sample volume of 62.5 mm^3 (125mm^2 area, 0.5mm thick) should raise the temperature by only 3 K, based on the estimated heat capacity of GAP (0.4 J/g/K). Therefore, it is difficult to imagine how a four-fold increase in sample thickness can significantly affect the probability of initiating a thermal explosion by the laser.

A sustained detonation depends on the rate of reaction behind the shock front, because it is this energy that must be transmitted to the front in the form of a pressure wave to sustain the shock. Hare and Dlott have demonstrated that some reactions occur on a nanosecond time scale behind a shock front in PMMA.[13] It is not known whether enough heat can be released on this time scale to sustain a shock, even in energetic materials. In principle, reactions that occur well behind the shock front can contribute energy to the shock front in a freely detonating solid. This is because the sum of the local sound speed and the particle velocity in the reaction zone is greater than the shock front velocity. If the thickness of sample is less than the characteristic reaction zone thickness in the freely detonating material, detonation can only be achieved by overdriving the system (e.g. with a high laser fluence). In other words, the fluence required to initiate detonation should decrease as sample thickness increases, but only to the point where the sample thickness is roughly equal to the reaction zone thickness in free detonation. This is consistent with the sensitivity behavior in Figure 6. It suggests that reaction zone of GAP in free detonation is greater than 2 mm.

Additional experiments on this system and other energetic materials will provide a clearer picture of the rates and mechanism involved in detonation. Such fundamental investigations of the chemistry of energetic materials are in important prerequisite to the rational design of less sensitive explosives.

CONCLUSIONS

Thin films of glycidyl azide polymer at 77 K have been subjected to laser-generated shock waves. Infrared spectra show substantial loss of azide functional groups after passage of the shock waves, presumably by elimination of molecular nitrogen and formation of nitrene radicals. The shock wave velocity has been measured by use of a velocity interferometer and shock pressures calculated. Shock-induced detonation has been observed in thick GAP samples.

ACKNOWLEDGMENT

This research is supported by the Air Force Office of Scientific Research.

REFERENCES

1. C. A. Wight, T. R. Botcher, J. Am. Chem. Soc. **114**, 8803 (1992).

2. T. B. Brill, D. G. Patil, J. Duterque, and G. Lengelle, Combust. Flame **95**, 183 (1993).

3. T. R. Botcher, C.A.Wight, J. Phys. Chem. **97**, 9149 (1993).

4. T. R. Botcher, C.A.Wight, J. Phys. Chem. **98**, 5441 (1994).

5. T. R. Botcher, D. Beardall, C.A.Wight, L. Fan, and T. J. Burkey, J. Phys. Chem. in press

6. M. Peyrand, S. Odiot, E. Oran, J. Boris, and J. Schnur, Phys. Rev. **B 33**, 250 (1986).

7. M. L. Elert, D. M. Deaven, D. W. Brenner, and C. T. White, Phys. Rev. **B 39**, 1435 (1989).

8. C. S. Coffey and E. T. Toton, J. Chem. Phys. **76**, 949 (1982)

9. S. F. Trevino and D. H. Tasi, J. Chem. Phys. **81**, 348 (1984)

10. F. J. Zerilli and E. T. Toton, Phys. Rev. **B 29**, 5891 (1984)

11. D. D. Dlott and M. D. Fayer, J. Chem. Phys. **92**, 3798 (1990).

12. A. Tokmakoff, M. D. Fayer, and D. D. Dlott, J. Phys. Chem. **97**,1901 (1993)

13. D.E.Hare and D.D.Dlott, Appl. Phys. lett. **64**, 715 (1994)

14. L. C. Yang, J. Appl. Phys. **45**, 2601 (1974).

15. P. E. Shoen and A. J. Campillo, Appl. Phys. Lett. **45**, 1049 (1984).

16. I.-Y. S. Lee, J. R. Hill, and D. D. Dlott, J. Appl. Phys. **75**, 4925 (1994).

17. L. C. Yang and V. J. Menichelli, Appl. Phys. Lett. **19**, 473 (1971)

18. J. Sakata and C. A. Wight, J.Phys. Chem. **99**, 6584 (1995).

19. J. N. Johnson and L. M. Barker, J. Appl. Phys. **40**, 4321 (1969).

20. I.-Y. S. Lee, J. R. Hill, H. Suzuki, D. D. Dlott, B. J. Baer, and E. L. Chronister, J. Chem. Phys., in press.

21. CRC Handbook of Chemistry and Physics, 74th ed. (CRC, Boca Raton, FL 1993).

22. R. W. Woolfolf, M. Cowperwaite, and R. Shaw, Thermochimica Acta **5**, 409 (1973).

23. Y. Haas, Y. B. Eliahu, and S. Welner, Combust. Flame **96**, 212 (1994).

24. J. K. Chen and T. B. Brill, Combust. Flame **87**, 157 (1991).

25. Y. Oyumi and T. B. Brill, Combust. Flame **65**, 127 (1987)

26. R. Shaw, J. Chem. Phys, **54**, 3657 (1971).

27. M. Cowperthwaite and R.Shaw, J. Chem. Phys. **53**, 555 (1970).

Heterogeneous Reaction of Boron in CHNO and CHNOF Environments Using High-Pressure Matrix Isolation

Jane K. Rice * and Thomas P. Russell ♣
Naval Research Laboratory, Chemistry Division, Code 6110, Washington, D.C. 20375-5320
* Presenting Author
♣ Author to Whom Correspondence Should Be Sent

Abstract:

We have developed a technique in which the decomposition of energetic materials can be initiated under high pressure conditions which resemble the pressures reached in the non-ideal detonation process. A gem anvil cell is cooled to cryogenic temperatures, 50 K, and remains in thermal contact with the cooling element throughout the experiment. Following initiation, the reaction products are rapidly cooled and quenched on the microsecond time scale and detected using FTIR spectroscopy. In the present study, binary mixtures of boron with energetic materials containing (CHNOF) and lacking fluorine (CHNO) are compared. The differences in the reaction products suggest that the presence of the fluorine substituent leads to a complete combustion of boron to B_2O_3. In the decomposition of binary mixtures lacking the fluorine substituent, the boron in appears to be unchanged following the reaction of the oxidizer. The observed products are compared to predict the affect of fluorine on the formation of boron combustion products in the two environments.

Introduction:

High energy density materials have been utilized as additives in numerous propellant and explosive applications. Controlling the energy release under the heterogeneous reaction conditions may be accomplished through the addition of solid fuel additives in propellant and explosive formulations. One solid fuel additive of interest for high energy high density applications is boron due to its high gravimetric heating value (13,800 cal/g). Only beryllium has a higher heating value than boron (15,890 cal/g). In addition, the potential for tailoring boron reactions under high pressure conditions is of interest in order to develop controlled, non-ideal detonations. However, the predicted energy release far exceeds the experimental value. Although important, the high thermodynamic heating value will not itself determine a desirable fuel for energetic material applications. The fuel must ignite, combust and sustain reaction in the desired pressure, temperature and time regime. Unfortunately, boron currently does not meet these criteria for energetic material applications. Several dominating factors restrict the usefulness of boron as a fuel additive for energetic materials applications. The high melting point and boiling point increase the heating times associated with boron ignition. In addition, an oxide layer inhibits the ignition of the particles and restricts the sustained combustion process. Therefore, full utilization of the thermodynamic energy release has not yet been achieved.

Efforts to enhance the release of energy from boron particle combustion have been in progress for many years. The delayed ignition and quenching followed by oxide layer evaporation produces a two stage reaction process for boron combustion. Several theories which describe the heterogeneous reactions of the first stage of boron combustion have been reported. [1-4] The latest model describes the dissolution of boron into the B_2O_3 (l) layer and has been demonstrated to accurately fit observations of boron annealing in the presence of O2 [4]. The primary problem of boron reaction is the formation of a thick oxide layer which quenches the combustion process. The liquid B_2O_3 layer encompasses the boron particle at temperatures below 2400 K. Once a temperature above 2400 K is achieved, the B_2O_3 oxide layer vaporizes exposing the pure boron material. The vaporization process initiates the second stage of boron combustion. One approach to improve boron ignition and eliminate two stage combustion is to add reactive fluorine species in the combustion process. In binary energetic mixtures, this is accomplished by introducing difluoramine (-NF2) substituents to the oxidizer structures mixed with boron particles. Gas

373

phase models of this approach predict the presence of several more exothermic gas phase channels which lead to higher temperatures and more efficient conversion of B_2O_3 (l) and boron to gas phase materials. [5,6] Two of these reaction are given below:

$$B_2O_3 \text{ (l)} + 3F2 \text{ (g)} \rightarrow 1.5 \, O2 \text{ (g)} + 2 \, BF3 \text{ (l)} \quad (1)$$

$$B_2O_3 \text{ (l)} + HF \text{ (g)} \rightarrow OBF \text{ (g)} + HOBO \text{ (g)} \quad (2)$$

Our interest in boron combustion is as a potential fuel structure for advanced development of non-ideal reaction processes produced during composite explosives. To date, the chemical reaction of these materials is investigated through extrapolation of decomposition and combustion studies at high temperatures and low pressures. The extrapolation of these results into the appropriate reaction regimes is only a good first approximation. There is limited direct knowledge of high-pressure combustion because probing this regime is a difficult problem of condensed phase chemistry. Recently, a new technique which combines high-pressure and thermal shock conditions with low temperature matrix isolation has been demonstrated.[7] The experiment is designed to model the temperatures, pressures and reaction times present in the deflagration to non-ideal detonation regime. In this work a thermally driven reaction of pure oxidizers and composite materials are studied. The reaction of pentaerythrytol tetra nitrate (PETN), petnaerythrytol dinitrate difluoramine (PETN-NF2) were studied at pressures up to 4.0 GPa. In addition, the mixture of PETN and PETN-NF2 with crystalline boron was studied up to 4.0 GPa. Comparison of the reaction products detected from the pure materials and the composite mixtures is employed as a first step to understand the affects of fluorine chemistry on boron reactions under extreme conditions. Preliminary results of this work will be presented.

Experimental:

The experimental methods and apparatus for obtaining high pressure matrix isolation from a sample subjected to a thermal jump by pulsed laser heating under static high pressures and cyrogenic temperatures is presented below. The experimental setup has been described in detail elsewhere and only a brief description is given: [7]

The high pressure apparatus is a Merrill Bassett gem anvil cell employing cubic zirconia anvils fabricated from stainless steel. The cell is designed for 180° transmission measurements. The small size of the cell is desired to enhance cryogenic cooling capabilities. A 3.0 μg sample is placed as a thin film (3-5 μm thick) between two NaCl salt windows. The typical sample diameter is 250-300 μm. This entire system, the two NaCl windows and the thin sample, are statically compressed to the desired starting pressure. After the sample has been statically compressed, the cell is mounted onto an APD displex refrigerator and cooled to 40-50 K inside a vacuum chamber. A temperature jump (T-Jump) in the sample is produced by the absorption of a single laser pulse initiating reaction. A single pulse (8 ns) at 532 nm from a Continuum Nd:YAG frequency doubled laser is employed. Typical laser energies delivered to the sample are ≈ 10 mJ. The typical beam diameters are ≈ 500 μm, giving a laser fluence of 0.5 J/cm2.

After the reaction mixture is quickly re-cooled to 40-50 K by conduction of heat into the surrounding material, thus arresting the reaction process, the products are then examined by FTIR spectroscopy. Infrared spectra are collected a) before laser initiation at low temperature, b) following laser initiation at low temperature and high pressure c) after warming the sample to room temperature at high pressure, and d) following the release of pressure and removal of the sample from the anvil cell.

Results and Discussion:

Figure 1 shows a series of FTIR spectra collected for the reaction of PETN-NF2 and crystalline boron inside a cubic zirconia anvil cell loaded at 1.7 GPa.

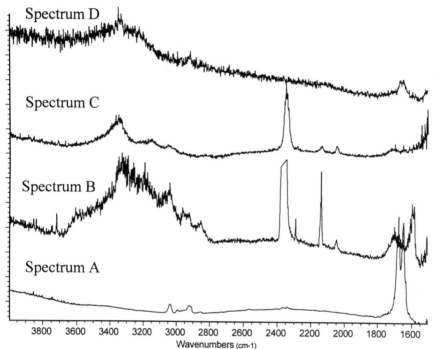

FIGURE 1: Series of spectra from the deflagration of PETN-NF2 in the presence of crystalline boron powder. Spectrum A is the infrared spectra of the PETN-NF2 composition at 50 K and 1.7 GPa. Spectrum B is that of the products isolated at 1.7 GPa and 50 K after the deflagration reaction is arrested. Spectrum C is the infrared spectra collected of the products at 1.7 GPa after the sample temperature has been warmed to room temperature. Spectrum D is the recovered residue spectrum from the deflagration reaction.

The products formed and isolated in spectrum B are identified by their infrared absorption frequencies. The tentative assignments for the observed frequencies are provided in table 1. The products observed at 1.7 GPa and 50 K are; CO_2, HOCN, CO, H_2O $(NO)_2$, NO_2 and two unidentified bands at 2141 and 2051 cm-1. These unidentified bands are consistent with a -BO stretching frequency of a matrix isolated -BO intermediate [8,9] but is not in close agreement with the assignment of BO, B_2O_2, B_2O_3. However, the unidentified bands are consistent with the observed -BO sym stretching frequency at 2141 and 2071 cm-1 reported previously by Snelson for OBF. [10] Therefore, the tentative assignment for the 2141 and 2050 cm-1 absorption is OBF. Infrared absorptions at 3074 and 3041 are tentatively identified as NH stretches and absorptions at 2961, 2928, and 2857 cm-1 are tentatively identified as CH stretches. In addition, a broadband absorption from 3613-2800 is observed. As the sample is warmed to room temperature the broad absorption observed in the 3000 cm-1 becomes ordered and three specific absorptions (3224, 3151,3201 cm-1) are detected. The observed CH stretches 2961, 2928, and 2857 cm-1, the NH stretches at 3074 and 3041 cm-1 and the NO_2 stretch at 1594 cm-1 disappear when the sample is warmed. All other species present at cold temperatures remain at room temperature and 1.7 GPa. Finally, after the pressure is released from 1.7 GPa to atmospheric pressure (spectrum D) the reaction products observed in spectra B and C disappear. Therefore, these products are either volatile products or reactive intermediates which escape once the pressure is released. Spectrum D shows the recovered solid residue from the PETN-NF2/boron deflagration process.

Table 1: IR bands and Tentative Assignments of the Observed Products in the Deflagration of PETN-NF2 in the presence of crystalline boron.

FREQUENCY CM -1	PETN-NF2 50 K	Post-Reaction 50 K	Post Reaction Room-Temp	RESIDUE
3720		HOCN	HOCN	
3607		HOCN, H2O	HOCN, H2O	
3613-2800		Broadband		
3357				NH Str
3224			NH Str	
3151			NH Str	NH Str, B_2O_3
3201			NH Str	
3074		NH Str		
3041		NH Str	NH Str	
3038	CH Str			
2992	CH Str			
2961		CH Str		
2928		CH Str		
2924	CH Str			
2857		CH Str		
2352		CO2	CO2	
2290		HOCN,HCN	HOCN,HCN	
2148		CO	CO	
2141		BO	BO	
2051		BO	BO	
1704		(NO)2	(NO)2	
1677	NO2			
1675				C-NO2
1649	NO2			C-NO2
1594		NO2		
1486				
1437				B_2O_3
1201				
1043				B_2O_3

Comparison of the residue spectrum D with the residue spectrum of pure PETN-NF2 at 1.7 GPa and a solid phase spectrum of B_2O_3 is shown in Figure 2.

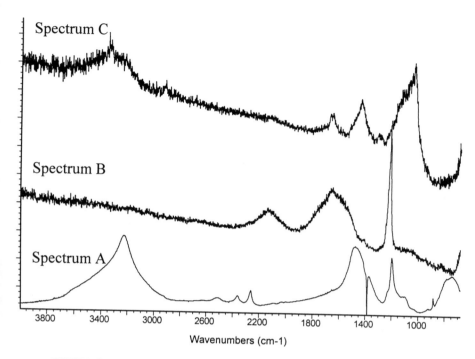

Wavenumbers (cm-1)

FIGURE 2: Comparison of the residue spectra from pure PETN-NF2 and PETN-NF2/boron deflagration at 1.7 GPa. Spectrum A is the solid phase spectrum of B_2O_3 in KBr. Spectrum B is the residue spectrum from the recovered PETN-NF2 deflagration at 1.7 GPa. Spectrum C is the residue spectrum from the deflagration of PETN-NF2/boron.

All samples are removed from the anvil cell for analysis. Several broadband absorptions are observed in the PETN-NF2/boron residue at 3357, 3151, 1675,1649, 1437, and 1043. Comparison to the solid phase spectra of B_2O_3 indicates that several of these absorptions are due to the strong absorption present in B_2O_3. In addition, optical evaluation of the remaining residue indicates that a white powdery substance is evenly dispersed throughout the sample. This is in contrast with the starting sample which is observed to be optically opaque from the black boron powder. No black powder is observed under microscopic evaluation in the recovered sample. Therefore, it B_2O_3 powder is present in the final products produced during the PETN-NF2/boron reaction. In addition, comparison to the recovered residue of pure PETN-NF2 indicates substantially different solid phase products are produced during the reaction with crystalline boron. The residue spectrum for pure PETN-NF2 has broadband absorptions at 2150, 1650, and a sharp absorption at 1100 cm-1.

The reaction of boron in PETN was attempted at 1.7 GPa to examine the effect of fluorine on boron combustion. A similar sequence of spectra were collected for the PETN/boron reaction as is presented for the PETN-NF2/boron reaction above. The reaction sequence for PETN/boron is not presented. The general results for this reaction are as follows. Initially, the sample is again optically opaque due to the solid boron present in the sample. The sample was precompressed to 1.7 GPa and T-jumped with a single laser pulse from the Nd:YAG laser. After the reaction is arrested, similar products are observed in the PETN/boron reaction as are seen in the PETN-NF2/boron reaction above. The major difference between the isolated products is the lack of -BO stretching frequencies in the PETN/boron reaction. Furthermore, no frequencies were observed in the solid residue spectrum to indicate the formation of B_2O_3. Finally, optical evaluation of the the recovered sample indicates that the sample is still optically opaque with a black solid present in the sample. The black solid in the recovered sample closely resembles the original solid boron initially present. This indicates that boron powder does not combust in the presence of PETN while it combusts completely in the presence of PETN-NF2. Therefore, the introduction of fluorine with the substitution of -NF2 into the oxidizer structure enhances the reactivity of solid boron in the sub-millisecond time frame provided by the experiment.

Conclusions:

The reaction of boron in a CHNO and CHNOF environment has been accomplished under high pressure conditions. In a CHNO environment, no reaction of the solid boron powder was observed in the sub-millisecond time frame provided by the isolation experiment. However, when reaction of boron occurs in a CHNOF environment substantial reaction of the boron powder is observed. Absorption corresponding to the strong absorption of B_2O_3 are observed in the residue and a polycrystalline white powder is observed optically in the recovered sample. In addition, isolation of two BO stretching frequencies are observed in the CHNOF environment which are not detected in the CHNO environment. The observed BO stretching frequencies are 2141 and 2050 cm-1 and are tentatively assigned to the intermediate OBF.

Acknowledgments:

The authors wish to acknowledge financial support from the Office of Naval Research and Naval Research Laboratory accelerated research initiative on the heterogeneous decomposition of energetic materials.

References:

1. Brzustowski, T. A. and Glassman, I., *Spectroscopic Investigation of Metal Combustion,* Heterogeneous Combustion, Progress in Astronautics and Aeronautics, **15**. Eds. Wolfhard, H. G., Glassman, I. And Green, L. Jr. Academic Press, New York, pp.41-73, (1964).
2. King, M. K.; J. of Spacecraft and Rockets, **19**, No. 4, pp 294-306, (1982).
3. Li, S. C. and Williams, F. A;;, Twenty-third Symp. (Int.) on Combustion, pp. 1147-1154, (1990).
4. Kuo, K. Presented ONR Boron Combustion Workshop, Penn State, June 6-8, (1995).
5. Brown, R. C. Kolb C. E. Yetter R. A. Dryer, F. L. and Rabitz H., Combustion and Flame, **101**, pp. 221-238 (1995).
6. Soto, M. and Page, M. Publication This Proceedings.
7. Rice, J. K. And Russell, T. P. Chem. Phys. Lett. **234**, pp.195-202 (1995).
8. Sommer, A. White, D. Linevsky, M. J. and Mann D. E., J. Chem. Phys. **38**, pp 87-98 (1962).
9. Weltner, W. Jr., and Warn, J. R.; J. Chem Phys. **37**, pp 292-303 (1962).
10. Snelson, A., High Temp. Sci. **4**, pp. 141-146 (1972).

EXPLOSIVE THERMAL DECOMPOSITION MECHANISM OF NTO

DAVID J. BEARDALL, TOD R. BOTCHER AND CHARLES A. WIGHT
Chemistry Department, University of Utah, Salt Lake City, UT 84112,
wight@chemistry.utah.edu

ABSTRACT

The initial step of the thermal decomposition of NTO (5-nitro-2,4-dihydro-3H-1,2,4-triazol-3-one) is determined by pulsed infrared laser pyrolysis of thin films. Rapid heating of the film and quenching to 77 K allows one to trap the initial decomposition products in the condensed phase and analyze them using transmission Fourier-transform infrared spectroscopy. The initial decomposition product is CO_2; NO_2 and HONO are not observed. We propose a new mechanism for NTO decomposition in which CO_2 is formed.

INTRODUCTION

NTO, shown in Figure 1, is an energetic material similar to RDX (hexahydro-1,3,5-trinitro-1,3,5-triazine), having a high density[1] and heat of formation.[2,3] However, NTO is much less sensitive to impact-induced ignition than RDX, which has prompted several recent investigations into the decomposition mechanism of NTO.[4-9]

Östmark, *et al.* used laser induced mass spectrometry to study NTO decomposition.[10,11] They observed formation of CO_2 and NO_2, and postulated that C-NO_2 bond cleavage is the initial step of thermal decomposition, followed by breakup of the azole ring. Menapace, *et al.* studied the decomposition of neat NTO and NTO dissolved in acetone or trinitrotoluene.[12] Their work indicated that the decomposition is bimolecular, beginning with H atom transfer and subsequent loss of HONO. The work by Oxley, *et al.*[13] is of special interest, in that they observed the formation of CO_2, N_2 and a polymeric solid which defied characterization. They also measured a primary deuterium kinetic isotope effect (DKIE) in the thermal decomposition of NTO, concluding that H atom transfer occurs in the rate determining step.

Of further interest is the work of Fan, Dass and Burkey,[14] in which they isotopically labelled NTO and found that N_2 formed during decomposition comes primarily from the adjacent nitrogen atoms in the ring. In addition, Brill and co-workers studied NTO thermal decomposition using Fast Thermolysis/FTIR[6] and found no evidence that NO_2 or HONO are generated. Instead, they report that CO_2 is the principal gas phase product.

With so little direct evidence for formation of NO_2 or HONO, the generally accepted decomposition mechanism (which requires the C-NO_2 bond be one of the first to break) must be called into question. This paper will address the issue of C-NO_2 bond cleavage and present an alternative mechanism to describe NTO decomposition.

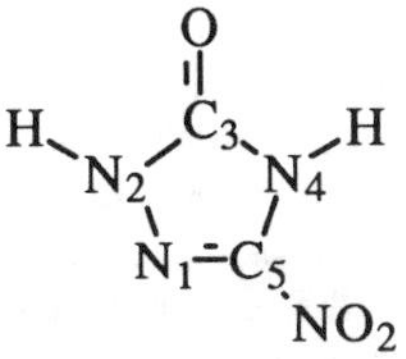

In previous work,[15-18] thin-film laser pyrolysis (TFLP) was used to determine that the initial step in the decomposition of RDX was cleavage of an N-NO_2 bond. TFLP is an elegant technique for trapping the initial decomposition products formed at high heating rates, which closely mimics actual detonation conditions. A thin (1-5 μm) film of explosive is deposited onto a thick (3 mm) transparent CsI window, cooled to 77K in vacuum,

Figure 1 - NTO ring numbering.

and irradiated using a pulsed CO_2 laser, which causes rapid (35 μs) heating to several hundred Kelvin and initiates NTO decomposition. As the film cools to 77K over several milliseconds, the decomposition products are trapped by a CsI cover window placed on the film, and products are identified by Fourier-transform infrared spectroscopy.

EXPERIMENTAL

Because NTO does not have strong absorbance bands accessible to a pulsed carbon dioxide laser, it is necessary to deposit a thin film of absorbing substrate onto the window before the NTO layer to convert the energy of the laser pulse into heat, which then diffuses into the NTO film and initiates decomposition. Three different absorbing substrates are used during the course of these experiments: dibasic potassium phosphate salt (K_2HPO_4), low molecular weight poly(vinyl chloride) (PVC, Aldrich Chemical Co.) and sodium tetrafluoroborate salt ($NaBF_4$). Layered samples (using K_2HPO_4, PVC or $NaBF_4$) and solid solutions (K_2HPO_4 or PVC only) are studied.

NTO isotopically labelled with ^{13}C at the carbonyl position (NTO-^{13}C) used in this work was synthesized by T. Burkey at the University of Memphis. Normal NTO (NTO or NTO-^{12}C, m.p. 262°C) was synthesized by us according to a procedure communicated to us by T. Burkey. The mass- and infrared spectra of NTO-^{12}C match those of the α-polymorph of NTO.[1]

All films studied in this work are deposited from solutions of the pure compound or mixtures onto polished CsI windows using an airbrush (Paasch type H with a #1 needle). All pure K_2HPO_4 films are deposited from methanol, as are K_2HPO_4/NTO solutions, and dried at 70°C after each deposition. Layered films of K_2HPO_4 and NTO are also dried at 70°C. Thin films of pure PVC and PVC/NTO solution are deposited from N,N-dimethylformamide (DMF) and placed in a beaker in a 140°C sand bath to drive off solvent. Layered films of NTO on PVC are similarly dried. $NaBF_4$ films are deposited from methanol and NTO films are deposited from methanol or tetrahydrofuran. Layered $NaBF_4$/NTO films are dried in a manner identical to PVC films. All films are mounted at the tip of a vacuum cell shown in Figure 2, evacuated overnight and cooled to 77K before pyrolysis. Spectra of the films are taken at 298K and 77K before and after pyrolysis using a Mattson RS-10000 spectrometer using 64 scans at 0.5 cm⁻¹ resolution.

A Pulse Systems Model LP140G carbon dioxide laser (35 μs pulse length) is used to irradiate all of the samples reported in this work. Laser fluence for any given experiment is determined by measuring the power of several laser pulses with a Scientech absorbing disk calorimeter and placing the sample at the required distance from a gold focusing mirror (1 m radius of curvature). At high fluences, the beam profile is smaller than the area of the window, so the beam is rastered across the sample in such a way that the entire sample area probed by infrared spectrometer beam is subjected to a single laser pulse.

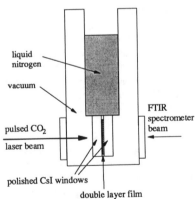

liquid nitrogen

vacuum

pulsed CO_2 laser beam

polished CsI windows

FTIR spectrometer beam

double layer film

Figure 2 - Schematic of vacuum cell.

RESULTS

Layered Samples

Initially, thin layered films of K_2HPO_4 and NTO were prepared as described above and pyrolyzed. Three decomposition products were detected: CO_2 (s, 77K, 2345 cm^{-1}, fluence ≥ 0.4 J/cm^2), N_2O (s, 77K, 2250 cm^{-1}, fluence >0.4 J/cm^2) and an unidentified solid (s, 77K, 2168 cm^{-1}, fluence ≥ 0.75 J/cm^2). Because CO_2 was an unexpected product, several control experiments were conducted to determine its source. The first consisted of pyrolysis at fluence up to 2.6 J/cm^2 of thin neat K_2HPO_4 films prepared as described above, sometimes yielding trace amounts of CO_2 from pyrolysis of methanol trapped in the film. In the second set of control experiments a layer of Teflon® was either sprayed (aerosol form) or stretched (teflon tape) between the salt and NTO layers. Subsequent pyrolysis yielded relatively large amounts of CO_2, coming presumably from the NTO.

PVC and NTO were layered on windows from DMF and methanol solutions, respectively. Pyrolysis of pure PVC films at fluence up to 3.1 J/cm^2 produced no detectable CO_2, while pyrolysis of layered films at 3.1 J/cm^2 produced measurable amounts of CO_2 in all cases. In no case was NO_2 detected.

Pyrolysis of ~1μm neat $NaBF_4$ films at 3.0 J/cm^2 produced no measurable CO_2. Experiments using NTO-^{12}C and NTO-^{13}C were carried out to eliminate ambiguity about the source of CO_2. Pyrolysis of layered films of $NaBF_4$ and NTO-^{13}C yielded >95% $^{13}CO_2$ (2279 cm^{-1}) at low laser fluence (~0.6 J/cm^2) and 90% $^{13}CO_2$ at high fluence (2.4 J/cm^2). Figure 3 shows that simultaneous pyrolysis of separate samples of NTO-^{12}C and NTO-^{13}C on $NaBF_4$ at 0.6 J/cm^2 produced exclusively $^{12}CO_2$ and $^{13}CO_2$, respectively.

Solid Solutions

Deposition and pyrolysis of single-layer, two component films of K_2HPO_4 and NTO was carried out. The mole fraction NTO in the prepared films (0.233, 0.066, 0.016, 0.0041 and 0.0046) was determined by measuring the integrated intensities of selected NTO and K_2HPO_4 bands and comparing them to reference bands in spectra of pure NTO or K_2HPO_4 films of known mass. All pyrolyses were at a laser fluence of 1.03 J/cm^2. The yield of CO_2 (molecules CO_2/ initial number NTO molecules) is plotted as a function of NTO mole fraction in Figure 4. Results of pyrolysis of NTO/K_2HPO_4 solid solutions are consistent with a trace amount of methanol remaining in the films before pyrolysis.

Deposition of thin films of NTO/PVC solution was carried out as described. Pyrolysis was at 3.2 J/cm^2. Mass percent NTO was 0.20, 0.11, 0.06, 0.03, 0.015 and 0.008. The yield of CO_2 relative to the amount of NTO present in sample films was calculated as described above and appears in Figure 5.

Gas Phase Decomposition Products

Gas phase decomposition products of

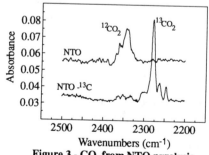

Figure 3 - CO_2 from NTO pyrolysis

NTO were collected after pyrolysis of thin films and crystals of NTO in an evacuated gas cell fitted with NaCl windows. Pyrolysis was carried out by multiple (~30) laser pulses at 3.0 J/cm². Gaseous CO_2 produced was always $^{12}CO_2$ or $^{13}CO_2$ from pyrolysis of NTO-^{12}C or NTO-^{13}C, respectively. Some samples of NTO-^{13}C were refluxed in and recrystallized from $^{18}OH_2$. Pyrolysis of these NTO-^{13}C samples (wet with $^{18}OH_2$) showed the presence of gas phase $^{18}OH_2$ in

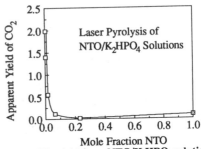

Figure 4 - CO_2 yield from NTO/K₂HPO₄ solution

addition to $^{13}CO_2$ in the gas cell. No evidence for the formation of $^{18}O^{13}C^{16}O$ or other isotopomers of CO_2 was observed.

CONCLUSION

In all experiments conducted, CO_2 was the initial decomposition product detected by infrared spectroscopy. Other products were detected, but only at higher energies. In experiments conducted just above the threshold for thermal decomposition, CO_2 was the only product detected. There was no evidence for formation of NO_2 (1617 cm⁻¹), N_2O_4 (1735 cm⁻¹) or HONO (1117 cm⁻¹). Experiments using NTO-^{13}C are ideal for detection of N_2O_4 because the carbonyl band shifts from 1717 cm⁻¹ to 1656 cm⁻¹, clearing a spectral region between 1715 cm⁻¹ and 1800 cm⁻¹ which makes possible ready detection of small amounts of N_2O_4.

CO_2 can be demonstrated to be the initial infrared active product formed during thermal decomposition by comparing the scaled intensity of product peaks as a function of laser fluence. Products formed concurrently with CO_2 will have a constant value as laser fluence changes, while products formed before or after CO_2 in the thermal decomposition will have monotonically decreasing or increasing slopes, respectively. Figure 6 shows that both N_2O and the product associated with the 2168 cm⁻¹ band appear after the CO_2 formation step, and that N_2O has a formation threshold much higher than that for carbon dioxide.

Experiments have shown that the CO_2 formed during thermal decomposition includes the carbonyl group carbon, and most likely the carbonyl oxygen as well. Possible sources of oxygen which participate in oxidation of the carbonyl group to CO_2 include the -NO_2 group in the same or a neighboring NTO molecule, or impurities in NTO film (most likely water, because NTO is recrystallized from water). Control experiments with a Teflon® barrier between K₂HPO₄ and NTO films eliminate the possibility of this salt oxidizing NTO.

The gas cell experiments were conducted to test whether water is a source of oxygen for CO_2 formation. The presence of $^{18}OH_2$ vapor in the cell without incorporation of the ^{18}O label in product CO_2 indicates that water participation in carbonyl oxidation is not significant.

Thus, we can clearly state that

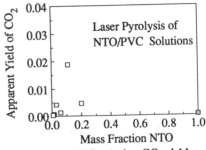

Figure 5 - PVC solution CO_2 yields.

carbonyl oxidation must be by interaction with a $-NO_2$ group in the same or a neighboring NTO molecule. The pyrolysis of solid solutions of NTO in K_2HPO_4 or PVC over a broad range of NTO concentrations can indicate whether or not the yield of CO_2 is dependent on the initial concentration of NTO in the sample. If CO_2 formation is independent of NTO concentration the reaction must be unimolecular, while an intermolecular reaction would show a concentration dependence. Considering all the data in Figure 4 and 5, we conclude that CO_2 formation is unimolecular.

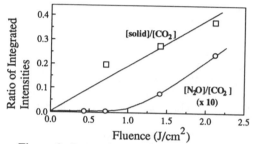

Figure 6 - Determination of product formation order.

A mechanism based on this data is shown in equation (1) below, beginning with ring opening and oxidation of the carbonyl group.

$$\text{(1)}$$

Formation of N_2 from the adjacent nitrogen atoms as described in equation (2) is consistent with the isotopic work done by Fan et al.

$$\text{(2)}$$

ACKNOWLEDGMENTS

This research is supported by the Air Force Office of Scientific Research. We thank Ted Burkey at the University of Memphis and Leanna Minier at Sandia/Lawrence Livermore National Labs for important discussions during the course of this work.

REFERENCES

1. K.-Y. Lee and R. Gilardi in <u>Structure and Properties of Energetic Materials</u>, edited by K.H. Liebenberg, R.W. Armstrong, and J.J. Gilman (Mater. Res. Soc. Proc. **296**, Pittsburgh, PA,

1993) p. 237.
2. J.P. Ritchie, J. Org. Chem. **54**, 3553, (1989).
3. A. Finch, P.J. Gardner, A.J. Head, and H.S. Majdi, J. Chem. Thermodynamics **23**, 1169 (1991).
4. E.F. Rothgery, D.E. Audette, R.C. Wedlich and D.A. Csejka, Thermochimica Acta, **185**, 235 (1991).
5. X. Yi, H. Rongzu, W. Xiyou, F. Xiayun and Z. Chunhua, Thermochimica Acta, **189**, 283 (1991).
6. G.K. Williams, S.F. Palopoli and T.B. Brill, Combust. Flame, **98**, 197 (1994).
7. T.J. Burkey and S.A. Shackelford, Thermal Decomposition of NTO and NTO/TNT Mixtures, Report No. 49620-88-C-0053, 1992.
8. K.V. Prabhakaran, S.R. Naidu and E.M. Kurian, Thermochimica Acta, **241**, 199 (1994).
9. G.K. Williams and T.B. Brill, J. Phys. Chem. **99**, 12536 (1995).
10. H. Östmark, H. Bergman, G. Åqvist, A. Langlet and B. Persson, in Proceedings of the 16th International Pyrotechnics Seminar, (1991) p. 874.
11. H. Östmark, FOA Report D-20178 2.3, National Defense Research Establishment, Sundbyberg, Sweden, 1991.
12. J.A. Menapace, J.E. Marlin, D.R. Bruss and R.V. Dascher, J. Phys. Chem. **95**, 5509 (1991).
13. J.C. Oxley, J.L. Smith, Z.Zhou and R.L.Mckenney, J. Phys. Chem. **99**, 10383 (1995).
14. L. Fan, C. Dass and T. Burkey, private communication.
15. C.A. Wight and T.R. Botcher J. Am. Chem. Soc., **114**, 8303 (1992).
16. T.R. Botcher and C.A. Wight, J. Phys. Chem., **97**, 9149 (1993).
17. T.R. Botcher and C.A. Wight in Structure and Properties of Energetic Materials, edited by D.H. Liebenberg, R.W. Armstrong and J.J. Gilman (Mater. Res. Soc. Proc.**296**, Pittsburgh, PA, 1993) p. 47.
18. T.R. Botcher and C.A. Wight, J. Phys. Chem., **98**, 5441 (1994).

Time Resolved Optical Spectroscopy to Examine Chemical Decomposition of Energetic Materials Under Static High Pressure and Pulsed Heating Conditions

Thomas P. Russell **+, Theresa M. Allen and Y. M. Gupta**,

* Naval Research Laboratory, Chemistry Division, Code 6110, Washington, D.C. 20375-5320
** Washington State University, Shock Dynamics Center, Pullman WA, 9164-2814
+ Author to Whom Correspondence Should Be Sent

ABSTRACT

The study of the deflagration or detonation reactions of energetic materials is challenging due to the high pressure, high temperature, and time domain under which the reactions occur. Experimental measurements, are presented that demonstrate the ability to continuously monitor the global reaction times and reaction sequences associated with chemical reactions under these conditions. Time resolved absorption spectroscopy is used in conjunction with a high pressure gem anvil cell to probe the real-time chemical processes during pulsed-heating. Samples are initiated by a rapid thermal jump induced by absorption of a single laser pulse. Time resolved absorption spectroscopy of 3,6 trinitroethylamine tetrazine reaction is demonstrated by the real time measurement of the decrease in the π-π* absorption at 110 ns temporal resolution during laser heating at pressures up to 3.5 GPa.

INTRODUCTION

Despite several decades of research, relatively little is known about the chemical processes associated with the deflagration or detonation regime. Historically, the chemical reaction determined from low pressure/slow heating experiments are extrapolated to the temperature and pressure regimes of interest. These data provide the basis for a more realistic estimation of the thermodynamic properties and product distributions used to describe the highly confined chemistry. However, this extrapolation enhances the uncertainty and error in the estimated properties.

Over the last several years, experimental approaches have been developed in an attempt to understand chemical reactions under extreme conditions (high temperature/high pressure). Static high pressure decomposition studies have provided kinetic and thermodynamic data on the decomposition of several nitramine compounds.[1-3] These studies do not provide any information about the reaction mechanisms or final products. The decomposition times are on the orders of minutes to hours. Static high pressure burn rate experiments provide a physical measurement of the affects of pressure on the combustion rate (burn rate) of the material. [4] However, no chemical information is provided. Detonation mass spectrometry has provided limited chemical data on reactions under detonation conditions. [5] The temporal evolution of products has been reported. Typically, small molecular weight species are detected and often multiple species with similar masses and poor resolution make product identification very difficult. Recently, high pressure matrix isolation has been reported. [6] The sample is reacted at high pressure and the product species are trapped by low temperature (50 K) freezing of the reaction sequence. Product species are identified but no temporal information is provided.

These experimental approaches have provided a good foundation for advancing chemistry under extreme conditions. With the development of fast spectroscopic techniques and additional advances toward the reproduction of the necessary experimental conditions, new approaches may now be developed. The experiments designed to probe this chemical regime must mimic the representative pressure, temperature, volume (density) and time conditions if the chemical understanding of this difficult reaction regime is to continue. This paper describes two optical spectroscopic techniques which are designed to approach the time, temperature and pressure conditions of this reaction regime, while providing unique chemical information not previously attainable. Time resolved absorption spectroscopy in conjunction with a high pressure gem anvil cell is presented.

Mat. Res. Soc. Symp. Proc. Vol. 418 © 1996 Materials Research Society

EXPERIMENTS

The experimental method and apparatus for obtaining time-resolved optical spectra from a sample subjected to pulsed laser heating under static high pressure are described below. The experimental setup has been described in detail elsewhere [7-12] and only a brief description will be given. The high pressure apparatus is an NBS diamond anvil cell designed for both transmission and reflection spectroscopic techniques. [9,10] Initial pressures are measured by the ruby fluorescence technique.[10-12]

High Pressure Cell

The high pressure cell used in the present work is a NIST anvil cell [9,10] and is fabricated from a high-temperature, high-strength superalloy, Inconel 718. The cell is designed for 180° transmission and reflection measurements. The cell employed here used single crystal cubic zirconia anvils. Single crystal cubic zirconia anvils have good optical quality for *in situ* visible spectroscopic measurements while permitting pressures in excess of 10 GPa. [11] The initial temperature in the sample is measured using a chromel-alumel thermocouple in contact with the nickel gasket which confines the sample under pressure. Pressure inside the anvil cell is measured by the ruby fluorescence pressure measurement technique [10,12] using a peak shift calculation or a line-shape model. The ruby fluorescence measurements are accurate to ± 0.05 GPa when they are made in a hydrostatic environment at room temperature. [12] The uncertainty of the pressure measurement is larger in NaCl, ± 0.15 GPa, due to the absence of the strictly hydrostatic conditions. The high pressure cell is mounted on a micrometer positioning device for alignment in the time resolved spectroscopy experiments.

Samples for UV/Vis spectroscopy are prepared by compressing a 3-5 μm sample film between two NaCl salt windows surrounded by a nickel gasket. The gasket hole diameter is 250 μm with a 200 μm thickness. A small ruby sphere, <15 μm, is placed within the salt window. The entire sample (NaCl salt windows and sample) is compressed inside the gem anvil cell (GAC) to the desired initial pressure.

Pulsed Heating and Time Resolved Spectroscopy

After the sample has been statically compressed, the experiment is carried out in two parallel steps. A thermal jump (T-jump) in the sample is produced by the absorption of a single laser pulse initiating the reaction. A single pulse (3.5-4 μs) at 514 nm from a flash lamp pumped dye laser (Cynosure) is used to heat the sample. Typical laser energies delivered to the sample are ≈ 1.0 mJ and ≈ 6.3 mJ for absorption and emission experiments, respectively. Typical values for beam diameter are ≈ 0.15-0.20 mm, giving a laser fluence of ≈ 4.5-5.0 J/cm^2.

Simultaneous with laser deposition, *in situ* spectroscopic probing of the reaction processes are examined by using either time resolved absorption or time resolved emission spectroscopy. The light transmitting through the sample is collected by the microscope objective and exits the microscope into a Kaiser holographic notch filter (514 nm) to eliminate any elastically scattered laser light. The remaining light is collected into a SPEX 1681 single spectrometer which disperses the light over the wavelength region between 350-650 nm. The light exiting the spectrometer is coupled to a Cordin 160 model no. 5B streak camera. The temporal distribution is accomplished by streaking the photoelectrons across a two dimensional phosphor screen. A 1024 x 256 LN$_2$ cooled Spectrum-One CCD camera collects the image of the two-dimensional streak camera phosphor screen. The entire process results in a series of spectra showing intensity as a function of wavelength and time. This experimental system is capable of in situ single-pulse time-resolved absorption at 110 ns temporal resolution during laser heating of the sample inside the GAC.

RESULTS AND DISCUSSION

Absorption Spectroscopy

A solid sample of 3,6 trinitroethylamine tetrazine (TNEAT) was loaded inside the GAC. The GAC sample absorption is determined by first collecting the transmission spectrum of a blank NaCl cell, followed by the transmission spectrum of a cell loaded with approximately 3.2 x 10^{-7} g of TNEAT. Two absorption bands, π-π* and the η-π*, are observed between 380 - 600 nm. The pressure dependence of the π-π* and the η-π* transitions was determined up to 4.0 GPa and the laser energy was adjusted to compensate for the change in absorption. Absorption spectra for laser heated TNEAT is accomplished by the following procedure. First, a reference spectra of the laser heated NaCl is recorded at the desired pressure. Second, the reaction spectra are recorded and are then compared with the reference spectra. Transmission data are converted to absorption as follows:

$$A(\lambda,t) = \log\frac{I_r(\lambda,t) - I_o(\lambda,t)}{I(\lambda,t) - I_o(\lambda,t)}. \tag{1}$$

A is the absorbance as a function of wavelength and time. I_r is the transmitted signal from the reference (NaCl), I_o is the background signal, and I is the transmitted signal through the sample.

The time resolved absorption spectra of TNEAT deflagration were collected during laser heating at a variety of pressures (atmospheric, 0.8 GPa, 2.2 GPa, and 3.5 GPa). The sample was photolyzed at 514 nm and the reaction was monitored by probing the spectral changes over the wavelength region 380-600 nm. The observed changes were collected in real-time with 110 ns temporal resolution at all pressures studied.

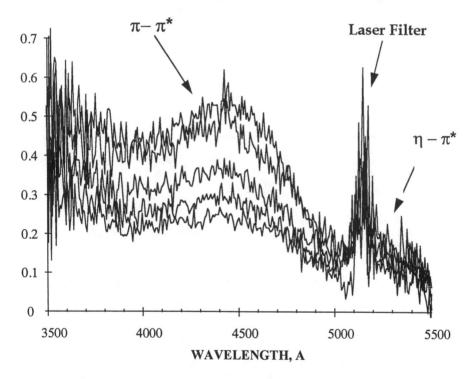

Figure 1. Time resolved UV-Vis absorption spectra of the laser photolysis of TNEAT at 0.8 GPa. For clarity the spectra presented are at a temporal resolution of 220 ns.

Figure 1 shows a typical series of spectra during photolysis. The selected spectra are the observed changes in the π-π^* transition at 450 nm with at 110 ns temporal resolution at 0.8 GPa during the laser heating reaction. No

387

change is observed initially. After 0.55 μs, a decrease in the absorption band is detected. A continuous decrease in absorption is observed for 0.8 μs. Once the decrease in absorption is no longer detected, the sample maintains a constant transmission. The recovered residue shows a featureless, broadband absorption between 380-500 nm, which is consistent with the observed increase in absorption. Figure 2 shows the pressure dependence of the TNEAT deflagration reaction by monitoring the change in absorption of the π-π* transition at 450 nm as a

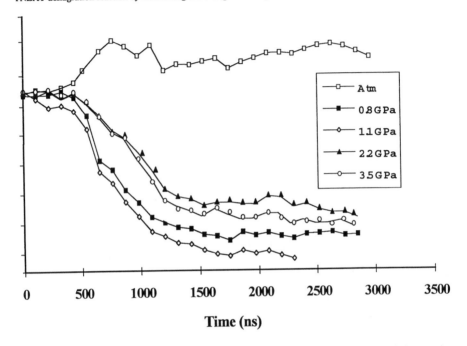

Time (ns)

Figure 2. Pressure dependence of the change in the π-π* transition at 450 nm. Note the increase in absorption at atmospheric pressure. Under high pressure conditions (0.8-3.5 GPa) a decrease in absorption is detected.

function of time. At atmospheric pressure, an absorbance increase is detected throughout the reaction process . Under high pressure conditions (0.8-3.5 GPa), the material reacts and the sample absorption decreases until a constant absorption is achieved.

The concentration of TNEAT is proportional to the measured absorbance. Therefore, the absorption change is a direct measure of the mole fraction (α) of reaction as a function of pressure. This is accomplished by taking the peak absorbance normalized to zero at $t = 0$. For decomposition at any specified time the change in absorption was subtracted from unity yielding, α, a term proportional to the mole fraction. A typical α vs. time curve is shown in figure 3 for each pressure studied. The general feature of the curves are similar. α is not linear as a function of time, nor does it follow any of the common kinetic laws. The curves can be identified with the initial, intermediate and final stage of a sigmoid (s-shaped) α vs. time curve found for thermal decomposition of a single solid in an autocatalytic-type reaction.[18-19] These curves are characterized by an induction phase (α < 0.2), and intermediate acceleratory or normal growth stage, and a final decay stage (α > 0.9). Our data shows an initial induction phase which appears to be pressure dependent with larger induction periods associated with higher pressures. However, the ignition delay decreases as the pressure increases up to 3.5 GPa.

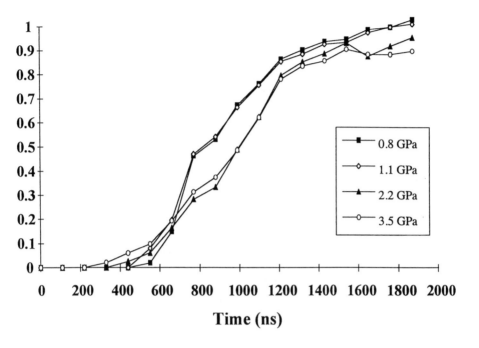

Figure 3. Typical α vs. time sigmoid-type rate curves at a variety of pressures (0.8, 1.1, 2.2 and 3.5 GPa), where α is the fraction of decomposed TNEAT during photolysis. Sigmoid curves such as those shown are characteristic of autocatalytic decomposition reaction of single solids. Curves are not fitted. The points are connected by lines.

CONCLUSIONS

The presented results are preliminary but qualitatively indicate the value of the high pressure time resolved absorption measurements. The measured absorption changes provide a real-time chemical probe of the deflagration reaction of TNEAT. The detected absorption changes in the TNEAT reactions indicate that at least two distinct reaction mechanisms are present. Below 0.8 GPa, the reaction produces an increase in absorption. Above 0.8 GPa, a decrease in absorption indicates a different reaction process. The mole fraction of reaction was accomplished as a function of pressure up to 3.5 GPa. The measured α vs. time curves are sigmoid or s-shaped. Sigmoid curves are characteristic of decomposing single solid materials, particularly highly energetic materials such as azides, fulminates, and permanganates. The induction part of the sigmoid curves decreases with increasing pressure, while the ignition delay decreases with increasing pressure.

Acknowledgments

We would like to thank Jeff Battaro and Rob Schmidt of Stanford Research Institute for the TNEAT sample. The authors wish to acknowledge financial support from the Office of Naval Research and Naval Research Laboratory accelerated research initiative on the heterogeneous decomposition of energetic materials.

REFERENCES

1. G. J. Piermarini, S. Block, and P. J. Miller,, *J. Phys. Chem,* 91 (1987) 3872. and G. J. Piermarini, S. Block, and P. J. Miller, *J. Phys. Chem.,* 93 (1989) 457.

2. P. J. Miller, G. J. Piermarini, and S. Block , *Applied Spectroscopy,* 38, (1984) 680.

3. T. P. Russell, P. J. Miller, G. J. Piermarini, S. and Block, *J. Phys. Chem.,* 97 (1993) 1993.

4. S. F. Rice and M. F. Foltz, *Combustion and Flame,* 87 (1991) 107.

5. N. C. Blais, *J. Energetic Mat.,* 6 (1987) 255.

6. J. K. Rice and T. P. Russell, Chem. Phys. Lett., 234 (1995) 195.

7. T. P. Russell, T. A. Allen, J. K. Rice, and Y. M. Gupta, J. De Physique, C IV, Vol 5 (1995), 553.

8. T. P. Russell, T. A. Allen and Y. M. Gupta, (to be submitted).

9. C. E. Weir, E. R. Lippincott, A. Van Valkenburg, and E. N. Bunting, *J. Research NBS,* 63A, 55 (1959).

10. Barnett J. D., Block S., and Piermarini G. J., *Rev. Sci Instrum. ,* 44 (1973) 1.

11. Russell, T. P. and Piermarini G. J. (to be submitted).

12. G. J. Piermarini , S. Block , J. D. Barnett, R. A. Forman, *J. Appl. Phys.,* 46 (1975) 2774. and Block, S. and Piermarini G. J., *Physics Today,* 29 (1976) 44.

13. P. W. Jacobs, F. C. Tompkins, In *Chemistry of the Solid State*; Garner W. E. Ed.; Butterworth: London 1955; Chapter 7.

14. D. Dollimore, J. Dollimore, D. Nicholson, In *Reactivity of Solids*; deBoer, J. H. Ed. Elsevier: Amsterdam, 1961; p 627.

TIME RESOLVED EMISSION STUDIES OF ALUMINUM AND WATER HIGH PRESSURE REACTIONS

C. A. Brown, T. P. Russell
Chemistry Division, Code 6110, Naval Research Laboratory Washington D. C. 20375

ABSTRACT

The detonation of underwater explosives is a complex problem involving a temporally dependent heterogeneous reaction regime of oxidizer reactions and high pressure metal combustion. For simplicity, underwater explosions may be described as a two stage reaction process. First, the oxidizing material detonates to produce species under extreme conditions of temperature (up to 5000 K) and pressure (up to 10 GPa). The chemical energy produced from this reaction is transferred to the bulk water as three forms of work: (1) shock, (2) heat, and (3) initial bubble formation. Second, the species produced by the oxidizer detonation form a high pressure and high temperature reactive fluid that surrounds the solid particles. The solid particles are primarily consumed while the pressure is decreasing from 10 GPa to 0.1 GPa at a reaction temperature in excess of 3200 K. The secondary reaction of the solid particles produces a lower energy shock and a pressure response that reinforces the initial energy delivered to the bulk water medium. The ability to tailor this late energy release between shock and bubble formation is dependent on the reaction time and chemistry of the solid particle under extreme conditions. We present a series of single-shot time resolved emission experiments that probe the reaction of aluminum particles under extreme conditions. The temporal behavior of the observed species is used to gain insight into the chemical reaction mechanism that leads to the formation of Al_2O_3 during underwater detonations.

INTRODUCTION

Underwater explosive formulations are a multiphase composition of materials designed to release a significant amount of energy at late time. This is accomplished by combining an organic oxidizer with solid aluminum particles supported in a polymer matrix. The detonation chemistry of the composite explosive is a heterogeneous process that involves rapid mechanical, physical, and chemical changes propagating under non-ideal conditions. Despite the importance of high pressure metal combustion, direct experimental measurements are lacking. If modeling capabilities are to be improved for a system that produces a non-ideal response, advanced chemical and physical experimental evaluations of the heterogeneous reactions are necessary under appropriate pressure, temperature, volume (density), and time conditions. We recognize that this will only be achieved when the chemical and physical sequences are probed under representative pressure, temperature, volume (density) and time conditions. Our experimental studies focus on the metal chemistry in a high density reactive fluid with a goal of improving the predictive capabilities associated with these non-ideal processes. This paper focuses on the reaction chemistry of metal particles in a water medium. Time resolved emission spectroscopy of the reaction of Al metal with water is presented as a function of pressure up to 0.8 GPa.

There are numerous reports of studies of Al reactions with oxidizers, including water. The articles that are most relevant to our work, which focus on the visible emission from aluminum reactions are briefly described. Brzustowski and Glassman measured the emission from thermally ignited aluminum in a General Electric Model M-2 photographic flash bulb in

1965.[1] They spectroscopically identified the three major products of the Al and O_2 reaction as atomic Al, molecular AlO, and a continuum emission (attributed to Al_2O_3 (s) particles). High pressure gas phase studies (up to 4.053 MPa) of aluminum combustion with oxygen have been reported by Driscoll, et. al., in shock tube studies.[2]   The Al/O_2 reaction produced the intermediate species Al, Al^+, AlO and a continuum emitter (attributed to Al_2O_2). AlO emission peaked at 1.4 msec after ignition with continuum emission appearing 2 msec after ignition and persisting throughout the experimental time sequence of 12 msec. The only published paper on emission observed from aluminum exploding wire combustion in water was performed by Jones and Brewster in 1991.[3]   While no time resolved spectra were obtained, three pressures were studied: 449 kPa, 101 kPa, 33.5 kPa. They spectroscopically identified Al, AlO, OH, AlH, and H. Continuum emission was not a major feature of the reported spectrum.

EXPERIMENTAL

The experimental technique is described in detail elsewhere,[4,5] therefore only a brief description is provided here. In these experiments, a sample is placed in the focal plane of a dye laser (Cynosure) beam that heats the sample with a single pulse(8 μs, 30-100 mJ), which is focused to 1 mm. The sample consists of either aluminum foils or aluminum particles. During subsequent reaction, the collected emission light is dispersed by a Spex 500M monochromator that is calibrated in the wavelength region being monitored. Light exiting the monochromator impinges on a Hamamatsu model C2830 Streak Camera that streaks the signal in time. Temporal resolution windows of 10 ns to hundreds of microseconds are employed for studying early reaction dynamics and to obtain global time information. A Spex Spectrum 1 CCD images the signal from the streak camera.

Three types of experiments are performed: Al foil reactions in air at ambient pressure, Al foil reactions in water at ambient pressure, and Al and water reactions in a cubic zirconia anvil cell, CZAC, at high pressure. In the ambient pressure experiments, the foils are mounted inside either quartz or glass cuvettes (10 mm). Since the light collection occurs on the side opposite that of heating, an amount of energy able to produce a 'window' in the material must be deposited on the sample in order to detect emission.

To conduct high pressure experiments, we use a cubic zirconia anvil cell that is a miniature Merrill Bassett cell made of 301 stainless, designed for 180° transmission measurements and operating pressures >10 GPa.[6,7] The gasket is Inconel 600 with a gasket-hole 250 μm (diameter) by ~200 μm (depth). Pressure inside the CZAC is measured using the ruby fluorescence measurement technique and is calculated on a peak shift model.[8,9]

RESULTS

The time resolved reaction of aluminum particles has been investigated under three experimental regimes. First, Al reactions in air were probed. These studies were followed by Al in H_2O (l) experiments at atmospheric pressure. Finally, Al reactions in H_2O were probed at pressures up to 0.8 GPa. Complete description of the reaction processes under each condition will be described in detail elsewhere.[10] Therefore, only a brief overview of the related reaction processes of Al and H_2O reactions is described here. The three major species spectroscopically observed in these reactions were atomic Al, molecular AlO and a continuum emitter. The atomic Al bands detected were the emission doublet at 394 and 396 nm. Four of the AlO (B→X) vibrational bandheads: (2,0) at 447 nm, (1,0) at 465 nm, (0,0) at 484 nm, and (0,1) at 508 nm were observed. A broadband continuum emission was observed over the entire spectral region investigated (380 nm - 600 nm).

The early and late time reactions of Al in air and water have been probed with a series of single-shot time-resolved optical emission experiments at different temporal windows ranging from 10 ns/spectrum to 1 msec/spectrum. The spectra collected permit the investigation of both the early reaction dynamics and the overall aluminum chemistry. The temporal and pressure dependence of the Al, AlO and a continuum emitter is presented to describe the reaction of aluminum metal in underwater detonations. Time-resolved spectra were collected under all experimental conditions, allowing comparison of the chemical lifetimes of the reaction intermediates as a function of reaction time and pressure.

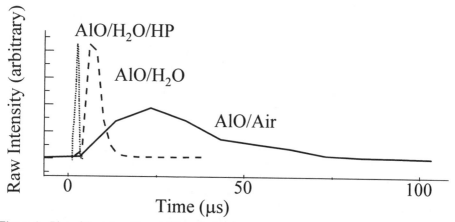

Figure 1. Plot of the intensities of AlO emission from Al reactions in air, water, and in water at pressure (<0.1 GPa). For comparison, the plots are referenced to the first appearance of emission (t=0).

A comparison of the AlO chemical lifetimes in the Al/Air, Al/H$_2$O, and Al/H$_2$O (at 0.1 GPa) reactions is shown in Figure 1. Note that the air and water spectra have the same experimental conditions and are thus directly comparable. The high pressure data were taken under different conditions, and while not directly comparable, qualitative trends can be observed. The AlO emission from Al/air reactions persists the longest, with a lifetime > 75 μs. The AlO emission from reactions in water appears and disappears much more quickly. A chemical lifetime of 15 μs is obtained. At 0.1 GPa, the observed AlO emission has an even shorter lifetime, 1μs. The Al chemical lifetimes (not shown) also show the same trend. Atomic Al emission and absorption are observed under all conditions. Both the emission and absorption are combined to determine a chemical lifetime. The Al atom chemical lifetime decreases as the conditions are changed from Al/air, Al/H$_2$O, and Al/H$_2$O at 0.1 GPa. Above 0.2 GPa, no Al or AlO emission is detected.

Figure 2 shows the pressure dependency of the Al and H$_2$O reaction at pressures up to 0.8 GPa. The intermediates, reported as a function of pressure, represent the initial species detected during the reaction with a time resolution of 1 μs. The relative percentages of Al, AlO and the continuum emitter were determined by integrating the spectral response of each species

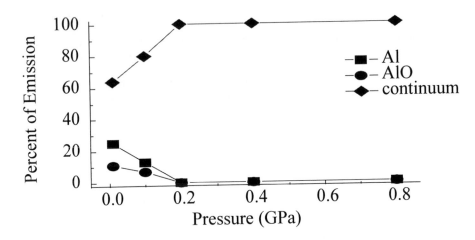

Figure 2. Pressure dependence of the relative abundance of the initial intermediate chemical species produced during the Al/H_2O reaction.

and the continuum emitter from the total emission detected. The contributions from unresolved species are not taken into account. No attempts were made to correlate observed emission intensities with absolute species concentrations. Therefore, each reported species abundance is correlated with the evolving spectral features and represents changing fractional concentrations of species relative to each other within a given experiment. Figure 2 shows the pressure dependence of the reaction intermediates up to 0.8 GPa. At atmospheric pressure, the initial chemical intermediates observed are Al, AlO and a continuum emitter. As the pressure increases to 0.1 GPa the relative abundance of Al and AlO decrease and the continuum contribution increases. Above 0.2 GPa, no Al and AlO are detected. The only emission detected is a continuum spectrum. Under all conditions, the most abundant emission observed is from the continuum.

DISCUSSION

The solid residue recovered from quenched underwater detonations is shown to consist of a combination of Al_2O_3 and Al metal.[11] Cross sectioning of the recovered particles shows a solid aluminum core with an Al_2O_3 coating.[11] The increase in particle volume due to the Al_2O_3 coating is directly related to the stoichiometric conversion of Al metal to the less dense Al_2O_3 (s). These results suggest that the Al metal reactions in underwater detonations proceed through a confined reaction mechanism during which the products are not able to diffuse from the reacting surface. Upon cooling, the reaction quenches and forms a particle consisting of a solid Al metal core with an Al_2O_3 coating.

Our experimental results, in conjunction with the recovery work from underwater detonations described in the preceding paragraph, start to provide a unique picture of the Al metal reaction scheme under extreme conditions. As the initial experimental conditions are changed from air to pressurized water, some general trends of the reactions are observed. The observed chemical lifetimes of Al and AlO decrease as the confinement of the system increases.

Furthermore, as the pressure is increased to 0.8 GPa, the relative abundance of the initial chemical intermediates decreases. Above 0.1 GPa, Al and AlO are no longer detected. Only continuum emission is observed. The increased confinement increases the frequency of collisions among the reactants. This accelerates the rate of formation of later time species (polyatomic Al species). As the reaction continues, the unreacted Al particle diameter shrinks, forming a larger heterogeneous reaction zone around the solid particle. Therefore, a chemical gradient between the particle surface and the water medium is produced. The initial chemical species (Al and AlO) remain near the surface of the particle and the reaction zone becomes dominated by the polyatomic Al species.

ACKNOWLEDGMENTS

The research was conducted with support from the Office of Naval Research and the Naval Research Laboratory. C. A. Brown is an NRL/NRC post-doctoral fellow. Some experiments were conducted in the laboratory of Y. M. Gupta at the Shock Dynamics Center in the Physics Department of Washington State University with the aid of T. M. Allen and G. Pangilinan and with the support of ONR.

REFERENCES

1. Brzustowski, T. A. and Glassman, I., *Heterogeneous Combustion, Progress in Astronautics and Aeronautics Series*, **15**, Wolfhard, I, Glassman, I, and Green, L. eds., New York: Academic Press, 1965, pp. 41-73.

2. Driscoll, J.F., Nicholls, J.A., Patel, V., Shahidi, B.K., and Liu, T.C., *AIAA Journal,* 1985, pp. 856-858.

3. Jones, M. R., and Brewster, M. Q., *J. Quant. Spectroc. Radiat. Transfer*, **46**, pp. 109-118, 1991.

4. Russell, T. P., Allen, T. M., Rice, J. K., and Gupta, Y. M., *J. Physique* (in press).

5. Russell, T. P., Allen, T. M., and Gupta, Y. M., *Chem. Phys. Lett.* (submitted).

6. Merrill, L. and Bassett, W. A., *Rev. Sci. Instr.*, **45**, pp. 290-294 (1983).

7. Russell, T. P. and Piermarini, G. J., (manuscript in preparation).

8. Block, S. and Piermarini, G. J., *Phys. Today*, **29**, 44-55 (1976).

9. Piermarini, G. J., Block, S., Barnett, J. D., and Forman, R. A., *J. Appl. Phys.*, **46**, pp. 2774-2780 (1975).

10. Brown, C.A. and Russell, T.P., manuscripts in preparation.

11. Carlson, D.W., Deiter, J.S., Doherty, R.M., and Wilmot, G., *Proceedings Tenth Symposium (International) on Detonation* Boston MA 1994.

DETONATION IN SHOCKED HOMOGENEOUS HIGH EXPLOSIVES

C. S. YOO, N. C. HOLMES, and P. C. SOUERS
Lawrence Livermore National Laboratory, University of California, Livermore, CA 94551,
yoo1@llnl.gov

ABSTRACT

We have studied shock-induced changes in homogeneous high explosives including nitromethane, tetranitromethane, and single crystals of pentaerythritol tetranitrate (PETN) by using fast time-resolved emission and Raman spectroscopy at a two-stage light-gas gun. The results reveal three distinct steps during which the homogeneous explosives chemically evolve to final detonation products. These are i) the initiation of shock compressed high explosives after an induction period, ii) thermal explosion of shock-compressed and/or reacting materials, and iii) a decay to a steady-state representing a transition to the detonation of uncompressed high explosives. Based on a gray-body approximation, we have obtained the CJ temperatures: 3800 K for nitromethane, 2950 K for tetranitromethane, and 4100 K for PETN. We compare the data with various thermochemical equilibrium calculations. In this paper we will also show a preliminary result of single-shot time-resolved Raman spectroscopy applied to shock-compressed nitromethane.

MOTIVATION

In contrast to better known macroscopic properties of high explosives (HEs), the energetic processes are poorly understood at an atomistic and/or molecular level. The detonation velocity, particle velocity, energetics of detonation, detonation products, thermal decomposition, material parameters controlling chemical sensitivity, etc, have well been explored for many energetic molecules. However, it is poorly understood even for a relatively simple system like nitromethane, how thermal shock energy transfers to HE molecules and dissipates through phonons and vibrons of HE lattice, what kind of transients are initially generated, and how these species evolve and eventually lead to detonation.

Plane shock wave experiments provide a well controlled way of investigating high explosives at well characterized shock states. Combined with time-resolved measurements, they provide an unque opportunity to monitor the changes in HE during detonation. The emphasis of shock wave research in the past, however, has been on understanding the continuum or bulk properties of HE such as detonation velocities, particle velocities, and burn rates. Although various thermochemical and phenomenological models have been developed to interpret these measurements [1], obtaining the description of detonation at a molecular level still remains a challenging problem. This is in part due to the lack of information regarding chemical species and kinetics associated with detonation. Therefore, a logical approach to this problem is a shock-wave experiment combined with fast time-resolved spectroscopy, being capable of characterization of various short lived transients and their kinetics [2-4]. Furthermore, recent advances in computational capability make such an understanding of detonation in microscopic details feasible.

The goal of this study is to get new physical insight into shock initiation and detonation chemistry by using fast time-resolved emission and Raman spectroscopy at a two-stage light-gas gun. The density, particle size, pores, defects, etc, strongly affect the energetics of HE processes. However, they often interfere with the events that are intrinsic to HE molecules. For example, the previous study performed in shocked HE powders [4] showed strong emission from voids and grain boundaries, rather than from HE molecules or transients that were in high density states. Therefore, this study will be performed in several homogeneous high explosives that minimize such effects, including nitromethane (NM), tetranitromethane (TNM), and single crystals of pentaerythritol tetranitrate (PETN).

SCIENTIFIC ISSUES IN DETONATION

Detonation in high explosives can be classified into three steps based on their spatial and temporal characteristics. (1) The primary step of thermal excitation and vibrational energy transfer. This process occurs in a sub-ps to one ns regime and proceeds in a highly non-equilibrium manner. During this period, hot spots are formed and highly vibrationally excited HE molecules dissociate to form transient species [5]. (2) The second step is one in which chemical kinetics of the primary species dominate. In this period, one to several 100 ns, the primary species strongly interact with each other and, as a result, they either recombine together or form the secondary products. The latter case typically proceeds exothermically in many energetic molecules. (3) Finally, the propagation step where the exothermic reaction eventually leads to detonation. In this period, ~ μs, the final detonation products are created, and the chemical equilibrium is achieved.

The primary detonation process, energy transfer and vibrational relaxation, occurs on a ps time scale. In the case that a fast (~ 5 Km/s) propagating shock wave thermally pumps HE molecules, the initiation reaction would then be completed within approximately 100-molecular layer determining the shock front. The mechanisms for energy transfer and vibrational activation in this region can be characterized by fast spectroscopy with a ps time resolution and be directly compared with theoretical results such as a molecular dynamics simulation carried out over the thickness of the shock front [6]. However, because these processes are completed within a 50 nm molecular layer, an order of the diffraction limit of visible light, shock wave measurements should be performed in thin films to avoid a dynamic uncertainty. A consequence is then the lack of any observable signal during the single, transient shock-wave event.

Detonation products of high explosives are reasonably well understood based on recovery experiments, bomb calorimetric measurements, and thermochemical analyses [7]. However, it is not so well understood how high explosive molecules evolve in time to final detonation products, a crucial issue that should be addressed for understanding shock-initiation as well as for engineering new energetic materials. For example, in a relatively simple high explosive like nitromethane, there have been at least six different mechanisms previously reported to explain how nitromethane initiates the energetic process, as summarized in Fig 1 [8-13]. They range from dimerization, a concerted type reaction mechanism typically occurring in condensed matter, to dissociation, a gas phase reaction. It simply reflects the infancy of our understanding of detonation in a molecular level. Obviously, the first step to this problem is to identify chemical species produced during the initiation period and to model the reaction in terms of the first principles in physics and chemistry such as *ab-initio* quantum chemical calculations. The kinetic measurements should be followed and the molecular based reaction models be developed.

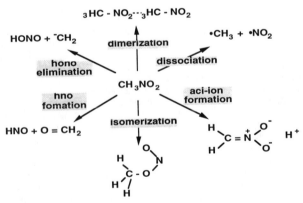

Figure 1. Various initiation reactions proposed for nitromethane at high pressures and temperatures.

Shock wave research on high explosives to date has been concentrated more on understanding the third propagation process. The energetics of these processes are relatively well known in many explosives, including the von-Neumann spike and *C-J* conditions, detonation wave velocity, material velocity, and thermochemical description of detonation. The detonation is traditionally conceived with the assumption of "thermal and chemical equilibrium". Neither is this assumption true in many cases, nor it is sufficient enough to appropriately describe the processes associated with the performance of high explosives. For example, most of thermochemical equilibrium models fail to describe the detonation velocity in many high explosives, particularly insensitive carbon-rich systems due to the effects like carbon coagulation [14]. Therefore, it is important to understand the mechanisms and kinetics for pre-, during, and post-detonation processes. In the next section, we show time-resolved temperature measurements in detonating high explosives, providing information on detonation kinetics. We then also show our preliminary results of single-shot time-resolved spontaneous Raman spectroscopy performed on shock compressed nitromethane.

SHOCK TEMPERATURES BY TIME-RESOLVED EMISSION SPECTROSCOPY

The shock temperature of high explosives is important for understanding shock-initiation and detonation processes; however, it has been one of less frequently measured thermodynamic variables. On the other hand, unlike other variables including pressure, volume, and energy, temperature strongly depends on thermochemical models which require experimental verification.

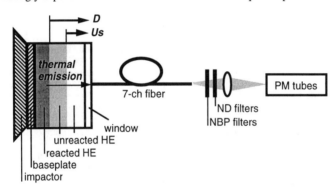

Figure 2. The experimental setup for time-resolved shock temperature measurements of transparent homogeneous high explosives.

Figure 2 shows the experimental setup to measure time-resolved temperatures of shock-compressed high explosives. The sample in a 6-8 mm thick cavity between a baseplate and a transparent window is compressed by using a two-stage gas-gun which can accelerate an impactor up to 8 km/s. Thermal emission from shock-compressed high explosive is time-resolved by using six photomultiplier tubes for which spectral windows (40 nm bandwidth) are set in the spectral range between 350 and 700 nm by using narrow beam pass filters. In this setup, thermal emission is collected from the central 3 mm area of HE, limited by the acceptance cone of an optical fiber bundle [15]. This results in fast time resolution, a few ns, primarily depending on the rise time of recording electronics. The temperature of the high explosive is then obtained by fitting this time-resolved thermal emission data to a gray body Planck function at 1 ns intervals during the event of interest, typically one μs long.

Figure 3 shows the time-resolved temperatures of NM, TNM, and (110) PETN single crystal [16] shocked to 10.7, 12.9 and 13.3 GPa, respectively. The temporal profiles can be characterized into i) a predetonation zone showing initial increases in temperature, ii) a superdetonation zone showing rapid temperature increases to the peak values, and iii) a normal detonation zone showing temperature decreases to steady state values.

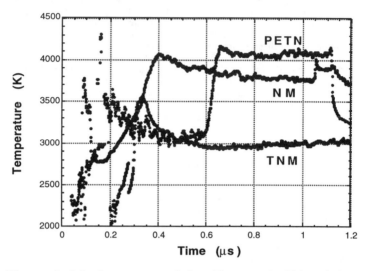

Figure 3.   Time-resolved shock temperatures of selected homogeneous high explosives.

The temperature changes in Fig 3 can be understood in the context of the thermal explosion model [17,18]. In homogeneous high explosives, the reaction is believed to initiate behind the shock front near the baseplate/HE interface as shown in Fig 2. This means that the detonation initially occurs from shock-compressed high explosives (superdetonation) and, then, from uncompressed high explosives (normal detonation) as the detonation front catches the shock front. Figure 4 illustrates various thermodynamic states through which homogeneous high explosive molecules evolve during detonation.

The initial shock state, i, is defined by the impedance match between an impactor and unreacted HE. The pressure of this shocked but unreacted HE increases to the von-Neumann spike, S^*, by chemical energy release behind the shock front. The chemical reaction initiates at S^*, and the products eventually evolve to the CJ^* state as the reaction is fully developed. These detonation products then expand isentropically to the condition matching the impedance of the impactor k, as long as the reaction front remains behind the shock front. At a later time, however, the detonation front catches the shock front, and the detonation occurs directly from unshocked HE at S, and the products evolve to the CJ state and to the impedance match condition, j, between the isentropically expanded HE products and the impactor.

The temperature, on the other hand, increases as the reaction progresses and peaks at the CJ state. The opacity of the reaction products is substantially higher than that of unreacted high explosive due to formation of graphite and other hydrocarbons at high temperatures [20]. Under these conditions thermal emission measured across the reaction zone (Fig 2) is mostly from the reaction front, and the measured temperatures in Fig 3 can be correlated to those at various thermodynamic states in Fig 4. The temperature will initially increase to that at S^* during predetonation and then to the CJ^* temperature of superdetonated products as the exothermic chemical reaction fully develops. The CJ^* temperature then further decays to the CJ, as the detonation front catches the shock front. The temperature should then remain at the CJ, as long as the impact condition is at` or below the CJ condition. The temperature changes observed in Fig 3 are clearly consistent with this model.

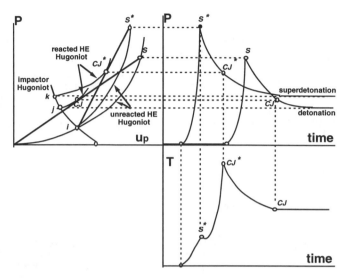

Figure 4. Thermodynamic states of the shock-compressed homogeneous high explosive and its reaction products, together with the pressure and temperature changes at the reaction front.

The *CJ* temperatures of these explosives can be obtained from the steady-state values in Fig 3 as discussed above. The results are summarized in Table I, together with the calculated *CJ* temperatures using themochemical equilibrium codes for comparison. Clearly, the calculated temperatures are sensitive to various models, and the current data could provide critical constraints for various models. The *CJ* temperature of PETN is higher than other two due to the abundance of oxygen. The *CJ* temperature of TNM is substantially lower than that of NM, likely due to the absence of hydrogen in TNM. It is also noticed that, unlike TNM and NM, PETN detonates at substantially lower pressures than the *CJ* pressure near 31 GPa, consistent with the previous observation that PETN is sensitive along the (110) direction [21].

Table I. The steady-state temperatures of shock compressed homogeneous high explosives in comparison with the calculated *CJ* temperatures [19].

HE	Pressure (GPa)		Temperature (K)				
	unreacted	reacted	Measured	CHEQ	CHEETA	BKWR	JCZ3
NM	107	128	3800	3750	3620	3000	3400
TNM[1]	129	137	2950	2640	2830	2200	2500
PETN[2]	133	191	4100	4430	4330	3300	4200

1. TNM used in the experiments is 98 % pure. The CHEQ and CHEETA calculations were also done in a mixture of 98 % TNM and 2 % NM.
2. (110) PETN single crystals were used.

DETONATION KINETICS BY SHOCK TEMPERATURES

In addition to the *CJ* temperature, the shock temperature data also provides information regarding the detonation kinetics. It is evident from Fig 3 that neither shock initiation nor detonation occurs instaneously. There is a relatively long induction period for HE to initiation; for example, the shock-initiation takes about 50 ns for NM and PETN and 200 ns for TNM. The transitions to the *CJ* * states occur in 20-50 ns in TNM and PETN but in 200 ns in NM. The peak temperatures then approach the steady-state values for the next 100-400 ns, representing the transition to detonation in uncompressed high explosives. The temporal structure of this measured temperature is now subject to the analysis by using various chemical models such as a reaction flow model [22].

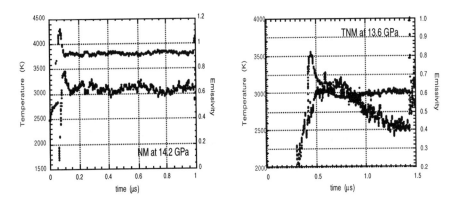

Figure 5.   The emissivity changes of shock compressed NM and TNM, together with their temperature changes.

In addition to the complicated structure in the predetonation period, there is an evidence for a weak time-dependent change in the post detonation period, long after the reaction is fully developed in TNM. This can be illustrated more clearly in the emissivity change. Figure 5 shows the emissivity changes in NM and TNM during detonation, plotted with respect to their temperature changes. The emissivity of NM reaches to a steady-state value after the initial rise, similar to the temperature change. A similar behavior of emissivity was observed in most of other experiments of NM and PETN single crystals. However, the emissivity change in TNM is quite different from those of NM and PETN. It increases to the peak value rather slowly and then decreases continuously for the next several 100 ns. This result implies that chemical changes occur even at long times, ~500 ns, after the detonation.

Similar non-equilibrium changes in post-detonation have also been previously recognized in insensitive high explosives [23] and have been conjectured to be due to carbon coagulation in carbon rich systems. The nucleation and growth of carbon particles typically takes several 100 ns to one µs time period [14,20]. However, because TNM is a carbon difficient system and is believed to form no carbon particles based on thermochemical calculations [19], we believe it is not the case of TNM. Therefore, the nature of this postdetonation process in TNM is not known. On the other hands, one may conjecture other possibilities, including phase separation [24] and/or chemical decomposition [25] of final detonation products, CO_2, H_2O, and N_2, at high pressures and temperatures. Clearly, further experimental and theoretical studies are needed to clarify this important issue.

402

SHOCK INITIATION BY TIME-RESOLVED RAMAN SPECTROSCOPY

The chemistry of shock compressed high explosives starts well before the detonation, typically a few 10 - 100 ns after being shocked (see Fig 3). In order to understand the predetonation chemistry, it is necessary to characterize transient species, reaction pathways, and kinetics, all of which may be obtained by using fast time-resolved Raman spectroscopy. The experimental setup for measuring time-resolved Raman scattering, shown in Fig 6, is similar to those used in the previous studies performed in the stepwised shock-wave loading experiments [26,27]. However, in the present study the experiments are designed to obtain time-resolved Raman data at the onset of detonation in single shock-compressed high explosives by using a 2-stage gas-gun. Details regarding the experimental setup will be presented elsewhere.

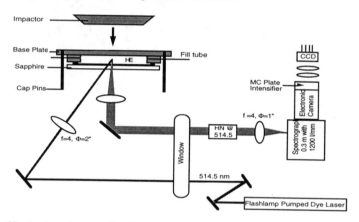

Figure 6. Single-shot time-resolved spontaneous Raman spectroscopy for detonating high explosives.

Preliminary Raman data from pure nitromethane shocked to 9.8 GPa is shown in Fig 7, as recorded on a 2-D multichannel plate intensifier. Five sharp vibrational features of NM, $\delta_s(NO_2)$ at 655 cm^{-1}, $\nu(CN)$ at 917 cm^{-1}, $\delta_s(CH_3)$ at 1379 cm^{-1}, $\nu_s(NO_2)$ at 1402 cm^{-1}, and $\nu_a(NO_2)$ at 1561 cm^{-1}, are evident, together with other broad features, which form the background and mostly arise from thermal emissions of NM. The pressure induced shifts toward the red are small but evident at least for three major bands, $\nu(CN)$, $\delta_s(CH_3)$, and $\nu_s(NO_2)$. The red shifts, "mode hardening", observed in this study range from 1.0 cm^{-1}/GPa in $\nu(CN)$ to 1.5-2.0 cm^{-1}/GPa in $\delta_s(CH_3)$ and $\nu_s(NO_2)$. These shifts are consistent with the previous single shock data [28], but are substantially smaller than that observed in the step-wise loading experiments [27]. It is probably because the step-wise loading results in substantially lower temperature and higher volume compression than the single shock loading to the same peak pressure.

The background in Fig 7 rapidly increases at approximately 200 ns after NM being shocked, t_n, but disappears as the laser intensity diminishes. It is also noted that the intensity of this broad feature is substantially higher at shorter wavelengths toward the laser wavelength at 514.5 nm. Therefore, this feature is not due to thermal emission, but is likely due to the elastic scattering or laser induced fluorescence, both of which could arise from chemical reaction of NM and/or transients. The disappearance of $\nu(CN)$ and the background enhancement were also observed previously in the sensitized NM experiments and were attributed to the CN bond scission and the fluorescence from transient species, respectively [27].

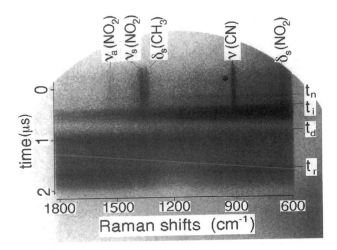

Figure 7.   Raman scattering data of nitromethane shocked to 9.8 GPa. Various events t_n, t_i, t_d, t_r, respectively, indicate the shock arrival at NM, initiation, detonation, and reshock time.

CONCLUSION

Understanding detonation chemistry requires advances in both experiments and theory. We have demonstrated that shock wave experiments combined with time-resolved spectroscopy are useful to study the shock initiation and detonation in homogeneous high explosives. In addition to the capability measuring the *CJ* temperature accurately, time-resolved emission spectroscopy also reveals the detonation kinetics, crucial for developing new theoretical reaction models. The kinetics in the predetonation period are complicated and depend strongly on pressure, temperature, time, and initiation reactions. We also found an indication of non-equilibrium behavior of TNM in the postdetonation period. The studies on shock-initiation are currently under way by using time-resolved Raman spectroscopy at a two-stage gas-gun.

ACKNOWLEDGMENTS

We thank Ervin See, Jim Crawford and Bruce Morgan at LLNL for the technical assistance. We also appreciate J. Dick at LANL for providing the PETN single crystals and for comments valuable to the study. The discussions with S. Sheffield at LANL and B. Nellis, D. Erskine, F. Ree and M. van Thiel at LLNL were very useful for the study. This work was performed under auspices of the U.S. DOE by the LLNL.

REFERENCES

1.   R. Cheret, Detonation of Condensed Explosives (Springer-Verlag, New York, 1993).

2.   Y.M. Gupta, J. De Phys. IV, **C4**-345, (1995).

3. D.S. Moore and S.C. Schmidt, Shock Waves in Condensed Matter-1987, S.C. Schmidt and N.C. Holmes, Eds. (North-Holland, Amsterdam, 1988), pp 35.

4. A.M. Renlund and W.M. Trott, *ibid*, pp. 547.

5. D.D. Dlott and M.D. Fayer, J. Phys. Chem. **92**, 3798 (1990).

6. D.W. Brenner, D.H. Robertson, M.L. Elert, and C.T. White, Phys. Rev. Lett. **70**, 2174 (1993).

7. P.C. Souers and J.W. Kury, Propellants. Explosives, Pyrotechnics **18**, 175 (1993).

8. S.A. Sheffield, R. Engelke, and R.R. Alcon, In-situ study of the chemically driven flow field in initiating homogeneous and heterogeneous nitromethane explosives in Proceedings of the Ninth International Symposium on Detonation, pp39, Portland, Oregon (1989).

9. R. Engelke, L.E. William, and C.M. Rohlfing, J. Phys. Chem. **90**, 545 (1986).

10. C.F. Melius, J. de Physique, **C4**-341 (1987).

11. J.J. Dick, J. Phys. Chem. **97**, 6195 (1993).

12. C.P. Constantinou, J.M. Winey, and Y.M. Gupta, J. Phys. Chem. **98**, 7767 (1994).

13. C.J. Piermarini, S. Block and P.J. Miller, J. Phys. Chem. **93**, 457 (1989).

14. M.S. Shaw and J.D. John, Shock Waves in Condensed Matter-1987, S.C. Schmidt and N.C. Holmes, Eds. (North-Holland, Amsterdam, 1988), pp. 503.

15. N.C. Holmes, Rev. Sci. Instru. **66**, 2615 (1995).

16. PETN single crystals oriented to (110) plane were obtained from J. Dick at the LANL. In the present experiments, shock wave propagates to the direction perpendicular to the (110) plane.

17. A.W. Campbell, W.C. Davis, J.B. Ramsey, and T.R. Travis, Phys. Fluid **4**, 511 (1961).

18. D.R. Hardesty, Combustion and Flame **27**, 229 (1976).

19. M. van Thiel, F.H. Ree, and L.C. Haselman, Accurate Determination of Pair Potentials for a CwHxNyOz system of Molecules: a Semiempirical Method, UCRL-ID-120096, LLNL (March, 1995).

20. N.C. Holmes, Rev. Sci. Inst. **64**, 357 (1993).

21. J.J. Dick, R.N. Mulford, W.J. Spencer, D.R. Pettit, E. Garcia, D.C. Shaw, J. Appl. Phys. **70**, 3572 (1991).

22. C.M. Tarver, Combustion and Flame **46**, 157 (1982).

23. A.N. Dremin, S. Savrov, and A.N. Amdrievskii, Comb. Expl. and Shock Waves, vol.1, 1965, pp 1.

24. F.H. Ree, J. Chem. Phys. **84**, 5845 (1986).

25. M.L. Japas and E.U. Frank, Ber. Bunsenges. Phys. Chem. **89**, 793 (1985).

26. C.S. Yoo, Y.M. Gupta, and P.D. Horn, Chem. Phys. Letts. **159**, 178 (1989).

27. G.I. Pangilian and Y.M. Gupta, J. Phys. Chem. **98**, 4522 (1994).

28. A.M. Renlund and W.M. Trott, <u>Shock Waves in Condensed Matter</u>- 1989, S.C. Schmidt, J.N. Johnson, and J.W. Davison, Eds. (Elsevier Science Pub; New York, 1990) pp 875.

DYNAMIC CHEMICAL PROCESSES OF DETONATIONS: UNDERWATER EXPLOSION AND COMBUSTION PROCESSES WITH THE USE OF WATER AS AN OXIDANT

V.G. SLUTSKY *, S.A. TSYGANOV, E.S. SEVERIN
*Semenov Institute of Chemical Physics, Russian Academy of Sciences, 4 Kosygin Str., Moscow 117977, Russia, kinet@glas.apc.org

ABSTRACT

The possibility of producing underwater unconfined explosions with the use of propylcarborane as a fuel and water as an oxidant was demonstrated experimentally. The specific TNT equivalent of the energy release in the explosion reaction between propylcarborane and water was found to be 1.5 times higher than conventional explosives. A possibility of organizing the combustion of initially unmixed propylcarborane and water was experimentally proven as well.

INTRODUCTION

Most conventional fuels and explosives are hydrocarbon-based materials which use oxygen or oxygen-rich compounds as oxidants. In contrast to hydrocarbon-based materials, metallic and organometallic compounds can use alternative oxidants (e.g., water) in combustion and detonation processes. The use of water as an oxidant and metallic/organometallic compounds as fuels allow one to increase energy characteristics of apparatus for underwater propulsion or explosion/detonation [1-3]. Nevertheless, combustion and detonation processes with the use of water as an oxidant remain insufficiently investigated.

The present paper is devoted to the experimental investigations of possibilities for organizing the combustion and explosion processes with the use of propylcarborane (PC) as a fuel and water as an oxidant. Propylcarborane $C_2B_{10}H_{11}-C_3H_7$ is a nontoxic liquid boron-based compound. The specific TNT equivalent of the heat effect for the PC reaction with water is 2.4 [4],

$$C_2B_{10}H_{11}-C_3H_7(L) + 15H_2O(L) = 5B_2O_3(L) + 17H_2 + CH_4 + C_4H_{10}, \qquad (1)$$
$$-\Delta H_{298} = 10.1 \text{ MJ/kg PC; TNT(calc.)} = 2.4.$$

The major reactant products are boron anhydride and hydrogen. Propylcarborane does not react with air and water under normal conditions without special initiation.

EXPERIMENTAL

To estimate the reactivity of carboranes in water vapor at elevated temperatures, we used gaseous carborane $C_2B_4H_6$ as an example. In our experiments, we measured ignition delays of the $C_2B_4H_6 + 12 H_2O$ mixture at 1000 - 1600 K and 3.0 ± 0.3 MPa. The experiments were carried out in a 40-mm × 40-mm × 3.4-m shock tube (T > 1300 K) and in a 80-mm-i.d. × 240-mm rapid compression machine (T < 1100 K). Both experimental facilities were preheated to 403 ± 3 K. In the shock-tube experiments, the ignition delay corresponded to the time interval between the arrival of the shock wave at the tube end plate and the beginning of a sharp pressure rise at the same point. In the rapid-compression-machine experiments, the ignition delay was determined as the time

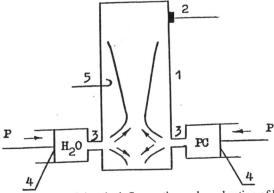

Fig. 1. Scheme of the setup for studying the inflammation and combustion of liquid carborane/water sprays: (1) combustion reactor, (2) pressure gauge, (3) injectors, (4) plungers, (5) plug points.

interval between the end of compression and the beginning of a pressure rise in the compressed volume. For comparison, we also investigated self-ignition of the stoichiometric propane/air mixture at 3.0 ± 0.3 MPa using the same experimental technique.

To prove a possibility of organizing the combustion of liquid propylcarborane/water sprays, we used the setup schematically shown in Fig. 1. The setup consisted of a 40-mm-i.d. $\times$ 140-mm (180 cm³ volume) confined combustion reactor (1) equipped with a quartz piezoelectric pressure gauge (2). Two injectors (3) with a 1-mm exit opening diameter were placed opposite each other at the bottom of the reactor. Interspaces between injectors (3) and pushed plungers (4) were filled with water and PC. Plungers were set in motion by gases from colloxylin powder weights burnt in gas generators. The speed of motion of plungers was registered with electromechanical gauges. Before

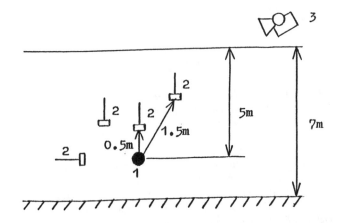

Fig. 2. Scheme of the installation for studying the underwater unconfined explosions: (1) charge, (2) membrane gauges, (3) video-tape recorder.

408

the experiment, the reactor was filled with the rich ($\varphi = 1.1$) hydrogen/air mixture under a pressure of 0.9 MPa. When pressing the "start" button, there was a spark discharge between plug points (5) that initiated hydrogen combustion. As a result of combustion, the reactor was filled with the water vapor/nitrogen mixture at a pressure of 7 MPa and temperature of 2600 K. This temperature is considerably higher than the combustion temperature (1600 K) of PC in water. That is why the combustion of colloxylin powder weights, setting plungers in motion, was electrically initiated with the 45-ms delay after the hydrogen burnout. This allowed us to investigate the inflammation and combustion of PC/water sprays at initial temperature of 1800 K due to heat losses during the delay time. In our experiments, we measured pressure in the reactor, starting from the initiation of hydrogen combustion. The pressure record made it possible to detect the inflammation and combustion of sprays.

The experiments, which proved a possibility of producing underwater unconfined explosions with the use of propylcarborane as a fuel and water as an oxidant, were carried out in a natural 7-m depth reservoir (Fig. 2). The charge of 0.3 kg total weight was placed at a depth of 5 m. The charge consisted of PC and a condensed explosive (CE) of a special shape. The CE played a role of both a dispergator and explosion/detonation initiator. The total trotyl equivalent (TNTt) of the explosive action of the charge was determined by deflections (δ) of membrane gauges and the maximum diameter (D_m) of the disturbance arising on the reservoir surface due to the explosion. The gauges were placed at distances of 0.5 to 1.5 m from the charge. The time history of the surface disturbance was registered by a video-tape recorder. The calibrated δ(TNTt) and D_m(TNTt) curves were initially obtained in explosions of a standard explosive. The TNT equivalent of the energy release in explosion reaction of PC with water was determined as a difference between TNTt and the trotyl equivalent of the CE part of the charge. The experiments were carried out for various CE/PC ratios.

RESULTS AND DISCUSSION

The experimental results of gas-phase self-ignition studies are shown in Fig. 3. Figure 3 demonstrates that the reactivity of the $C_2B_4H_6/H_2O$ gaseous mixture is close to that of the

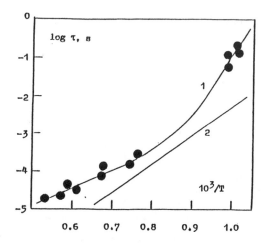

Fig. 3. Ignition delay times for the stoichiometric $C_2B_4H_6/H_2O$ (1) and C_3H_8/air (2) gaseous mixtures at 3.0 ± 0.3 MPa.

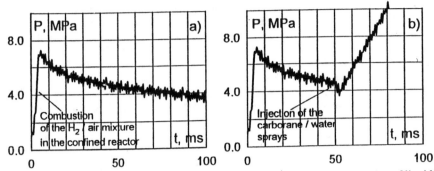

Fig. 4. Pressure records in the combustion reactor without (a) and with (b) injection of liquid carborane/water sprays.

propane/air mixture. This result indicates a principal possibility of producing the combustion and explosion/detonation of carborane/water mixtures.

The experimental results, which proved the possibility of organizing the combustion of liquid carborane/water sprays, are displayed in Fig. 4. Figure 4a represents the pressure record in the combustion reactor during (first 5 ms) and after the H$_2$/air mixture combustion. Figure 4b represents the pressure record corresponding to the injection of PC/water sprays into the reactor. The intensive pressure rise after the injection corresponds to the inflammation and combustion of liquid carborane/water sprays. This result revealed the possibility of organizing the combustion of initially unmixed liquid carborane and water.

The treatment of the explosion experiments with the charges consisted of PC and CE has shown that, starting with a certain CE/PC ratio, the specific TNT equivalent of the energy release in the explosion reaction between PC and water becomes equal to the thermodynamically predicted one (Eq. (1)),

$$TNT_1(experimental) = 2.4. \qquad (2)$$

This result 1) proves the possibility of producing underwater unconfined explosions of initially

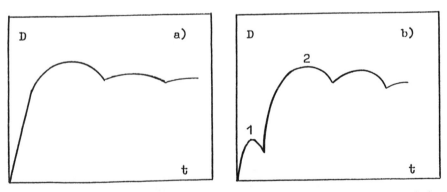

Fig. 5. Time histories for diameters of disturbances arising on the surface of the reservoir due to underwater unconfined explosions of propylcarborane-based charges: (a) corresponds to a single-stage regime of explosion, (b) corresponds to a double-stage regime of explosion.

unmixed PC and water, and 2) indicates that the thermodynamically predicted energy release of the PC/water reaction was achieved in those explosions. A typical time history, D(t), for the diameter of the disturbance arising on the surface of the reservoir due to the above charge explosion is shown in Fig. 5a. The D(t) behavior is similar to that due to underwater explosions of condensed explosives. This means that the explosion of the CE part of the charge and the explosion reaction between PC and water took place practically simultaneously. We called the observed regime of the underwater unconfined explosion as "Single-Stage Regime".

Varying the CE/PC ratio, we have found the second regime of the underwater unconfined explosion, which we have called "Double-Stage Regime". A typical D(t) behavior for this regime is shown in Fig. 5b. The first disturbance (relatively small in diameter) corresponds to the CE explosion. The second one (relatively large in diameter) corresponds to the unconfined PC/water explosion. The specific TNT equivalent of the PC/water explosion, which was determined from the maximum diameter of the second disturbance, exceeded 5.0 in some our experiments,

$$TNT_2(\text{experimental}) \geq 5.0. \qquad (3)$$

The TNT_2 obtained in the experiments exceeds the thermodynamically predicted value. This effect is known for unconfined explosions in air [5], and is due to a relatively slow decay of blast waves formed by unconfined explosions in comparison with the attenuation of blast waves formed by condensed explosives.

CONCLUSIONS

1. The possibility of organizing the combustion of initially unmixed propylcarborane and water has been demonstrated experimentally.

2. The possibility of producing the underwater unconfined explosions with the use of propylcarborane as a fuel and water as an oxidant has been experimentally proven; the thermodynamically predicted energy release of the propylcarborane/water reaction was achieved in those explosions.

3. Two regimes for dynamics of the energy release were found for underwater unconfined explosions of the propylcarborane-based charges: a single-stage regime is characterized by a prompt explosion of a condensed explosive part of the charge and the explosion reaction between propylcarborane and water; the two above types of explosions are separated in time for a double-stage regime.

4. Propylcarborane was found to be a promising fuel for underwater combustion and explosion/detonation processes with the use of water as an oxidant.

ACKNOWLEDGMENTS

This work was sponsored in part by the Office of Naval Research under grant number N68171-95-C-9076 and in part by the Russian Foundation for Basic Research under grant number 93-03-4345.

REFERENCES

1. Kenneth K. Kuo and Roland Pein, Combustion of Boron-Based Solid Propellants and Solid Fuels, CRC Press, Boca Raton, 1993, pp. 1-510.

2. H. Scmidt, J. Lineberry and L. Crawford, AIAA-87-0382, 25th AIAA Aerospace Sciences Meting, Reno, NV, Jan 1987.

3. S. Tsyganov, V. Slutsky, E. Severin and E. Bespalov in Combustion, Detonation, Shock Waves (Proceedings of the Zel'dovich Memorial), edited by S. Frolov (Semenov Inst. of Chem. Phys. Moscow, 1994), 2, pp. 83-86.

4. V. Pepekin, Yu. Matyushin, V. Kalinin, Yu. Lebedev, L. Zakcharkin and A. Apin, Izvestia AN SSSR, Ser. Chem. No. 2, pp. 268-273 (1971), in Russ.

5. W. E. Baker, P. A. Cox, P. S. Westine, J. J. Kulesz and R. A. Strehlow, Explosion Hazards and Evaluation, Elsevier Sci. Publ. Comp., New York, 1983, pp. 104-190.

A REACTIVE FLOW MODEL WITH COUPLED REACTION KINETICS FOR DETONATION AND COMBUSTION IN NON-IDEAL EXPLOSIVES

PHILIP J. MILLER
Engineering Sciences Branch, Research and Technology Division
Naval Air Warfare Center, Weapons Division
China Lake, CA 93555-6001

ABSTRACT

A new reactive flow model for highly non-ideal explosives and propellants is presented. These compostions, which contain large amounts of metal, upon explosion have reaction kinetics that are characteristic of both fast detonation and slow metal combustion chemistry. A reaction model for these systems was incorporated into the two-dimensional, finite element, Lagrangian hydrodynamic code, DYNA2D. A description of how to determine the model parameters is given.

INTRODUCTION

The ignition and growth concept of shock initiation and detonation wave propagation in nearly ideal heterogeneous solid explosives has been described by Lee and Tarver [1]. It has been applied through the use of the two-dimensional, finite element, Lagrangian hydrodynamic code, DYNA2D [2], to a variety of PBX-type explosives and propellants [3,4]. The reactive flow hydrodyamic computer code model consists of: an unreacted explosive equation of state, a reactive product equation of state, a reaction rate law that governs the chemical conversion of exposive molecules to reaction product molecules, and a set of mixture equations to describe the states attained as the reactions proceed. The unreacted equation of state is derived from shock Hugoniot data. The reaction product equation of state is derived from expansion data, such as that obtained in a cylinder test. The reaction rates are usually inferred from embedded gauge and/or laser interferometric measurements of pressure and/or particle velocity histories. The ignition and growth reactive flow model has been used to analyze a great deal of one- and two-dimensional shock initiation and self sustaining detonation data. In this reactive flow formulation, the unreactive and product equation of state are both JWL (Gruneisen) forms:

$$P = A\exp(-R_1 V) + B\exp(-R_2 V) + \omega C_v T/V , \qquad (1)$$

where P is pressure; V is relative volume; T is temperature; and A, B, R_1, R_2, ω (the Gruneisen coefficient); and C_v (the average heat capacity) are constants. The ignition and growth reaction rate law is the three term form for ignition, growth and completion as described in reference [4].

Composite explosives containing significant amounts of aluminum (Al) and ammonium perchlorate (AP) often release a substantial amount of their energy late in the explosion process. The detonation velocity and pressure are substantially lower than in ideal explosives. But delaying the energy release enhances the damage in underwater, air and internal blast applications. Energy release can persist in the Taylor rarefaction region for hundreds of microseconds after the detonation state. This late energy release introduces a time-dependency into the problem of determining the equation of state (EOS) [5-7]. As a result, the EOS appears to depend on the size of the charge and the degree of confinement, making it difficult to characterize these explosives using traditional small-scale experiments, such as the cylinder test,

413

the wedge test, etc, where the tests appear to be not scalable. Figure 1. shows cylinder expansion tests of varying diameters for a typical (aluminum/ammonium percholarate/nitramine) underwater explosive. These results, clearly demonstrate the effect of the late reactions on the energy release.

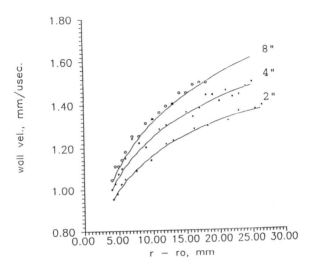

Figure 1. Calculated (solid lines) and Observed 2, 4, and 8 Inch Cylinder Tests for an Underwater Explosive Having Significant Energy Release After the CJ-State. The Data are Scaled to 1" Tests.

In a previous report a JWL EOS, containing a slow energy release rate term, was introduced in the DYNA2D hydrocode [9], and calibrated by fitting the experimental data of the non-scalable results obtained from different diameter cylinder expansion tests. The resulting EOS reproduced the experimental data of large-scale tests of underwater explosions [10,11].

In that paper, the ignition and growth reactive flow model was not considered, but, the JWL EOS for the products of reaction was modified to the form

$$P = A(1-\omega/R_1V)\exp(-R_1V) + B(1-\omega/R_2V)\exp(-R_2V) + \omega(E+\lambda Q)/V , \qquad (2)$$

where $\qquad d\lambda/dt = a(1-\lambda)^{1/2}p^{1/6}$.

The late energy release term λQ, primarily a result of the metal combustion processes, was added to the traditional JWL EOS, yielding $\omega(E+\lambda Q)/V$. The new parameters a and Q were added to the list of adjustable parameters A,B,R_1,R_2,ω, and E_0 in the traditional JWL EOS. These are usually determined by fitting the results of hydrocodes to the cylinder test data, and to measured detonation pressures and velocities. Here, E_0 is the energy content of the explosive that sustains the detonation CJ conditions. Q is the additional energy release after the CJ plane. The adjustable parameters are first evaluated using the TIGER chemical equilibrium code [8], then fine tuned by fitting the early and late time regimes of the varying size cylinder tests. This model did not consider the unreactive explosive equation of state, but did consider the equation of state at detonation front and that of the final state, when combustion was complete, where the difference between the latter two states was the energy released after the detonation sonic plane.

In this paper, the slow metal combustion and energy release kinetics are combined with the ignition and growth reactive flow model for detonation. The resultant energy release rate equations for detonation and combustion are coupled and depending on their relative rates are demonstrated to greatly effect the gas expansion and shock dynamics of an explosive. Thus, the model can be extremely useful for designing and the understanding of non-ideal explosives in specific applications.

REACTIVE FLOW MODEL FOR DETONATION AND COMBUSTION PROCESSES

The model developed here is based on the work of Guirguis and Miller [9-15]. A somewhat similar approach has been reported by Merrill and co-workers [16]. The model is based on a coupling of the global kinetics for the fast reactions that determine the detonation state and the slow metal combustion kinetics that determine the late time energy release after the CJ state. Whereas, these reaction mechanisms represent two different time regimes depending on the rate of reaction of the late reactions, considerable overlap of the energy release can occur. The model consists of: an unreacted explosive equation of state, an intermediate reactive product equation that describes the state of the system at the Chapman-Jouguet (CJ)-like surface (here the flow is sonic and only a fraction of the total energy contributes to sustaining the detonation front), a final reactive product equation of state (includes the energy released from the total of the detonation and combustion reactions), a reaction rate law that governs the chemical conversion of the explosive molecules to the reaction product molecules of the CJ state, a reaction rate law that governs the chemical reactions between the products of the detonation state and the remaining unreacted species (mostly metal) of the system, and a set of mixture equations to describe the states attained as the reaction proceed.

The actual chemical kinetics and energy release mechanisms, involving energy and heat transfer between the solid and gaseous products of reaction, will not be considered here. A strictly empirical approach, based on experimental data and on some thermochemical equilibrium calculations, will be used to determine the model for specific explosive compositions. The reaction model that was in used formulating the above approach is

$$\text{Reactants} \xrightarrow{\text{ } k_\lambda{}^{\text{global}} \text{ }} \text{Products} \xrightarrow{\text{ } k_\alpha{}^{\text{global}} \text{ }} \text{Products.} \qquad (3)$$
$$\text{(expl.)} \qquad\qquad \text{(detonation)} \qquad\qquad \text{(combustion)}$$

The reaction rate for the fast reactions involving the detonation process are given by the usual expressions used in the ignition and growth model.

$$d\lambda/dt = I(1-\lambda)^b(\rho/\rho_0-1-a)^x + G_1(1-\lambda)^c\lambda^dP^y + G_2(1-\lambda)^e\lambda^gP^z , \qquad (4)$$

where λ is the fraction reacted; t is time; ρ_0 is the initial density; ρ is the current density; p is the pressure in Mbars; and I, G_1, G_2, b, x, a, c, d, y, e, g, and z are constants. The first term in Equation (2) controls the initial rate of reaction ignited during shock compression and is limited to fraction reacted $\lambda \leq \lambda_{igmax}$. The second term in Equation (2) is used to simulate the relatively slow growth of hot spot reactions during low pressure shock initiation calculations, and the third term is used to rapidly complete the shock to detonation transition in those calculations. The reaction rate for slow energy release after the CJ state is given by

$$d\alpha/dt = \lambda a P^{1/6}(1-\alpha)^{1/2} \qquad (5)$$

where α is the fraction of reacted material in the final state that did not react in the intermediate state. This kinetic term is based primarily on the metal reacting with oxygen containing species from the detonation state. A more complete description of this rate equation is given below. This reaction is coupled to the detonation reactions since it only occurs from the products of detonation and unreacted components of the composition. The model assumes pressure and temperature equilibrium and that the relative volumes are additive.

$$V_{Total} = (1-\lambda)\ V_{expl} + (1-\alpha)\lambda V_{det.} + \alpha V_{comb.} \qquad (6)$$

This reactive rate model for energetic materials containing significant amounts of metal, using equations (4) through (6), was than incorporated into the two-dimensional, finite element, Lagrangian hydrodynamic code, DYNA2D.

RATE EQUATION FOR THE COMBUSTION PROCESSES

To determine rate equation for the late energy release that occurs after the detonation state, the assumption is made that the energy release kinetics are rate controlled by the combustion of the metal. Previous experiments on aluminum combustion were usually conducted at atmospheric pressure [17]. Some observed aluminum particles burning in a flame. Others measured the oxidation rate of thin sheets of aluminum in an oxidizer gas. Both types of studies have shown that the aluminum reaction is diffusion limited. As the thickness of the Al_2O_3 layer formed on the surface gets thicker, the rate of diffusion of the oxidizer gas through the barrier layer decreases, causing a decrease in the reaction rate. During the oxidation of the metal sheet, the surface area is constant. As a result, the oxidation is described by the simple kinetic law, $\lambda = (kt)^{1/2}$, where λ is the mass fraction of Al_2O_3 and k is a constant that depends on the temperature and the pressure of the oxidizer gas. In the combustion of aluminum particles, burning progresses from the surface towards the center, leaving an increasingly thicker layer of Al_2O_3. In addition to the reduction in the rate of diffusion, the area of the interface at which the reaction occurs gets smaller, yielding $[1-(1-\lambda)^{1/3}]^2 = kt/r^2$, where r denotes the particle radius. A number of these expressions exist for various reaction conditions and geometric configurations, but the correct one, that is obeyed when an irregularly shaped aluminum particle burns in the high-pressure and temperature environment of the detonation products, is unknown. Here, the general rate expression $d\lambda/dt = a(1-\lambda)^{1/2}p^{1/6}$ was chosen, where the constant a includes the dependency on the particle size, $(1-\lambda)^{1/2}$ is for the geometrically decreasing reaction surface area, and the pressure dependence $p^{1/6}$ is based on both theoretical analysis and experimental measurements [18]. The temperature dependence, usually in the form $e^{-\Delta H/RT}$, is assumed included in the constant a, because the aluminum combustion in the detonation products environment occurs within a narrow range of relatively high flame temperatures. The temperature of this reaction has been measured [19] in the explosive gases of 1/2 gram of an AP/Al/RDX explosive in a highly confined detonation experiment. The Al particles with diameter of 20 microns were observed to burn for about 700 microseconds at approximately 2600 K until the pressure dropped to well below 1/2 Kbar.

416

APPLICATION OF THE MODEL AND SUMMARY

The information required to determine this model for a specific energetic compostion includes, (1) detonation velocity and pressure, (2) ignition and growth rate parameters to reach the intermediate detonation state, (3) rate parameters for complete reaction to the final state, and (4) the equations of state of the unreactive material, the CJ-like detonation state and the final state of the system when combustion is complete. All of this data input can be derived from experiment, although the use of thermochemical equilibrium calculations can be helpful. The detonation velocity and unreactive Hugoniot can be obtained by the usual experimental methods. The energies of the detonation state and the final state, where reaction is completed, are easily calculated using chemical equilibrium codes by assuming little or no metal reaction contribution to the detonation reactions. This will not be true when the size of the metal particles approaches 1 micron or less. In that case a comparision of the observed and calculated detonation velocities are used to estimate the extent of metal reaction in the detonation state. The equation of state of the intermediate detonation state products and the rate parameters for the late energy release may be obtained by fitting the modified JWL equation of state, equation (2), to two different size cylinder expansion tests, where the time dependency of reactions are evident. By setting $\lambda=0$ in the resultant equation, the intermediate state JWL is found. This equation of state could also be obtained directly from the chemical equilibrium codes by assuming no aluminum reaction, however, the use of these codes to determine release isentropes are notorius for not being able to reproduce experimental results. The final products JWL equation of state can be found by either two methods, (1) using the equation of state of the detonation state found above, but letting, in this case, $\lambda=1$, or (2) using the thermochemical equilibrium code for the case where all the reactant components are allowed to react in the detonation state. Use of either method gives essentially the same results, because the energy release resulting from the late chemical reactions only become significant at large volume expansions and here the equation of state is dominated by the total chemical energy of the system and they are the same in both cases. Finally, the reactive rate parameters for the ignition and growth rate equations used to obtain the intermediate state may be obtained by the usual methods of fitting embedded gauge stress-time profiles or by using the simplified SIG method described elsewhere in these proceedings [20].

The use of this model and variations of it have been applied to ammonium perchlorate, aluminum, and nitramine underwater explosive and propellant systems. The model was applied to the cylinder expansion tests of a typical underwater explosive and the calculated and observed results are shown in Figure 1. The calculated effect of the late energy release is demonstrated on the shock wave produced in water from a 55 lb. explosive charge, Figure 2. Clearly, even though the ideal explosives have a considerable larger initial pressure, the effect of late metal reactions is to produce a larger shock in the water at increasing distances from the explosive. The performance of underwater explosives, in particular, can be fully understood only by considering the time dependence of the metal reaction energy release. Another effect of the late chemical energy release is shown in Figure 3. Here, the maximum size of a bubble produced in water from the detonation of an underwater explosive is compared to that of ideal explosives. The figure demonstrates that the inclusion of late metal reactions in the calculations after the detonation state results in a maximum bubble radius that is twice the size of the typical ideal explosive. The model has been and is being used to aid in the design of new explosive compositions, containing newly developed energetic ingredients, by the parametric investigation of varying the relative rates of reaction of the early and late reactions to determine the effects on explosive performance in specific applications (unpublished work). It has been used for aiding in the engineering design of new warhead concepts to take advantage of the delayed energy release.

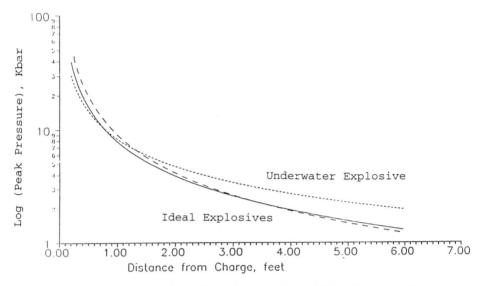

Figure 2. Calculated Peak Pressures in Water Resulting From a 55 lb. Charge. The Results Compare a Typical Underwater Explosive to Those of Ideal Explosives.

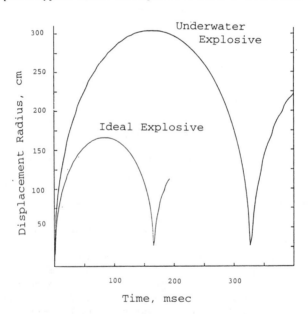

Figure 3. Calculated Bubble Radii Produced from the Explosion of 11 lb. Charges in Water. The Results Compare Underwater Explosives to that of an Ideal Explosive.

ACKNOWLEDGMENTS

The Author acknowledges the continued support of Dr. Judah Goldwasser, Program Officer for the ONR 6.2 Undersea Weaponry program. The Author particularly thanks Dr. Raafat Guirguis of the Naval Surface Warfare Center, White Oak Laboratory, for his help in modifing the DYNA2D hydrocode and for his originality of ideas upon which this paper is based.

REFERENCES

1. E. L. Lee and C. M. Tarver, Phys. Fluids 23, 2362(1980).

2. J. O. Hallquist, User's Manual for DYNA2D-an Explict Two-Dimensional Hydrodynamic Finite Element Code with Interactive Zoining," Lawrence Livermore National Laboratory Report UCID-18756, July 1980.

3. C. M. Tarver and J. O. Hallquist, Seventh Symposium (International) on Detonation, Naval Surface Weapons Center NSWC MP-82-334, Annapolis, MD, 1981, p.488.

4. C. M. Tarver and L. G. Green, Ninth Symposium (International) on Detonation, Office of the Chief of Naval Research, OCNR 113291-7, Portland, OR, 1989, p.701.

5. J. W. Kury, H. C. Hornig, E. L. Lee, J. L. McDonnel, D. L. Ornellas, M. Finger, F. M. Strange, and M. L. Wilkins, "Metal Acceleration of Chemical Explosives," Fourth (International) Symposium on Detonation, ACR-126(1965).

6. C. L. Mader, Numerical Modeling of Detonations, University of California Press (1979).

7. H. M. Sternburg and L. Hudson, "Equations of States for Underwater Explosives", Proceedings of the 1987 International Symposium on Pyrotechnics and Explosives, Beijing, China, China Academic Press.

8. M. Cowperthwaite and W. H. Zwisler, Users Guide of the TIGER Computer Program, Stanford Research Institute, Menlo Park, California, (1984).

9. P. J. Miller and R. H. Guirguis, "Experimental Study and Model Calculations of Metal Combustion in Al/AP Underwater Explosives," in Structure and Properties of Energetic Materials, Edited by J. Gilman, R. Armstrong and D. Liebenberg (Mater. Res. Soc. Proc., MRS publication Vol. 296, page 299, Pittsburgh, PA, 1993.)

10. P. J. Miller, "A Small-Scale Experiment to Characterize Underwater Explosives," 1993 JANNAF Propulsion Systems Hazards Subcommittee Meeting, CPIA publication 599, page 55, Fort Lewis, WA.

11. P. J. Miller and R. H. Guirguis, "Effects of Late Chemical Reactions on the Energy Partition in Non-Ideal Underwater Explosions", Proceedings of the Joint AIRAPT/APS Conference on High Pressure Science and Technology, June, 1993, Colorado Springs, CO., page 1417, 1993.

12. R. H. Guirguis and P. J. Miller, "Time-Dependent Equations of State for Aluminized Underwater Explosives," in the Tenth Symposium (International) on Detonation, ONR , Arlington, VA, OCNR 33395-12, page 665, 1993.

13. P. J. Miller, "Metal Acceleration by Composite Explosives Containing Aluminum," Bulletin of American Physical Society, Vol.39, page 926, 1994. Presented at APS Meeting on Condensed Matter in Pittsburgh, PA April 1994.

14. P. J. Miller, "Reaction Kinetics of Aluminum Particles in Detonation Gases," JANNAF PSHS Proceedings, San Diego, CA, CPIA publication 615, 1994, page 413.

15. R. H. Guirguis, "Relation Between Early and Late Energy Release in Non-Ideal Explosives," Proceedings of the 1994 JANNAF PSHS, San Diego, CA, CPIA publication 615, page 383.

16. C. Merrill, A. Nichols, and E. Lee, "Explosive Response of HTPB Propellants to Impact and Shock" 1994 JANNAF PSHS Meeting Proceedings, San Diego, CA, CPIA publication 615, 1994 page 343.

17. W. E. Brown, D. Dollimore, and A. K. Galwey, in Comprehensive Chemical Kinetics, edited by C. H. Bamford and C. F. H. Tipper (Elsevier Scientific Publishers), New York, 1980), p.68.

18. T. B. Grimley, in Chemistry of the Solid State, edited by W. E. Garner (Butterworths Scientific Publications, London, 1955), p.336.

19. P. J. Miller, Unpublished Results, This Work.

20. P. J. Miller, "A Simplified Method for Determining Reactive Rate Parameters for Reaction Ignition and Growth In Explosives," This Symposium of the 1995 Fall MRS Meeting.

HIGH PRESSURE EQUATION OF STATE FOR Fe₂O₃

FRANK J. ZERILLI, HERMENZO D. JONES
Energetic Materials Research and Technology Department, Naval Surface Warfare Center Indian Head Division, 10901 New Hampshire Avenue, Silver Spring, MD 20903-5640

ABSTRACT

A Debye-Gruneisen equation of state for hematite (Fe_2O_3) was developed for use in studying the high pressure solid state reaction $Al + Fe_2O_3 \rightarrow Fe + Al_2O_3$. A simple, single Curie temperature Curie-Weiss ferromagnetic term was used to account for the complex antiferromagnetic contributions to the entropy and specific heat. In the absence of information on the variation of Curie temperature with pressure, it is taken to be constant, so that the antiferromagnetism makes no contribution to the pressure-volume equation of state.

INTRODUCTION

In order to study the effects of shock pressures in the solid state reaction

$$2\,Al + Fe_2O_3 \rightarrow 2\,Fe + Al_2O_3$$

a Debye-Gruneisen equation of state was developed for solid Fe and solid Fe_2O_3. The equation of state for solid, liquid Al, and solid Al_2O_3 were already available and a high pressure equation of state for iron obtained separately. In this work we describe the equation of state for Fe_2O_3, including its complex antiferromagnetism. Superimposed on the antiferromagnetism is a weak ferromagnetism which disappears at approximately the antiferromagnetic Curie point[1]. There is some slight disagreement as to the temperature at which these transitions occur[2], but the consensus is that the temperature is in the range 950 to 1000 K.

DEBYE-GRUNEISEN EQUATION OF STATE

A Debye-Gruneisen equation of state with a single Debye temperature is used to describe Fe_2O_3. The equation of state may be derived from the Helmholtz free energy[3]

$$A = E_0 + NkT \left\{ \frac{9}{8}\frac{\theta}{T} - D\left(\frac{\theta}{T}\right) + 3\ln(1 - e^{-\theta/T}) \right\} \tag{1}$$

where $D(x)$ is the Debye function

$$D(x) = \frac{3}{x^3} \int_0^x \frac{z^3}{e^z - 1} dz \tag{2}$$

and the Murnaghan[4] form

$$E_0 = \frac{B_0 V_0}{n(n-1)} \left\{ \left(\frac{V_0}{V}\right)^{n-1} - (n-1)\left(1 - \frac{V}{V_0}\right) - 1 \right\} \tag{3}$$

has been chosen for the cold compression isotherm. If γ/V is constant where

$$\gamma \equiv -\frac{\partial \ln \theta}{\partial \ln V} \tag{4}$$

is the Gruneisen parameter, then the Debye temperature is

$$\theta = \theta_0 e^{\gamma_0 (1 - V/V_0)}. \tag{5}$$

Entropy, Internal Energy, Pressure

The entropy, internal energy, and pressure may be derived from the Helmholtz potential in the usual way.

$$S = -\left(\frac{\partial A}{\partial V} \right)_T = -\frac{A - E_0}{T} + Nk \left\{ \frac{9}{8} \frac{\theta}{T} + 3 D\left(\frac{\theta}{T} \right) \right\} \tag{6}$$

$$E = A + TS = E_0 + NkT \left\{ \frac{9}{8} \frac{\theta}{T} + 3 D\left(\frac{\theta}{T} \right) \right\} \tag{7}$$

$$p = -\left(\frac{\partial A}{\partial V} \right)_T = -\frac{dE_0}{dV} - NkT \left\{ \frac{9}{8} \frac{\theta}{T} + 3 D\left(\frac{\theta}{T} \right) \right\} \frac{1}{\theta} \frac{d\theta}{dV} \tag{8}$$

The pressure may be written in the Gruneisen form

$$p = p_0 + \frac{\gamma}{V}(E - E_0) \tag{9}$$

where

$$p_0 = -\frac{dE_0}{dV} = \frac{B_0}{n} \left\{ \left(\frac{V_0}{V} \right)^n - 1 \right\}. \tag{10}$$

Note that

$$-V \frac{dp_0}{dV} = B_0 \left(\frac{V_0}{V} \right)^n \tag{11}$$

so that B_0 is the bulk modulus at $T=0$, $p=0$.

Hugoniot

The Hugoniot[5,6] equation for the conservation of energy across a shock front is

$$E_H - E_1 = \frac{1}{2}(p_H + p_1)(V_1 - V) \tag{12}$$

and substituting Eq. (9) in Eq. (12) we obtain

$$E_H(V) = \frac{p_0(V) + p_1 - \frac{\gamma}{V} E_0(V)}{2 - \frac{\gamma}{V}(V_1 - V)}(V_1 - V) \tag{13}$$

and

$$p_H(V) = p_0(V) + \frac{\gamma}{V} \{E_H(V) - E_0(V)\}. \tag{14}$$

Specific Heat, Thermal Expansion

The specific heat at constant volume is given by

$$C_V = T\left(\frac{\partial S}{\partial T}\right)_V = NkT\left\{4D'(x) - \frac{3}{e^{x}-1}\right\}\left(\frac{\partial x}{\partial T}\right)_V \tag{15}$$

where

$$x = \frac{\theta}{T}, \quad \left(\frac{\partial x}{\partial T}\right)_V = -\frac{x}{T} \tag{16}$$

and

$$D'(x) = \frac{3}{e^{x}-1} - \frac{3}{x}D(x). \tag{17}$$

The specific heat at constant pressure is

$$C_P = T\left(\frac{\partial S}{\partial T}\right)_P = C_V\{1+\alpha\gamma T\} \tag{18}$$

where

$$\alpha = \frac{1}{V}\left(\frac{\partial V}{\partial T}\right)_P. \tag{19}$$

FERROMAGNETIC CONTRIBUTION TO THE FREE ENERGY

The complex ferromagnetism-antiferromagnetism in hematite is modeled with a single ferromagnetic contribution to the Helmholtz free energy. From Wannier's[7] description of Curie-Weiss ferromagnetism, we can derive the form of the ferromagnetic contribution to be

$$A_{ferr} = N_f k\left\{\frac{1}{2}T_c\lambda^2 - T\ln\left[2\cosh\left(\frac{T_c}{T}\lambda\right)\right]\right\} \tag{20}$$

where N_f is the number of spin 1/2 magnetic moments μ, k is Boltzmann's constant, T_c is the Curie temperature (a function of volume V), and

$$\lambda = \frac{M}{N_f\mu} \tag{21}$$

where M is the magnetization. The thermodynamic equilibrium condition for the magnetization is

$$\left(\frac{\partial A_{ferr}}{\partial \lambda}\right)_{T,V} = N_f kT_c\left\{\lambda-\tanh\left(\frac{T_c}{T}\lambda\right)\right\} = 0, \tag{22}$$

thus λ satisfies the equation

$$\lambda = \tanh\left(\frac{T_c}{T}\lambda\right). \tag{23}$$

The ferromagnetic contributions to the entropy, internal energy, pressure, and specific heat are:

$$S_{ferr} = N_f k\left\{\ln2-\frac{(1+\lambda)}{2}\ln(1+\lambda)-\frac{(1-\lambda)}{2}\ln(1-\lambda)\right\} \tag{24}$$

$$E_{ferr} = -\frac{1}{2}N_f k\lambda^2 T_c \tag{25}$$

$$P_{ferr} = N_f k\lambda^2 \frac{dT_c}{dV} \tag{26}$$

$$C_{Vferr} = N_f k \frac{(1-\lambda^2)\lambda^2}{1-(1-\lambda^2)\frac{T_c}{T}}\left(\frac{T_c}{T}\right)^2 . \tag{27}$$

If T_c is constant, then $p_{ferr} = 0$, and the contribution to the specific heat at constant pressure C_{Pferr} is equal to C_{Vferr}.

OBTAINING EQUATION OF STATE PARAMETERS

The density, molecular weight, longitudinal and transverse sound speeds, thermal expansion coefficient, and specific heat for hematite are listed in Table 1. In Table 2 are listed constants which may be derived from the data in Table 1. The third column in Table 2 shows the equation from which the given constant is calculated. A range of values is given for the Debye temperature. The lower limit (357 K) results from assuming that the Debye cutoff frequency is determined by the acoustic

Table 1. Density, molecular weight, longitudinal and transverse sound speeds, thermal expansion coefficient, and specific heat for hematite (Fe_2O_3).

ρ_0	(kg m^{-3})	5007	a
MW	(kg mol^{-1})	0.1596922	b
c_l	(m s^{-1})	7780	a
c_t	(m s^{-1})	4020	a
α	(K^{-1})	2.42 x 10^{-5}	c
c_P	(J K^{-1} kg^{-1})	649.8	b

[a]*LASL Shock Hugoniot Data*, S. P. Marsh, ed., University of California Press, Berkeley, 1980.
[b]*JANAF Thermochemical Tables, Third Edition* (J. Phys. and Chem. Ref. Data, 14,supp. 1, 1985), American Chemical Society and American Institute of Physics, New York, 1986.
[c]B. J. Skinner, *Handbook of Physical Constants, Rev. Ed.*, S. P. Clark, Jr., ed., Geological Society of America, New York, 1966.

Table 2. Some derived constants for hematite (Fe_2O_3).

c_s (m s^{-1})	6244	$c_l^2 - \frac{4}{3}c_t^2$
κ_s (GPa)	195.2	$\rho_0 c_s^2$
G (GPa)	80.92	$\rho_0 c_t^2$
ν	0.318	$\frac{1}{2}\frac{(c_l^2-2c_t^2)}{(c_l^2-c_t^2)}$
E (GPa)	402.6	$2G(1+\nu)$
θ_D (K)	357-610	$\frac{h}{k}\left(\frac{9N}{4\pi V}\right)^{\frac{1}{3}}\left[\frac{1}{c_l^3}+\frac{2}{c_t^3}\right]^{-\frac{1}{3}}$
γ	1.45	$\frac{\alpha\kappa_s}{\rho c_P}$

424

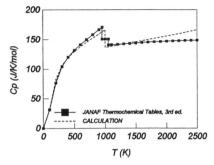

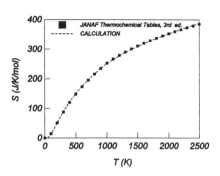

Figure 1. Specific heat at constant pressure (0.1 MPa) for hematite.

Figure 2. Entropy for hematite at 0.1 MPa compared to JANAF Thermochemical data.

modes of the molecular lattice while the upper limit (610 K) is calculated on the basis of a lattice with five times the number of molecules having 1/5 the molecular weight.

$N_f k$ and T_c are estimated from discontinuity in C_p as illustrated in Fig. 1. Using the bulk modulus κ_s to estimate an initial value for B_0, guessing an initial value for n, values for θ_0 and γ_0 are found by fitting entropy vs. temperature (Fig. 2). Using this value of γ_0, new values for B_0 and n are found by fitting the Hugoniot (Fig. 3). This process is iterated until a consistent set of parameters is found. The values of the parameters found by this method are listed in Table 3. The Debye temperature obtained by this procedure agrees with the Debye temperature calculated from the sound velocities under the assumption that hematite is an atomic solid (610 K, Table 2).

As illustrated in Fig. 3, calculations with these parameters match the Hugoniot very well up to about 50 GPa. Above 50 GPa there appears to be a phase transition which is not accounted for in this equation of state. The calculated specific heat and entropy are compared, in Figs. 1 and 2, respectively, to the data from the JANAF Thermochemical Tables, showing good agreement for the specific heat and very good agreement for the entropy. The calculated Gibbs energy and enthalpy also match the data very well as shown in Figs. 4 and 5.

Table 3. Parameters for the equation of state of Hematite.

B_0	(GPa)	224.5
n		1.5
γ_0		2.41
θ_0	(K)	632
T_c	(K)	1000
$N_f k$	(J K^{-1} mol^{-1})	20.0

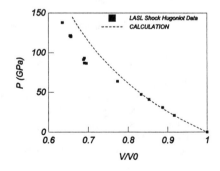

Figure 3. Calculated Hugoniot for hematite compared to experimental data.

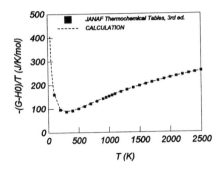

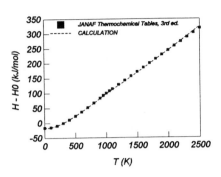

Figure 4. Calculated Gibbs energy for hematite at a pressure of 0.1 MPa, compared to JANAF Thermochemical data.

Figure 5. Calculated enthalpy for hematite at a pressure of 0.1 MPa, compared to JANAF Thermochemical data.

CONCLUSION

Treating the solid Fe_2O_3 as an atomic solid with a single Debye frequency and accounting for the complex ferromagnetic behavior with a single, pressure independent Curie temperature gives a good representation of its equation of state up to pressures of the order of 50 GPa.

ACKNOWLEDGEMENTS

This work was supported by the NSWC Energetic Materials Research and Technology Department ATT Program.

REFERENCES

1. A. Aharoni, E. H. Frei, and M. Schieber, "Curie Point and Origin of Weak Ferromagnetism in Hematite", *Phys. Rev.* **127**, 439 (1962).
2. J. Lielmezs and A. C. D. Chaklader, "Reversible Thermal Effect in α-Fe_2O_3 at 690°±5°C", *Phys. Rev. Letters* **13**, 866 (1965).
3. G. Liebfried and W. Ludwig, in *Solid State Physics*, edited by F. Seitz and D. Turnbull, Academic, London, 1961, Vol. 12, p. 275.
4. F. D. Murnaghan, *Proc. Nat. Acad. Sci.* **30**, 244 (1944).
5. H. Hugoniot, "Sur la propagation du mouvement dans les corps et specialement dans le gaz parfaits", *J. de l'ecole polytechnique* **58**, 1-125 (1889).
6. R. Courant and K. O. Friedrichs, Supersonic Flow and Shock Waves, Springer-Verlag, New York, 1976, pp. 116ff.
7. G. H. Wannier, *Elements of Solid State Theory*, Cambridge University Press, Cambridge, 1959, pp. 94ff.

DETONATION AND DEFLAGRATION PROPERTIES OF PYROTECHNIC MIXTURES

KATSUMI TANAKA
National Institute of Materials and Chemical Research,
Tsukuba Research Center, Tsukuba, Ibaraki 305, Japan, tanaka@nimc.go.jp

ABSTRACT

Theroretical calculation of detonation and deflagration properties of pyrotechnic mixtures have been performed including report charges and display charges. Calculation were performed with the KHT (Kihara-Hikta-Tanaka) code[1]. KHT results are compared with a modified version of the TIGER code[2] which allows calculation with 900 gaseous and 600 condensed product species at high pressure. Detonation properties computed by KHT and BKWS (Becker-Kistiakowskii-Wilson) give favorable agreement with experimental results of detonation velocity measurements. Hydrodynamic computation by one dimensional Lagrangian hydrodynamic code using the isentrope given by KHT constant volume explosion, indicated that experimental results for blast wave measurement for 30kg and 50kg of report charge were an incomplete reaction. Underwater detonation experiments with explosive charge of 25g, however, indicates a more energetic nature than the KHT prediction. This scale effect indicates complicated slow reactions and a number of condensed phase deflagration products of powder mixtures such as aluminum or titanium with oxidizers such as potassium perchlorate or nitrate salts as suggested by Hobbs et al[3].

INTRODUCTION

The study of explosion properties of pyrotechnic mixtures have been used to obtain the safety separation distance between manufacturing plants and inhabited residence. A recent accident which occurred in unapproved buildings of a fireworks company in Japan showed unexpected, large damages to surrounding buildings. Several experiments have been performed for report charges to study the detonation or deflagration properties of firework mixtures.

DETONATION/DEFLAGRATION PROPAGATION

Detonability of report charges and display charges have been studied by two steel tube tests. One is a 50/60 mm steel tube test based on the method of TRANSPORTATION OF DANGEROUS GOODS/TEST AND CRITERIA recommended by the UN. Another is a 30/34 mm steel tube test with a steel witness plate to measure the plate dent and confirm the detonation propagation. A report charge of Al/Ti/KP(potassium perchlorate) (15/15/70 weight percent) powder mixtures of initial density of 0.9 g/cm3 gives a dent of approximately 0.5 mm in depth but no dent was observed with the display charge. Detonation velocity has been measured by resitance wire to continuous change of electrical resitance or two contact pins. Both tests have shown the detonation propagation velocity of 1000 to 2000 m/s. The average detonation velocity was approximately 1500m/s. A high epxlosive booster charge of 50 g was used for the UN classification test for the CLASS 1 substances. A pentolite charge of 20 g or electric fuse was used for the 30/34 mm steel tube test to initiate a sample. The continuous detonation velocity measurements have suggested unsteady detonation propagation.

Table 1A Assumed detonation products to KHT calculation:
KCl,O2,Cl2,AlCl3,SO2,SO,S,KCl(s),Al(s),Al2O3(s)

	D(m/s)	P(GPa)	T(K)	Qd(J/g)	Qis(J/g)
KP/Al/S;10/3/1	2054(1800)	1.12	6410	7218	
KC/Al/S;10/3/1	1753(1700)	1.82	6463	7399	
KC/Al;90/10	2398	1.33	3528	3386	1548
KC/Al;85/15	2288	1.22	4641	4926	1674
KC/Al;80/20	2004	0.94	5635	6462	1477
KC/Al;75/25	1507	0.52	6526	7998	925
KC/Al;70/30	fail				

Table 1B
KCl,O2,Cl2,AlCl3,SO2,SO,S,Al,AlO,Al2O,Al2O3(s)

	D(m/s)	P(GPa)	T(K)	Qd(J/g)	Qis(J/g)
KP/Al/S;10/3/1	2427(1800)	1.95	7132	6508	
KC/Al/S;10/3/1	3452(1700)	1.75	7341	6554	
KC/Al;90/10	3369	2.75	3528	3386	1603
KC/Al;80/20	2994	2.93	5635	6462	3214
KC/Al;75/25	2565	0.52	6526	7998	3394
KC/Al;70/30	2565	1.80	8192	8270	3478
KC/Al;65/35	2776	1.33	3528	7144	3373
KC/Al;60/40	2957	1.22	4641	5666	3285
KC/Al;50/50	2938	0.94	5635	2704	2143
KC/Al;40/60	fail				

Table 1C
KCl,K2CL2,O2,Cl2,AlCl3,SO2,SO,S,Al,AlO,Al2O,Al(s),Al2O3(s)

	D(m/s)	P(GPa)	T(K)	Qd(J/g)	Qis(J/g)
KP/Al/S;10/3/1	2222(1800)	1.25	7147	6625	2967
KC/Al/S;10/3/1	2044(1700)	1.47	7353	6683	2750
KC/Al;90/10	2548	1.89	3486	2419	1783
KC/Al;80/20	2496	2.01	6351	5503	3193
KC/Al;75/25	2247	1.68	7459	7068	3407
KC/Al;70/30	1874	1.21	8397	8538	3511
KC/Al;65/35	2155	1.56	7638	7500	3398
KC/Al;60/40	2393	1.86	6557	6005	3289
KC/Al;50/50	2465	1.83	3979	3047	2172
KC/Al;40/60	fail				

KP(Potassium Pechlorate);KClO4,   KC(Potassium Chlorate);KClO3
();measured by Hatanaka et al[4].   D; Detonation velocity
P; Chapman Jouguet pressure   Qd: Heat of detonation for C-J composition
Qis; Blast wave energy(TNT 4520 J/g)

428

Hobbs and author[3] studied the equilibrium calculations of firework mixtures using BKWS-TIGER which can reproduce approximately the detonation velocity of report charges composed of aluminum, titanium and potassium perchlorate(KP) where the major component of detonation products are $Al_2O_3(l)$, $Ti_3O_5(l)$, $KCl(l)$ for condensed phase and O_2,O,KCl,KO,TiO_2, AlO and ClO for gaseous products. A large number of possible detonation products and phase changes of condensed products for report charges and display charges makes equilibrium calculations difficult. Phase changes under high pressure state are still unknown. In the Chapman Jouguet calculations by KHT, 14 species of detonation products were assumed as shown in Table 1. In Table 1A, formation of solid KCl was assumed. Tables 1B and IC assume no condensed phase KCl. The detonation temperatures of report charges and display charges are well higher than boiling point of KCl under normal condition. The assumption of K_2Cl_2 as a product reproduce the detonation velocity measurements. Tiger-BKWS also reproduces detonation properties of report charges for the oxygen-rich mixtures. However, Tiger-BKWS calculation which assumes the chemical equilibrium state could not predict the explosive hazard for fuel-rich explosives, where the major detonation products predicted by BKWS are in condensed phase for pyrotechnic mixtures. Then total quantity of predicted gaseous species are negligibly small for fuel(metals)-rich pyrotechnic mixtures. The explosive hazards for fuel-rich pyrotechnic mixtures should be in non-chemical equilibrium state. The experimental and calculated detonation properties of the pyrotechnic mixtures are similar to the gaseous detonation. Little change of detonation velocities was predicted with the variation of initial densities by both KHT and BKWS.

BLAST WAVE MEASUREMENTS

Field experiments[5] of blast waves for typical report charge for titanium, aluminum and potassium perchlorate(Al/Ti/KP 15/15/70 weight percent) showed the TNT equivalence of about 50 % for a charge weight of 30 and 50kg. Also, blast wave properties of H-6 (RDX/TNT/Al/Wax 45/30/20/5) high explosive charge of 100kg was measured. The report charge was confined in a cardbord drum and initiated by electric squib. The pressure of blast wave was measured by piezo-electric transducers at several locations. Measured blast wave profile for aluminum loaded explosives of H-6 compares well with numerical calculation by the OBUQ code, which is a one dimensional Lagrangian hydrodynamic code for blast waves and undewater detonations. The used equations of state of detonation products are calculated by KHT code where the isentropes through the chapman-Jouguet point are assumed. The report charges tested give lower overpressure than numerical calculation. This result indicates that the several tens microns of aluminum of particle size included in report charges reacted partially. Case given in Table1A predicts lower overpressure. The time of arrival of the shock wave are shown in Fig.1 and Fig.2 for measurements and calculation. For the case of H-6, the difference of arrival time is considered to be an 3 to 5 ms initiation delay by commercial electric detonators. However, for the case of report charges, the difference of time arrival between calculation and experimetal measurement is approximately 20 ms. This indicates a slow deflagration propagation rate. Experimental results show the increase of TNT equivalence at far distance indicating successive reaction behind the shock wave.

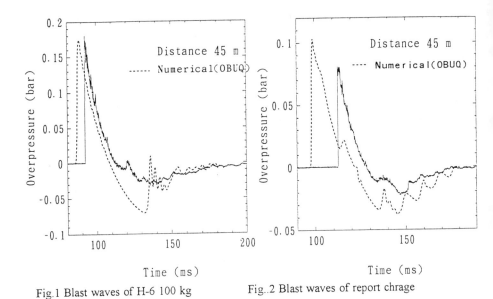

Fig.1 Blast waves of H-6 100 kg   Fig..2 Blast waves of report chrage

UNDERWATER EXPLOSION

The underwater detonation for the Al/Ti/KP(15/15/70 weight percent) system showed a bubble energy of 2 to 2.5 kJ/g compared to the 1.85 kJ/g of TNT. The bubble energy represents the expansion work by explosion products for underwater explosion. Shock energy of this report charge was 0.35 to 0.39 kJ/g comparing with 0.84 to 0.88 kJ/g for TNT. The underwater detonation tests showed the bubble energy of energy measurement of 3 to 3.5 kJ/g for Al/potassium chlorate(KC) which is nearly twice of bubble energy of TNT. Hydrodynamic calculation by OBUQ gives lower explosion energy than experimental results. Bubble energy was estimatd by both bubble pulse of oscillation time and maximum bubble radius. Both estimations give an agreement within a few percent of an experimental error which indicates the instantaneous reaction of report charges.   Explosion energies of Al/KC report charges were measured as a function of aluminum content by Aoti et al[6]. In the Al/KC system, 50 wt.% of atomized aluminum gives maximum bubble energy. A display charge composed of KP/barium nitrate/Al/sulfur(38/8/46/8 weight percent), which has no detonability, also showed a bubble energy of 3 to 3.5 kJ/g, this was estimated by bubble pulse duration and a shock energy of 0.16 to 0.18kJ/g. KHT and BKWS cannot predict these extremely large energy of deflagaration products.

DISUCUSSION

A most of heat of reaction is generated by solid or liquid alumina which gives high reaction energy and consumes most of oxygen of report charges and display charges. Thus, the total amount of gaseous products are small and reduce the detonation/deflagaration preesure.

Then the blaswave effects for pyrochnic mixtures with high reacion energy are relatively low due to the less work than TNT energy performed by an expansion of gaseous products which is main energy of blast waves and underwater detonation. The several thousands Kelvin of detonation or deflagration temperature makes many species vaporize. Experiments using small charges show nearly complete reaction, while large scale experiments indicate incomplete reaction. Blast wave for 30 to 50 kg of weakly confined report charges were measured within 200 ms, while the energy of small scale experiments for 25 g of report charges were measured within 100 ms. An extraordinary long duration of bubble pulse observed in display charges can be considered a result of slow reaction. However, the large bubble energy observed in small scale underwater experiments of report charges is still unknown. Observed bubble energy corresponds to nearly the total energy of chemical reaction. The measurement of reaction products will give the solution of the mechanism of high blast performance. The reaction of water and deflagration products is under the consideration in underwater explosion. Also, the effect of heat transfer into water from a bubble should be considered in future work.

CONCLUSION

The detonation and deflagration properties of report charges and display charges are studied. The high reaction temperature makes most of products vaporize, except aluminum oxide, which gives high blast wave effects. An increase of TNT equivalence at far distance for blast wave experiments of the KP/Ti/Al pyrotechnic mixtures suggests the continuous reaction after explosion.

REFERENCE

1. K.Tanaka,"Detonation Properties of High Explosives Calculated by Revised Kihara-Hikita Equation of State,"Proceedings Eighth Symposium(International) on Detonation, Albuquerque, 548,NM(1985)

2. M.L.Hobbs and M.R.Baer,"Calibrating the BKW-EOS with a Large product Species Data Base and Measured C-J Properties," Tenth Symposium(International) on Detonation, Boston,MA(1993)

3. M.L.Hobbs, K.Tanaka, T.Matsunaga and M.Iida, "Equilibrium Calculations of Firework Mixtures," 3rd(Beijing) International Symposium on Pyrotechnics and Explosives, Beijing China(1995)

4. S.Hatanaka,A.Miyahara,T.Hayakawa and Y.Hirosaki,"Steel Tube Test for Pyrotechnic Mixtures,"Japan Explosive Society Spring Symposium,Tokyo,Japan(1994)

5.M.Iida,Y.Nakayama,T.Matsunaga,S.Usuba,Y.Kakudate,M.Yoshida,K.Tanaka and S.Fujiwra, J.National Chemical Laboratory for Industry 85,No.6, 193(1990)

6. T.Aoti,A.Miyake and T.Ogawa, National University of Yokohama, private communication

SHOCK INDUCED ENERGETIC PHASE TRANSFORMATIONS

RICHARD D. BARDO, DAN AGASSI
Materials Department, Naval Surface Warfare Center, White Oak, Silver Spring, MD 20903

ABSTRACT

Ultrafine diamond dust and the metastable fullerenes C_{60} and $B_{24}N_{36}$ offer new pathways of energy release which are unattainable in conventional energetic materials. Theoretical and experimental studies indicate that the high bare-surface strain energies of these small particles are released in rapid (< 1 μsec) phase changes. For diamond, C_{60} and $B_{24}N_{36}$, strain energies of 1000, 1210 and 1245 cal/g, respectively, may be released during transition to graphite, diamond and hexagonal BN under pressures exceeding 100 kbar. It is shown that the behavior of these nanoscale particles under high shocks may be traced to their intrinsic five-fold and icosahedral symmetry, which leads to an incompatibility of global and local driving forces for space filling under high compression. From a correlation function analysis, it is shown that shear can induce long range correlations of fluctuations which arise from geometrical frustration, thereby reducing the activation energy for phase change. For the first time, critical stresses and symmetry rules are determined for such phase changes in aggregates of nanoscale particles.

INTRODUCTION

Energy release from a conventional explosive involves the simultaneous breaking and formation of chemical bonds in organic molecules.[1] On the other hand, the fullerenes C_{60} [2] and $B_{24}N_{36}$ [3] and ultrafine diamond dust have high bare surface energies which can significantly enhance reactivity and offer new pathways of energy release which are unattainable with organics. It is known experimentally, for example, that C_{60} compressed in a diamond anvil cell undergoes an explosive phase transition in less than 1 μsec.[4] The fullerene molecules represent the limiting species of the series of ultrafine diamond and diamond-like particles. The diameters range from 7Å for fullerenes to 100Å for diamond. These small sizes make the particles particularly attractive for use as energetic materials themselves or as additives to conventional explosives to provide energy enhancement. Ultimately, in addition to their intrinsic high surface energies, the shapes of the particles become important at high compression. It will be shown in this paper that shapes of five-fold symmetry determine critical pressures for phase transition and the types of products generated.

Ultrafine particles are effective scavengers of atoms such as hydrogen which lower the surface energy. This is the case for the covalent solids such as diamond, in which the high bare surface energy results from dangling hybrids. In fact, computations by Badziag, et al.[5] indicate that < 30Å diamonds coated with H atoms can be more stable than graphite. In addition to scavenging, the diamond grains produced in detonations can undergo further conversion to graphitic or amorphous carbon, due to high residual temperatures which are not quenched sufficiently rapidly.[6] On the other hand, the prevention of adsorption on diamond and any real enhancement of energy production remains an unresolved issue. It is known, for example, that the incompressibility of hard particles can increase detonation velocity and CJ pressure alone.[7]

These potential problems with diamond dust would appear to make fullerenes an attractive alternative, since bare particles are routinely generated. However, they are known to produce stable $C_{60}H_{36}$ species, which have carbon-bonded H atoms protruding outside or inside the cage of C_{60}.[8] The smallest, stable, hydrocarbon cage particles are pentagonal dodecahedrane ($C_{20}H_{20}$), and the first symmetrical fragments of a diamond crystal structure, adamantane ($C_{10}H_{16}$).

THEORY

Surface Energy and Five-fold Symmetry

The equilibrium shape of a crystal is of direct practical interest only for very small crystals, since for a large crystal a significant alteration in shape can be achieved only by transporting a large number of atoms through a large distance. The Gibbs conditions and the Wulff theorem are the most important approaches to consideration of the equilibrium shapes of particles. Gibbs postulated that the total surface free energy G_s must be a minimum at fixed volume (size),[9]

$$\delta G_s = \delta \int \gamma \, dS = 0, \tag{1}$$

where γ is the specific surface free energy (in erg/cm^2) and the integral is over the surface of the body. Based on the Gibbs approach, Wulff proposed that the velocities of crystal growth in the directions of normals to the faces are proportional to the surface free energies of these faces.[10,11] Generally, then, the face with the slowest growth rate will survive in competition with other faces and become the dominant feature in the final morphology. For example, in diamond $\gamma_{(111)} < \gamma_{(100)}$ so that (111) planes dominate the shape of the crystals. Thus, octahedral diamond comprises a majority of well-shaped natural and/or high-pressure/high-temperature synthesized diamonds. On the other hand, nonequilibrium conditions such as those encountered in chemical vapor deposition (CVD) and detonations yield truncated octahedra such as single crystal or twinned cubo-octahedra, iscosahedra, and decahedral Wulff-polyhedra [12]

This diversity of shapes is a function not only of the atomic structure of the particle surface, but also of all the interactions between the surface and the gas phase environment, especially any impurities present such as hydrogen. In spite of these complications, the smallest particles tend to have icosahedral or decahedral shapes. This is seen in part from the consideration of the ratio of the surface energy to the elastic energy, which is proportional to the particle radius R and to the surface-to-volume ratio S/V.[11] Of the five regular polyhedra, the icosahedron and pentagonal dodecahedron have the values of S/V at 3.97/l and 2.69/l, where l is the edge length. Thus, nanoscale particles will tend to have five-fold symmetry. This will be the case of any early nucleation stage where the growth process is limited. Indeed, high resolution electron microscopy (HREM) shows the existence of such particles of ~ 1 µm size produced under CVD conditions[12] and of 10 nm size generated by the arc-discharge method,[13] which is also used to form fullerenes with five-fold symmetry. For CVD, substrate temperatures >1000 °C appear to favor the icosahedral structure.[11] Crystallites of > 5 µm size can also have near five-fold symmetry, which, however, results from twinning so that no real icosahedral phase exists.[13]

The universality of the presence of "five-fold" symmetry in metal grains, cells, and bubbles has long been recognized.[14] This is due to the conflict between the local and overall (global) solutions of the surface energy requirements. In the case of random contacts of particles of approximately uniform size or when grains differ in size, pentagons are by far the most common grain face. For a uniform distribution of grains, polygons with other than four to six sides will be rare, with grain shapes approaching the tetrakaidecahedron in shape.[14]

To satisfy local energy requirements, an icosahedron forms when the decrease in surface energy surpasses its increase in elastic and twin energies. For diamond, Figure 1 displays the transformations which can minimize the surface energy by rehybridization of the tetrahedral C-C bonds. The indicated distortion of the FCC lattice corresponds to a tetrahedral bond angle change

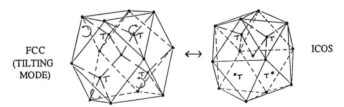

FCC
(TILTING
MODE)

ICOS

Figure 1. Transformation of diamond FCC lattice to icosahedral clustering, corresponding to shear rehybridization. Each structure contains the minimum number of C atoms required to form the icosahedron. Extended structures are obtained by adding shells. The actual C-C bonds are arranged around the atoms indicated by T. The other lines serve to outline the shapes.

of > 3^0 and a corresponding change in C-C bond length and bond strength. Thus, the transformation from left to right corresponds to minimization of the local free energy, whereas the reverse process represents minimization of the global free energy. In a compacted mass of nanoscale icosahedra, the five-fold symmetry prevents the filling of space and simultaneous minimization of the global and local free energies. Consequently, at moderate shear stresses and strains, there will be competition between the tendencies to minimize both energies, resulting in fluctuations between the two structural types. These fluctuations may be associated with the "vibrational" modes inherent in each structure. The tilting mode, indicated in Figure 1 by circular arrows in the FCC structure is one such mode, and the breathing mode is another. For a so-called "multilayer icosahedron" with diameter approaching 50Å, its internal local order is nearly that of a crystal, despite the five-fold symmetry.

Critical Stresses

High bare surface energies of the nanoscale particles are required for energetic phase transformations to occur or to be of practical significance. These energies G_s are given in Table 1 for nanoscale diamond, and for C_{60}, and $B_{24}N_{36}$.

Table 1. Calculated Surface Energies and Critical Stresses for Phase Change.

Particle	G_s (in cal/g)[a]	$v_B - v_A$ (in cm³/g)	P_C (in kbar)[b]
Diamond	1000	0.02-0.09	456-2050
C_{60}	1210	0.31-0.35	141-159 (170-200)
$B_{24}N_{36}$	1245	0.31-0.35	146-165

[a] Calculated values from References [2] and [3]. Calculated bare surface strain energies are estimated to be higher than empirical values by 100-300 cal/g.
[b] Known experimental values[16] for nonhydrostatic compression are given in parentheses.

Of particular interest here is the formation of gas and the release of 1245 cal/g in the reaction[3]
$B_{24}N_{36}$ ------> 24 BN + 6 N_2 .

Minimization of the total surface free energy at some critical stress P_C will allow space filling to occur and energy to be released. This situation is shown schematically by the potential energy curves in Figure 2. The interacting assemblage of icosahedral grains is best approximated by the wide, steep-walled, and flat-bottom potential (labeled B), and the FCC lattice by a narrow, quadratic-like potential (labeled A). The point of intersection of the curves locates the activation energies in the absence of applied stress. The steepness of the walls of the icosahedral potential corresponds to the observed stiffness (hardness) of nanoscale compacts, indicating the need for high stress levels to cause a phase transition. The flatness corresponds to the existence of low-frequency acoustic phonon modes which characterize icosahedral structures. A square-well potential has been used by Narasimhan and Jaric[15] to show that, for certain well widths, monatomic icosahedral quasicrystals can be highly robust and more stable than cp, hcp, fcc, or bcc crystal lattices up to very high stress levels.

The quadratic potential profile in Figure 2 is determined by the elastic deformation associated with shearing in the FCC lattice. Such shear corresponds to rehybridization in Figure 1 and, for diamond is represented by the total energy change per bond, where C_{44}, ε, and a are, respectively, the elastic constant for shear deformation, the strain, and the lattice constant.

$$\delta E_s = C_{44} \, \varepsilon^2 \, (a^3/8) \qquad (2)$$

For bond angle and bond length changes of 3^0 and 2%, respectively, $\varepsilon = 0.08$. Then, since $C_{44} = 48.6 \times 10^{11}$ erg/cm^3 and $a = 3.56$Å for diamond, Equation (2) gives $\delta E_s = 110$ cal/g.

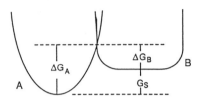

Figure 2. Intersecting potential curves for FCC lattice (A) and interacting icosahedral particles (B). G_s denotes the higher surface energy of the nanoscale particles.

The Gibbs free energy changes ΔG_A and ΔG_B in Figure 2 are related to the surface free energy G_s by $\Delta G_A = \Delta G_B + G_s$. For a first order phase transition to occur in the presence of an applied critical stress P_C, the potential curves in Figure 2 must be shifted to the extent that the internal energies satisfy $E_A = E_B$. As a result, the point of intersection of the curves will be shifted down to give a much a lower activation barrier approximately equal to the value of δE_s above. From the equality of the free energies G_A and G_B, it is readily shown that, at low temperature T,

$$G_s = P_C(v_B - v_A), \tag{3}$$

where v_A and v_B are specific volumes for the space-filling diamond lattice and compact of icosahedral particles, respectively. For FCC, octahedral shaped diamond and for icosahedral polyhedra, $v_A = 0.47\,l_A^3$ and $v_B \geq 12.18\,l_B^3$, respectively. The volume difference in Equation (3) includes the free volume for tightest packing of icosahedral particles. Also, for the golden mean $l_A/l_B = (1 + \sqrt{5})/2$, $0.76 \leq v_A/v_B \leq 0.92$. Thus, since the specific volume for diamond is of order $v_A \cong 0.28$ cm^3/g, $0.30 \leq v_B \leq 0.37$ cm^3/g, and $0.02 \leq v_B - v_A \leq 0.09$ cm^3/g, where the ranges correspond to random packings of the particles. Calculation of P_C gives the results shown in Table 1. Similar calculations are done for C_{60} and $B_{24}N_{36}$, giving the results displayed in Table 1. It should be emphasized that P_C in Equation (3) corresponds to nonhydrostatic compression resulting from the required presence of shear deformation, Equation (2). Therefore, the stress levels indicated in Table 1 should also apply to shocked states as well, and indeed, this is found to be the case for the fullerenes.[16]

<u>Symmetry Selection Rules</u>

As mentioned previously, it is impossible to fill Euclidean space with particles of five-fold symmetry. This has three important consequences. First, there is an incompatibility of global and local driving forces. Second, the crystalline long-range-order (LRO) and nanoscale short-range-order (SRO) requirements must be balanced by geometrical frustration and fluctuations, such as the situation represented in Figure 1. Third, long-range correlations of the geometrical fluctuations, which must exceed the range of the SRO interactions, give vibrational modes corresponding to crystalline or LRO interactions. These requirements for a phase transition, which were introduced previously as the equality of G_A and G_B and of E_A and E_B to derive Equation (4), allow maximum sampling between the two types of structures.

For a phase change to occur, the correlation function $C = \langle \delta G(\bar{r}_i)\,\delta G(\bar{r}_j) \rangle$ must be finite as $|\bar{r}_i - \bar{r}_j| \longrightarrow \infty$. Here, $\delta G(\bar{r}_i)$ is the free energy fluctuation corresponding to a geometrical change in structure i. For FCC diamond in Figure 1, the corresponding geometrical fluctuations are represented by LRO tilting and breathing phonon modes. In the case of icosahedra, the appropriate counterpart fluctuations are represented by SRO phason modes. The coupling of these phonon and phason modes allows the phase transition to occur. This coupling is represented by

$M = \sum \sum |<F_{\alpha r}{}^i \psi_r{}^i | V | F_{\alpha s}{}^j \psi_s{}^j >|^2 \rho$, where the products $F\psi$ pertain to phonon or phason wavefunctions F and electronic wavefunctions ψ, V is a free energy coupling term, and ρ is a phonon or phason density of states. The summations are over all the states of the coupled system. The expressions for C and M are related to each other through complicated equations[17] which will not be reproduced here. For the present purpose, it suffices to indicate here that C vanishes only when the important terms in M are zero.

The Landau free energy coupling term V may be taken to be the one used in connection with elastic phase transformations considered by Bak.[18] While he also studied first-order phase transitions, his approach is quite different from the present one, which rigorously treats the quantum mechanical behavior of nanoscale particles. Since elastic deformation leading to first-order transitions requires elastic free energy terms to at least third-order in the icosahedral strains ξ_i, terms such as the product $\xi_i \xi_j \xi_k$ in V must be considered.

While evaluation of M is difficult, the straightforward use of group theory determines the specific symmetries leading to the important nonzero matrix elements in M. Now the symmetry of FCC diamond belongs to the tetrahedral point group T, and that of nanoscale diamond and fullerenes belongs to the icosahedral group I. Group T is a subgroup of group I. Since the only representations of I which correspond to elastic phase transitions are Γ_4 and Γ_5, the above third-order products of the strain transform as the symmetric part of the representations [$\Gamma_m{}^3$], m = 4,5.[18] Of particular importance here are the strains associated with the four-dimensional representation Γ_4. They are associated with deformations along three mutually perpendicular twofold axes of an icosahedral diamond particle or fullerene, leading to FCC diamond of T symmetry. This is the situation shown in Figure 1. Distortions along either of the three-fold or five-fold axes are associated with the five-dimensional representation Γ_5, which may yield a hexagonal phase of D_{6h} symmetry, such as graphite, and other symmetries such as D_{5d} and D_{3d}.

The symmetrical part of the representation [$\Gamma_4{}^3$] includes the invariant representation Γ_1 and 4 Γ_4 and 5 Γ_5. The representation Γ_1 corresponds to the bulk modulus and does not contribute to the transformation. Since the functions F must transform as either the phonon mode Γ_3 or the phason mode $\Gamma_{3'}$, the wavefunction products $F\psi$ also transform as either the representation Γ_3 or $\Gamma_{3'}$, if the important electronic functions correspond to ground states which transform as the totally symmetric representation and if there are no electronic excitations. This will be the case at shock stress levels of interest to the detonation community. Now the direct products $\Gamma_3 \times \Gamma_3$, $\Gamma_3 \times \Gamma_{3'}$, and $\Gamma_{3'} \times \Gamma_{3'}$, which represent phonon-phonon, phonon-phason, and phason-phason couplings, respectively, decompose to give $\Gamma_1 + \Gamma_3 + \Gamma_5$, $\Gamma_4 + \Gamma_5$, and $\Gamma_1 + \Gamma_{3'} + \Gamma_5$, respectively. From this analysis, then, the strict conditions for nonzero matrix elements are that (1) V transforming as Γ_4 couples only phonons to phasons and (2) V transforming as Γ_5 couples phonons to phonons, phonons to phasons, and phasons to phasons. Therefore, condition (1) shows that diamond (arising from strains of Γ_4 symmetry) will be obtained only by phonon-phason coupling, where the phonon mode belongs either to an FCC diamond "seed" in an aggregate of nanoscale particles or to another icosahedral particle in the aggregate. In the latter case, the phonon mode is acoustic only and corresponds to center-of-mass or translational motion only. The early preferential excitation of such low-energy modes is to be expected in a shock wave.[1] On the other hand, condition (2) shows that graphite and other carbon species (arising from strains of Γ_5 symmetry) will be generated by all couplings. However, phonon-phonon couplings are related to elastic scattering and have no phase change contributions. Moreover, higher-energy phason-phason couplings are expected to play a relatively minor role, leaving phonon-phason interactions as the primary contributors for energetically producing FCC diamond, diamond-like phases, and graphite, or from $B_{24}N_{36}$, hexagonal BN and N_2.[3]

SUMMARY

The universality of five-fold symmetry appears in nanoscale structures, regardless of the nature of the atomic bonding. Here, surface reconstruction in the traditional sense involving Jahn-Teller distortions and Fermi-level pinning in larger particles appears to play a limited role as particle size diminishes in the covalently bonded systems, with icosahedral symmetry playing an increasingly important role. On the other hand, surface adsorption on diamond and its possible detrimental effect on energetic materials needs to be investigated further. However, the fullerenes, particularly highly-energetic $B_{24}N_{36}$, have bare surface strain properties of possible practical value. The potential energetic output, the ability to produce N_2 gas for useful work, and the high calculated critical pressure for phase transition make $B_{24}N_{36}$ a possible high-output, low-sensitivity energetic material. The synthesis of this material is now being studied.

The existence of particles of icosahedral symmetry was shown to allow the development of rigorous symmetry selection rules for phase change. For this purpose, the quantum mechanical behavior of particles of decreasing size was exploited. It was found that the coupling between phonon and phason modes gives the predominant interaction in a nanophase aggregate, thereby laying the groundwork for further modeling and the development of a predictive capability.

REFERENCES

1. R. D. Bardo, Int. J. Quantum Chem. S20, p. 455, (1986); R. D. Bardo in Proceedings of the Ninth Symposium (Int.) on Detonation, Vol. 1, OCNR 113291-7, 28 Aug-1Sep 1989, p. 235.

2. M. A. Wilson, L. S. K. Pang, G. D. Willett, K. J. Fisher, and I. G. Dance, Carbon 30, p. 675 (1992).

3. R. D. Bardo, C. T. Stanton, and W. H. Jones, Inorg. Chem. 34, p. 1271(1995).

4. T. P. Russell, Naval Research Laboratory, private communication.

5. P. Badziag, W. S. Verwoerd, W. P. Ellis, and N. R. Greiner, Nature 343, p. 244 (1990).

6. H. Hirai, K. Kondo, and T. Ohwada, Carbon 31, p. 1095 (1993).

7. R. Guirguis, NAVSWC TR 91-716, Silver Spring, MD., 1991.

8. R. E. Haufler, J. Phys. Chem. 94, p. 8634 (1990).

9. J. W. Gibbs, Collected Works, Longmans, Green and Co., New York, (1928).

10. G. Wulff, Zeit. Krist. 34, p. 449 (1901).

11. C. Herring in Structure and Properties of Solid Surfaces, edited by R. Gomer and C. S. Smith (The University of Chicago Press, 1953), p. 5.

12. W. Zhu, A. R. Badzian, and R. Messier in SPIE 1325, Diamond Optics III, p. 187 (1990).

13. M. Miki-Yoshida, R. Castillo, S. Ramos, L. Rendon. S. Tehuacanero, B. S. Zou, and M. Jose-Yacaman, Carbon 32, p. 231 (1994).

14. D. W. Thompson, On Growth and Form, 2nd Ed., (Cambridge, 1942).

15. S. Narasimhan and M. V. Jaric, Phys. Rev. Lett. 62, p. 454 (1989).

16. C. S. Woo and W. J. Nellis, Science 254, p. 1489 (1991); T. Sekine, Proc. Japan Acad. 68, Ser. B, p. 95 (1992).

17. R. D. Bardo in Shock Compression of Condensed Matter, edited by S. C. Schmidt, J. N. Johnson, and L. W. Davison (North-Holland, Amsterdam, 1989), p. 595.

18. P. Bak, Phys. Rev. B 32, p. 5764 (1985).

DETONATION SYNTHESIS OF NANO-SIZE MATERIALS, J. Forbes, J. Davis, and C. Wong, Naval Surface Warfare Center, Indian Head Division, Silver Spring, MD 20903

ABSTRACT

The detonation of explosives typically creates 100's of kbar pressures and 1000's K temperatures. These pressures and temperatures last for only a fraction of a microsecond as the products expand. Nucleation and growth of crystalline materials can occur under these conditions. Recovery of these materials is difficult but can occur in some circumstances. This paper describes the detonation synthesis facility, recovery of nano-size diamond, and plans to synthesize other nano-size materials by modifying the chemical composition of explosive compounds. The characterization of nano-size diamonds by transmission electron microscopy and electron diffraction, X-ray diffraction and Raman spectroscopy will also be reported.

INTRODUCTION

Nano-size materials have properties uniquely different from those of the bulk. For example, composites made from nano-particles are tougher and harder than bulk materials of the same material and ceramics that are brittle as bulk materials have ductility when made of nano-materials[1]. Interest in nano-materials has increased efforts to find synthesis methods for them.

This article reports on synthesis of nano-size diamonds obtained by the detonation of explosives. This method for synthesis of nano-size diamond has been reported by a number of researchers. A representative sampling of these publications[2-6] is given in the references.

The yield and particle size of the diamonds from detonation synthesis are primarily dependent on the pressure and temperature realized in the detonation process. A lesser particle size dependence is found as a function of explosive compositions and pre-detonation atmospheres[6] surrounding the explosive.

EXPERIMENT AND RESULTS

The present apparatus for the detonation synthesis of nano-size materials is a closed steel sphere with nominal wall thickness of 2.2 cm. Detonations of explosives up to 450 g weights have been done in this facility. All the products of detonation are contained inside the sphere. A sealed door allows access to the interior of the sphere. The sphere also has a gas collection system attached to sample the gas atmosphere before and after detonation of the test explosive.

Cylinders of Pentolite (50/50 PETN/TNT) weighing 450 g were detonated inside this metal sphere filled with air, carbon dioxide, argon, or nitrogen at one atmosphere of pressure. The cast(C) or pressed(P) Pentolite charges were nominally 5.1 cm diameter by 15 cm long cylinders. The results of these experiments are given in Table I. The solid products of detonation form a soot which is collected and chemically reduced by mineral acids. Chemical reduction of the soot for this study was performed by S. Eidelman, SAIC.

Transmission electron microscopy (TEM) and electron diffraction, X-ray diffraction and Raman spectroscopy were used to characterize the reduced soot. The characterizations were performed at NRL. The TEM was obtained with a Philips CM-30 transmission electron

Table I. Summary of Pentolite Experiments

Expt. no.	HE TYPE	DENSITY (g/cm$^{3)}$	CHARGE WEIGHT (g)	CHARGE ENVIRONMENT	WEIGHT OF COLLECTED RESIDUE (g)
002	Pentolite (P)	1.56	408	AIR	SMALL AMT. [NO DIAMONDS FOUND AFTER REDUCTION]
005	Pentolite (C)	1.65	398	81%CO$_2$/ 19%AIR	33.6
006	Pentolite (C)	1.65	403	69%ARGON/ 31%AIR	5.4
007	Pentolite (C)	1.45	410	96%NITROGEN/ 4%OXYGEN	55.5
008	Pentolite (C)	1.60	411	AIR	1.6

microscope. The X-ray diffraction pattern was obtained with a Philips APD1700 automated powder diffractometer system using Cu K-alpha radiations. Raman spectra were obtained with a micro-Raman spectrometer at 514.5 nm excitation.

The electron diffraction of the chemically reduced soot from detonation of Pentolite in CO_2 is shown in Figure 1. Diamond is identified by the characteristic diffraction rings. A micrograph of similar powder obtained with dark field technique was shown in Figure 2. Single and clustered particles were present. Single particles from about 2 to 10 nm in size were visible. Clusters an order of magnitude larger are present. The sizes of diamond particles obtained from detonation of Pentolite in air were similar.

The identity of diamond in these powders was also confirmed from characteristic X-ray and Raman lines. The findings are summarized in Table II. The average particle sizes were obtained from relative X-ray diffraction line widths. The percentage of reduced sample was determined from the reduced and starting weights. The percentage yield of diamond in the acid-reduced soot was determined from relative X-ray diffraction line intensities. A Raman spectrum of the acid reduced soot is shown in Figure 3. The diamond line was much broadened and the magnitude of the shift reduced. The shift was between 1320 and 1327 cm^{-1} as compared to the value of 1332 cm^{-1} of natural diamond.

Non-diamond phases of carbon and some impurities were present in the purified samples. Notably, X-ray diffraction lines identified silicon dioxide as being present. The source of this oxide was not known, but could originate from the steel walls of the sphere[7] and/or from residual sand left in the sphere after sand blasting prior to the first experiment.

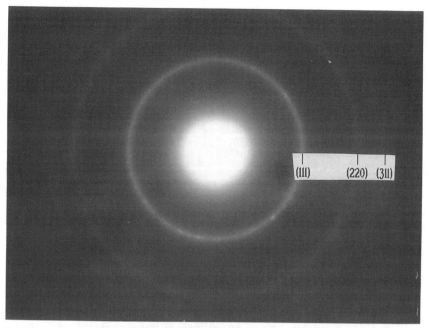

Figure 1. Electron diffraction rings from nano-diamonds
created by detonation of Pentolite in CO_2

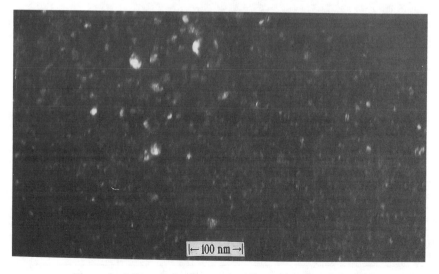

|← 100 nm →|

Figure 2. Micrograph of nano-size diamonds using dark field
technique from detonation of Pentolite in CO_2

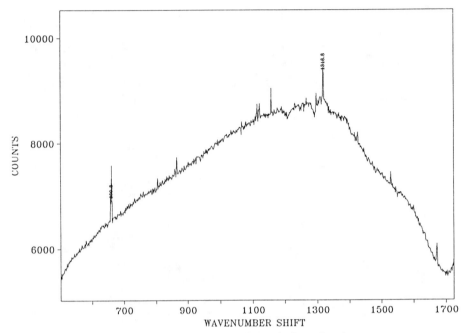

Figure 3. Raman spectrum of chemically reduced soot
from detonation of Pentolite in CO_2

Table II. X-ray diffraction and Raman scattering results

	TECHNIQUE	AIR	CO_2
SIZE (nm)	X-RAY DIFFRACTION LINE BROADENING	6.2	4.8
IMPURITY	X-RAY DIFFRACTION LINES	SiO_2	SiO_2
% DIA. IN PURIFIED POWDER	X-RAY DIFFRACTION LINE INTENSITIES	23	13
% PURIFIED POWDER IN SOOT	WEIGHING POWDERS	14	20
RAMAN SHIFT (cm^{-1})	RAMAN SCATTERING	1320-1327	1320-1327

DISCUSSION AND CONCLUSIONS

Identity of diamond in acid-reduced detonation soot was established by electron diffraction, X-ray diffraction and Raman shifts are characteristic of diamond. Particle sizes of 4 to 6 nm were estimated from X-ray line widths and more accurately from transmission electron micrographs. The average particle sizes agree with those obtained by researchers using different explosives[2-6]. The magnitude of the Raman shift and line width for our nano-size diamonds is also in agreement with other researchers[8].

The Raman shift for natural stress-free diamond[8] is 1332 cm^{-1} and line width is 3 to 4 cm^{-1}. The Raman line broadening and reduction in magnitude for the nano-diamonds were probably due to the very severe mechanical conditions these powders were formed under. The exact nature of the Raman line broadening is not known, but could be due to both the small particle sizes and defects present. More thorough chemical reduction techniques would lower the number of inclusion impurities and defects in the nano-diamonds.

The yield of diamond in the acid-reduced powders was low. No attempt was made in this limited study to improve this yield. Use of a more powerful carbon containing explosive and pressurizing an inert gas prior to detonation would increase the yield considerably[6].

The facility described in this paper can be used to produce nano-size materials other than diamonds. The composition of explosive compounds can be modified to contain constituents of the desired materials. Future experiments will use an azide in combination with a silicon compound to attempt production of nano-size silicon carbide. This approach is similar to that proposed by DeVries[9]

ACKNOWLEDGEMENTS

This work was jointly funded by ONR-332 and NSWC/IH independent research funds. S. Eidelman (SAIC) suggested the creation of this facility to provide data for his ONR-332 contract on the synthesis of nano-size materials. C. Vold, L. Troilo and J. Butler, and R. Vardiman of NRL did the X-ray diffraction, Raman spectroscopy, and TEM, respectively.

REFERENCES

1.   R. Dagani, C&EN, p 18, Nov. 23, 1992.

2.   V. M. Titov, V. F. Anisichkin, and I. Yu. Mal'kov in Ninth Symposium (International) on Detonation, Portland, OR., Aug 1989, pp. 407-416, Office of the Chief of Naval Research publication OCNR 113291-7.

3.   N. Roy Greiner, D. S. Phillips, J. D. Johnson, and Fred Volk, Nature, **333**, p 440, (1988).

4.   E. A. Petrov, G. V. Sakovich, and P. M. Brylyakov, Sov. Phys. Dokl. **35**(8), p 765, (1990).

5.   A. L. Vereschagin, G. V. Sakovich, V. F. Komarov, and E. A. Petrov, Diamond and Related Materials, **3**, p 160, (1993).

6.   V. L. Kuznetsov, A. L. Chuvilin, E. M. Moroz, V. N. Kolomiichuk, Sh. K. Shaikhutdinov, Yu. V. Butenko, and I. Yu. Mal'Kov, <u>Carbon,</u> Elsevier Science Ltd, Great Britain, p. 873, (1994).

7.   D. L. Gur'ev, E. V. Lazareva, and L. I. Kopaneva, Fizika Goreniya i Vzryva, <u>19</u>, No.2, p 110 (1983).

8.   M. Yoshikawa, Y. Mori, H. Obata, M. Maegawa, G. Katagiri, H. Ishida, and A. Ishitani, Appl. Phys. Lett. **67,** p 694, (1995)

9.   R. C. DeVries, <u>Diamond and Related Materials,</u> **4,** p 1093, (1995).

METAL/METAL EXOTHERMIC REACTIONS INDUCED BY LOW VELOCITY IMPACT*

Diana L. Woody, Jeffery J. Davis, Philip J. Miller
Naval Air Warfare Center Weapons Division, China Lake, California 93555-6001

ABSTRACT

This paper discusses experimental results from an effort conducted to discern the basic mechanism of reactions in porous metal/metal compositions under rapid plastic flow conditions. Small-scale impact tests were performed on various intermetallic mixtures: $3CuO + 2Al$, $Fe_2O_3 + 2Al$, $Ni + Al$, and $5Ti + 3Si$. The addition of polytetrafluoroethylene (Teflon) to the metal/metal mixtures has been demonstrated to affect the extent of the reactions. Real-time emissivity and species evolution measurements of the reacting materials were used to discern the chemical reactions occurring under rapid plastic flow conditions.

INTRODUCTION

Metal/metal and metal/metal oxide materials have exhibited the ability to generate highly exothermic reactions capable of occurring in the same time frame as detonations in conventional energetic materials [1,2]. Until recently, these materials were assumed to initiate exclusively under shock conditions. However, work done by Woody and Davis [3,4,5] with materials under mechanical impact conditions, as well as by Nestereenko and Meyers [6,7] with materials under explosive compaction conditions, has demonstrated the role that rapid plastic flow can play in the reactions of these mixtures. Horie has modeled their shock reaction based upon a critical plastic flow [8].

Small-scale experiments have been performed to observe the effect of rapid plastic flow produced by impact reactions on porous metal/metal compositions [3]. This paper explores the effect of the addition of polytetrafluoroethylene (Teflon) to several metal/metal and metal/metal oxide mixtures including $3CuO + 2Al$, $Fe_2O_3 + 2Al$, $Ni + Al$, and $5Ti + 3Si$. Under impact conditions of 13 m/s, self-propagating high temperature synthesis (SHS)-like reactions have been observed. A two-color infrared detector and spectrometer were used for real-time emission measurements of the reacting materials.

EXPERIMENT

A drop weight impact machine was used to induce a plastic flow in the test samples. The impact machine used for these experiments consisted of an anvil, an accelerated guided drop weight, a base, and a release triggering device. The impact machine is described fully in another publication [4]. Elastic shock cords were used to accelerate the drop weight to obtain impact velocities of 13 m/s. At these velocities, the plastic flow has been measured to be approximately 60 m/s. The impact of the drop weight on the anvil was planar to within 2 mrad. The samples used for the tests weighed 0.2 g and were in loose powder form prior to impact. Because the samples were loose powders, their initial porosity was not measured. The light emanating from the impacted sample came from its surface.

Teflon was mixed with the metal mixtures described above by the mortar and pestle method. The specifics of the materials are listed in Table I.

---

* Approved for public release; distribution is unlimited.

TABLE I. Characteristics of Materials Used in This Study.

Material	Manufacturer	Material Data
Titanium (Ti)	CERAC, Inc.	325 mesh, 99.5% purity, 20 µm or less
Silicon (Si)	CERAC, Inc.	325 mesh, 99.5% purity, 10 µm or less
Aluminum (Al)	CERAC, Inc.	325 mesh, 99.97% purity, 5 - 15 µm
Teflon	DuPont	7A, 35 µm, 60% crystallinity
Iron Oxide (Fe_2O_3)	CERAC, Inc.	325 mesh, 99.97% purity, ≥15 µm
Copper Oxide (CuO)	CERAC, Inc.	325 mesh, 99.97% purity
Nickel (Ni)	CERAC, Inc.	325 mesh, 99.5% purity, 10 - 20 µm
Garnet Paper	Norton, Inc.	180A Garnet A511

The two-color infrared detector consisted of a HgCdTe element juxtaposed to an InSb element. Each element measured 0.101 by 0.101 cm with an active area of 0.010 cm^2. The elements were housed in a liquid-nitrogen-cooled Dewar and kept at an operating temperature of 77 K. The InSb element's spectral response was from 2 to 5 µm. The HgCdTe element was capable of detecting wavelengths from 5 to 12 µm. The signal from each infrared detector element was transmitted as a voltage through an initial voltage amplifier and then transferred to a LeCroy digital oscilloscope.

The spectrometer system was used to observe the emissions generated by the gases produced upon reaction of the impacted sample. This system consisted of a SPEX 1877 spectrometer, a Tracor Northern TN-6312 1024 element array detector, and a TN-6500 controller. Although this system permits time-resolved spectroscopy measurements in the extended visible range from 200 to 800 nm, the gratings were set for viewing from 430 to 640 nm. Three spectra were taken during a single impact. The sampling time for each spectrum was 800 µs. After 10.24 ms, the detector was able to take another spectrum for 800 µs. The list of elements and molecules in Table II is based upon known lines for the elements and molecules [9].

TABLE II. Results Obtained with the Spectrometer for Some Compositions Tested.

Material Tested	Spectrometer Data
5Ti + 3Si/Teflon	TiO, C-O, C-C
Ti/Teflon (35 µm)	TiO, C-O
3CuO + 2Al	AlO, CuO,
3CuO + 2Al/Teflon	C-O, C-C, AlO, CuO, and CuF
Fe_2O_3 + 2Al/Teflon	FeO, AlO, Al_2O_3, C-O, C-C

RESULTS AND DISCUSSION

The reaction of the materials to impact was quantitatively defined from the infrared detector's signals. The extent of the reaction was quantified from the area under the infrared emission curve. The most extensive reactions in these experiments were defined as the largest peak emissions and the shortest time-to-peak emission registered by the two-color infrared detector. The peak emission and time-to-peak emission for some of the experiments are given in Table III.

Qualitative signs of reactions consisted of such parameters as a visible light emission, an audible signal, and characteristics of the recovered sample's surface. The most exothermic samples exhibited a visible flash and sustained burning after impact up to 1 to 3 seconds in duration; charring was also found on the recovered sample. It was observed that the burning duration could be varied by changing the percentages of Teflon added to the mixtures. The results obtained with the spectrometer are given in Table II, and the results obtained with the infrared detectors are given in Table III.

TABLE III. Results Obtained with the Infrared Detectors for Some of the Compositions Tested.

Material and Composition	Radiation From InSb Detector, mV	Radiation From HgCdTe Detector, mV	Time-to-Peak Emission, μs
5Ti + 3Si	29	1.41	2950
(99%) 5Ti + 3Si (1%) Teflon (35 μm)	16	6.4	2950
(90%) 5Ti + 3Si (10%) Teflon (35 μm)	3547	55	56
(80%) 5Ti + 3Si (20%) Teflon (35 μm)	3664	68.9	56
(80%) Ti (20%) Teflon (35 μm)	2890	64.5	31.5
3CuO + 2Al	3250	164.1	16.25
(90%) 3CuO + 2Al (10%) Teflon	3090	163.8	27
(75%) 3CuO + 2Al (25%) Teflon	3000	158.1	44
(50%) 3CuO + 2Al (50%) Teflon	62.5	10.2	750
Fe_2O_3 + 2Al	22.7	3.9	2300
(99%) Fe_2O_3 + 2Al (1%) Teflon	3280	94.5	3075
(90%) Fe_2O_3 + 2Al (10%) Teflon	3500	154.7	36
Ni + Al	34.8	10.9	1340
(90%) Ni + Al (10%) Teflon	63.3	7.7	810
(50%) Ni + Al (50%) Teflon	62.5	10.2	750

Note: Percentages are by mass.

5Ti + 3Si

As shown in Table III, impact of the 5Ti + 3Si mixture generated a small emission, as recorded by the infrared detectors. The small emission coupled with the longer time-to-peak emission indicated a heating of the impacted metals due to dynamic compaction rather than a chemical reaction. X-ray diffraction of the recovered samples reinforced our conclusion that reaction did not take place in that system. The addition of Teflon to 5Ti + 3Si considerably increased the peak emissions recorded by the infrared detectors and reduced the time-to-peak emission two orders of magnitude (from milliseconds to microseconds). A detailed study has been performed on the effect that varying the percentage by weight of Teflon added to various 5Ti + 3Si mixture has upon the exothermic reactions observed [10].

3CuO + 2Al

An extensive exothermic reaction was observed for the neat 3CuO + 2Al mixture. The reaction was characterized by a large infrared emission from the two detector elements, a visible flash, and sustained burning. The spectrometer registered strong emission lines for the neat mixture. The 3CuO + 2Al mixture was the only material in our study that showed reaction without the addition of Teflon. As can be seen in Table III, increasing the percentage by weight of Teflon added to the 3CuO + 2Al mixture progressively decreased the peak emission and increased the reaction time registered by the infrared detectors.

$Fe_2O_3 + 2Al$

As shown in Table III, the Fe_2O_3 + 2Al mixture generated a relatively small emission upon impact. These results were similar to those obtained upon impact of the 5Ti + 3Si mixture. The addition of Teflon to the Fe_2O_3 + 2Al mixture considerably increased the peak emissions upon impact. Increasing the percentages of Teflon added to the Fe_2O_3 + 2Al mixture had a significant effect on the extent of the exothermic reaction and the signature of the infrared emission curve. The time-to-peak emission was decreased two orders of magnitude (from milliseconds to microseconds) for the mixture containing 10% Teflon. The spectrometer data are shown in Table II.

Ni + Al

Table III shows that, although the addition of Teflon to the Ni + Al mixture increased the infrared emission recorded by the detectors and decreased the reaction time, the overall increase in exothermic release was not as substantial as that observed during the impact of the 5Ti + 3Si/Teflon and Fe_2O_3 + 2Al/Teflon mixtures. The qualitative signs of the reaction mentioned in the preceding paragraphs of this paper were not present under these impact conditions.

CONCLUSIONS

The results of these experiments show that the variation in the amount of Teflon added to the metal/metal and metal/metal oxide mixtures can significantly alter the extent, initial time of reaction, and the duration of the exothermic reaction upon rapid shear conditions. The addition of Teflon also influences the likelihood of a material to produce a sustained burning reaction. The 5Ti + 3Si, and Fe_2O_3 + 2Al mixtures exhibited relatively similar responses to the addition of varying percentages of Teflon. Under the impact conditions discussed in this paper, the highly exothermic reaction of 2Al + 3CuO did occur under plastic flow conditions without the addition of Teflon. However, when Teflon was added to the 3CuO + 2Al mixture, the exothermic reactions decreased in intensity.

ACKNOWLEDGMENTS

The author gratefully acknowledges the Office of Naval Research (6.1 Independent Research program, Judah Goldwasser, cognizant technology area manager) for sponsoring and supporting the work reported.

REFERENCES

1. N.N. Thadhani, *J. Appl. Phys.* 76, p. 2129 (1994).

2. R.A. Graham, *Solids Under High-Pressure Shock Compression,,* Springer-Verlag, 1993.

3. J.J Davis and D.L. Woody, "Reactions in Neat Porous Metal/Metal and Metal/Metal Oxide Mixtures Under Shear Induced Plastic Flow Conditions," *Proceedings of Explomet Conference,* El Paso, TX, August 1995.

4. D.L. Woody, J.J. Davis, and J.S. Deiter, *Proceedings of American Physical Society*, Seattle, WA, August 1995.

5. D.L. Woody, J.J Davis, and P.J. Miller, "Impact Induced Solid Sate Metal/Metal Reactions," *Proceedings of JANNAF Hazards Meeting*, San Diego, CA, August 1994.

6. V.F. Nestereenko, M.A. Meyers, H.C.Chen, and J.C. LaSalvia, *Applied Physics Lett.* 65 (24), December 1994.

7. V.F. Nestereenko, M.A. Meyers, H.C.Chen, and J.C. LaSalvia, *Metallurgical and Materials Transactions A*, Vol. 26A, 1995.

8. Y. Horie, "Kinetic Modeling of Shock Chemistry," presented at Explomet 95, El Paso, TX, August 1995.

9. R.W. Pearse and A.G. Gaydon, *The Identification of Molecular Spectra*, Chapman and Hall (1965).

10. D.L. Woody and J.J Davis, *J. Appl. Phys.*, in progress.

AUTHOR INDEX

SUBJECT INDEX